Cell Biology

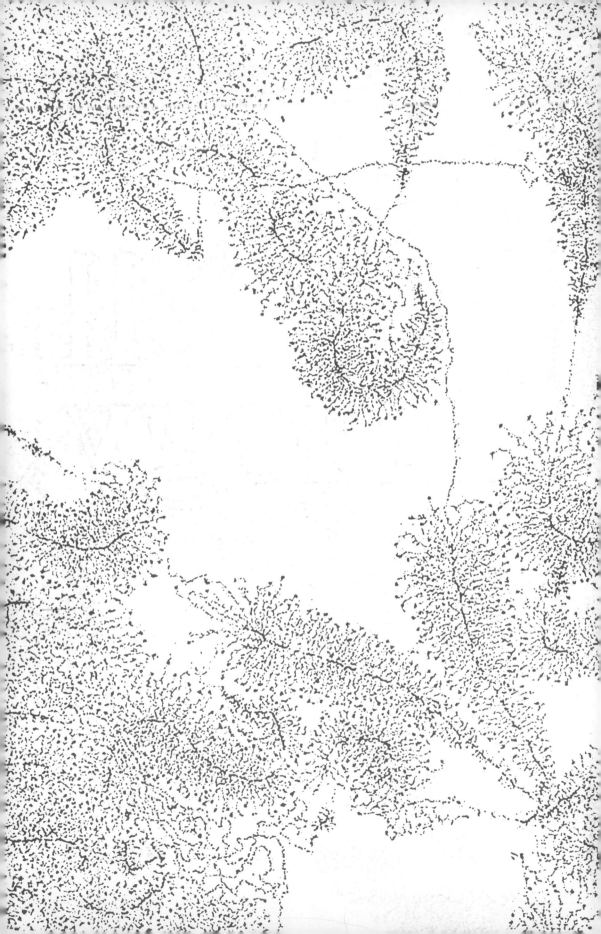

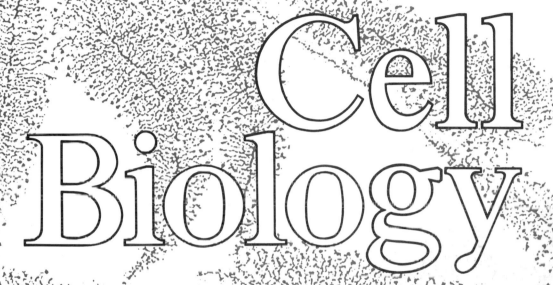

E. J. Ambrose

Professor of Cell Biology, University of London
Chester Beatty Research Institute

Dorothy M. Easty

Senior Lecturer in Cell Biology
Chester Beatty Research Institute

Cell Biology

Nelson

THOMAS NELSON AND SONS LTD

36 Park Street, London WIY 4DE   P.O. Box 25012 Nairobi   P.O. Box
21149 Dar es Salaam   P.O. Box 2187 Accra   77 Coffee Street San Fernando
Trinidad

Thomas Nelson (Australia) Ltd 597 Little Collins Street Melbourne 3000/
Thomas Nelson and Sons (South Africa) (Proprietary) Ltd 51 Commissioner
Street Johannesburg / Thomas Nelson and Sons (Canada) Ltd 81 Curlew
Drive Don Mills Ontario / Thomas Nelson (Nigeria) Ltd P.O. Box 336 Apapa
Lagos

Illustrations by K. G. Farrall

First published in Great Britain 1970
Copyright © E. J. Ambrose and Dorothy M. Easty 1970

17 761002 6

Phototypeset by Oliver Burridge Filmsetting Crawley Sussex
Printed by Fletcher & Son Ltd Norwich

Contents

PART 3   The life of cells

PART 4   Cellular dynamics

PART 5   Early and simple forms of life

Preface

Many universities now include a course in cell biology either as an introductory course for biology and medical students or as part of a combined biology course. It is recognized that a study of cell biology can provide a unifying background to biology in a way which cannot be achieved by a specialist course in other biological subjects. The striking progress made in the understanding of cell function at the molecular level in recent years has also stimulated interest.

While the importance of molecular biology is recognized and strongly emphasized in this book, it is only of relevance to the integrated function of whole cells. This is the approach we have adopted.

Some students wishing to study cell biology will have taken courses in physics and chemistry and have had no previous experience of biology. Chapter 1 is intended to provide a general introduction to cell biology for such students. Others will have studied biology but will have little previous knowledge of the rather specialized type of chemistry which is needed to follow the recent developments in molecular biology and biochemistry. Chapter 2 is intended as an introduction for such students. Both chapters include a summary of the practical methods used in cell biology, molecular biology, and biochemistry which the student will encounter in practical courses.

Some students may wish to omit Chapter 1 or 2 or both and commence with Part 2, which describes the structure and function of the various cellular components which are found almost universally in micro-organisms, and in plant and in animal cells. These include the cell nucleus, mitochondria, cell membranes, and ribosomes. The cell nucleus is rather a special case since it plays a key role in cytology. The chromosomes, seen at metaphase in cell division, have been widely studied; we have preferred to describe the interphase nucleus first as a functioning organelle because it is in the interphase nucleus that the genes of the chromosomes are actively engaged in the control of cellular synthesis. During mitosis, the chromosomes are carried as passengers to the two daughter cells. Mitosis is an integrated activity of the whole cell.

Having described the function of the cellular components

in interphase cells, Part 3 deals with the integrated function of growing cells, in the cell cycle, in mitosis, in the behaviour of germ line cells and the genetics of bacteria, and plant and animal cells. In Part 4 attention turns to the dynamic function of whole cells, cytoplasmic and whole cell movements, cellular interactions, and differentiation in development. Developmental biology provides a link between cell biology and general biology, and an attempt is made to show the relevance of a cellular approach in relation to courses in general biology. In Part 5 we consider the formation of biological structures from biochemical building units, and discuss the orgin of the latter.

We should like to express our thanks to Sir Gavin de Beer, F.R.S., at whose instigation this work was undertaken, and to Professor M. Abercrombie, F.R.S., for continuous help during the preparation; to Professor E. N. Willmer, F.R.S., Professor J. M. Thoday, F.R.S., Dr D. H. Northcote, Dr Robert Cox, Dr A. D. Greenwood; also to Dr G. C. Easty, Dr R. W. Tindle, Mrs Salley Wareham, and many other colleagues of the Chester Beatty Research Institute, who have given us so much helpful advice or have read manuscripts of chapters; and to Mr Navin Sullivan and Mr Martin Lewin of the publishers for their continual help and stimulation.

E.J.A.
D.M.E.

*We don't see words in nature
but always only the initial letters of words,
and when we set out to spell,
we find that the so-called new words
are in their turn
merely the initial letters of others.*

LICHTENBERG

PART 1 Introductory cell biology

Cell biology can be approached either from a background of the biological or the physical sciences, because to a considerable extent it is a common meeting place for biologists and specialists in molecular studies. In Chapters 1 and 2 we therefore provide background reading both for physicists and chemists and for biologists.

In Chapter 1 we begin with a historical introduction to the subject of cell biology. Then we describe the sizes, shapes, and structures of cells in general terms, and also discuss how they associate to form tissues in multicellular organisms. How cells move, grow, and divide is also briefly described.

Chapter 2 contains those particular physical and chemical concepts which are relevant as background reading in connection with cellular biochemistry and molecular biology. Particular emphasis is placed on the properties of long-chain molecules (macromolecules) which form a major part of cell structures.

Laboratory work in cell biology must of course form a major part of any course. Some of the optical, biophysical, and biochemical methods used are described at the end of Chapters 1 and 2.

1 Properties of cells

Omnis cellula e cellula. (Every cell is derived from a cell.)—
Virchow, 1855

However brilliant a scientist may be, and however careful and systematic his observations, he is always limited by the apparatus and techniques that are available at the time. No man saw the moons of Jupiter until Galileo developed the first astronomical telescope in 1609, and set in train the revolution that overthrew the Ptolemaic system and established the Copernican view of the universe.

In biology, it was the invention of the microscope which led to the discovery that organisms are composed of the individual units now known as cells. The simple microscope was invented by Galileo in 1610; and the first compound microscope was made by Robert Hooke. Hooke examined thin slices of cork (Fig. 1–1) under his microscope in 1665, and saw that the cork was composed of box-like compartments which he called cells. He wrote:

> But judging from the lightness and yielding quality of the cork ... that possibly, if I use some further diligence, I might find it to be discernible with a Microscope, I with the same sharp Pen-knife, cut off from the former smooth surface an exceeding thin piece of it, and placing it on a black object plate, because it was itself a white body, and casting the light on it with a deep plano-convex Glass, I could exceedingly plainly perceive it to be all perforated and porous, much like a Honeycomb ...
> ... these pores, or cells, were not very deep, but consisted of a great many little Boxes ...
> I no sooner discern'd these, which were indeed the first microscopical pores I ever saw, and perhaps that were ever seen . . . , but methought I had with the discovery of them, presently hinted to me the ... reason of all the Phaenomena of Cork; ...

At about the same time as Robert Hooke was working in London, a Dutch draper called Antony van Leeuwenhoek was building himself a simple microscope, with lenses of short focal length which he ground in his spare time. With this microscope he observed single-celled organisms in a drop of pond water. He wrote:

> To my great surprise, I found that it contained many very small animalcules, the motions of which were very pleasing to behold. The motion of these little creatures, one among another, may be likened to that of a great number of gnats or flies disporting in the air.

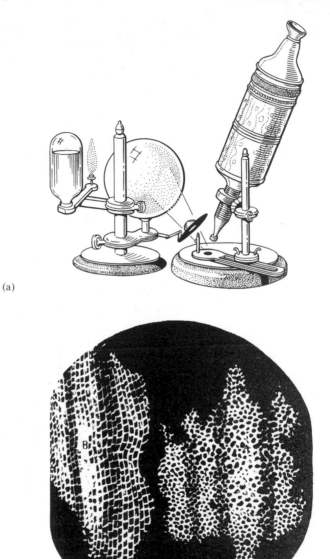

(a)

Fig. 1–1 **(a)** The microscope used by Robert Hooke. It is a compound microscope with an objective and eyepiece lens. The sphere contains a solution of clear brine. Overall magnification about × 270. **(b)** Robert Hooke's drawing of cork. Hooke called the compartments 'cells'; they are in fact plant cell walls

(b)

Leeuwenhoek was also the first to observe the nuclei of living cells, in the red corpuscles of salmon blood, although he did not of course appreciate the significance of what he saw. He noticed, too, that there were bacteria present in the tartar deposits which form on teeth. His work with the microscope led him to doubt the theory, commonly held at that time, that living creatures could be born from dead matter. Leeuwenhoek felt certain that microbes 'as well as all other small living creatures are produced from their like by means of egg, seeds or spawn, according to the nature implanted in them at their first generation'.

These early experimental observations by microscopists were the beginnings of cell biology. By the beginning of the 19th century it was known that all of the various organs of animals were made up of tissues such as muscle, bone, cartilage, and fat; and similarly that the stems, roots, leaves, and other organs of higher plants were also composed of distinct tissues. Dutrochet, in the early years of the 19th century, put forward a unifying theory of the composition of tissues, based on careful studies of both plant and animal material. He concluded that all of these tissues were composed of 'globular cells of an extreme smallness', although there could be many diversities of shape and structure. He further suggested that these cells he observed were held together in the tissues 'only by cohesion'. Dutrochet's ideas represented an important advance, and are still very much in line with present-day thinking on the organization of cells in tissues.

The work of Schleiden on plant cells and of Schwann on animal cells, both published in 1839, led to a clearer definition of cells and their function. Schleiden concluded from his observations on plants that although each cell of every tissue he observed could be looked upon as an independent unit, it must also contribute to the life of the whole organism to which it belonged. Schwann worked on a variety of animal cells at the same time, and came to similar conclusions. The deductions of the two microscopists formed the basis of what came to be known as the *cell theory*. This simply stated more clearly than had been done before that all living matter, from the simplest of unicellular organisms to very complex higher plants and animals, is composed of cells, and that each cell can act independently but also functions as an integral part of the complete organism.

During the next decade, biologists were concerned with the problem of how new cells are formed. Both Schleiden and Schwann thought that minute new cells might be formed inside older ones, or that perhaps cells were formed simply by crystallization from fluid. But studies of developing embryos showed that during growth cells duplicate themselves, one parent cell giving rise to two daughter cells. This process came to be known as *cell division*. These observations were summarized by Virchow in 1855 in the Latin statement given at the head of this chapter, which simply means that all cells are derived from other living cells. Virchow wrote:

Where a cell exists there must have been a pre-existing cell, just as an animal arises only from an animal and the plant only from a plant. The principle is thus established, even though strict proof has not been produced for every detail, that throughout the whole series of living forms, whether entire animal or plant organisms or their component parts, there rules an eternal law of continuous development or continuous reproduction.

About the middle of the 19th century biologists also began to realize that what was actually contained inside cells must be of fundamental importance. Purkinje, in 1839, gave the name 'protoplasm' to the contents of cells. It became clear, however, as structures were observed inside cells, that there could be no single substance which could be called protoplasm. Leeuwenhoek, as mentioned above, had observed that there is a region of different appearance inside salmon red blood cells. Fontana, in 1781, described oval bodies, which must from his description have been nuclei, in eel skin cells. In 1833, Robert Brown observed a dense circular region inside plant cells which he called a *nucleus*. After studying several species of flowering plants he found that a nucleated cell appeared to be common to all of them. Gradually, it became clear that the nucleus is always involved in cell division. Its precise role, however, remained uncertain until the development in the latter part of the 19th century of dyes which specifically stained nuclear material, and made it possible for microscopists to observe the behaviour of the nucleus more clearly.

It is now known that cells contain many different structures which perform a wide range of functions (Fig. 1–3). There are a great many chemical compounds involved in these processes and the interior of the cell is in a state of continuous change and activity. We hope to show how modern ideas in cell biology have developed from the cell theory of the 19th century. But we should like at this stage to give a word of warning. It is true that a study of the biology of cells provides an excellent background for a unified view of general biology, which it would be impossible to achieve in any other way, just as the study of atomic behaviour which developed from the basis of Dalton's atomic theory has provided the background for present-day chemistry and physics. But in the case of biology, too much emphasis has been placed in the past on the structure and function of individual cells, and not sufficient on the way in which cells interact in the development and functioning of whole organisms. This has certainly been due partly, as in other aspects of science, to experimental limitations. It is important, however, to look upon an organism as an integrated whole, with properties which cannot be explained in terms of those of the individual cells.

There is also a more recent danger, which has arisen from the great advances made in molecular biology in the last twenty years or so. Chemists and physicists have sometimes tried to interpret biology largely in terms of the deoxyribonucleic acid (DNA) molecules of the cell nucleus. The DNA molecule is a little like a reference library, in that it carries very important information needed by cells in order to carry out various functions. But DNA operates as a source of information inside a living cell, in which the various substances

micron = 10^{-6} m

angstrom 10^{-10} m or

10^{-4} micron

and activities are balanced with extraordinary precision. When DNA is isolated from cells, and so separated from the substances which carry its information to other parts of the cell, it no more fulfils a function than do the dusty books of a reference library when they are not used as a source of information by enquiring readers.

Lastly, in experimental work in cell biology, the importance of working with living organisms, tissues, or cells cannot be overemphasized. Fixed and stained preparations can only provide limited information about structure in relation to function. We shall try at all stages to consider the functional aspects of all structures observed within the living cell.

1–1
Sizes and shapes
of cells

In this book two units of measurement are used—the *micron* or micrometre (μ), which is 1/1,000,000 of a metre (10^{-6}m), and the *angstrom unit* (Å) which is 10^{-4} microns or 10^{-10}m. The micron is a convenient unit for expressing the sizes of cells, while the much smaller angstrom unit is convenient for recording the sizes of the atoms and molecules of which cells are composed. The nanometre, which is 10^{-9}m, or 10 Å, is now an internationally accepted unit.

Shown in Fig. 1–2 are the sizes of some typical cells found amongst micro-organisms (including bacteria, moulds, and so on), plants, and animals. The logarithmic scale indicating size is necessary in order to show on one diagram the great range of cell sizes. For comparison, the sizes of a hydrogen atom, a haemoglobin molecule, and a virus are marked at the lower end of the scale. (Viruses, strictly speaking, cannot be classed as cells because they consist of little more than a nucleic acid core surrounded by a coat of protein molecules and they cannot exist independently of a host cell in which to multiply.) The smallest free-living organisms are the *mycoplasmas*. These minute cells, the smallest of which are only 1,000 Å in diameter, have been isolated from soil and have been identified as the infective agent in various animal infections. They are sometimes called pleuropneumonia-like organisms or PPLOs, since Iwanowsky first demonstrated their existence in 1892 by showing that the infective agent in pleuropneumonia could pass through filters which were capable of trapping bacteria. The structure of mycoplasmas and other cells shown in Fig. 1–2 will be discussed later in this chapter.

Next on the scale are the *bacteria*, which vary greatly in size from a diameter of 5,000 Å for the cocci up to about 20 μ in length for some of the filamentous bacteria. The shape of bacterial cells also varies widely: cocci are spherical; bacilli are rod-shaped; spirilla, as their name implies, have a corkscrew shape; and filamentous bacteria are thread-like. Bacteria are ubiquitous, occurring in fresh and salt water, in

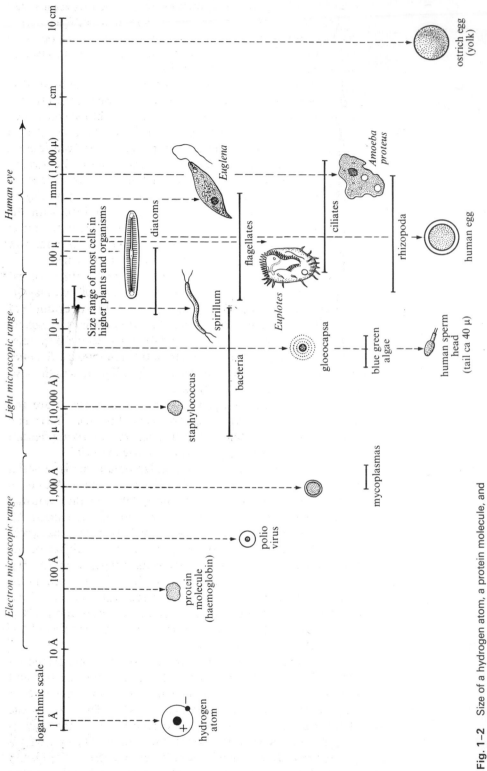

Fig. 1-2 Size of a hydrogen atom, a protein molecule, and a virus compared with the size range of cells of micro-organisms, plants and animals

the soil, and in plants and animals, where their existence may be helpful, harmful, or innocuous. The many different types occur so abundantly that according to one estimate the total weight of bacteria alive at any one time may amount to twenty times the weight of all other living material.

Within the size range of bacteria are the cells of *blue-green algae*. These occur as single cells of approximately 10 μ in diameter (slightly larger than a human red blood cell, which is 7–8 μ in diameter) but are generally loosely associated in colonies. Some filamentous colonies may be several centimetres in length. Blue-green algae occur in fresh water and in oceans and in such widely differing environments as rock surfaces in the desert and cold lakes in Antarctica. Their capacity to proliferate in relatively barren regions, combined with the fact that they may be carried for large distances in the air in the form of spores, means that control of their growth has now become a major problem in certain parts of the world.

Many larger unicellular organisms are motile; for example the *flagellates* and *ciliates* move by means of cytoplasmic projections (flagella and cilia, see Section 1–6). An example of a flagellate is *Euglena*, commonly found in freshwater ponds, soil, and mud. Most of the many different species are spindle-shaped; the largest reach a length of 0·5 mm (500 μ). *Euplotes* is a ciliated organism, approximately 120μ in length, also commonly found in pond water.

The unicellular organisms known as *diatoms* are the most important primary food producers of the sea. These may be up to 100 μ or more in length. Their most striking feature is an outer enclosing case made of silicate material.

One of the largest of unicellular organisms is *Amoeba proteus*, which is about 1 mm (1,000 μ) in length. Amoebae occur mainly in fresh water and are notable for their unusual method of locomotion by means of pseudopodia (Section 1–6).

So far only unicellular organisms have been considered. Most of the cells which make up the tissues of higher plants and animals fall into a very narrow size range, from 20–30 μ (Fig. 1–2). Probably the largest single cell produced in an animal body is the yolk of an ostrich egg, which is about 5 cm in diameter. By comparison, the human ovum is about 200 μ across and the head of the human spermatozoon is about 5 μ long, with a 30–50 μ tail.

The lower limit of cell size is probably determined by the smallest volume into which the minimum number and size of essential cellular components may be fitted in order for independent cellular existence to be possible. The size of some of the essential cellular macromolecules is such that myco-plasmas must be very near to this lower limit. The upper limit to cell size is determined by several factors. (Egg cells form a special class because they contain nutrients for the developing embryo.) The nucleus directs and controls the many activities

PROPERTIES OF CELLS

taking place in the cytoplasm, so that the relation between nuclear and cytoplasmic volume is obviously important; there must be a limit to the amount of cytoplasm which a particular nucleus can control. Some larger unicellular organisms, such as the amoeba *Chaos chaos*, are multinucleate, which overcomes this particular problem. There is also the question of the necessary diffusion of substances from one part of the cell to another, which becomes increasingly difficult with larger cells. Then there is the important consideration of the cell and its environment. The larger a cell becomes, the lower is its surface area to volume ratio; and the more difficult is communication between the cell and its surroundings. The surface area of some cells is increased by, for instance, the formation of long thin protuberances or, conversely, invaginations of the cell surface.

1-2
Structure of cells

With mycoplasmas and blue-green algae there is no clear distinction between the nucleus and cytoplasm (Fig. 1-3a). The nucleic acids, which direct and order the various cell processes such as the synthesis of protein molecules, are all distributed fairly uniformly throughout the cell interior. In bacteria, however, there is a more clearly defined nuclear region, while in the cells of all higher plants and animals the nuclear material is surrounded by a membrane and is clearly visible as a separate compartment in both living and fixed cells (Figs 1-3b and 1-3c). All cells are surrounded by an outer membrane, known as the plasma membrane, which encloses and protects the cell contents.

Recent research in biochemistry and molecular biology has shown that the structure and function of cellular molecules are remarkably similar throughout the plant and animal kingdoms. The four principal classes of cellular molecules are nucleic acids, proteins, lipids, and carbohydrates. The chemical structures of these substances will be considered in Chapter 2.

Nucleic acids are long-chain molecules: they are built of repeating units of comparatively small molecules, joined end to end in a chain. This simple construction gives rise to many interesting physical and chemical properties, which will be described in later chapters. There are two types of nucleic acid: deoxyribonucleic acid (DNA) and ribonucleic acid (RNA). DNA plays a supremely important role in cellular inheritance, in growth and development, in cell division, and in the synthesis of protein molecules and other cell constituents. RNA plays an essential role in the synthesis of proteins.

Proteins are also long-chain molecules, but are composed of different building units (amino acids). There are many different kinds of protein; perhaps the most important are the enzymes, which act as biological catalysts and facilitate the many biochemical reactions occurring inside the cell. Proteins are important structurally in animal and, to a lesser extent, in

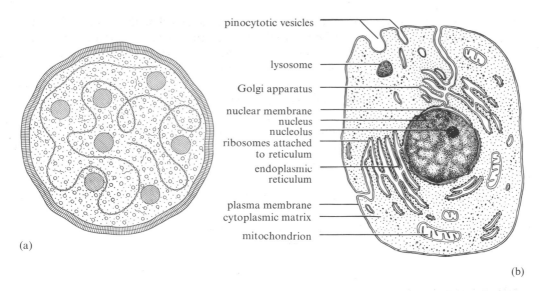

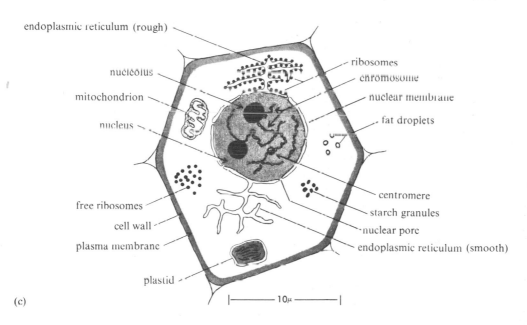

Fig. 1-3   **(a)** Mycoplasmas are the smallest known living cells. This is a schematic diagram of *Mycoplasma gallisepticum*. The double strands are DNA, the large spheres are ribosomes, the short threads are soluble RNA, and the small hollow spheres are soluble protein. The outer coat is a phospholipid–protein bilayer. (With permission from H. J. Morowitz and M. E. Tourtellotte, The smallest living cells. Copyright © March 1962 by Scientific American Inc. All rights reserved.) **(b)** Generalized representation of an animal cell showing the subcellular components. (Reprinted by permission of Dodd, Mead and Company, Inc. from *Biology in Action* by N. J. Berrill. Copyright © 1966 by Dodd, Mead and Company, Inc.) **(c)** Generalized representation of a plant cell showing structures similar to those of animal cells, but in addition plastids (including chloroplasts), starch grains, and the outer cell wall. (With permission from J. D. Watson, *Molecular Biology of the Gene*. W. A. Benjamin Inc., New York, 1965.)

plant cells. Cellular membranes, such as the plasma membrane and nuclear membrane, are composed of a complex of lipid (see below) with protein, which gives them their unique properties. In multicellular organisms, certain specialized cells secrete (produce and export) proteins which are necessary for the functioning of other parts of the organism.

The actual synthesis of proteins is carried out with the help of RNA molecules, which play an intermediate role between the DNA molecule carrying the necessary information and the synthetic process itself. This process takes place at the site of units composed of RNA and protein, known as *ribosomes*. In the case of cells which do not have a nuclear membrane, such as mycoplasmas, the ribosomal units are freely dispersed throughout the interior of the cell (Fig. 1–3a). When a nuclear membrane is present, however, the ribosomes are mostly contained in the cytoplasm. They may be freely dispersed, or, particularly in cells which secrete protein to the surrounding medium, they may be attached to an assembly of interconnected membranes known as the *endoplasmic reticulum* (Figs 1–3b and 1–3c).

The *lipids* are another important class of biological substances. Both the intracellular membranes and the outer plasma membrane of cells contain lipid molecules. These are basically long-chain compounds, but the chains are much shorter than those of nucleic acids and proteins and apart from charged end groups are hydrocarbon in nature, which means that they are insoluble in water. It is the insolubility of lipids which gives membranes the capacity to enclose and protect—for instance, the plasma membrane encloses the cytoplasm and protects the cell from the external environment, and the various intracellular membranes protect regions of the cell which have special functions to perform. Because membranes are complexes of lipid with protein they have very interesting properties concerning the diffusion and transport of various substances. As well as performing an important role, the plasma membrane permits the inward passage of various substances (metabolites) which are necessary for cellular function, and the outward passage of unwanted or harmful substances. Most cells contain a reserve of lipid molecules, which are stored as granules in the cytoplasm.

Carbohydrates or *sugars* form the last major class of biological substances. Glucose, a simple sugar molecule, is the source of energy for most cells. It is produced during photosynthesis by plant cells and is stored in the form of long-chain molecules—*glycogen* in animal cells and *starch* in plant cells. *Cellulose*, the chief component of the wall surrounding the outer plasma membrane in the cells of higher plants, is also a long-chain molecule formed from glucose units. The rigid outer cell wall of bacteria and the more gelatinous one of algae consist mainly of carbohydrate. There are carbohydrate molecules on the surface of most mammalian cells, and they appear to play a part in the specific interactions between cells.

Almost all cells in plant and animal organisms contain *mitochondria* in the cytoplasm. These are compartments, surrounded by a lipid membrane, with a thread-like or rounded form; they contain *cristae*, which are internal sub-divisions (Figs 1–3b and 1–3c). Mitochondria are the sites of energy production in cells and contain enzymes which are necessary for the extraction of energy from the breakdown products of glucose and other foodstuffs, and for then making it available for cellular use. In cells such as bacteria and mycoplasmas, which contain no mitochondria, these particular enzymes appear to be associated with the plasma membrane.

The main intracellular difference between plant and animal cells is that the former contain the green pigment *chlorophyll* and can carry out *photosynthesis*, while the latter do not. The chlorophyll is contained in a special compartment known as a *chloroplast*, which is again surrounded by a lipid membrane and contains highly organized disk-like structures called *grana* interconnected by a membrane system. In chloroplasts, energy from sunlight is trapped by the chlorophyll molecule, and used to make sugars from simple inorganic molecules. Certain bacteria, the Rhodobacteriaceae, are also photosynthetic, and here the green chlorophyll colour may be obscured by the presence of a red or purple pigment. In blue-green algae and in photosynthetic bacteria the structures involved in photosynthesis are much smaller than in other algae and all higher plants, consisting of small bodies known as *chromatophores*. All other cells, including all animal cells, which cannot carry out photosynthetic activity, require a supply of an energy-rich substance (usually glucose) from the environment.

This brief outline of the main components of animal and plant cells is summarized diagrammatically in Figs 1–3b and 1–3c. Although thread-like strands of DNA are indicated in the nucleus of the plant cell, these are not in fact visible in so-called 'resting' cells (cells not actively engaged in division). Resting nuclei in both plant and animal cells have the much more uniform appearance shown in Fig. 1–3b, the only clearly detectable organized structure being the *nucleolus*, which contains most of the nuclear RNA.

1–3
Cells in
association

Although a unicellular organism may show a surprising degree of structural differentiation and function, there is a limit to the possibilities. All higher organisms are multicellular, and in this way overcome many of the limitations imposed by size and shape in single cells. There is considerable division of labour, with many cells sharing the functions carried out by a single cell in a unicellular organism. This results in some cells specializing in certain functions, which leads to greater efficiency. A multicellular organism has also a better insurance against accidents; it can survive the death of some of its component cells, whereas an accident that causes, for instance, the puncturing of a cell wall and membrane may be the end for a unicellular organism.

The way in which an association of like cells can lead to new properties and behaviour may be seen by examining the cellular slime mould *Dictyostelium* (Fig. 1-4a). Slime moulds are classed as motile fungi or as Protozoa. They are not truly plant cells as they cannot photosynthesize, nor, on the other hand, are they sufficiently motile to move in pursuit of food. They flourish only where food materials are available, for

Fig. 1-4   (a) Life cycle of the cellular slime mould *Dictyostelium*. Dry spores change into amoeboid cells, which aggregate around individual cells. A motile aggregate and slug are formed; the latter becomes stationary and develops a stalk which gives rise to a new sporangium. **(b)** The free-swimming plant *Volvox*. A colony is shown with daughter colonies embedded in the matrix, and a section through the surface. (With permission from W. T. Keeton, *Biological Science*. W. W. Norton, New York, 1967. Modified from *College Botany* by H. J. Fuller and O. Tippo. Copyright © 1949 by Holt, Rinehart, and Winston, Inc. Adapted and reprinted by permission of Holt, and Winston, Inc., Publishers, New York.)

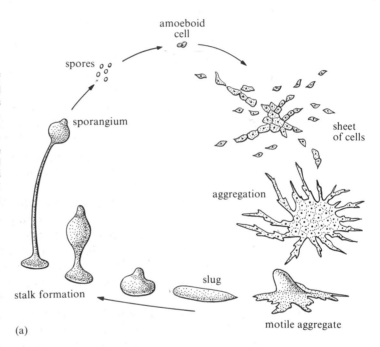

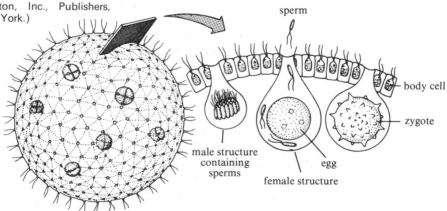

example in decaying vegetation. *Dictyostelium* has a life cycle which involves both cell aggregates and also cells moving independently in an amoeboid way (Section 1–6). When a spore settles, it splits open and releases a single amoeboid cell. This feeds on bacteria, and if it is near such a food supply it will grow and divide. This process continues until a sheet of independent single cells is formed. When the population of cells is sufficiently large, movement of cells begins to form a number of separate centres of aggregation. A substance called *acrasin*, secreted by one or more cells, and with the capacity to stimulate secretion in more distant cells, appears to be responsible for this process of attraction. These aggregates are still motile and may have a length of more than 2 mm, consisting of very many cells. Finally a dense mass or *slug* is formed, and migration ceases. The slug gradually becomes erect, the anterior cells forming a stalk which is encased in a cellulose sheath. The stalk grows by means of further cells gliding up the outside. The last cells to glide up form a round mass at the top, the *sporangium*, inside which they differentiate into spore cells, and the whole process is repeated. There are various questions which are still only partly answered: for instance, what enables an aggregate to move as a unit? How does the slug convert into a stalked sporangium? Does the position of individual cells in the slug determine their final positions in the sporangium? What is clear is that a group of like cells in association can behave in a very different way from the individual cells.

Colonies of cells also occur in simple plants. An example is the multicellular *Volvox*, a freshwater alga. Each aggregate consists of a hollow sphere, about the size of a pin head, composed of about 500 to 50,000 cells embedded in a gelatinous matrix. These cells are very similar to the unicellular alga *Chlamydomonas*, possessing a nucleus and cytoplasm, a chloroplast, an eyespot, and two flagella which enable them to be motile. In *Volvox*, each cell is linked to its neighbours by fine cytoplasmic strands and is so arranged that the flagella project outwards beyond the gelatinous layer (Fig. 1–4b). The *Volvox* sphere acts as a single organism. It has polarity, in that when it swims through the water one pole of the sphere is always foremost; as well as moving forward, the sphere also spins. This implies considerable coordination of the beating of the flagella of individual cells, since otherwise only random motion would result. Thus the cells are organized into a working unit which moves quite differently from the individual cells. There is some slight degree of cell specialization, as a few cells are larger than the rest and are usually found at the rear of *Volvox* when it is swimming. These cells are specialized for reproduction, which may be asexual (vegetative) or sexual; some daughter colonies still embedded in the matrix of the parent colony are shown in Fig. 1–4b.

In aquatic plants, supported and surrounded by water, the chief requirement is an adequate light supply for photo-synthesis, which means that the plants cannot live too far below the surface. So there is no problem for non-motile organisms in shallow water; they may also exist in deeper water if they possess a means of anchoring themselves at a level not too far below the surface. Plants of this kind do not have to be complex multicellular organisms with many specialized cells. For instance, the sea lettuce *Ulva* (a green alga) consists essentially of two rows of cells (Fig. 1–4c) embedded in a mucilaginous material in a thin expanded sheet, the *thallus*, which is about 30 cm in length and anchored by branching outgrowths of specialized cells. The cells of the thallus are arranged with the nucleus on the inner side, and the single choroplast of each on the outer side. Since the sheet is thin, water and minerals are rapidly transported and light is easily transmitted to the choroplasts. Any cell can change from a vegetative to a reproductive cell; in the smaller diagram of Fig. 1–4c the two end cells are producing reproductive cells.

By comparison, the problems of plants growing on land are considerable. They are no longer supported by water and they must still obtain and store water to prevent the tissue from drying up. The general pattern in higher plants involves the growth of roots downwards into the soil to ensure the supply of water and minerals, then a stiff stem or trunk, above which are supported the leaves, which need as much light as

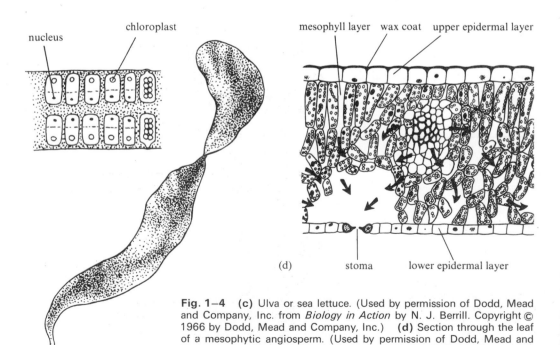

nucleus   chloroplast   mesophyll layer   wax coat   upper epidermal layer

(d)   stoma   lower epidermal layer

(c)

Fig. 1–4 **(c)** Ulva or sea lettuce. (Used by permission of Dodd, Mead and Company, Inc. from *Biology in Action* by N. J. Berrill. Copyright © 1966 by Dodd, Mead and Company, Inc.)   **(d)** Section through the leaf of a mesophytic angiosperm. (Used by permission of Dodd, Mead and Company, Inc., from *Biology in Action* by N. J. Berrill. Copyright © 1966 by Dodd, Mead and Company, Inc.)

possible for photosynthesis. The stem is essentially a continuous two-way transport system between the roots and the leaves. A high degree of cellular differentiation is clearly necessary in such a complex organism. Figure 1–4d shows a diagrammatic section through the blade of a leaf. The upper layer consists of epidermal cells covered by a wax coat, which prevents loss of moisture by evaporation. Light passes through the epidermis to the thick mesophyll layer of chloroplast-containing cells, where photosynthesis takes place. A cross-section of a vein, or vascular bundle, providing water and mineral salts from the roots, is also shown (the direction of movement from the vein is indicated by arrows). Finally there is the lower epidermal layer, containing openings known as stomata, the size of which is regulated by special guard cells to control the entry of carbon dioxide (required for photosynthesis) and of water.

In animals there is a very large variation and an enormous range of sizes, from simple multicellular organisms to exceedingly complex ones. Like the unicellular Protozoa, animals have to obtain other organisms as a source of food. This they do either by moving about in pursuit of prey, or by manipulating their environment so that food is brought to them. The whole organism has to be directed to a particular way of life from amongst those possible in its environment, and to ensuring the health and activity of its constituent cells as well as to coordinating their functions.

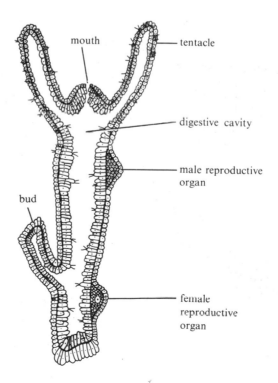

mouth

tentacle

digestive cavity

male reproductive organ

bud

female reproductive organ

Fig. 1–4 (e) Section through *Hydra*, a coelenterate. (With permission from Biological Sciences Curriculum Study, *Biological Science: Molecules to Man*. Houghton Mifflin Company, Boston, 1963.)

Sponges are an example of animals which are composed of relatively few cell types and show a low grade of organization. They are essentially an arrangement of epithelial cells united together on a supporting base, usually of horny or siliceous material which forms a filtration system for extracting the necessary foodstuffs from water. Sponges are stationary, remaining attached to a sea or river bed while water passes over them.

More cell types and a greater degree of organization are seen in *Hydra*, one of the invertebrate coelenterates. It occurs in fresh water and has a supple hollow body, about 1 cm in length (Fig. 1–4e). Its base attaches to a substrate, and at its opposite end is the mouth, which is surrounded by a ring of tentacles and is the only opening in the body, both for taking in food and rejecting waste material. The body and tentacles are basically two cell layers thick, and consist of musculo-epithelial cells, separated by a colloidal layer. The outer layer forms the body wall and the inner one lines the digestive cavity; many of the cells are flagellated. Because these epithelial cells contain contractile muscle fibrils, the tentacles and body can shorten and re-lengthen, or undulate, as required. This muscular activity is controlled by a nervous system; sensory cells placed between the epithelial cells are connected by a network of nerve cells lying between the two cell layers. Sting cells present in the body wall are for paralysing prey, which is then fed into the mouth by means of the tentacles. The prey is slowly drawn inside, where cells in the mouth region secrete digestive enzymes which cause it to

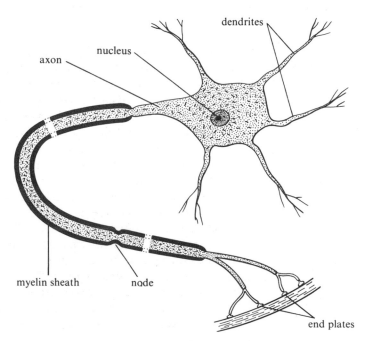

Fig. 1–4 **(f)** Motor neuron cell of vertebrate.

disintegrate. Reproduction is by budding of the two-layered epithelium from the body wall, or in certain conditions by sexual reproduction. The entire organism, which consists of less than ten basically different cell types, shows a considerable degree of organization; it is even able to move by turning somersaults.

As in plants, a multicellular animal requires considerable adaptation to exist out of water. First, the body must have adequate support, which has resulted in the development of a skeletal structure; then there is the problem of locomotion, involving the development of limbs, which lift the body clear of the ground; other changes are necessitated by the requirements of respiration, a circulatory system, and the conservation of moisture. The mammal is the most advanced type of animal, and in higher mammals the degree of cell differentiation and interaction reaches a very high level indeed. An example of a highly differentiated cell is the motor neuron cell shown diagrammatically in Fig. 1–4f. The function of a nerve cell is to transmit a stimulus from one place to another. Extensions known as dendrites carry signals towards the cell body, and the axon carries them away from it. The axon is covered by a layer of myelin and the nodes which occur along the axon play an important part in the speed of conduction. When the signal reaches the motor end plates it is transmitted to a muscle cell and sets in train the contraction of the muscle.

Man possesses the most highly developed brain of all the mammals and has very many different specialized cells in the various organs and tissues, the functions of which must all be organized and coordinated to ensure the efficient working of the whole complex organism.

1–4 Why cells need energy

A continuous supply of energy is necessary just to maintain the high degree of organization within cells. (As the second law of thermodynamics states, isolated systems tend spontaneously towards a greater degree of disorganization.) When the flow of energy to a cell ceases, intracellular organization breaks down and the cell dies. The energy supply to a cell is used to do work of various types, all of it being directed towards the maintenance and reproduction of the cell's own structure and function. For instance, chemical work is involved in the synthesis of cellular materials such as proteins, and in particular for all of the various synthetic processes taking place during growth and cell division and in the repair and replacement of different tissues in multicellular organisms. Osmotic and electrical work in cells is concerned chiefly with controlling the passage (in both directions) of molecules and ions across the plasma membrane (Section 2–14). Mechanical work is required for various cell movements, such as muscle contractions, or the beating of cytoplasmic processes such as flagella, or again for the movements involved during cell division.

All organisms depend on photosynthesis as a source of energy, either directly as in plants, or indirectly in animals, which ingest as food the various carbohydrates, fats, and proteins produced by plant cells by the utilization of light energy. The chemical breakdown of these foodstuffs by the respiratory metabolism (Section 2–8) of cells releases energy. This is used, as is the energy derived from photosynthesis, in the synthesis of a phosphate compound, adenosine triphosphate or ATP (Section 2–9). ATP can then act as an energy donor for the wide variety of biochemical reactions inside cells. ATP is thus a means by which the energy derived from sunlight or from food metabolism may be trapped within cells in a controlled way and released when required.

1–5
How cells grow
and multiply

When single cells reach a certain size, they divide to form two daughter cells. The size reached before division varies widely from species to species. The actual process of cell division, however, known as *mitosis*, is remarkably similar for all cell types. The various stages involved in mitosis are described in Chapter 9, and here all we need say is that mitosis involves a division of the parent cell so that the two daughter cells are more or less equal. It is clear that a highly organized nuclear division must be involved, with identical replicas of the DNA genetic material of the parent cell passing to the daughter cells. If this were not so, the genetic information would be different in the daughter cells, and the cell character would not be preserved. Duplication of all the other cellular materials, RNA, protein, lipid, and carbohydrate molecules, must also occur, which involves division of the cytoplasm as well as of the nucleus. Since many of the bodies being duplicated are relatively numerous in the cytoplasm (e.g., mitochondria, ribosomes), exactly equal distribution in the daughter cells is not essential.

Cell division is the way in which reproduction occurs in most simple unicellular organisms, such as mycoplasmas, amoebae, single-celled algae, and, generally, in bacteria (see below). With the exception of amoebae, these cells do not possess an organized nucleus surrounded by a membrane, so that the process is somewhat simpler than that of mitosis. Again, however, it is essential that exact copies of the parent cell DNA are passed on to each daughter cell. Reproduction in rod-shaped bacteria is shown diagrammatically in Fig. 1–5. The cellular components are duplicated, and the cell increases in size lengthwise. When it has reached a certain length, and new cell wall material has been formed in the middle, the cell divides transversely to form two daughter cells. Although reproduction of an asexual type is mostly restricted to microorganisms, it can occur in some multicellular organisms. It may take place by means of an outgrowth from the parent, known as 'budding'. This is the way in which new cells are formed in the unicellular fungus, yeast. The bud grows and develops until it has the form and function of the parent cell,

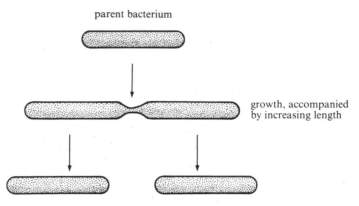

parent bacterium

growth, accompanied by increasing length

Fig. 1-5 Reproduction by simple division in bacteria

division, resulting in two identical daughter cells

and then becomes detached. A similar process can take place in *Hydra* (Fig. 1–4e), a bud of the two-layered epithelium forming on the side of the body and developing to form a complete new individual. The formation of runners in certain plants—for example, the strawberry—is again a closely related process. In this case the outgrowth remains attached to the parent even after it has taken on the adult form.

The capacity to regenerate certain tissues, possessed by most embryonic animals, is still present in some mature animals and plants and is a type of asexual reproduction. This is true of planarians (flatworms), which seldom exceed 0·5 cm in length. If a region of the body of a planarian becomes constricted, and the section of the body on the other side of this restriction is pinched off (Fig. 1 6), then this section develops a new head and new internal organs. Similarly, cuttings from plants such as roses will develop roots when placed in soil and will eventually grow into a complete new plant. An 'eye' of a potato can give rise to a potato plant, and a single cell of

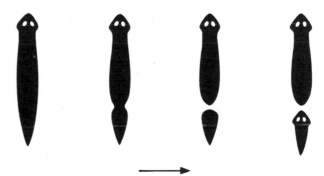

Fig. 1-6 Growth of a new individual from the pinched-off tail in flatworm

carrot root tissue has the capacity to grow, undergo mitotic divisions, and develop into a carrot plant.

In most highly developed plants and animals, however, reproduction is sexual. A type of mating can also occur in bacteria (see Section 10–6); this is known as *conjugation* and involves the transfer of DNA from the male bacterium into the female. A similar process can occur in many ciliates. In this case each cell forms two nuclei, one of which acts as a female and the other as a male during conjugation. In higher organisms, however, sexual reproduction always involves the formation of specialized male and female germ cells, or *gametes*, which unite to form a *zygote*, from which the new individual develops. This arrangement leads to considerable advantages from the evolutionary point of view; it allows variations in the composition of the genetic material of individuals, and so organisms can adapt to a changing environment. Gamete cells are *haploid*, which means that they contain half the amount of genetic material possessed by the ordinary *diploid* cells of the body. The formation of gametes involves a special process of reduction division called *meiosis*, described in detail in Chapter 10. When the gametes fuse, the zygote so produced contains the diploid amount of genetic material, half of it being derived from one parent and half from the other.

In some very simple organisms the gametes that fuse to form a new individual appear to be identical. An example of

Fig. 1–7 Life cycle of the green flagellate, *Chlamydomonas*. Individual cells associate in pairs and fuse together to form a zygote, which becomes encapsulated. Meiotic divisions occur and the capsule opens to release new biflagellate free-swimming individuals. These can divide mitotically (asexual reproduction) but eventually they fuse again to form a new zygote

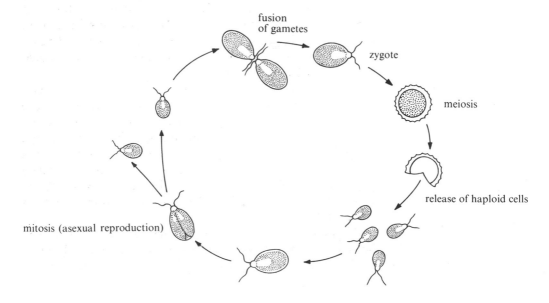

fusion of gametes

zygote

meiosis

release of haploid cells

mitosis (asexual reproduction)

this is the flagellated green alga *Chlamydomonas* (Fig. 1–7). The zygote formed from apparently identical gametes becomes encysted, undergoes meiosis, and liberates haploid cells, each of which can reproduce asexually indefinitely. Eventually conditions are suitable for haploid cells to function as gametes, which pair off, fuse, and form a new generation of zygotes.

In the majority of higher plants and in many lower organisms, male and female gametes occur in the same parent. For instance, it was mentioned earlier that reproduction in *Volvox* may be sexual. The specialized reproductive cells can give rise to a number of male gametes or to a single large female gamete and in certain species both types occur in the same individual. This has obvious advantages for survival when the chances of a meeting between two individuals with gametes of the opposite sex are poor. This arrangement also occurs in some lower animal organisms, for example some species of *Hydra* (Fig. 1–4c), where male and female reproductive organs can occur on the same individual. In *Hydra* and in *Volvox* reproduction can be either asexual or sexual, depending upon the environmental conditions—again, a clear advantage for survival.

The main difference between the reproductive process in most plants and in animals is that in plants meiosis does not result immediately in gamete formation. Instead, the haploid cells so formed go through a period of growth and development, resulting from mitotic division, at the end of which true gametes are formed. The length of this haploid growth phase varies greatly in different plants. Mosses are an interesting example of plants which spend most of their life cycle in the haploid condition. The diploid organism resulting from the zygote formed by the fusion of male and female gametes is not self-sufficient, but remains attached to the haploid parent and is dependent on it for part of its food supply. When it is sufficiently mature meiosis takes place, resulting in haploid spore cells, which under favourable conditions will develop into haploid moss plants (cf. reproduction in the mould *Neurospora,* Section 10–5).

The most highly developed plants are the flowering ones, in which the diploid condition is the usual state of the mature organism. The haploid stage of growth is very much reduced, and there is a self-sufficient diploid stage. Other adaptations which have taken place involve highly efficient ways by which the male gametes of one individual can reach the female gametes of another, and also seed formation, by which the young plant developing from the zygote is provided with food and protection.

In some lower animals male and female gametes may also be produced on the same individual. In higher animals, however, particularly in vertebrates, the usual situation is for an individual to produce, by meiotic division inside a specialized organ, either male gametes (*sperm*) or female gametes (eggs or *ova*). Each sperm is highly motile, possessing a tail (or flagellum) which assists transport to the non-motile ovum. In most aquatic animals, fusion of egg and sperm takes place after both types of gamete have been shed into the surrounding medium, and so the process is one of *external fertilization*, which is clearly uncertain and inefficient; many sperm may not come into contact with an egg, and many eggs may not be fertilized. There is considerable loss, too, because the developing young are not well protected and die or are eaten.

In most higher animals, *internal fertilization* of the ovum occurs within the female, a much more efficient process. In reptiles and birds the fertilized egg when laid is enclosed in a protective shell, providing an aqueous environment and a food supply for the developing embryo. The most highly developed reproductive pattern is that of mammals, where the embryo develops in a well-protected state inside the body of the female, and is provided with nourishment by her for some time after birth.

1–6
How cells move

Even for photosynthetic unicellular organisms it may be an advantage to be able to move to obtain the optimum conditions of sunlight. Organisms which do not photosynthesize have to be able either to move in search of food, or to make the food come to them.

Cellular locomotory mechanisms may be divided into three main groups: those which depend upon the activity of specialized structures on the cell surface, such as flagella or cilia; those which involve a streaming movement of the cellular cytoplasm, as in amoeboid movement; and, in higher animal organisms, those concerned with muscle fibre contractions in specialized structures of muscle cells bound together by connective tissue, or in sheets of musculo-epithelium.

Cilia and flagella are whip-like processes extending outwards from the cell surface. They are actively contractile, which enables them to perform a rhythmical beating or whip-like movement capable of moving a single cell along. In multicellular organisms this movement causes liquid to move over the cell surface. Cilia and flagella are found in all animal organisms, from Protozoa to man, and also on the cells of certain lower plants. They were first observed on Protozoa, and their presence on the epithelia of higher animals was observed about the mid-19th century. Electron microscopy has shown that there is no fundamental difference in the structures of cilia and flagella, apart from their length. They are either short in relation to the cell size, and numerous, in which case they are known as cilia, or they are relatively long and few (often only one) and in this case are known as flagella.

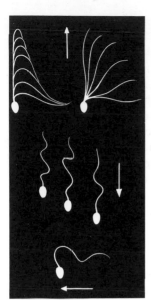

Cilia may be 5 μ to 10 μ in length and flagella up to about 150 μ. Their shafts consist of a regular arrangement of protein fibres (see Section 11–4) encased in an outer membrane continuous with the cell membrane. Movement is due to a beating or power stroke of the shaft, in which it is extended fairly rigidly, followed by a recovery stroke in which it is swept back by bending at the base (Fig. 1–8). Flagella, being considerably longer than cilia, can carry out more varied movements, but these are mostly variations of the same basic stroke. It is clear that highly coordinated contractions of the fibres (see Section 11–4) must be involved in this beating process, but the actual mechanisms are not clearly understood. There must also be a conduction of impulses, as in muscle cells; the necessary energy supply appears to come from the region at the base of the shaft.

The group of flagellated unicellular organisms contains a variety of different types, ranging from those which contain chlorophyll and carry out photosynthesis, like *Euglena* (Fig. 1–9), and those which are much more animal-like in their behaviour, such as *Trypanosoma.* Others, the dinoflagellates, have an armoured cell wall made up of plates of cellulose fitted together. The reproductive cells of lower plants, usually only the male, and the spermatozoa of animals are also each equipped with a flagellum. Electron microscopy has shown that the sperm tail has basically the same design as the cilium and flagellum. A spiral twist of the tail, producing a propeller effect, is particularly noticeable in spermatozoa.

Cilia are effective as a means of locomotion only in very small organisms (an example of a ciliated unicellular organism is *Euplotes,* Fig. 1–2). In more complex multicellular organisms their function is to move materials along. An example is the mantle cavity of bivalve molluscs such as the clam (Fig. 1–10). The rhythmic action of cilia on the gills draws water containing

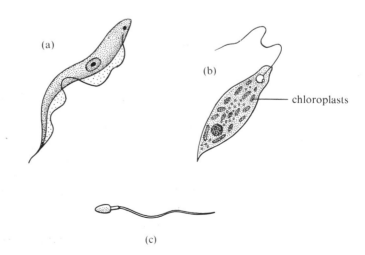

(a)

(b)

chloroplasts

(c)

Fig. 1–9 (a) *Trypanosoma,* length ca. 50μ. (b) *Euglena,* length ca. 200μ. (c) Human sperm, length ca. 50μ.

microscopic food particles into the cavity. These are sorted out by a ciliary mechanism and directed to the mouth, while the strained water escapes via the exhalant siphon. In man, the trachea is lined with ciliary epithelium; in this case the rhythmic beating together of cilia on neighbouring cells filters out unwanted impurities such as dust particles, and washes the cell surfaces clean by keeping a liquid film moving past them. The various sensory cells in most animals are modified ciliated or flagellated cells. Here the direction of the impulse is reversed, since it is in response to an external and not an internal stimulus. In the complex light-sensitive rod cells of the vertebrate retina the membrane and fibre structure of cilia have been modified to receive light and convert it into an electrochemical stimulus.

The second of the three types of movement mentioned above is the cytoplasmic streaming seen in amoebae and many other Protozoa, in certain slime moulds, and in the vertebrate white blood cells called macrophages. A simpler type of cytoplasmic streaming occurs in plant cells. In amoebae, movement takes place by continual changes in shape of the

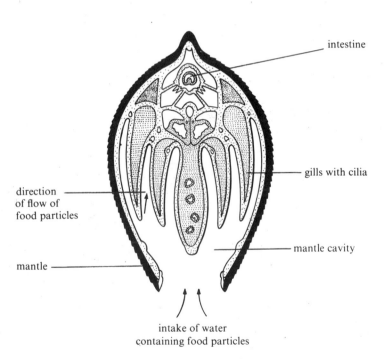

direction
of flow of
food particles

mantle

intestine

gills with cilia

mantle cavity

intake of water
containing food particles

Fig. 1–10 Vertical cross-section of a bivalve mollusc, showing mantle cavity

cell. As the result of cytoplasmic streaming, projections of the cytoplasm known as *pseudopodia* are continually being formed (Fig. 1–11a). These are simple extensions of a very flexible cell membrane, any portion of which can be extended outwards. Pseudopodium formation means that the cell is being continuously built up at the front or advancing end, and at the same time being drawn in and shortened at the rear end. It is only the inner *endoplasm* which flows forward in the direction of movement (Fig. 1–11b). When it reaches the front of the cell, the endoplasmic material flows to the sides of the cell and becomes transformed into the stiff, non-moving *ectoplasm*. At the rear of the cell, the opposite process is taking place, ectoplasmic material changing to fluid endoplasm which then forms part of the forward-moving stream. The forces involved and the precise mechanisms are highly complex, and various theories have been proposed to explain them (see Chapter 11).

In complex animal organisms the cells most specialized for movement are muscle cells. Their effectiveness depends upon the contractile properties of muscle fibres, which result from changes in form and shortening of the proteins of which the fibres are composed. Thanks to electron microscopy, considerable advances have been made in relating muscle behaviour to events at the molecular level. These will be discussed in detail in Chapter 11. Muscle tissue is generally classified as either striated or smooth, depending upon its appearance under the light microscope, the former having

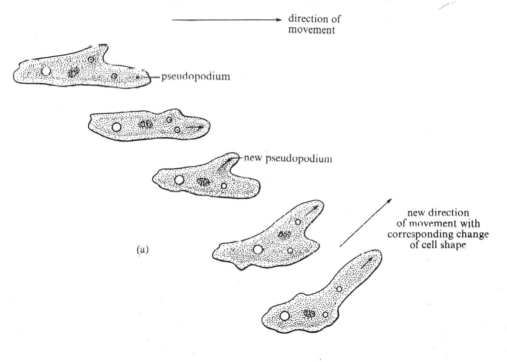

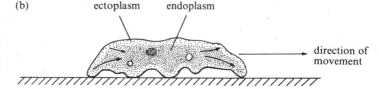

Fig. 1–11 Movement in an amoeba. **(a)** Formation of pseudopodia. **(b)** Transverse section showing movement of cytoplasm

striations of microscopic dimensions (Fig. 1–12). This also corresponds to a difference of function; muscles involved in limb movements, which act quickly and are under voluntary control, are striated, while muscles of the gut, arteries, and uterus, which are slow-acting and involuntary, are smooth. The sheets of smooth muscle which line tubular structures such as the digestive tract are particularly good at moving bulky material of a type that cilia, only able to move particles or liquid within their own short reach, could not deal with. These sheets of muscle are capable of a wavelike action, known as *peristalsis*, in which a wave of muscle relaxation passing along the tube is followed by a wave of contraction. By this means, material is pushed along inside the tube; this is the way in which blood flows through arteries, and in which food is transported in the digestive system of higher animals. It is also the basis of some forms of locomotion; for instance, the flatworm has a flattened ventral surface which facilitates movement on a substrate. Movement takes place

Fig. 1–12 Electron micrograph of striated rabbit muscle (magnification × 21,000). Individual fibrils lie in east–west direction. Each fibril is traversed by a regular banded structure of A-band, I-band, Z-line, and H-zone. (With permission from H. E. Huxley.)

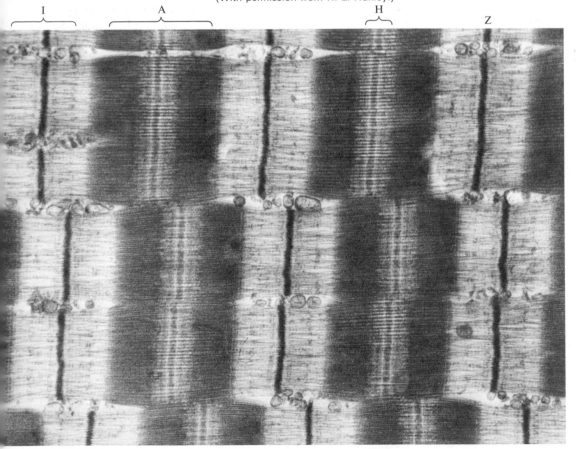

partly by waves of peristalsis passing along the muscle sheets from head to tail, and partly by rhythmical beating of the cilia of the ventral surface against a slime track secreted by specialized gland cells.

The locomotion of animal tissue cells as observed *in vitro* is a complex process involving undulating movements of the plasma membrane and interactions between membranes of different cells. It will be considered further in Chapters 11 and 13.

1–7
Life-span of cells

Unicellular organisms can achieve a certain kind of immortality; if they are not destroyed by some accident, they will eventually go into mitosis and produce two daughter cells, from which in turn new daughter cells will arise. In the laboratory cells can grow and divide under optimal conditions, with plentiful supplies of nutrients and adequate space to accommodate the growing cell population. Under natural conditions, however, the environment is rarely so favourable. Limiting factors for a particular unicellular organism may be the light and temperature conditions; there may be a shortage of nitrogen for protein synthesis, or of phosphorus for energy processes, so that cells compete with one another for limited resources and many may die. Motile cells have the advantage of being able to move to new surroundings when there is overcrowding. But because of competition immortality at the molecular level of unicellular organisms is seldom realized in practice. An organism must change with its environment if it is to survive.

In multicellular organisms the situation is very different, because the organism as a whole is able to outlive the life-span of its individual cells. Apart from the processes of wound healing and repair, which take place in response to a particular injury, individual cells die and are replaced continually throughout the life of the organism. In mammals and birds, the high body temperature means that there is a high cellular metabolic rate which tends to shorten the life of the individual cells, so that the difference between their life-span and that of the whole animal is more marked than in other organisms. Exceptions to this are the germ line cells and those of the developing zygote (see Chapter 12). Turnover in the central nervous system and the muscular system is very slow; the cells in the brain and spinal cord are not replaced, so that those present in a young animal must last a life-time. This may be connected with the fact that nerve cell replacement would interfere with the maintenance of stable nerve circuits. It is possible that in many other organs most cells of the young adult last the entire life-span.

There is a continuous turnover of certain cell types, however, even in young animals, the processes of cell death and new cell growth taking place at the same time. With a young organism still in its growth phase the production of new cells must be in excess of the death of old cells. When the organism reaches maturity, however, a balance is attained between the birth rate and death rate of cells in the various tissues. Each cell type has its own rate of turnover; the white phagocytic cells of the blood exist individually for approximately 1 to 3 weeks, red blood cells live for about 120 days, while bone and cartilage cells are relatively long-lived and their replacement is very slow. The processes of growth and replacement are regulated by various growth controlling mechanisms (Chapter 9), which operate to maintain the structure and coordinated function of the various parts of the organism. In malignant disease (cancer) there is a breakdown in this process—a cell (or cells) escapes from the usual growth control restrictions and is able to grow and divide, and to move more freely than normal cells (Chapter 13). This results frequently in the ability to invade and damage normal tissues, and to spread by way of the blood or lymphatic systems to other regions of the organism. Malignant cells may arise in any tissue in which cell proliferation takes place. One theory is that they are due to somatic mutations—that is, to mutations in cells other than the germ cells of an individual (Chapter 10). The number of mutated cells present in various tissues must increase with the age of the individual.

Failing disease or injury, a multicellular organism dies from the process of *ageing*, which takes different forms in different types of cell. Ageing is still imperfectly understood, but it seems to be connected with the high degree of specialization required for carrying out specific functions, which is known for many cell types to lead to loss of capacity for cell division. The decrease in rate of production of new cells when an organism reaches maturity, so that a balance between cell death and cell production is achieved, has been mentioned already. However, there is a continued progressive decrease in the production rate with increasing age of the organism, although the life-span of individual cells does not appear to alter significantly. This means that not all dying cells are being replaced, and there is a loss of weight and loss of function in tissues and organs; for instance, in animals the heart does not pump so much blood in a given time as it did when the individual was a young adult, the blood is not oxygenated so efficiently as it passes through the lungs (which have also lost some of their capacity), the actual blood volume diminishes, partly because tissues retain less water and partly because the production of red blood cells slows down, and so on. All of this means that a greater burden is thrown on the remaining cells.

Hormonal changes affecting adversely the functioning of various organs, the accumulation of waste material from various metabolic processes inside certain cells, loss of elasticity in elastic fibres, and an increase in the amount of collagen in connective tissue—all play a part in ageing, resulting in such well-known changes as stiffening of the joints and hardening of the arteries. No new nerve cells are produced, but there is a replacement of intracellular material at the molecular level, and this process also becomes slower with advancing age. In the absence of death from injury or disease, the changes brought about by ageing eventually result in death due to the breakdown of some vital structure, such as the rupture of a blood vessel or the occlusion of an artery.

The reasons why the various processes of ageing take place are not yet clear. One theory is that in mammals, at least, radiation may be a contributory factor, particularly by X-rays and cosmic rays. Since exposure to radiation is known to decrease the life of an individual, perhaps this is just due to a speeding up of changes which would anyway have occurred at a later time. However, laboratory work on mice has indicated that radiation seems to alter the order in which different causes of death appear, so it would appear not to be a simple acceleration process.

Certainly, there is an enormous range in the potential life-span of different species of both plants and animals, and there does seem to be a relationship between continued growth throughout the lifetime of an individual and relatively few symptoms of ageing. For instance, in plants there are the giant sequoia and redwood trees at one end of the longevity scale, living to thousands of years and always continuing to grow, whereas annual plants largely cease to grow soon after the flowering period when the fruit begins to form, after which they soon age and die. Among animals, the process of ageing appears to be accompanied by obvious external changes in warm-blooded animals, whereas in some long-living animals such as the sturgeon (with a recorded life-span of more than 100 years) and the tortoise (a life-span of more than 150 years), in which continuous growth during life occurs, the process of ageing is so slow as to be almost undetectable. These observations have led to the suggestion that an animal capable of almost indefinite growth might not age at all.

Another suggestion is that the changes characteristic of ageing are inherent in the genetic make-up of an individual, and that the environment can only speed up or slow down the process. It was mentioned earlier that the number of somatic mutations in an organism must increase with age. When mammalian cells are cultured *in vitro* their lifetime is restricted to a certain number of cell generations, at the end of which they either die out or sometimes spontaneously develop into what is called an established cell line. The latter can be propagated indefinitely, and its cells usually have a different chromosome complement from that of the original animal, and show other differences in culture. Human and chick fibroblasts appear not to be able to develop into cell lines (Hayflick, 1966), but after a certain number of generations at constant growth rate there is a slowing down for several generations and the cells finally die out. Since this slowing down is accompanied in human fibroblasts by an increased incidence of chromosomal abnormalities and a change in morphology, it has been compared to ageing processes in organisms.

To understand all of the processes mentioned in this section it is necessary to know a great deal more about the processes of development in general because the changes due to ageing are fundamentally a continuation of a course of development which began in the embryo stage (as in the increase in number and thickness of collagen fibres in connective tissue, and the thickening and loss of elasticity of elastic fibres). The problems of cell death (which plays an essential role in the development of the animal embryo, see Section 13–7), cell growth, malignancy, and ageing are all different aspects of cell development, about which much remains to be known.

In this chapter the importance of considering cell biology as an introduction to general biology has been emphasized. But even in such complex problems as the ageing of organisms the importance of cellular changes has become clear. The continued life of organisms depends on the maintenance of the function of the individual cells. In the following chapters attention will be concentrated entirely on these cellular functions. In Chapters 12 and 13, which deal with the cell in development, we shall return to the function of cells and their control by other cells in the organism. This book must clearly have a limited scope and cannot deal with more than a few of these aspects of cellular function in higher organisms.

**1–8
Experimental
methods:
ways of looking
at cells**

The importance of new experimental methods in cell biology was emphasized in the introduction to the chapter. These techniques can only be learnt by practical experience in the laboratory. In the section which follows, the general principles of some of the more important experimental methods used are described; the practical details have not been dealt with at length.

Compound microscope. Objects less than 50 μ across can only be seen under the microscope. There are many kinds of microscope, each suited to a particular purpose, of which the most generally used is the compound microscope.

Modern laboratory microscopes have been developed from the original compound microscope of Hooke. The path of rays through such a microscope is shown diagrammatically in Fig. 1–13. In the modern microscope the objective lens is composed of several elements; this ensures that variation in focal length with wavelength (chromatic aberration) and spherical aberrations are eliminated. High resolving power is achieved by use of a large aperture (large cone of convergent rays) in the condenser (Fig. 1–14).

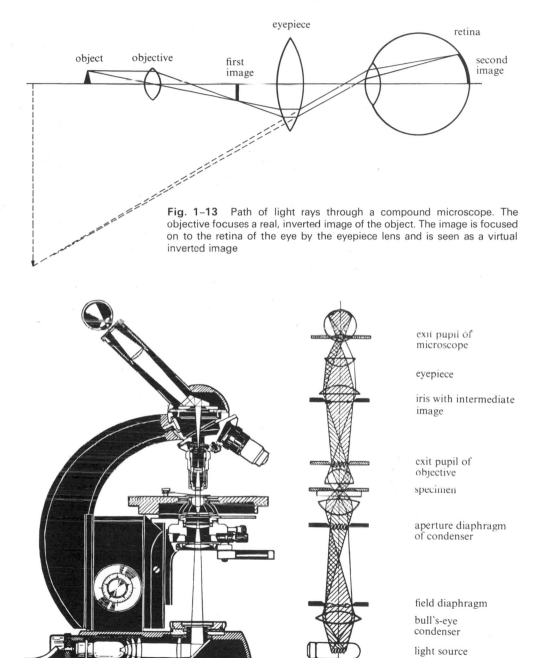

Fig. 1–13   Path of light rays through a compound microscope. The objective focuses a real, inverted image of the object. The image is focused on to the retina of the eye by the eyepiece lens and is seen as a virtual inverted image

Fig. 1–14   Sectional diagram of a modern research microscope. Path of rays arranged for high definition (Köhler illumination). (Reproduced by permission of Carl Zeiss, from R. Schenk and G. Kistler, *Photomicrography*. Chapman and Hall, London, 1962.)

Because living cells are almost transparent, it is extremely difficult to see much structure in the optical microscope. Certain so-called vital stains, which do not kill the cell, can be added to the medium. Neutral red, for example, does not kill cells at low concentrations. But most early microscopy was carried out on cells which had been fixed and stained before examination. A fixative is a chemical substance which makes proteins coagulate. Most of the organelles, such as the nucleus, chromosomes, and mitochondria are preserved in their original shape after fixing. The fixative can be an acid, such as acetic acid, an organic solvent such as alcohol or acetone, or a water-soluble substance which reacts with certain groups on the protein molecules of the cytoplasm. Formaldehyde is a fixative of this kind. After fixation the cells can be washed and they remain stable in water. A dye solution is then added which stains the various cellular components. The type of stain chosen depends on the particular organelle which is of interest. For example, Feulgen stain is used to demonstrate the presence of DNA, and an excellent contrast between DNA (green-staining) and RNA (red-staining) is obtained with a mixture of methyl green and pyronin. Carmine, eosin, and crystal violet are other well-known stains used in cytology. After staining, the cells may be washed in alcohol, transferred to xylene and mounted beneath a coverslip in Canada balsam.

The study of cells inside whole tissues has been almost entirely on such fixed and stained specimens, because it is not possible to see the cells at any depth below the surface in whole organisms. In this case the fragment of tissue is embedded after fixation in the following way. A bath of molten wax is prepared; the specimen is removed from the xylene and transferred to the wax bath. When cold the specimen is transferred to a microtome for sectioning. In some cases a monomer, which sets later to form a solid plastic block, is used in place of molten wax. The sections are cut as shown in Fig. 1–15. The sections float off onto the water surface and may be picked up by raising a glass slide from below the surface of the water. They can then be stained and mounted in the same way as single cell suspensions. This is a very brief outline of the technique which is widely used as a routine in histology and pathology laboratories.

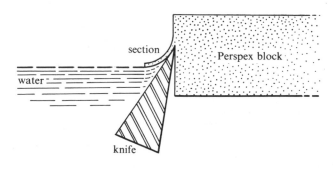

Fig. 1–15 Cutting of sections embedded in plastic for the electron microscope. A glass knife is used and the sections float off on the surface of the water

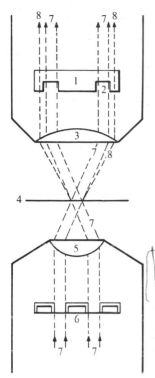

Fixed and stained preparations are still used for a variety of studies but it is now recognized that certain parts of the cell cannot be fixed satisfactorily.

Phase-contrast microscope.   For studies of cell movements it is of course only possible to work with healthy and living cells. With the aid of a compound microscope, Canti carried out some beautiful studies in the 1920s, but the development of new kinds of microscope has greatly increased the possibilities. The techniques still provide a high degree of contrast even when only a small change of refractive index occurs in various regions of the specimen. The phase-contrast principle was worked out by Zernicke in 1940. If a ray of light enters a sheet of water or glass of uniform thickness and refractive index it is transmitted without deviation, as shown in Fig. 1–16 (rays 7). In a preparation of living cells there may be sudden changes of refractive index between the inside and the outside of the cell, or between the nucleus and cytoplasm. An enlargement of the ringed region which includes the specimen is shown in Fig. 1–17. An annulus is used to give a hollow cone of rays, since full cone illumination would cause direct and diffracted rays to be superimposed. The region in the centre of Fig. 1–17 has a low refractive index and the outside a high refractive index. The transmitted rays are diffracted and will not enter the annulus 2; this region will appear dark. Diffracted and undiffracted rays strike different parts of the phase plate, where the phase relationship as well as the amplitude can be altered. Diffraction patterns are produced by interference and reinforcement of light waves. Variation in brightness occurs at boundaries of different refractive index in the specimen, which gives the characteristic halo appearance.

Interference microscopy.   It is interference between two beams of light which gives the familiar colours observed in thin films of oil on the surface of a road. If two beams of light from the same source are in phase, as shown in Figs 1–18a and 1–18b,

Fig. 1–16 Path of rays through a phase contrast microscope. The condenser contains an annular diaphragm; a similar diaphragm lies above the condenser. Rays deviated by the object do not pass through the objective diaphragm. 1 phase plate; 2 annular groove; 3 objective lens; 4 object plane; 5 substage condenser; 6 annular diaphragm; 7 direct rays; 8 diffracted rays

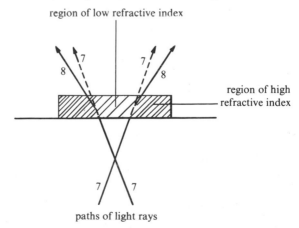

region of low refractive index

region of high refractive index

paths of light rays

Fig. 1–17 Path of rays through an object which has regions of low and high refractive index: 7 direct rays; 8 diffracted rays

they will supplement each other and produce a region of brightness as shown in Fig. 1–18c. When out of phase, as in Figs 1–18d and 1–18e, they will neutralize each other and the field will appear dark. Microscopes based on this principle

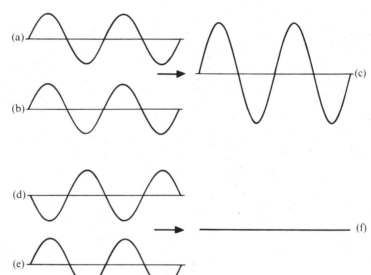

Fig. 1–18 Two plane light waves which can combine to give interference effects. Waves (a) and (b) are in phase and give rise to wave (c) with double the amplitude. Waves (d) and (e) are out of phase and give rise to zero amplitude (f)

have been produced by Merton, Ambrose, Dyson, Smith, and others. In the Smith (Baker) system the two rays which give rise to interference are separated at the specimen. One ray passes through the object and the other beside the object. The rays are brought together again above the objective as shown in Fig. 1–19. In this case the contrast depends on the refractive index and thickness of the specimen, not the gradient of refractive index. This image is a more exact representation of the true shapes of cells than that obtained by phase contrast. A change in contrast giving rise to a change of optical path $\Delta\lambda$ is quantitatively related to refractive index and thickness according to the following formula:

$$\Delta\lambda = t(\mu_s - \mu_r)$$

where t is the thickness, μ_s is the refractive index of the specimen, μ_r is the refractive index of the surrounding medium. $\Delta\lambda$ can be measured with the analyser attached to the interference microscope. This enables the dry mass (weight of protein, etc.) of individual cells and even separate organelles to be measured. The sensitivity is extremely high; very small differences in mass can be measured.

These quantitative measurements are carried out using

a monochromatic light source. If a white light source is used, parts of the cell which differ in thickness and refractive index appear in contrasting interference colours, which facilitates the observation of different regions.

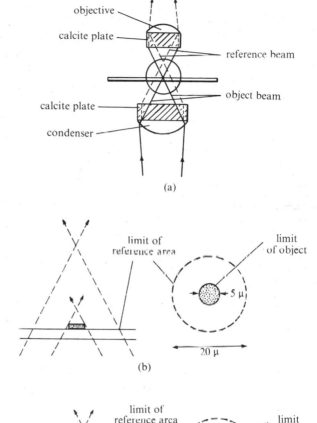

(a)

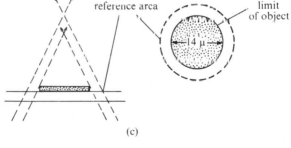

(b)

(c)

Fig. 1-19   (a) Path of rays through the double focus (Baker) interference microscope. **(b)** Outer cone of reference beam and small cone of rays passing through the object. **(c)** Similar, but with a large central cone for large objects

Ultraviolet microscopy.   The visible region of the spectrum extends from 6,500 Å (deep red) to 4,500 Å (violet). Wavelengths shorter than 4,000 Å are called ultraviolet (u.v.). In this region some cellular components such as nucleic acids and proteins absorb particular wavelengths. If u.v. microscopy is used in conjunction with visible light microscopy, much useful information may be obtained. Caspersson and others have used this method to measure the nucleic acid content of cell nuclei. Quartz lenses or mirrors must be used to transmit the ultraviolet light. The method has limited use for living cells because the cells are rapidly killed by ultraviolet radiation. It is important to use filters to protect the eyes, which can be damaged by ultraviolet radiation.

Fluorescence microscopy.　When certain chemical substances are irradiated with ultraviolet light they absorb the radiation and emit visible light. Objects which absorb these chemicals within living cells can therefore be observed as fluorescent areas when illuminated with ultraviolet light. The method is extremely sensitive and can detect minute quantities of material. It is particularly useful for studying how proteins and other molecules enter or are adsorbed on to cells. The proteins are labelled by coupling with the molecules of a fluorescent dye.

Electron microscopy.　*Transmission microscopy.*　The resolving power (or capacity to detect) of the light microscope is limited to objects of approximately the same dimensions as the wavelength of the light used. Objects between 2,000 Å and 3,000 Å in diameter can just be detected but may not be sharply resolved (Fig. 1–2). The wavelength of electrons which travel in a high vacuum in a high-voltage field can be so short that objects of 10 Å diameter or less can be resolved.

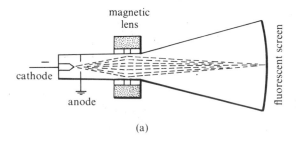

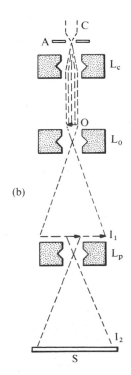

Fig. 1–20　Transmission electron microscope. **(a)** The use of a magnetic lens to focus a fine beam of electrons on a fluorescent screen. **(b)** Path of electrons through an electron microscope; C cathode; A annulus; L_c condenser lens; L_o objective lens which forms an image, I_1; L_p lens which forms magnified image, I_2, on a fluorescent screen photographic plate, S.

Everything must be maintained in a high vacuum to enable the electrons to travel with a high velocity without colliding with oxygen or nitrogen atoms in the air. Electrons travel in the cathode-ray tube and magnetic fields produced by coils act as lenses, as shown in Fig. 1–20a. The optical system of a complete microscope is shown in Fig. 1–20b. The final image is formed on a fluorescent screen or photographic plate.

Specimens are prepared by sectioning. An extremely delicate microtome is used with a glass knife, formed by fracture of a glass plate. The sections must be little more than 100 Å thick for high-resolution work. Fixation of cells for electron microscopy is extremely exacting in view of the fine detail observable under high resolution. The best fixative has proved to be osmic acid, which reacts almost instantaneously with cellular components. The sudden fixation maintains the structure of the cytoplasm extremely well. Various pictures taken this way are shown later in this book. An aldehyde (glutaraldehyde) has proved to be the best fixative for certain fibrous structures in the cytoplasm. No really satisfactory fixative has yet been found for the interphase nucleus.

Scanning electron microscopy.   The scanning electron microscope is a comparatively recent development. A very fine pencil of electrons scans the surface of the specimen, much as a screen is scanned in a television tube (Fig. 1–21a). Secondary electrons are emitted from the surface at the points where the beam is scanning. These are collected by the positively charged

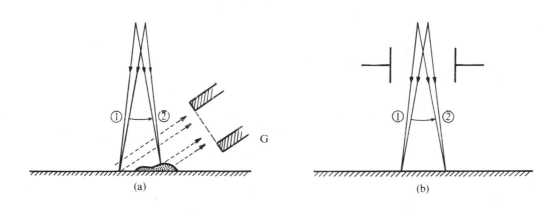

(a)   (b)

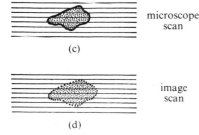

microscope scan

(c)

image scan

(d)

Fig. 1–21   Scanning electron microscope. **(a)** Scanning pencil of electrons which moves from 1 to 2 across the specimen; G is positively charged grid which collects electrons. **(b)** Scanning pencil of electrons which moves from 1 to 2 across a fluorescent screen in a television tube. **(c)** Form of scan on specimen. **(d)** Form of scan on television image.

grid (G). The signal from the grid is transferred to a television tube which scans in synchrony with the beam in the microscope (b). An image of the specimen is produced on the screen as shown in Figs 1–21c and 1–21d. The scanning microscope produces an image in much the same way as the image seen with the naked eye. The eye collects light which is reflected from solid objects somewhat like the grid in the scanning microscope. The image formed by this microscope therefore has a remarkable three-dimensional appearance, as in Fig. 11–24. Cell shapes are best preserved by plunging cells immediately into liquid propane (kept in liquid nitrogen) at −180°C. They may be subsequently dried after substitution of water with alcohol at −70°C.

Polarizing microscope. Polarized light is produced when the light waves lie in one plane. In Fig. 1–18, for example, light will be polarized if it vibrates in planes parallel to the plane of the paper. Light of this kind is produced by a calcite prism which splits the light into two beams, or by selective absorption of one of the polarized components in a plastic material in which the molecules are orientated parallel to each other. This is shown diagrammatically in Fig. 1–22a. Here a block of

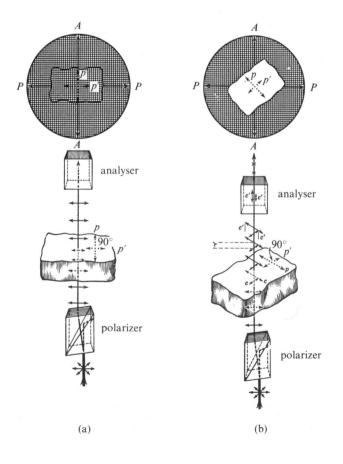

Fig. 1–22 (a) Passage of polarized light through an optically anisotropic substance. The polarizer and analyser are set in the crossed position. The specimen is arranged with optical axes parallel to the plane of the polarized beam (dark field— position of extinction). *A–A* and *P–P*, reference axes; *p* and *p'* are mutually perpendicular planes in which vibration might occur. (b) The same with the specimen set at 45° (position of brightness). *c* and *c* are the mutually perpendicular components of the incident vibrations; *c'* and *c'* and the components which pass through the analyser; *r* is the retardation of one of these components. (With permission from E. M. Chamot and C. W. Mason, *Handbook of Chemical Microscopy*, vol. 1. John Wiley, New York, 1958.)

(a)   (b)

plastic is shown lying in the beam of polarized light, with the long-chain molecules parallel to the direction p'. In such a material the refractive index is higher for light polarized so that it vibrates along the molecules (p') than for light which vibrates at right angles to the length of the molecules. In Fig. 1–22a the light is shown as vibrating along the length of the molecules. Under these conditions, the field appears dark in the microscope because the analyser, when set at right angles to the phase of vibration of the light, does not transmit the waves. But when the specimen is set so that the long-chain molecules lie at an angle of 45° to the direction of vibration of the incident light (Fig. 1–22b) a component of the light passes the analyser and the specimen is illuminated. This bright appearance of objects in crossed polarizers is known as birefringence or double refraction. The method can be extremely sensitive if the polarizers are made to produce a high degree of polarization. The polarizing microscope proved extremely useful for the study of the formation of the spindle in dividing cells and other structures in living cells.

Studying cell movements. Although direct observation with the phase-contrast or polarizing microscopes is useful for the study of cell movements, recording these on film allows them to be analysed subsequently. This can be done by making a long exposure on a film whence a track of the moving cell is

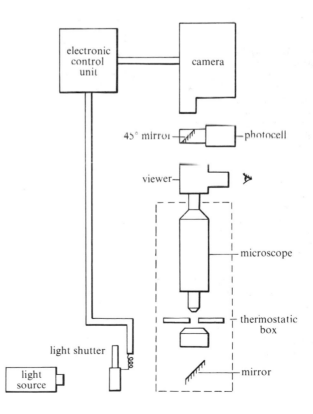

Fig. 1–23 Arrangement of equipment for time-lapse cinemicrography

obtained. But the best method is to record on ciné film. A schematic diagram of a time-lapse ciné apparatus is shown in Fig. 1–23. The microscope is mounted in a heated box, thermostatically maintained at 37°C. The camera is mounted above the microscope. A viewer is provided in the eyepiece which transmits part of the light to the camera and part to the viewing telescope. This enables the specimen on the stage to be focused by the operator while the camera is operating. A photocell is also provided to measure the light intensity so that the correct exposure on the film can be provided. The electronic control unit provides the time-lapse exposures on the film. For example, with an interval of 10 seconds between exposures, the cell movements will be accelerated 10×16 (160) times when the film is projected at 16 frames per second.

Cell culture methods.   Cell cultures are required for much of the modern work in biochemistry, molecular biology, and cell physiology. The most commonly grown cells are bacteria or mammalian tissue cells, although amphibian cells and plant cells are now being used more extensively. The methods used in bacterial culture are comparatively simple. Basic medium requirements are:

A source of carbon and nitrogen from which the bacteria can synthesize organic compounds.

An energy source. This is usually a substance which yields energy on oxidation which can be harnessed by the bacterial cell. Some bacteria can utilize one compound such as an amino acid as a source of carbon, nitrogen, and energy.

Mineral salts. Phosphorus, sulphur, calcium, and potassium are needed for bacterial growth. Traces of magnesium, manganese, copper, zinc, and cobalt are also required.

Growth factors. Certain bacteria are not able to synthesize some of the essential biochemical intermediates required by the cell. These must be supplied in the culture medium.

The acidity or pH of the medium must be controlled at between pH 6 and 8.

The optimum temperature varies with the species, some at 15–20°C, others at 30–37°C.

The cells can be grown in large rotating bottles containing a suspension of bacteria. Sometimes the culture medium is made up in agar, a gel which sets to form a sheet on the bottom of a Petri dish. In this case a very dilute suspension of bacteria is used and colonies are formed from single cells on the surface of the agar.

The culture medium for mammalian cells is much more complex than the medium required for bacteria. This is because the cells are less well equipped to synthesize essential biochemical intermediates. In the early days of tissue culture the medium was a mixture of serum and chick embryo extract rich in various factors required for growth. In recent years synthetic media have been developed which contain a large number of components. For example, Parker's 199 synthetic medium contains many different metabolites. Even with these complex media it is generally necessary to provide some foetal calf serum or horse serum to provide a supplement to produce cell growth.

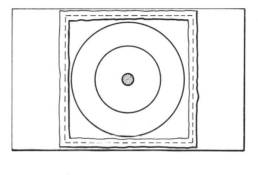

Fig. 1-24　The Maximow double coverslip method for explant culture, viewed from above and in cross section

The simplest technique for growing avian or mammalian cells is the *explant* method (Fig. 1-24). A small fragment of tissue is placed on a coverslip and covered with a drop of chick plasma and growth medium. The plasma sets to form a clot. The coverslip is mounted in a cavity slide with the clot attached to the under surface. The space in the cavity provides a source of oxygen and allows for exchange of carbon dioxide. The cells grow out from the tissue fragment as a sheet within the plasma clot.

For preparing large numbers of cells in culture, the *monolayer* method is extensively used (Fig. 1-25). The chopped tissue is treated with dilute trypsin, an enzyme which breaks up the intercellular cement and produces a suspension of separated cells. The trypsin is then removed by washing in a centrifuge and the cell suspension mixed with culture medium. The cells settle on the glass, attach to the surface of the vessel, and grow to form a monolayer. When the culture has covered the glass surface, the cells may be harvested by removal with trypsin. They can then be used to prepare fresh monolayers. By this method, large numbers of cells can be grown from a small mass of starting material.

Although monolayer cultures are useful for the study of cell growth, the cells tend to lose many of their characteristic

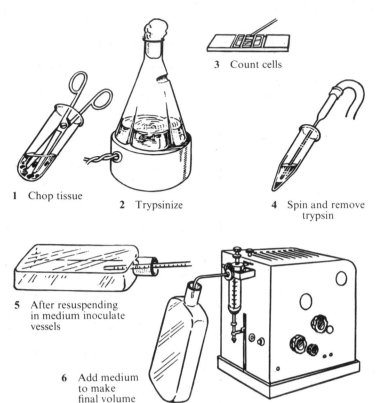

3   Count cells

1   Chop tissue

2   Trypsinize

4   Spin and remove
    trypsin

5   After resuspending
    in medium inoculate
    vessels

Fig. 1-25 Procedure for
the preparation of a cell suspension from fresh tissue,
for the monolayer method of
culture. (With permission
from J. Paul, *Cell and Tissue
Culture*. E. & S. Livingstone,
Edinburgh, 1970.)

6   Add medium
    to make
    final volume

properties under these conditions. To preserve the *in vivo*
characteristics of cells and tissues, the *organ* culture method
is generally used. This is now not used so much for cultures
of whole organs but for fragments of tissue approximately
1·5 mm in size. (With larger fragments there are problems in
obtaining adequate gas exchange.) These fragments are placed
on a raft or permeable sheet such as Millipore material, as in
the Trowell method shown in Fig. 1–26.

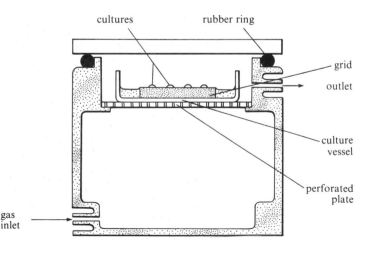

cultures   rubber ring

grid

outlet

culture
vessel

perforated
plate

Fig. 1-26 Chamber for
organ culture; Trowell's type
II chamber

gas
inlet

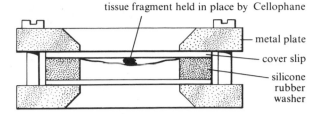

tissue fragment held in place by Cellophane

metal plate

cover slip

silicone rubber washer

Fig. 1–27 Rose chamber for time-lapse cinemicrography. Cross-section of assembled chamber

For studies of the movements and behaviour of cells by the time-lapse method, the cells are grown on glass coverslips which are then sealed in a chamber such as that shown in Fig. 1–27, which was designed by Rose. This consists essentially of a sandwich, the outer layers of which are two sheets of metal, bolted together, with circular holes to permit microscopy. The inner layers of coverslip, on one of which the cells are growing, form the top and bottom of the actual culture compartment and the sides are formed of special silicone rubber seals. The conditions of culture may be altered by injecting various substances through these rubber seals by means of a hypodermic needle. When the needle is withdrawn, re-sealing takes place automatically. Rose found that the cells showed better differentiation if they were covered by a strip of Cellophane.

Bibliography   Borradaile, L. A., L. E. S. Eastham, F. A. Potts, and J. T. Saunders, *The Invertebrata*, 4th edn. Cambridge University Press, London. 1963.

Brachet, J., The living cell. *Scientific American*, September 1961.

Butler, J. A. V., *Inside the Living Cell*. Allen and Unwin, London, 1959.

Chamot, E. M. and C. W. Mason, *Handbook of Chemical Microscopy*, vol. 1, 3rd edn. John Wiley, New York, 1958.

Comfort, A., The life-span of animals. *Scientific American*, August 1961.

Jensen, W. A., *The Plant Cell*. Macmillan, London, 1964.

Loewy, A. G. and P. Siekevitz, *Cell Structure and Function*, 2nd edn. Holt, Rinehart, and Winston, New York, 1969.

Mercer, E. H., *Cells and Cell Structure*. Hutchinson, London, 1961.

Paul, J., *Cell and Tissue Culture*, 4th edn. E. & S. Livingstone, Edinburgh, 1970.

Schenk, R. and G. Kistler (translator F. Bradley), *Photomicrography*. Chapman and Hall, London, 1962.

Willmer, E. N. (Ed.), *Cells and Tissues in Culture, Methods, Biology, and Physiology*, vols 1, 2, and 3. Academic Press, New York, 1965, 1966.

2 Basic chemical and physical concepts

The origins of a chemical study of living organisms can be found in the writings of the 16th century alchemist, Theophrastus Bombastus von Hohenheim (1493–1541), who called himself Paracelsus. Chemists of the 16th century were either practical men engaged in the making of pottery, glass, and so on, or alchemists who were seeking to transmute baser metals into gold. Paracelsus studied both these aspects of chemistry and attempted to apply them to medicine; from this arose the subject known as iatrochemistry (medical chemistry). The first scientific approach came from the work of a Swedish pharmacist, Carl Wilhelm Scheele (1742–1782). Amongst other preparations, he isolated citric acid from lime juice, uric acid from urine, and glycerol from animal fats. The importance of work on respiration in relation to cellular chemistry, particularly by Lavoisier, will be referred to in Chapter 6. What is now called organic chemistry was originally looked upon as biochemistry when it was thought that living matter differed fundamentally from non-living or inorganic matter. The synthesis of urea by Friedrich Wöhler in 1828 from inorganic materials, of acetic acid by Herman Kolbe in 1845, and particularly the work of Marcellin Berthelot (1827–1907), showed that the so-called 'organic' compounds owed their rather special properties to the nature of the carbon atom. But the name 'organic chemistry' remained. The name 'biochemistry' came to refer more generally to the chemical dynamics of living systems, particularly through the work of Claude Bernard (1813–1878) and Louis Pasteur (1822–1895).

The concept which began to emerge at this time was the essential biochemical unity of all living matter. Rudolf Virchow (1821–1902) stated *Omnis cellula e cellula* (Every cell is derived from a cell), and Pasteur also declared *Omne vivum e vivo* (All living things come from living material). It was a major advance when Hans and Eduard Buchner isolated the first enzyme in 1897. They discovered that the use of high temperatures or various solvents destroyed the activity of the substance they were attempting to isolate, and so they literally squeezed the liquid out of ground-up yeast cells with a hydraulic press. They obtained a solution of an enzyme, *zymase*, which catalysed the fermentation of glucose to produce alcohol, just as did intact yeast cells. The establishment of biochemistry as a recognized and separate academic subject was largely due to the work of Sir Frederick Gowland Hopkins in England in the early 1900s. He suggested in 1906 that animal diets required the inclusion of certain accessory factors (later called vitamins), and that the absence of these factors resulted in deficiency diseases.

2–1 Chemistry of carbon

Carbon atoms are the major component of biologically important molecules and the chemistry of carbon, or organic chemistry, is of supreme importance in determining many of the properties of biochemical compounds. Few other elements are required to make these compounds, the chief ones being hydrogen, oxygen, and nitrogen; other elements used are phosphorus and sulphur, and some metals. From these few elements, many different compounds are formed, by virtue of the fact that carbon atoms can combine with each other to form stable long chains and rings, many of which are extremely complex. These compounds are stable because the carbon atom is tetravalent and forms *covalent bonds* with itself and with other of the elements mentioned above. An *ionic* (or electrovalent) bond is an electrostatic bond between two oppositely charged ions. The bond is formed by the transfer of an electron (or electrons) from one atom to the other (Fig. 2–1a) which brings about the stable configuration, as in the inert gases, of eight electrons in the outer shell. In a *covalent* bond, on the other hand, a pair (or pairs) of electrons is shared between the two atoms concerned, each contributing equally to the bond (Fig. 2–1b). Since an appreciable amount of energy is required for a single electron transfer in an ionic bond, it is rare for more than three electrons to be transferred. In the carbon atom there are four electrons in the outer shell, which means that from

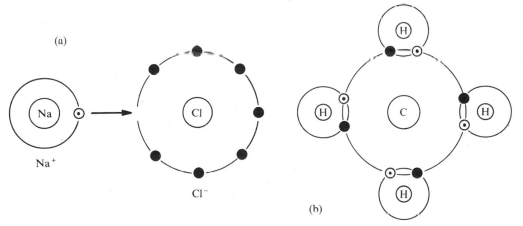

Fig. 2–1 **(a)** Ionic bonding. In NaCl, the Na atom contributes the single electron in its outermost shell to the Cl outer shell, which contains 7 electrons: ● electron from Cl; ⊙ electron from Na. **(b)** Covalent bonding. Electronic configuration of methane, CH$_4$; outer shell only of C atom depicted: ● electron from C; ⊙ electron from H

energy considerations the formation of ionic bonds is highly unlikely.

The fact that carbon is invariably tetravalent means that it forms four covalent bonds, which completely fill its outermost electron shell with the stable configuration of eight electrons (the shared electrons may be looked upon as a cloud of negative charge situated somewhere in the region between the two nuclei, and under the partial control of both). Since covalent bonds result from the sharing of electrons between two atoms, they always have a definite spatial orientation; in the case of carbon, the four bonds are arranged tetrahedrally around the central carbon atom (Fig. 2-2). It is the combination of these properties—the tetravalency of the carbon atom resulting in the formation of a stable outer electron shell and the highly symmetrical arrangement of its four covalent bonds—which makes carbon compounds so stable. Another factor which contributes to the unique behaviour of carbon is that it combines with almost equal facility with hydrogen, oxygen, and nitrogen, which adds considerably to the variety of organic compounds.

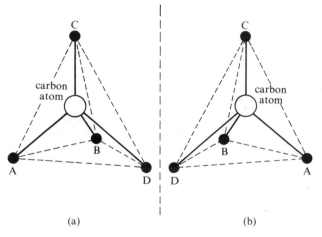

Fig. 2-2 Asymmetric arrangement of four different groups about a carbon atom. The two arrangements are in the form of mirror images

(a)   (b)

When a carbon atom is linked to four different atoms or groups, the molecule possesses *asymmetry* (Fig. 2-2). There are two possible arrangements in space of the same atoms linked together in the same ways; these are known as *stereoisomers*. Figure 2-2 shows that the two isomers are mirror images of each other. They may be distinguished because in solution they rotate the plane of polarized light, one isomer causing it to rotate in a clockwise direction (by definition *dextrorotatory* and given the symbol *d*) and the other making it rotate in an anticlockwise direction (*laevorotatory*, with the symbol *l*); for this reason they are known as *optical isomers*. As a result of asymmetric substitution of carbon atoms, a still greater variety of organic compounds derived from the same elements is possible.

The best-known ring or cyclic compound of carbon is

benzene, a six-membered ring compound, C_6H_6 (Fig. 2-3), which occurs biochemically as a side-chain of various amino acids (Fig. 4-1a). The structure of benzene used to be written as one of the two forms shown in Fig. 2-3a, with alternating double and single bonds, which were believed to interchange by a rapid oscillation of the double bonds. It is now believed on the basis of chemical and physical evidence that benzene has a *resonance* structure, which is a *hybrid* (Fig. 2-3b) of the two forms shown in Fig. 2-3a. A resonance hybrid is a distinct structure, intermediate in character between the possible *canonical* forms, and always of greater stability than they are. The connection between resonance structures and increased stability will be referred to in Section 2-9.

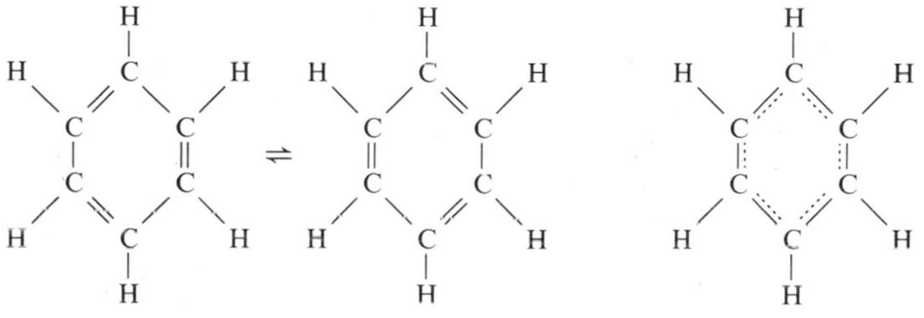

Fig. 2-3 Structure of benzene ring. **(a)** Resonating canonical forms. **(b)** Hybrid structure

(a)   (b)

The most important ring compounds in cellular biochemistry are heterocyclic, containing either oxygen or nitrogen as well as carbon in their ring structure. An example is pyridine which, in the form of the vitamin nicotinic acid or nicotinamide (Fig. 2-4), plays an important role in the functioning of the coenzyme nicotinamide adenine dinucleotide. Other important rings are the purines and pyrimidines (Fig. 2-10), which are constituents of the nucleotide sub-units from which nucleic acids are built.

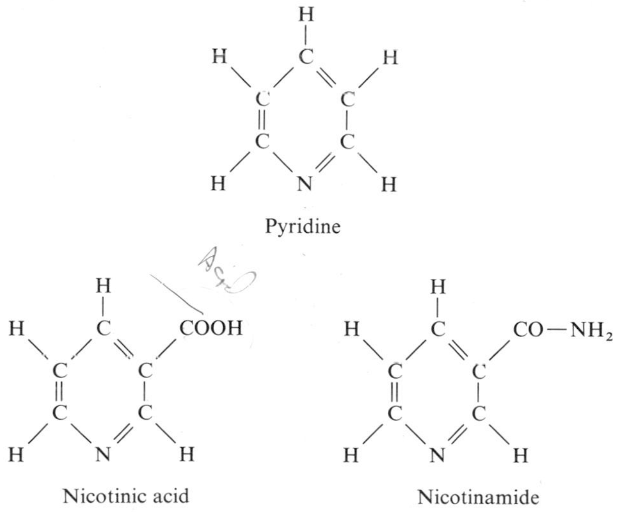

Pyridine

Fig. 2-4 Heterocyclic ring compounds

Nicotinic acid   Nicotinamide

2–2
Carbohydrates

As their name implies, carbohydrates or sugars are composed chiefly of carbon, hydrogen, and oxygen; some carbohydrate derivatives of biological importance also contain nitrogen and sulphur.

The *monosaccharides* are simple sugars which do not give smaller sugar units on hydrolysis. The most important sugar occurring in cells is glucose, which is a hexose (i.e., it consists of a chain of six carbon atoms). As shown in Fig. 2–5, there is an aldehyde group $-C\diagdown^H_O$ at carbon atom 1, and five hydroxyl (—OH) groups attached to the other five carbon atoms. Glucose exists in two forms—the open chain structure shown in Fig. 2–5a and the ring structure of Fig. 2–5b; this is a property common to all monosaccharides, due to the fact that the tetrahedral bond angles of the carbon atom result not in a linear but a bent carbon chain. The ends of the molecule tend to approach one another and, in glucose, the aldehyde group forms a semi-acetal compound with the OH group on the fifth carbon atom (Fig. 2–5b). This is known as a pyranose ring. The stereochemistry of monosaccharides is such that the ring formed is either five- or six-membered; a seven-membered ring would involve too much strain.

In fructose, a keto $\diagup^{}_{\diagdown}C{=}O$ containing sugar, the ring forms between the keto group and either the hydroxyl group on C atom 6 to give the pyranose form, or the hydroxyl group on C atom 5 to give what is called the furanose form (Fig. 2–6d).

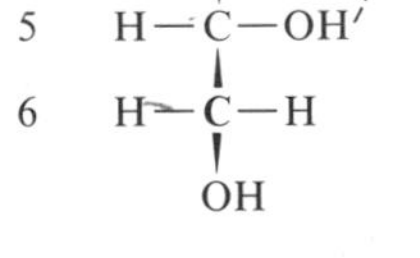

Glucose: open chain
(a)

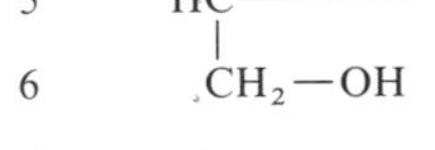

Glucose: ring structure
(b)

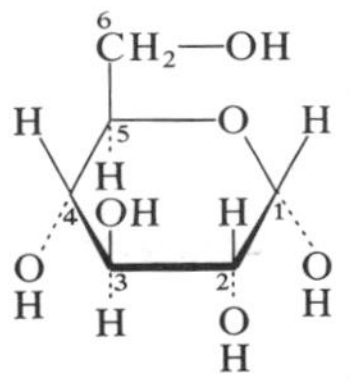

Glucose: Haworth representation
(c)

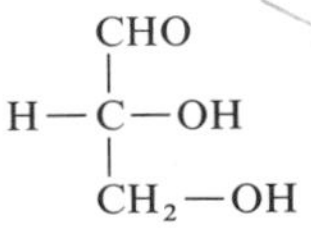

D-Glyceraldehyde
(d)

Fig. 2–5 (a), (b), and (c) Structure of glucose. **(d)** D-Glyceraldehyde is the reference compound to which all other compounds containing asymmetric groups are referred.

In pentose (five-carbon) sugars such as ribose, a five-membered furanose ring is formed. In the complex sugars (see below) it is always the cyclic form of the simple sugars which is present. A useful way of representing the ring structures of sugars was proposed by Haworth (1927), and is shown in Figs 2–5c, 2–6c, and 2–6e. The pyranose or furanose ring is considered to be in a plane perpendicular to the plane of the paper; thus in gluco-pyranose, carbon atoms 2 and 3 are in front of the paper, and carbon atom 5 and the ring oxygen lie behind the plane of the paper. The substituent groups are either above or below the plane of the ring. This is the method of representation which will be used in general in this book.

There are various carbon atoms in the simple sugars which are asymmetrically substituted, so that a number of optical isomers must exist. There are four asymmetrically substituted carbon atoms in the open chain structure of glucose, which means (since the substituted groups at each atom may be arranged in two ways) that there must be $2 \times 2 \times 2 \times 2$ (i.e. 2^4, or 16) possible different arrangements of groups on these 4 carbon atoms which will all give compounds with the same molecular formula, i.e. there must be sixteen different hexoses. These are labelled D- or L- according to a convention based on the configuration of glyceraldehyde. If the OH group on the highest asymmetrical C atom (carbon 5 in the case of glucose) is on the right when the aldehyde group is at the top of the formula, as in dextro-rotatory glyceraldehyde (Fig. 2–5d), they are labelled D. If it is on the left they are designated L.

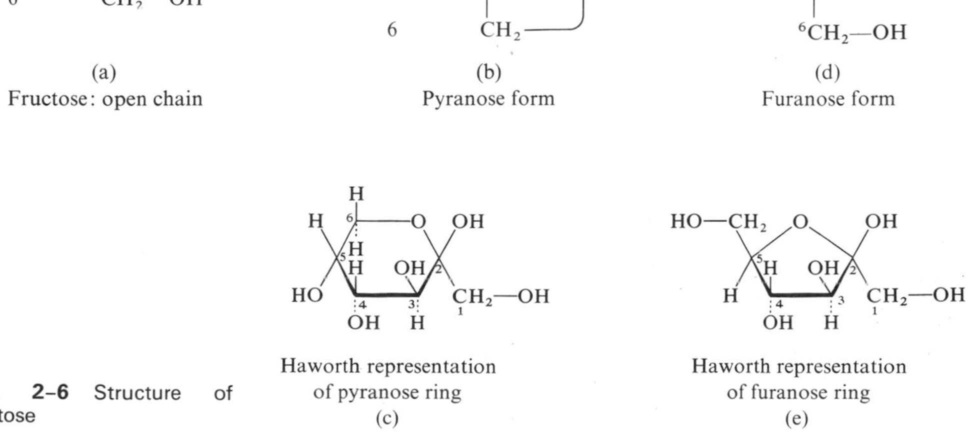

(a)
Fructose: open chain

(b)
Pyranose form

(d)
Furanose form

Haworth representation
of pyranose ring
(c)

Haworth representation
of furanose ring
(e)

Fig.   2–6   Structure   of
fructose

Glycodic bond
of sugar lin Roge

by losing N₂ Obstain
OH bonds
CC

According to this convention, both glucose and fructose are labelled D. This classification is based on configuration only, and D configuration is not always accompanied by dextro-rotatory capacity, nor L by laevorotation.

More complex sugars also play an important part in cellular biochemistry. Oligosaccharides (from the Greek, *oligos*, meaning 'few') are by definition compound sugars which on hydrolysis yield two to six molecules of simple sugars. They are usually crystalline and soluble in water, and have a sweet taste. The smallest oligosaccharides are the disaccharides, which yield two simple sugars on hydrolysis. For example, the disaccharide sucrose (Fig. 7–26), which consists of a glucose molecule and a fructose molecule joined by a glycosidic or sugar linkage, in which a molecule of H_2O is eliminated between two OH groups on adjacent sugar mole-cules. The elimination takes place between the hydroxyl of the semi-acetal group of glucose and the hydroxyl group at carbon atom 2 of fructose, the fructose being in the furanose form. The glycosidic bond plays a similar role in carbohydrate chemistry to that of the peptide bond in protein chemistry. With sucrose, the linkage between the two monosaccharides is known as an α-glycosidic bond, since glucose is here in the α configuration, with the OH group at carbon atom 1 orientated below the plane of the ring, as in Fig. 2–5c. There is also a β-glucose, in which the OH at carbon 1 is above the ring, and the H atom below; this will form a β-glycosidic linkage with another sugar, as in cellobiose (Fig. 7–27b).

Polysaccharides are compound sugars consisting of a large number of monosaccharides which are linked by glyco-sidic bonds, and are released when the polysaccharide is hydrolysed. Starch, for example, consists of many linked glucose units. Polysaccharides generally have very high molecular weights and are amorphous, insoluble, and tasteless.

2–3
Amino acids

Amino acids are the building units of proteins, and have the general formula:

$$H_2N-\underset{\underset{H}{|}}{\overset{\overset{R}{|}}{C}}-COOH$$

where the nature of the group R depends on the amino acid. There are twenty different amino acids commonly found in the proteins in cells (Fig. 4–1a). (Other amino acids, such as ornithine and γ-aminobutyric acid, also occur inside cells in addition to those found in proteins.) Since the NH_2 group is always attached to the C atom next to the carboxyl group (the α-carbon) they are known as α-amino acids. Except for glycine, the simplest amino acid, where R is equal to H (Fig. 4–1a), the α-carbon is always asymmetrically substituted, so there can be two optical isomers for each chemical formula. As with sugars,

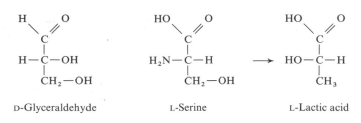

Fig. 2-7 Steric relation-ship of a naturally occur-ring amino acid, L-serine, to D-glyceraldehyde. When serine was converted to lactic acid it was found to give the L-form, so serine itself has the L-configura-tion with respect to D-glyceraldehyde

D-Glyceraldehyde   L-Serine   L-Lactic acid

amino acids are designated D- or L- according to their con-figuration with reference to that of D-glyceraldehyde. All of the amino acids of proteins have the L-configuration, as in serine (Fig. 2-7). D-Amino acids do occur in nature, however, being synthesized by certain plants and micro-organisms. Several antibiotics produced by bacteria—for example, the penicil-lamine found in penicillins—contain D-amino acids. (These antibiotics are not proteins, but contain smaller units known as peptides, which consist of amino acids linked by peptide bonds in the same way.) Also, D-glutamic acid is present in the capsular substance of the anthrax bacillus.

Figure 4-1a shows that certain of the protein amino acids, including threonine, contain two carbon atoms which are asymmetrically substituted, so that there are four possible isomers. However, only one of the four is found to occur naturally. Two of the amino acids, methionine and cysteine, contain sulphur, and some have a benzene ring on one side-chain—for example, phenylalanine—and some, such as tryptophan, contain a heterocyclic carbon-nitrogen ring.

The amino acids are generally soluble in water, and quite insoluble in nonpolar organic solvents such as ether. They also have fairly high melting points, and melting is often followed by decomposition. This suggests structures with highly charged polar groups, rather than the uncharged molecule of the general formula given at the beginning of the section. Further informa-tion about the structure of amino acids in solution has been gained from studying their behaviour as electrolytes. Since amino acids contain both an acidic carboxyl group and a basic amino group, they should be *amphoteric*—that is, capable of reacting with both alkalis and acids. Studies on alanine showed that it does not migrate in neutral solution if an electric field is applied. If alkali is added to the solution, alanine moves to the anode, i.e., it becomes negatively charged, and similarly if acid is added it becomes positively charged and moves to the cathode. It was suggested by Bjerrum in 1923 that amino acids in solution are best represented by the zwitterion formula,

$$\overset{+}{N}H_3-\underset{\underset{R}{|}}{\overset{\overset{COO^-}{|}}{C}}-H$$

with a negative charge on the carboxyl group due to the migra-tion of a proton to the NH_2 group, which has thus acquired a positive charge; a zwitterion is essentially an internal salt. When alkali is added to a neutral solution of an amino acid, it is neutralized by the proton on the amino group:

$$\overset{+}{N}H_3-\underset{\underset{R}{|}}{\overset{\overset{COO^-}{|}}{C}}-H \xrightarrow{\quad OH^-\quad} NH_2-\underset{\underset{R}{|}}{\overset{\overset{COO^-}{|}}{C}}-H$$

Similarly, added acid is neutralized by the acceptance of a proton by the carboxyl group:

$$\overset{+}{N}H_3-\underset{\underset{R}{|}}{\overset{\overset{COO^-}{|}}{C}}-H \xrightarrow{\quad H^+\quad} \overset{+}{N}H_3-\underset{\underset{R}{|}}{\overset{\overset{COOH}{|}}{C}}-H$$

Figure 2–8 shows the process diagrammatically.

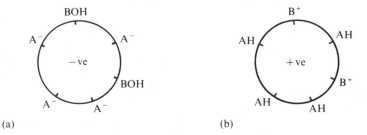

Fig. 2–8 Protein molecules with negative (A⁻) and positive (B⁺) charges: **(a)** in alkaline solution, **(b)** in acid solution

(a)   (b)

Amino acid solutions thus act as *buffers*—that is, they resist any change in the pH of their environment (see below). Studies on solid amino acids indicate that in this state also they may exist as zwitterions, which accounts for their ready solubility in water (non-ionizable organic compounds are not in general water-soluble).

Peptides and *proteins* are formed from amino acids linked together by *peptide bonds*. In this type of bond, the carboxyl group of one amino acid is linked to the amino group of a second:

$$NH_2-\underset{\overset{|}{R^1}}{\overset{\overset{R^1}{|}}{C}H}-COOH+NH_2-\underset{\overset{|}{R^2}}{\overset{\overset{R^2}{|}}{C}H}-COOH \longrightarrow NH_2-\underset{\overset{|}{R^1}}{\overset{\overset{R^1}{|}}{C}H}-CO-NH-\underset{\overset{|}{R^2}}{\overset{\overset{R^2}{|}}{C}H}-COOH+H_2O$$

Peptide bond formation in living cells is a complex process, and will be discussed in Chapter 4. When two amino acids are linked in this way, the product is called a dipeptide; three amino acids form a tripeptide, and a large number give a long-chain compound called a polypeptide. Proteins consist of one or more polypeptide chains, some of which are exceedingly complex and may have molecular weights of many millions. Proteins contain relatively few free α-amino and α-carboxyl groups, since these are involved in peptide bond formation. So the behaviour of proteins as electrolytes depends upon the ionizable groups (both basic and acidic) of the side-chains of the constituent amino acids. Proteins in solution will

migrate in an electric field, the direction of movement depending upon the nature and number of these various ionic groups.
Clearly, there must be a pH value (see below) different for
each protein, at which movement will not occur in either
direction; this is known as the *isoelectric point* (Fig. 2–9).

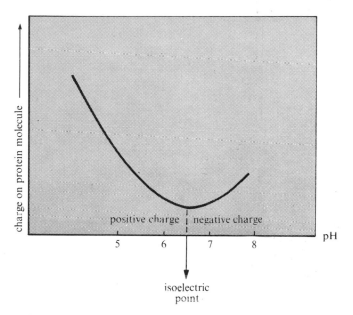

Fig. 2–9 Diagram demonstrating the isoelectric point
of a protein molecule at a
specific pH value

Because of their high molecular weights and large numbers of ionic side-chain groups, proteins have a very powerful
buffering capacity, which is of considerable importance in
biological systems. For example, the pH of mammalian blood
is controlled within the narrow range of pH 7·3 to pH 7·5
mainly by the buffering capacity of haemoglobin. The variety
of groups present on side-chains also means that proteins can
combine with a range of low molecular weight substances and
can act in a transport and storage capacity in biological
systems.

All of the amino acids in proteins possess the L-configuration about the α-carbon atom, so the structure of proteins
has molecular asymmetry. Since all the known enzymes
which catalyse biological reactions are proteins, and since
enzymes show considerable steric specificity (Section 7–1),
asymmetry of protein structure is of considerable biochemical
importance.

2–4
Nucleotides

Nucleotides are the building units of nucleic acid molecules.
Each nucleotide consists of three sub-units—a basic nitrogen-
containing ring compound, a pentose sugar, and a phosphate
group. The basic part of the molecule is always either a substituted pyrimidine ring (Fig. 2–10a), a heterocyclic six-
membered ring, or a substituted purine (Fig. 2–10b), which is a
pyrimidine fused with a five-membered heterocyclic ring.

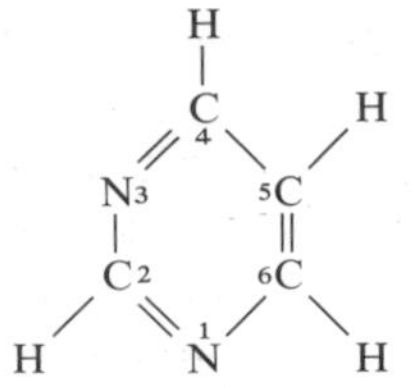

(a) Pyrimidine

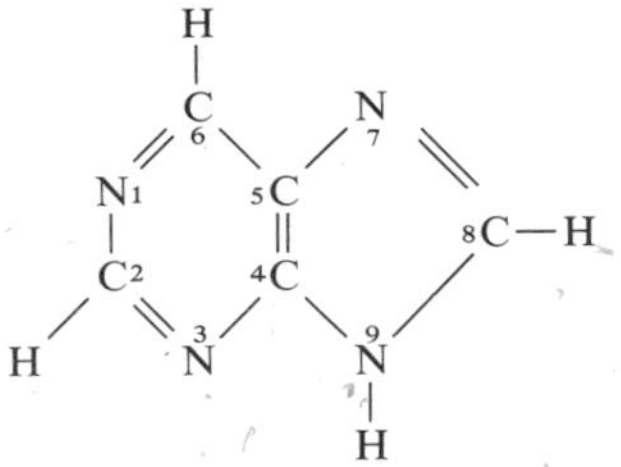

(b) Purine

Fig. 2–10 Heterocyclic ring structures: (a) pyrimidine, (b) purine

Both DNA and RNA contain two purines and two pyrimidines. The numbering of the atoms is shown in Fig. 2–10. When these compounds contain oxygen, as in uracil, a substituted pyrimidine present in RNA, there is an equilibrium between the *keto* (lactam) form and the *enol* (lactim) form (Fig. 2–11). It is the keto form which is predominant at physiological pH, and this is the form which combines with the pentose sugar by means of a glycosidic bond.

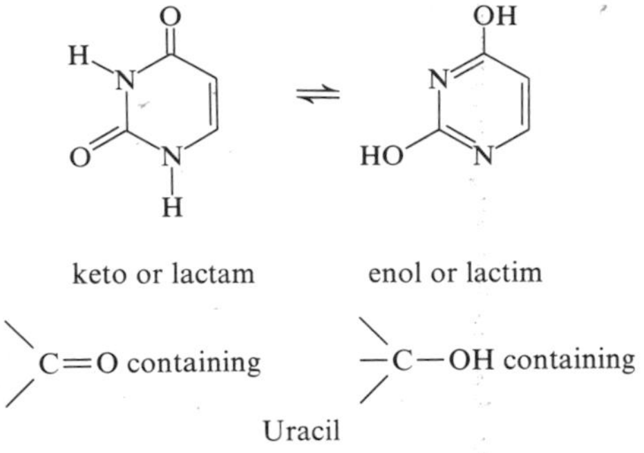

keto or lactam   enol or lactim

$\diagdown$C=O containing $-$C$-$OH containing

Uracil

Fig. 2–11 (above)   Isomeric structures of uracil

The pentose sugar in nucleic acids is always *ribose*; in RNA it is D-ribose and in DNA, as its name implies, it is deoxy-D-ribose (Fig. 2–12). It is always the OH on the C-1 carbon which is the point of attachment of the base. This is linked to the 1-nitrogen atom in the case of pyrimidines, and to the 9-nitrogen atom in purines (Fig. 2–12). To avoid confusion, the carbon atoms of the sugar ring are labelled 1′, 2′, and so forth. The combination of ribose with a purine or pyrimidine base is known as a *nucleoside* (which is a riboside or deoxyriboside of the base concerned). A *nucleotide* is a phosphate ester of a nucleoside; phosphoric acid, H_3PO_4, forms an ester

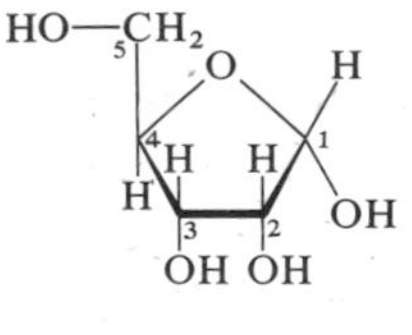

D-Ribose

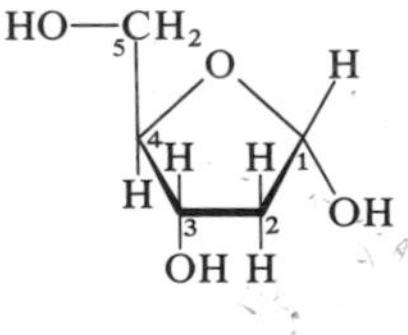

2-Deoxy-D-ribose

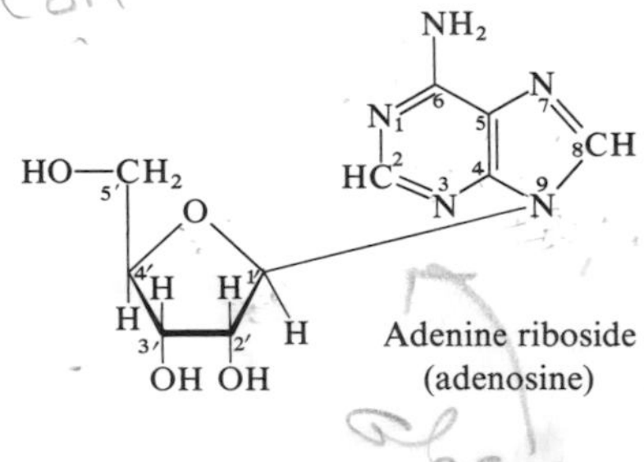

Adenine riboside (adenosine)

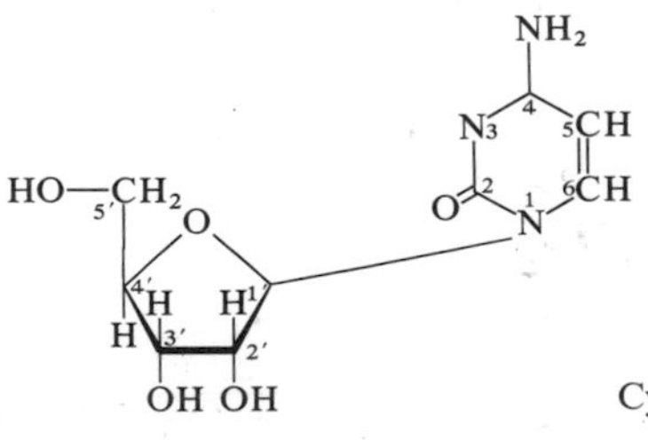

Cytosine riboside (cytidine)

Fig. 2–12 Structure of ribose, deoxyribose, and two nucleosides

linkage with one of the free hydroxyl groups on the ribose (or deoxyribose) component (Fig. 2-13). In nucleic acids, the nucleotide units are held together by phosphate diester linkages; one molecule of phosphate joins two nucleoside units by forming an ester linkage with the hydroxyl on the 3′ carbon atom of one sugar component and another ester linkage with the 5′ hydroxyl of the sugar of the adjacent nucleoside (Fig. 3-3).

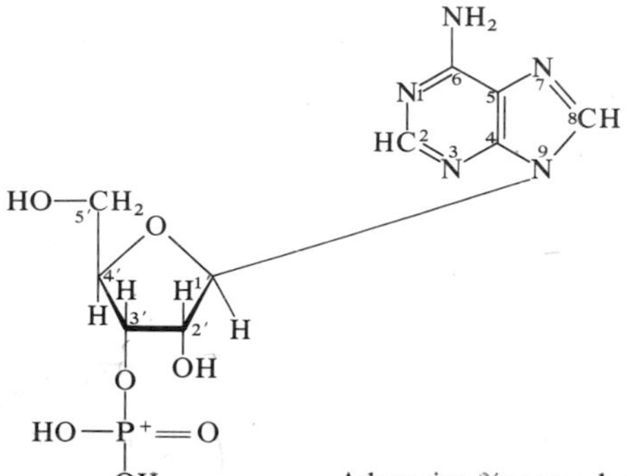

Adenosine 3′-monophosphate

Fig. 2-13 Structure of a nucleotide

Diphosphate and triphosphate esters of nucleosides also exist. Adenosine triphosphate (ATP) and adenosine diphosphate (ADP) play an important part in cellular processes requiring the release of energy, because of their ability to donate and accept phosphate groups (Fig. 7-13).

2-5 Lipids

Cellular lipids are esters (of general formula $R^1 . COOR^2$) of long-chain fatty acids with an alcohol. The fatty acids have the general formula $R^1 . COOH$, where R^1 is a hydrocarbon chain and the alcohols have the formula R^2OH, where R^2 may be a different hydrocarbon chain. The commonest fat-building alcohol in both plant and animal cells is glycerol (Fig. 2-14), but in animals cholesterol (Fig. 7-39) forms the alcohol part of the molecule in certain important lipids.

Neutral fats, or triglycerides, are formed by ester linkages of three fatty acids to all three available hydroxyl groups of glycerol (Fig. 2-14).

In phospholipids, such as lecithin (phosphatidylcholine) (Fig. 7-37), the molecule contains a charged phosphate group and a nitrogenous base in addition to glycerol and (usually) fatty acids. Phospholipids may be looked upon as derivatives

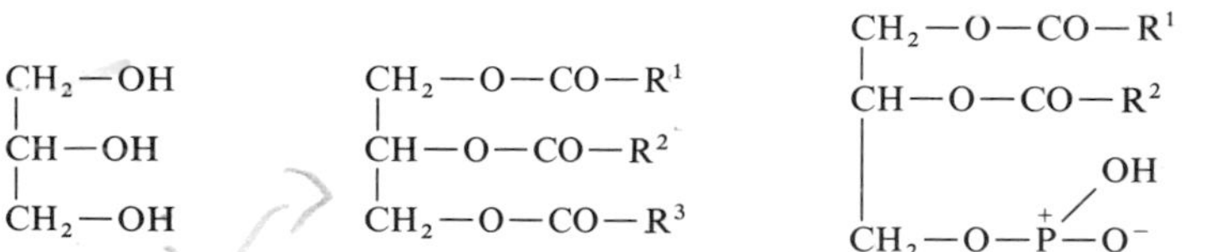

Fig. 2-14 Structure of glycerol, a triglyceride, and a phospholipid

Glycerol

Triglyceride: ester of glycerol with 3 fatty acids, R^1COOH, R^2COOH and R^3COOH

Phosphatidic acid

of phosphatidic acid (Fig. 2–14) and are sometimes called phosphatides.

The predominantly hydrocarbon, and so hydrophobic, nature of lipids is characterized by their very low solubility in water and high solubility in organic solvents.

2–6 Hydrogen ion concentration and pH

We use pH values to express the acidity or alkalinity of a solution. In water, there is always a small number of dissociated H^+ ions (or, more correctly the hydrated ion H_3O^+) and OH^- ions. This may be written as the reversible dissociation

$$H_2O \rightleftharpoons H^+ + OH^-$$

which, as indicated by the heavy arrow, has an equilibrium lying very much to the left. Applying the law of mass action (which states that the rate of a chemical reaction is proportional to the masses of the reacting substances) to this eqilibrium, we have, at constant temperature.

$$\frac{[H^+][OH^-]}{[H_2O]} = K$$

where K is a constant, and $[H^+]$ and $[OH^-]$ are the concentrations at equilibrium of the ions in gramme-ions per litre, and $[H_2O]$ that of the undissociated water in molarity. (Strictly, activities should be used, but these approximate to molar concentrations in extremely dilute solutions—see below.) Since water is only very slightly dissociated, even a large relative change in the concentration of ions makes very little percentage difference to the concentration of undissociated water molecules, which may therefore be regarded as constant. So we write:

$$[H^+][OH^-] = K_w$$

where K_w is the *ionic product* for water. Since equal numbers of H^+ and OH^- ions are formed in pure water, $[H^+] = [OH^-]$. This has been determined experimentally and found to be about 1×10^{-7} gramme-ions/litre at 25°C.

$$K_w = [H^+][OH^-] = (1 \times 10^{-7})(1 \times 10^{-7}) = 1 \times 10^{-14} \text{ (at 25°C)}$$

The hydrogen ion concentration is equal to the hydroxyl ion concentration only at neutrality. The ionic product, however, is always constant at constant temperature. If the hydrogen ion concentration rises above 10^{-7} gramme-ions/litre, the solution is termed *acidic*, and the hydroxyl ion concentration is less than 10^{-7} gramme-ions/litre; if the hydrogen ion concentration is less than 10^{-7} gramme-ions/litre the solution is called *alkaline* or *basic*, and the hydroxyl ion concentration is greater than 10^{-7} gramme-ions/litre. To characterize the acidity or alkalinity of a solution it is only necessary to state the hydrogen ion concentration, since the product of hydrogen and hydroxyl ion concentrations is always the constant, K_w.

Wide variations in hydrogen ion concentration are possible, so it is convenient in practice to use a logarithmic scale. The accepted one is the negative logarithm to the base 10 of the hydrogen ion concentration in gramme-ions/litre. This is called the pH of a solution. So, at neutrality, $[H^+] = 10^{-7}$ gramme-ions/litre, thus

$$pH = -\log_{10}[H^+] = -\log_{10}(10^{-7}) = 7.$$

The sign is reversed simply for convenience, to give a positive scale of readings. In an acidic solution the pH will be less than 7—for instance, where $[H^+]$ is as high as 10^{-3} gramme-ions/litre, the pH is $-\log_{10}(10^{-3}) = 3$. Conversely, in an alkaline solution the pH will be greater than 7; thus if $[H^+]$ is 10^{-9} gramme-ions/litre, the pH is $-\log_{10}(10^{-9}) = 9$.

In summary, acidic solutions with a high hydrogen ion concentration have a low pH, and alkaline solutions with a low hydrogen ion concentration have a high pH (Fig. 2–15). Measurements of pH are usually made quickly and accurately by use of a pH meter (see Section 2–15).

Fig. 2–15 pH of acidic and alkaline solutions

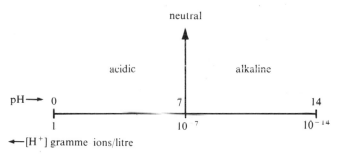

2–7
Energy changes in biochemical reactions

The processes involved in cellular biochemistry form an energy cycle. Foodstuffs such as sugars, lipids, and proteins act as a potential source of energy, undergoing degradation (sometimes called *catabolism*) inside cells with an accompanying release of energy. The energy released is utilized, by the intermediary of a few energy-rich compounds, in the synthesis (or *anabolism*) of various cellular components and in other cellular activities for which work is required, such as mechanical work in cell movements and muscle contractions, electrical work in the conductance of nerve impulses, work against osmotic forces, and so on. In the anabolic process of photosynthesis in green plants, energy from the sun is utilized to convert CO_2 and H_2O into glucose, with the release of oxygen.

The cell in which all of these biochemical processes take place does not exist in isolation, insulated from its surroundings. If heat is liberated in a cellular reaction it will be lost to the surroundings and the temperature conditions inside the cell will remain almost constant. The same argument applies to changes in pressure; cellular reactions generally take place at atmospheric pressure, and if there is a change in volume then interaction with the environment leads to a return to the original pressure conditions. For example, if there is an increase in volume because of gas being formed from solids or liquids, then work will be done against the environment to restore atmospheric pressure inside the cell. In looking at the energetics of biochemical reactions, therefore, we have to consider the energetics of reactions at *constant temperature* and *pressure*. Most cellular reactions are enzyme-catalysed and reversible (see Chapter 7), that is they may be written as

$$A + B \rightleftharpoons C + D.$$

Employing energetics, the most useful measure is the maximum work obtained from a reversible reaction under conditions of constant temperature and pressure. This, the *change in free energy* of the reaction, is symbolized as ΔF. If the free energy of the products of a reaction is less than the free energy of the initial reactants, there has been a decrease in free energy, and ΔF is negative $(-\Delta F)$; such reactions are termed *exergonic* (Fig. 7–3). If the reverse is true, there is an increase in free energy and a positive ΔF, the reactions in this case being termed *endergonic*. In practice, the energy made available in a cellular exergonic reaction is used to drive other biochemical reactions which are endergonic.

In the reversible reaction

$$A + B \rightleftharpoons C + D$$

at equilibrium conditions the rate of the back reaction is equal to the rate of the forward reaction. Reaction rates are proportional to the product of the activities of the reactants. (Activities must be used for non-ideal solutions; in very dilute solutions activities approximate to concentrations expressed in molarities.) So at equilibrium we may write:

$$\frac{(C) \times (D)}{(A) \times (B)} = K$$

Where $(A) \ldots$, etc. are the activities of the various reactants and products at equilibrium, and K is the *equilibrium constant*. Clearly, the further the reaction proceeds to the right, the higher will be the equilibrium constant.

An expression which is derived thermodynamically is that for the *standard change in free energy*, ΔF^0. This describes the maximum useful work which can be obtained on the conversion of unit activities of reactants to unit activities of products (at pH 0; at any other pH the standard free energy change is written as $\Delta F'$, and pH should be indicated). It is related to the equilibrium constant in the following way:

$$\Delta F^0 = -1363 \log_{10} K \text{ (at 25°C)}$$

This relationship is used to determine in the laboratory values of ΔF^0 for reversible reactions by measuring the con-

centrations of reactants and products at equilibrium. Reactions with a negative standard change in free energy have a K value greater than 1, and those with a positive ΔF^0 have a K value of less than 1.

Any disturbance in the conditions at equilibrium will alter the relative concentrations of the reactants and products; thus an increase in the concentration of A or B, or a decrease in the concentration of C or D, will drive the reaction further to the right.

The energy available from exergonic cellular reactions is used to drive various endergonic reactions by a coupling process, using *common reactants*. For example, the reactions $A \rightleftharpoons B$ and $B \rightleftharpoons C$ are coupled to give

$$A \rightleftharpoons B \rightleftharpoons C$$

If the reaction $A \rightleftharpoons B$ has a positive ΔF and would normally lie to the left of the equation at equilibrium, and the reaction $B \rightleftharpoons C$ has a negative ΔF, then energy available from the latter reaction is used to do work to enable the first reaction to proceed. The continual removal of B to form C means that the reaction $A \rightleftharpoons B$ goes continually further to the right. If the equilibrium constant of the reaction $B \rightleftharpoons C$ is high—that is, if there is a large negative free energy change—the final result will be the very efficient conversion of A to C via the intermediate B, which will finally remain only in very small quantities. In Fig. 2–16, the negative ΔF_2 for the $B \rightleftharpoons C$ reaction is greater than the positive ΔF_1 for the $A \rightleftharpoons B$ reaction, with the result that the final free energy level of C is lower than that of A, and the free energy change for the reaction $A \rightleftharpoons C$ is also negative.

Note, however, that although certain biochemical reactions are thermodynamically feasible, their rates may be very slow indeed in the absence of an enzyme to catalyse them, and few, if any, of the reactions of living cells are in thermodynamic equilibrium. Also, when substances are not miscible, they separate into two or more phases. A living organism is not homogeneous, but a heterogeneous multiphase system. A cell itself is also a multiphase system, and such a system, consisting of many interrelated reaction processes, is not in a state of

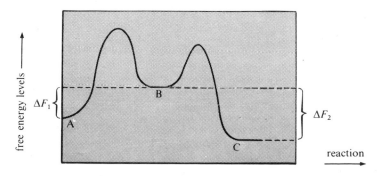

Fig. 2–16 Energy levels in chemical reactions

thermodynamic equilibrium, but in a 'steady state' in which the rates of chemical synthesis and degradation are balanced. Any alteration in environment, such as the addition of a metabolite or the removal of a reaction product, will cause the balance of the various intracellular reactions to be altered so as to preserve the steady state. Consequently, the accurate determination of ΔF for a biochemical reaction is very difficult. Even the determination *in vitro* of the ΔF of a homogeneous reversible reaction involves considerable difficulties, and various simplifying assumptions have to be made (see Section 2–15). Most of the ΔF values quoted in biochemical literature must therefore be looked upon as approximations, involving errors in the region of ± 10 per cent. The concept of the standard free energy change has, however, had many valuable applications to biochemical reaction systems.

2–8 Oxidation– reduction reactions

The term *oxidation* was used in the past to describe reactions in which substances combined with oxygen, e.g.

$$2Mg + O_2 \longrightarrow 2MgO$$

and *reduction* was used to describe the reverse process, that is removal of oxygen, e.g.

$$PbO + H_2 \longrightarrow Pb + H_2O$$

Later, the term oxidation was extended to cover reactions such as the change of ferrous ion to ferric ion,

$$Fe^{2+} \rightleftharpoons Fe^{3+} + e$$
(reduced)   (oxidised)

This is clearly an oxidation reaction, corresponding to a change in the oxides from FeO to Fe_2O_3.

Now a much broader definition is used—that *oxidation* is a progress involving a *loss of electrons*, and that an oxidizing agent is one which habitually accepts electrons. *Reduction* is regarded conversely as involving a *gain of electrons*, and substances which have the characteristic of giving up electrons to other substances are called reducing agents. The two processes are therefore complementary, oxidation always being accompanied by reduction; they are simply two aspects of the same change—the transfer of electrons from one substance to another. They are therefore given the name reduction–oxidation reactions or *redox* reactions.

Within the group classed as reducing agents, some compounds have a greater tendency to give up electrons than others. Similarly, in the class of oxidizing agents the electron affinity of different substances varies widely. To compare quantitatively these differing tendencies, we express the tendency of a substance to lose electrons as a numerical value (which may be positive or negative). On this basis substances may be arranged in a series, known as the *redox* series, according to their oxidation–reduction potential. This is determined

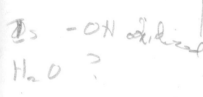

by measuring the electrode potential of various systems compared to that of hydrogen, which is taken as an arbitrary standard. The hydrogen electrode by definition has a potential of zero when an inert metal electrode dips into a solution at pH 0 (i.e., a solution of unit activity with respect to protons, see Fig. 2–15) in equilibrium with hydrogen gas at a pressure of 1 atmosphere. The reaction is

$$\tfrac{1}{2}H_2 \rightleftharpoons H^+ + e$$

The potential of any oxidation–reduction system can be determined if the system can accept or donate electrons reversibly at a metal electrode. The system is connected, via its electrode, to a standard hydrogen electrode (or other electrode which has been standardized) and the potential difference between the two electrodes measured by means of a potentiometer. With a metal, for instance, the electrode consists of the metal placed in a solution of its own ions (one of its salts); thus a rod of zinc could be placed in a solution of zinc sulphate. Some of the metal atoms tend to give up electrons to the metal and go into solution as positively charged ions, while at the same time some metallic ions in solution tend to take up electrons from the metal and become deposited as neutral atoms·

$$Zn \rightleftharpoons Zn^{2+} + 2e$$

Whichever tendency is greater determines whether the element becomes negatively or positively charged compared with the solution. Once equilibrium is finally reached, however, which means that the activities of the oxidized and reduced forms are equal, a constant potential difference will be maintained, provided the conditions are unchanged. The potential set up (compared to a hydrogen electrode) when an electrode is in contact with a molar solution of its own ions at 25°C is defined as the standard electrode potential of the element, E_0. This value is of great use in oxidation–reduction reactions, since it describes the relative affinities for electrons of various systems, and is then known as the *standard redox potential*. For zinc, the potential is negative compared to hydrogen (see the short table of standard redox potentials in Fig. 2–17).

System	Standard redox potential, E_0 (volts)
$K \rightleftharpoons K^+ + e$	$-2{\cdot}92$
$Ca \rightleftharpoons Ca^{2+} + 2e$	$-2{\cdot}87$
$Zn \rightleftharpoons Zn^{2+} + 2e$	$-0{\cdot}76$
$\tfrac{1}{2}H_2 \rightleftharpoons H^+ + e$ (pH 7)	$-0{\cdot}41$
(pH 0)	$0{\cdot}00$
$Fe^{2+} \rightleftharpoons Fe^{3+} + e$	$+0{\cdot}77$
$2H_2O \rightleftharpoons O_2 + 4H^+ + 4e$	$+1{\cdot}24$
$Cl^- \rightleftharpoons \tfrac{1}{2}Cl_2 + e$	$+1{\cdot}36$

Fig. 2–17 Standard redox potentials of common elements

Those substances which have a high positive potential are stronger oxidizing agents, or have a greater tendency to take up electrons, than those with a low positive potential, and these in turn have a greater tendency than those with a negative potential. So the series is one of increasing electron affinity. Included in the table is a system in which two different ions of one element, Fe^{2+} and Fe^{3+}, are in equilibrium with one another. As shown, this has a positive standard redox potential, which means that the ferric ion will act as an oxidizing agent, and be reduced by H_2, or, theoretically, by the reduced form of any system with a lower value of E_0 than its own. The redox series is thus of great value in predicting which reactions are theoretically possible.

Redox potentials are of considerable importance in biochemistry; many reactions involving electron transfer occur in living cells. One very important process is the oxidation of glucose by cells, which yields a considerable amount of energy. The overall reaction may be written as:

$$C_6H_{12}O_6 + 6O_2 \longrightarrow 6CO_2 + 6H_2O \qquad (\Delta F' = -685,000 \text{ cal})$$

This is the reverse of the photosynthetic process, in which light energy from the sun releases electrons which then take part in a series of reactions, the final result of which is the synthesis of organic compounds from carbon dioxide and water. As in photosynthesis, the oxidation of glucose does not take place in a single step, but in a series of gradual steps. The various reactions involved will be considered in Chapter 6. As may be seen from the free energy change shown above, the overall reaction is highly exergonic. The energy made available is released in small packets at various stages in the series of reactions.

Certain compounds play an extremely important role in this process, because they can exist in an oxidized or a reduced form, and so can act as a means of transferring electrons. For instance, the coenzyme nicotinamide adenine dinucleotide, NAD (Figs 7–14 and 7–15), takes part in its reduced form, NADH, in the respiratory chain in mitochondria which is a series of coupled oxidation–reduction reactions (Section 6–2). Iron-containing compounds known as cytochromes play a key role in this process; in their oxidized form the iron is in the ferric state, and in their reduced form it is in the ferrous state. During this series of reactions, energy is released from the NADH molecules in small packets, and is coupled to the formation of the energy-storing compound ATP from ADP. From start to finish the process is highly efficient. The standard free energy change for the oxidation of glucose is −685,000 cal. From the oxidation of one molecule of glucose, 38 molecules of ADP are converted into ATP. Each molecule of ATP has a standard free energy of formation of about +8,900 cal.; for 38 molecules, therefore, the free energy change is +338,000 cal. This means that glucose as a source of energy in cells is utilized with an efficiency of between 40 per cent and 50 per cent, a remarkably high value compared with that at which most man-made machines work.

In the table of Fig. 2–17, two values of the redox potential of hydrogen are given; one is the arbitrary zero at pH 0, and the other is the value at pH 7, which is approximately -0.4. In biochemical reactions the latter value, at physiological pH, is the important one. Figure 2–18 gives some standard redox potentials at pH 7 for various important biochemical systems.

System	E_0 at pH 7
$\frac{1}{2}H_2 \rightleftharpoons H^+ + e$	-0.41
$NADH \rightleftharpoons NAD^+ + H^+ + 2e$	-0.32
Lactate $\rightleftharpoons$ Pyruvate	-0.19
$\begin{array}{ll} CH_3 & CH_3 \\ | & | \\ CH{-}OH & C{=}O + 2H^+ + 2e \\ | & | \\ COO^- & COO^- \end{array}$	
Cytochrome $a \rightleftharpoons$ Cytochrome a (reduced) (oxidized)	$+0.29$
$Fe^{2+} \rightleftharpoons Fe^{3+} + e$	$+0.771$
$2H_2O \rightarrow O_2 + 4H^+ + 4e$	$+0.816$

Fig. 2–18 Standard redox potentials of some biochemical systems at pH 7

The usual method of determining these is by potentiometric titrations—that is, by starting either with the completely reduced form of a system and adding oxidizing agent to it, determining the potential difference at various stages of oxidation by reference to a standard electrode, or by carrying out this process in reverse, adding reducing agent to the completely oxidized form. The type of curve obtained is shown in Fig. 2–19. When 50 per cent of the substance is present in the

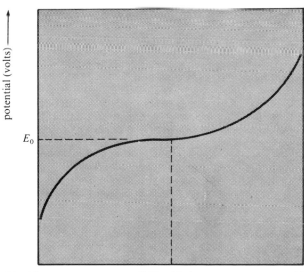

Fig. 2–19 Potentiometric titration curve

potential (volts) ⟶

E_0

reduced form   50% oxidation   oxidized form

reduced form and 50 per cent in the oxidized form, there is a point of inflection on the curve. The potential difference at this point is E_0, the standard redox potential.

In certain biochemical systems a direct potentiometric measurement is not possible. However, if a system of unknown potential reacts reversibly with a system of known potential, the equilibrium constant can be calculated by determining the concentrations of the various reactants at equilibrium. It may be shown thermodynamically that the change in redox potential of the combined systems is related to the logarithm of the equilibrium constant, and by this means it is possible to calculate the E_0 of the first system.

From Fig. 2–18, $NADH \rightleftharpoons NAD^+$ has a negative E_0, and thus its reduced form NADH will be expected to act as a very good reducing agent inside cells. Since the system

$$2H_2O \rightleftharpoons O_2 + 4H^+ + 4e$$

has a high positive E_0, we can predict that the oxidation of NADH by O_2 is probable thermodynamically. Note, however, that a favourable redox potential does not necessarily mean that a particular reaction will take place, but simply that it is thermodynamically feasible. Often the presence of a specific catalyst is required before the reaction will proceed at an appreciable rate. In fact, with the oxidation of NADH, the reaction will only take place in the presence of appropriate enzymes:

$$NADH + H^+ + \tfrac{1}{2}O_2 \xrightarrow{\text{enzymes}} NAD^+ + H_2O$$

This gives a negative change of free energy, since it is a thermodynamically possible reaction, ΔF being of the order of $-52,000$ cal.

2–9 Adenosine phosphate esters and energy-releasing reactions

There is one compound which often acts as a link between cellular endergonic and exergonic processes. This is adenosine triphosphate (ATP) which acts as both a store and source of energy.

ADP and ATP are the di- and tri-phosphate esters respectively of adenine ribonucleoside (Fig. 7–13); there is also a monophosphate, AMP. In all three compounds the CH_2OH group of ribose forms an ester link with the phosphate group of phosphoric acid, H_3PO_4. As an abbreviation the phosphate group is often written as Ⓟ, so that ATP is written as adenosine—Ⓟ—Ⓟ—Ⓟ.

ATP is hydrolysed in the presence of a suitable enzyme, or by dilute mineral acids, or alkalis. The hydrolysis involves the loss of the terminal Ⓟ, leaving ADP:

adenosine—Ⓟ—Ⓟ—Ⓟ + H_2O $\rightleftharpoons$ adenosine—Ⓟ—Ⓟ + H_3PO_4

The standard free energy of this reaction at pH 7 is about $-8,000$ cal, which means that there is a large release of energy.

Is hydrolysis
oxidation or
reduction ??)
,
exergonic —
hence oxidation?

There are two reasons for this. First, in each of the phosphate groups of ATP the oxygen atom, because of its tendency to acquire electrons, assumes a negative charge, which induces a positive charge on the neighbouring phosphorus atom (Fig. 7–13). This means that in the ATP molecule energy is required to overcome the electrostatic repulsion of the like positive charges on the phosphorus atoms, and to hold the molecule together. Thus when one phosphate group is removed by hydrolysis, this energy is released. Second, several resonance forms exist for both the reactants and the products of the hydrolysis; the more resonance forms which are possible, the greater is the stability (see Section 2–1). Among the reaction products, inorganic phosphate has the resonance structures:

$$
\underset{\overset{\|}{O}}{HO-\overset{\overset{O^-}{|}}{P^+}-O^-}
\qquad
\underset{\overset{|}{O^-}}{HO-\overset{\overset{O^-}{|}}{P^+}=O}
\qquad
\underset{\overset{|}{O^-}}{HO-\overset{\overset{O}{\|}}{P^+}-O^-}
$$

The number of resonance structures for ATP (and ADP) is clearly less, since two (one in the case of ADP) of the oxygens take part in the phosphate bond structures. There is thus greater stability, and hence lower energy values, on the right-hand side of the reaction, $ATP \rightleftharpoons ADP + \textcircled{P}$

Similar changes take place in the hydrolysis of the second phosphate group in ADP to give AMP:

$$\text{adenosine}-\textcircled{P}-\textcircled{P}+ H_2O \rightleftharpoons \text{adenosine}-\textcircled{P}+ H_3PO_4$$

This reaction also has a high negative free energy change, ΔF at pH 7 being $-6,500$ cal. On the other hand, the hydrolysis of the third phosphate in AMP has a much smaller free energy change ($\Delta F = -2,200$ cal). As in other phosphate esters, AMP has a number of resonance structures, leading to greater stability and lower free energy.

There are other biochemical compounds which on hydrolysis release large amounts of energy, and many of these contain phosphate—for example, phosphoenolpyruvic acid is hydrolysed to pyruvic acid with a ΔF of $-12,800$ cal:

$$
\underset{\overset{\|}{CH_2}}{\overset{\overset{COOH}{|}}{C}-O-PO_3H_2} + H_2O \longrightarrow \underset{\overset{|}{CH_3}}{\overset{\overset{COOH}{|}}{C}=O} + H_3PO_4
$$

One of the factors contributing to this high release of energy is the fact that, in the phosphate compound, pyruvic acid is held in the unstable enol form, while on hydrolysis it reverts to the much more stable keto form, at a lower energy level.

ATP is used in numerous cellular reactions for the storage of energy, by means of its formation from ADP. Subsequently energy is released by re-conversion of ATP into ADP.

Occasionally the expression high-energy (or energy-rich) phosphate bond is used, but this is very misleading. It is not the bond energy of the terminal phosphate group which is high; the ΔF value refers not only to the entire ATP molecule, but specifically to its hydrolysis, and the reaction products of its hydrolysis.

2–10 Biological catalysts

Some chemical reactions require a catalyst. In the manufacture of sulphuric acid by the contact process, for instance, sulphur dioxide is oxidized to sulphur trioxide at high temperature in the presence of a vanadium pentoxide catalyst:

$$2SO_2 + O_2 \longrightarrow 2SO_3$$

The molecules of sulphur dioxide and oxygen are adsorbed side by side on the surface of the catalyst. This adsorption reduces the potential energy barrier (or energy of activation) of the reaction (Fig. 7–3). The barrier can be regarded as a hurdle; the effect of the catalyst is to reduce the height of the hurdle so that more molecules at a given temperature can jump over. The catalyst itself is unaltered chemically at the end of the reaction, and does not alter the course or the free energy change of the reaction.

Biological catalysts or *enzymes* are always protein molecules, and are synthesized inside cells. The function of an enzyme is to speed up a thermodynamically possible cellular reaction to a rate which is suitable in the context of the various biochemical processes which take place in cells. Enzymes are extremely specific, and thousands of different kinds are required to catalyse the reactions inside cells. The polypeptide chains of enzymes are coiled up in a specific fashion to form round molecules of about 50–100 Å in diameter. In certain regions a localized group of amino acids (for example, A, B, and C, in Fig. 2–20a) has a particular three-dimensional arrangement; this is represented diagrammatically in Fig. 2–20b. This group provides the specific *active site* of an enzyme, on which the molecules taking part in a particular reaction are adsorbed, and thus plays a role similar to that of vanadium pentoxide in the synthesis of sulphur trioxide. Each of the many different kinds of enzyme molecule possesses an active site and the ability to catalyse a particular biochemical reaction.

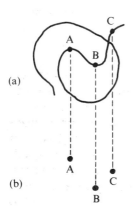

(a)

(b)

Fig. 2–20 Active site of an enzyme molecule. **(a)** Folding of a linear polypeptide chain in a specific manner to give rise to a corpuscular enzyme molecule. **(b)** Arrangement of three amino acid side chains, A, B, and C, in specific positions on the surface of the molecule provides an active site for the catalytic activity of the enzyme

2–11 Properties of long-chain molecules

The long-chain molecules of nucleic acids, proteins, and polysaccharides, and the somewhat shorter molecules of lipids, possess interesting physical properties. Synthetic long-chain polymers, such as nylon and Terylene or Dacron, have been extensively prepared in laboratories and are widely used in the textile industry. Synthetic polymer research has been of considerable value in biology, since the general properties of synthetic polymers are the same as those of polymers of biological origin.

The simplest biological polymers are formed from a single repeating unit (or monomer). Cellulose, the important structural polysaccharide of plants, is a good example. It consists of glucose molecules joined together by 1,4 linkages (see Section 7–4 and Fig. 7–27b). Cellulose chains may contain up to 20,000 glucose residues joined end to end and have molecular weights of up to 2 million. Starch is another polysaccharide which is formed from glucose monomers (Fig. 7–27a).

Next in complexity are the nucleic acids, which are composed of four different monomer units, three of which are the same for both DNA and RNA, while the fourth differs slightly in the substitution of the pyrimidine base concerned. There are variations in the order in which the nucleotide units are joined together, which makes for considerable variety in their composition.

Proteins are by far the most common polymer molecules found in cells, constituting over 90 per cent of the dry mass of most cells. They are formed from the twenty common amino acids shown in Fig. 4–1a. The possible variation in amino acid sequence in polypeptide chains is therefore immense.

Protein molecules are held together by several different types of bond. In a covalent bond, the pair of electrons which form the bond is shared between the two atoms concerned. If, however, one of these atoms has a greater affinity for electrons than the other, the electrons (or electron charge cloud) will not be shared equally, but will be displaced from a position mid-way between the two nuclei towards the nucleus of the atom which has the greater electron attracting properties. This asymmetric distribution of charge will give rise to a marked polarity, the best measure of which is its dipole moment, which is proportional to the effective charges set up and their distance apart. When hydrogen combines with certain strongly electron attracting elements such as oxygen or nitrogen (as in water or ammonia), the polarity in the covalent bond is so marked that in the liquid state the positive pole on one molecule aligns itself towards the negative pole of the adjacent molecule, as in

$$
\begin{array}{c}
H^{\delta+} \\
\diagup \\
H^{\delta+}\cdots\cdots\ O^{\delta-} \\
\diagup \qquad\qquad \diagdown \\
O^{\delta-} \qquad\qquad H^{\delta+} \\
\diagdown \\
H^{\delta+}
\end{array}
$$

As a result of dipole–dipole interaction the molecules are associated in pairs, or even in chains. The molecules are said to be joined by *hydrogen bonds*, arising from the very small size of the hydrogen atom, and its ability to approach closely to the electronegative elements. Hydrogen bonds are much weaker than covalent or ionic bonds and are easily broken by thermal agitation. In solutions of polypeptide and polynucleotide chains, hydrogen bonds can form between $>$NH and $>$C $=$ O groups and play an extremely important role in determining the spatial arrangement of the molecules. They provide as it were the gossamer threads which maintain the organization of long-chain molecules in specific configurations, and also the relation of these forms to associated water molecules. The structures they maintain are nevertheless labile, due to the low energy involved in the formation of hydrogen bonds. For this reason the three-dimensional structure of protein molecules can be changed extremely easily.

There are even weaker forces of interaction between molecules, known as *van der Waals forces*. Their exact nature is uncertain, but they are residual forces which exist after the primary chemical affinities between atoms have been satisfied. Van der Waals forces are believed to result from the interaction of rapidly moving electrons in adjacent molecules. They occur for example between the hydrocarbon chains of lipids:

$$
\begin{array}{ccccccc}
 & C & & C & & C & \\
\diagdown\diagup & & \diagdown\diagup & & \diagdown\diagup & & \diagup \\
CH_2 & & CH_2 & & CH_2 & & CH_2 \\
 & & & & & & \longleftarrow \text{van der Waals forces} \\
CH_2 & & CH_2 & & CH_2 & & CH_2 \\
\diagup & & \diagdown\diagup & & \diagdown\diagup & & \diagdown \\
 & C & & C & & C &
\end{array}
$$

They are mainly responsible for the interactions between non-polar side chains of proteins, and help to maintain specific molecular structures.

2–12 Solutions of long-chain molecules

Colloidal properties. In a colloidal system, a substance in one phase is dispersed homogeneously in the form of very fine particles, droplets, or bubbles (depending on whether the disperse phase is solid, liquid, or gaseous) in a dispersion medium of different phase. It is therefore distinct from a true solution, in which only one phase is present. In biological material we are concerned with *sols*, in which a solid is dispersed in a liquid medium, and *gels*, in which a liquid is dispersed in a solid medium. A colloidal sol is therefore intermediate in its properties between a true solution and a coarse suspension; there is in fact a gradual transition between the three states as the size of the particle increases. In a sol, the particles are so small that their own thermal agitation, combined with the energy transmitted to them by bombardment with molecules of the dispersion medium, partially counteracts the effect of gravity, which means that the rate of settling out is not appreciable.

It is of great importance biologically that protein molecules can act as colloids. (Starch, agar, and fat globules

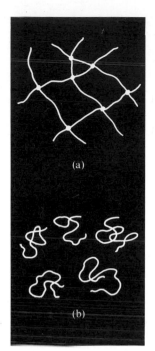

(a)

(b)

Fig. 2–21 **(a)** Structure of a gel showing a network of long-chain molecules which are surrounded by water molecules. **(b)** Long-chain molecules which are coiled up in solution as a sol

also form colloidal material.) Proteins often form a true solution in aqueous media, but as they are very large molecules they behave physically like colloidal particles. Since most of the dry weight of cells is protein, this has a profound effect upon cellular properties. Cellular contents (protoplasm, as this used to be called) form a complex solution, containing as well as proteins such substances as inorganic salts, simple sugars, and amino acids, and at the same time a heterogeneous colloid. The colloids of living material show a considerable affinity for the dispersion medium—that is to say, they are *lyophilic*, or, as the medium is water, *hydrophilic*. The colloids often become extensively solvated (hydrated), forming a close association with solvent molecules by means of weak bonds. A colloidal suspension of this type is much more stable than one consisting of a hydrophobic dispersed phase. Typical of lyophilic colloids is that they can exist in either the sol or the gel state and can easily change from one state to the other. Gelatin, a protein derived from collagen, is a well-known example of this.

In the gel state the long-chain molecules are held together at specific points by hydrogen bonds or other weak forces of interaction (Fig. 2–21a), and form an orderly three-dimensional interconnecting network throughout the dispersion medium, so that both phases are in effect continuous. A gel is a semi-solid, and possesses high viscosity, elasticity, and the ability to resist a shearing force. In the sol state, which is essentially fluid, the colloidal particles are dispersed randomly throughout the medium (Fig. 2–21b). Changes between the gel and the sol state are brought about by changes in temperature (below 34°C gelatin forms a gel, but above this temperature it is converted into the sol state), pressure, concentration, pH, salt concentration, in some cases possibly by supplying energy in the form of ATP (see Chapter 1), and by agitation of the solution. In the latter instance, gel-like solutions of certain long-chain molecules which become fluid on shaking or stirring are said to possess the property of *thixotropy*. Apparently, when the solution is at rest, weak associations form between the molecules and maintain the gel state. When the solution is stirred or shaken, however, these associations are destroyed, and conversion into the sol state occurs. On standing, the gel situation is again restored.

Viscosity. The viscosity of a solution can be measured by the resistance to movement of an object through it, and is generally expressed relative to the behaviour of water under the same conditions. Syrup containing sugar molecules at high concentration is a good example of a solution of high viscosity. Nucleic acids and proteins can form viscous solutions at extremely low concentrations; with these long-chain molecules, the viscosity depends in a complicated way on the chain length and the molecular weight. As can be seen from Fig. 2–22a, the

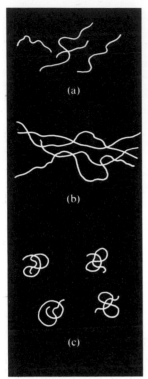

Fig. 2-22 Effects of molecular size and shape on the viscosity of solutions. **(a)** Low viscosity of a solution of short-chain molecules with few intermolecular contacts. **(b)** High viscosity of a solution of long-chain molecules with many intermolecular contacts. **(c)** Low viscosity of a solution of coiled up long-chain molecules with few intermolecular contacts

probability of contact or entanglement between molecules of low molecular weight and low chain length is less than between the same number of molecules of high chain length and high molecular weight (Fig. 2-22b). Many proteins, such as enzymes, do not exist in the form of thread-like molecules; because of interactions between the amino acid side-chains the long molecules become coiled up into very compact configurations which do not exhibit a high viscosity in solution (Fig. 2-22c).

Flow birefringence. If a solution of a long-chain molecule is placed between two sheared surfaces, as in Figs 2-23a and 2-23b, there is a tendency for the molecules to line up parallel to each other (and to the shearing surfaces). This can happen when a solution is placed between two parallel plates or between two coaxial cylinders. The refractive index parallel to the chains of the molecules is usually higher than at right angles to the molecules (Fig. 2-23b). Objects of this type may be observed in a polarizing microscope using crossed polarizers (Fig. 1-22), and appear as bright areas, in the regions where the molecules are lined up parallel to each other, on a dark field. These regions are said to show *birefringence* or double refraction. When lining-up of the molecules occurs because of a flow gradient resulting from shearing, the property is known as *flow birefringence*. Some alignment can also occur when a solution of long-chain molecules is allowed to evaporate slowly on a solid surface (Figs 2-24a and 2-24b). Here again, the refractive index parallel to the molecular chains is higher than that at right angles to them.

All of the properties of long-chain molecules described above are of major importance in determining the structure and behaviour of the cytoplasm of living cells. The fact that cellular contents can exist in sol-gel states explains how considerable integrity of structure is maintained in a system composed largely of water. Different regions or structures may be in different sol-gel states at various times, interconversion

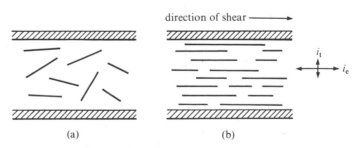

Fig. 2-23 Flow birefringence of solutions of long-chain molecules. **(a)** Solution of long-chain molecules between two parallel surfaces. **(b)** Parallel orientation of molecules due to application of shearing force: i_e refractive index parallel to molecules; i_t refractive index at right angles to molecules

being determined by alterations in the local environment. Continuous sol–gel interchanges in amoeboid movement have already been mentioned (Section 1–6) and will be discussed further in Chapter 11.

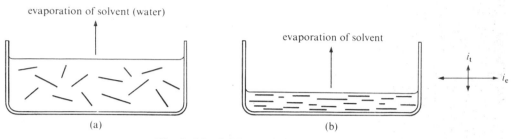

Fig. 2–24 (a) Long-chain molecules in solution. (b) Self orientation of molecules due to evaporation of solvent: i_e refractive index parallel to molecules; i_t refractive index at right angles to molecules

2–13
Behaviour of lipid molecules

In Chapter 1 we mentioned that lipid molecules play a highly important role in the organization of living cells because they form a major component of the outer or plasma membrane, as well as the membranes which divide the cell into compartments, such as the nuclear, mitochondrial, and lysosomal membranes. Certain general physicochemical characteristics of lipids are of importance in considering cell function.

Lipid–water mixtures. Because neutral lipids or triglycerides (Fig. 2–14) occur as distinct droplets of lipids in the cell cytoplasm, a study of the behaviour of lipid–water mixtures is of interest. If a neutral liquid hydrocarbon such as paraffin oil is shaken with water it becomes dispersed into minute droplets. When left to stand, these droplets coalesce to form a separate layer of lipid on top of the water. The water and oil are immiscible, since the water molecules are attracted to each other by stronger forces than those between water and lipid molecules, and similarly the lipid molecules are more strongly attracted to other lipid molecules than to water molecules. Some small intracellular molecules may tend to dissolve in both oil and water. For instance, citric acid

$$CH_2.COOH$$
$$|$$
$$C(OH).COOH$$
$$|$$
$$CH_2.COOH$$

is extremely soluble in water, by virtue of its polar acidic and hydroxyl groups. Its methyl ester, however,

$$CH_2.COO\ Me$$
$$|$$
$$C(OH).COO\ Me$$
$$|$$
$$CH_2.COO\ Me$$

is much less polar in character, and therefore much less

soluble in water. But the three hydrocarbon methyl groups are soluble in lipid. So if an aqueous solution of the methyl ester is placed below a layer of lipid, some of the trimethyl citrate molecules will enter the lipid layer, until eventually an equilibrium distribution is reached. If C_1 is the concentration of trimethyl citrate in water and C_2 the concentration in lipid, we write $C_2/C_1 = K$, where K is called the lipid/water *partition coefficient*. The more soluble in lipid a molecule is, the higher is the concentration C_2, and the higher the value of K. The K value for various small cellular molecules is of great importance in determining how easily they enter cells and how they are distributed within cells.

Polar lipids at interfaces.   Phospholipids contain a phosphate group and a nitrogenous base; for instance, lecithin or phosphatidylcholine has the formula

$$CH_2OCO \wedge\wedge\wedge\wedge\wedge\wedge\wedge\wedge\wedge\wedge (R^1)$$
$$|$$
$$CHOCO \wedge\wedge\wedge\wedge\wedge\wedge\wedge\wedge (R^2)$$
$$|$$
$$| \qquad\quad O^-$$
$$| \qquad\quad |$$
$$CH_2-O-\overset{+}{P}-OCH_2CH_2\overset{+}{N}(CH_3)_3$$
$$\qquad\quad |$$
$$\qquad\quad OH$$

where R^1 and R^2 are long carbon chains.

Here the basic part of the molecule is choline, an amino-alcohol with the formula $CH_2OH.CH_2N(CH_3)_3OH$. Phospholipids form an important part of cell membranes. The phosphate part of the molecule is highly polar and water soluble, while the hydrocarbon chains R^1 and R^2 are lipid soluble. They are similar in this respect to soap molecules, which are the Na and K salts (polar region) of long-chain fatty acids (hydrocarbon part of molecule).

Molecules of this type are of great interest to surface chemists. When a solution of such a molecule in alcohol is placed on a water surface, the alcohol is quickly dissolved in the water, while the molecules of polar lipid arrange themselves at the air–water interface with the polar portion in the water (Fig. 2–25a) and the hydrophobic lipid part projecting into the air. If thin waxed barriers are placed on the surface on either side of the lipid molecules, and the layer is compressed, the molecules are eventually brought into a closely packed assembly (Fig. 2–25b). The surface pressure, as measured by the force required to compress the layer further, rises steeply at this point of close packing. Polar lipids have a similar tendency to pack in this way at an oil–water interface; this in fact is more relevant to the situation in a cell, where there are many lipid membrane–water interfaces. Experimental work is, however, much simpler using the air–water system, and most of the early work by Rideal, Schulmann, and others involved this interface.

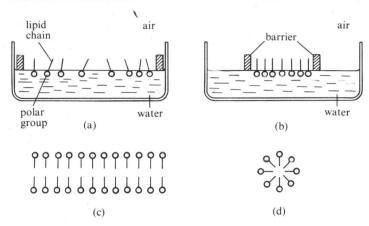

Fig. 2-25 Behaviour of polar lipids in water. **(a)** Polar lipids at an air–water interface; the lipid portions are in the air and the polar portions are in the aqueous phase. **(b)** The same, after compression between two barriers. **(c)** Bilayer formed by phospholipids, in which the polar ends of the molecules are in the aqueous phase while the non-polar regions are associated in the bilayer. **(d)** A spherical micelle which is similarly arranged with the polar ends of the molecules in the aqueous phase

When completely surrounded by water, polar lipid molecules tend to form a lipid bilayer as shown in Fig. 2-25c. The polar groups face the aqueous phase on both sides, while the non-polar hydrocarbon chains face and interpenetrate each other. Spherical micelles may also form (Fig. 2-25d, see also Fig. 5–5). The relation of these structures to intracellular membranes and to the cell plasma membrane will be discussed in Chapters 5 and 8.

2–14 Osmotic phenomena The membranes in living cells produce certain physico-chemical phenomena of importance in interpreting the behaviour of biological systems. If a small volume of a concentrated solution of some substance such as glucose is introduced into a large volume of water, the glucose rapidly becomes dispersed in the water until a uniform equilibrium concentration is reached between the two solutions. Certain natural and artificial membranes are *semipermeable*—that is, they are permeable to water but not to solutes dissolved in the water. For instance, if a glucose solution of high concentration C_A were placed on the left-hand side, A, of a semipermeable membrane, M (Fig. 2–26a), with a low concentration C_B of glucose on side B, the tendency would be for glucose molecules to pass from A to B until $C_A = C_B$. However, because the membrane is not permeable to glucose, but permeable to water, water molecules instead will pass from B to A to compensate for the difference in concentration. This flow of water through a semipermeable membrane is called *osmosis*. It can be prevented by applying pressure, P, to A, where P is known as the *osmotic pressure*. Then

$$P = RT(C_A - C_B)$$

where R is the gas constant, T is the absolute temperature, and C_A and C_B are the molar concentrations of solute.

The osmotic pressure is determined not by the nature of the solute molecules but simply by the difference in their molar concentration on either side of the semipermeable membrane.

Certain small ions, such as Na^+ and Cl^-, can pass through some semipermeable membranes; some biological membranes are permeable to one type of small ion but not to another. Often in biology a solution of a substance consisting of two diffusible ions is separated by a membrane from another solution containing a salt with a non-diffusible ion, such as a protein ion. Consider sodium chloride (Na^+ and Cl^-) at a concentration C_A and separated by a semipermeable membrane

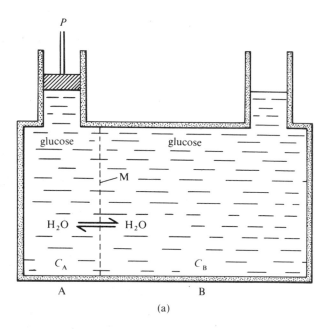

(a)

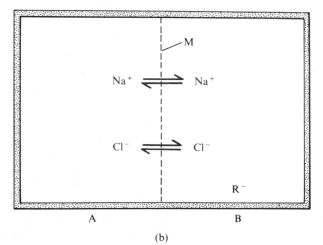

(b)

Fig. 2–26. (a) Osmotic pressure differences produced at a semipermeable membrane: C_A high concentration of glucose; C_B low concentration of glucose. The membrane is permeable to water but not to glucose molecules. **(b)** Donnan equilibrium. The membrane is impermeable to the negatively charged protein molecules, R^-. The Na^+ and Cl^- ions diffuse through the membrane, resulting in a difference of osmotic pressure and of electrical potential across the membrane

from a solution (Na$^+$ and R$^-$) of the sodium salt of a negatively charged protein molecule, where R$^-$ is the negative protein ion, at a concentration C_B (Fig. 2–26b). The Na$^+$ and Cl$^-$ ions may pass freely through the barrier while the R$^-$ ions are not diffusible. Some sodium and chloride ions will pass through the membrane and at equilibrium, as shown by Donnan in 1911, the concentrations of these will be unequal on the two sides of the membrane. If we suppose that under equilibrium conditions a concentration x of Na$^+$ and Cl$^-$ has diffused through the membrane, then the concentrations are

Na$^+$	Cl$^-$		Na$^+$	R$^-$	Cl$^-$
$C_A - x$	$C_A - x$		$C_B + x$	C_B	x

Donnan showed that

$$[Na_A^+][Cl_A^-] = [Na_B^+][Cl_B^-]$$

where [ ] indicates concentrations (although here also activities should be used, except in dilute solutions). Therefore,

$$(C_A - x)(C_A - x) = (C_B + x) \cdot x$$

This relationship, which indicates the ionic imbalance across the membrane, is known as the *Donnan equilibrium*. On side B protein ions are also present (at a concentration C_B) which produce a colloidal osmotic pressure.

Any ionic imbalance across a membrane results in an electrical potential across the membrane. The potentials are quite small, having a maximum value of about 70–100 millivolts (mV). But when the thinness of cell membranes (about 70–100 Å) is taken into account, the electric field as measured in volts/cm is enormous. These strong electric fields may be major factors in controlling many cellular functions.

2–15
Experimental
methods

The experimental methods described here are the specialized tools of biochemistry, molecular biology, and other biological disciplines. They consist chiefly of·

Preparative methods used in the separation of cellular components and in protein chemistry.

Analytical methods used for the determination of structure and other properties.

Measurements of respiration, metabolism, pH, and so forth.

Preparative methods. *Cellular homogenization.* The first stage in the preparation of subcellular components is to disintegrate the cells. This is often done in a Potter homogenizer which consists of two concentric cylinders separated by a narrow gap. The cell suspension passes through the gap in which the cells are subject to a shearing force sufficient to rupture the cell membranes.

Another method is to irradiate the cells with ultrasonic

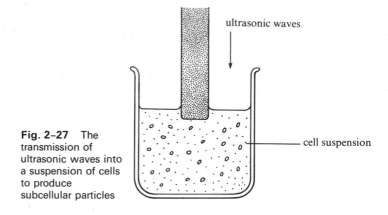

Fig. 2–27 The transmission of ultrasonic waves into a suspension of cells to produce subcellular particles

waves. These are generated by a piezo-electric crystal and are transmitted to a stainless steel rod placed in the solution containing the cell suspension, as shown in Fig. 2–27. The ultrasonic waves are transmitted through the water and generate vibrations which rupture the cell.

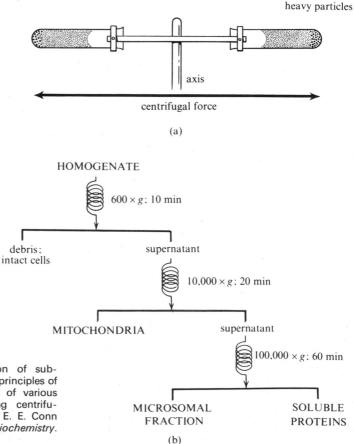

Fig. 2–28 The centrifugation of subcellular components. **(a)** Basic principles of centrifugation. **(b)** Separation of various subcellular components during centrifugation. (With permission from E. E. Conn and P. K. Stumpf, *Outlines of Biochemistry*. John Wiley, New York, 1964.)

Centrifugation.   Once a suspension of subcellular particles has been obtained, it is separated into various fractions as a first step in purification. The usual methods exploit differences in mass or density between the particles. Much of this work is now carried out with various types of centrifuge. A centrifuge consists of a pivoted cylinder which rotates at high speed about a vertical axis (Fig. 2–28a). The high rotational speeds produce centrifugal forces which may reach as much as 100,000 times the force of gravity; these high-speed machines need to be protected with armour plating for safety. The heavier particles are deposited first on the outer wall of the tube (Fig. 2–28a). A procedure for the separation of cellular fractions by successive centrifugation is illustrated in Fig. 2–28b.

Density gradients.   With large particles, an alternative method to centrifugation is to allow them to settle under gravity in a density gradient. The principle is illustrated in Fig. 2–29. First, two solutions—for example, solutions of sucrose of high and of low density—are allowed to flow into the base of a column, beginning with the low-density solution. As the column fills, the relative rates of flow are changed so that the upper part of the column is of low density and the density

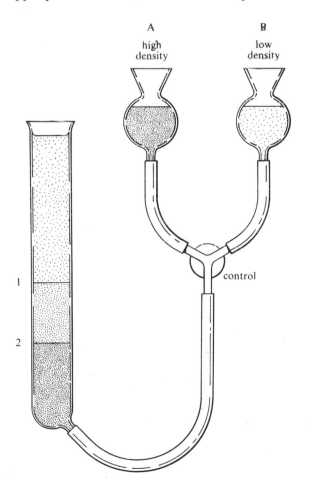

Fig. 2–29 The formation of a density gradient by the controlled flow of two solutions of different density. Two types of particles finally separate and lie at different levels—the less dense type at 1, and the more dense type at 2

increases progressively down the column. Such density columns are extremely stable. Now a suspension of particles or macromolecules is introduced into the column. The particles slowly separate according to density and lie at particular heights in the column.

Density gradients are of great value, particularly in the molecular biology of nucleic acids. Methods have now been developed whereby such gradients can be formed in rotating centrifuge tubes (zonal centrifuge). Particles again reach equilibrium at positions where their density equals that of the solution. They can then be drawn off continuously and collected as separate fractions.

The density gradient method is useful for separating mitochondria, endoplasmic reticulum, lipid particles, and so forth. The various fractions sediment in layers according to density.

The analytical ultracentrifuge works on the same principle as the preparative centrifuges, but it is fitted with an optical system for the analysis of moving boundaries and is described below.

Purification of macromolecules.   *Dialysis.*   This is often the first step in purification. It depends on the fact that some membranes are permeable to water molecules, salt molecules, and other small molecules found in cells, but are impermeable to macromolecules. The most commonly used membrane is made of Cellophane and is purchased in the form of tubes which can be knotted at the ends. If protein solution is to be purified, this is poured into the bag which is then knotted top and bottom and placed in a beaker of distilled water or saline —usually in a cold room to prevent deterioration of the biological material. The small molecules diffuse out through the membrane leaving the macromolecules inside the dialysis bag. The surrounding water may be replaced several times to achieve a more complete removal of unwanted small molecules.

Salting-out methods.   After fragmentation of cells many of the protein molecules remain in solution, and are found in the supernatant in the centrifuge. One of the oldest and still most widely used methods for purifying mixtures of proteins is the salting-out method. Proteins are amphoteric molecules carrying both negatively charged acidic groups and positively charged basic groups. It often happens that particular proteins exhibit characteristically distinct degrees of solubility in salt solutions of varying concentration. Certain proteins may be precipitated out of the solution and separated, leaving the other proteins in solution. Because the molecules are amphoteric their solubility is also dependent on pH. This too can be exploited for purification purposes.

Electrophoresis.   Biological macromolecules carry charged groups and therefore the molecules in solution carry a net electric charge (except at the isoelectric point). If two electrodes are placed in a solution of this kind and an electric field is applied, the positively charged molecules move towards the cathode, while the negatively charged molecules move towards the anode. The rate at which the molecules move depends on the magnitude as well as the sign of the charge. A molecule with a double charge will move at twice the speed of a molecule with a single charge. This difference provides the basis for the separation of various types of protein in solution.

In the early days (Tiselius, 1937), the proteins were allowed to move freely in solution and electrophoresis was carried out in a U-tube, as shown in Fig. 2–30. Nowadays,

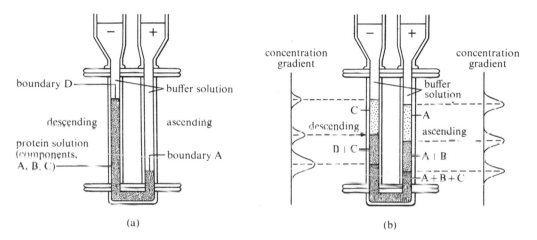

(a)   (b)

Fig. 2–30 The U-tube of an electrophoresis cell. **(a)** Setting-up of compensated boundaries. **(b)** Separation in ascending and descending boundaries

electrophoresis is usually carried out in a supporting medium of paper or gel (Fig. 2–31). This prevents rapid diffusion and preserves extremely sharp bands after separation. With paper electrophoresis, a small drop of the protein solution is placed at one end of the paper and the electric field is then applied, using an intermediate salt or buffer solution to make contact between the electrodes and the paper. When gel is used, it is often prepared in a vertical tube, and a drop of protein placed on top. Remarkably good separation of proteins can be attained with acrylamide gel (Fig. 2–31).

An even more sophisticated method of separation exploits the fact that proteins do not carry a net charge at the isoelectric point. If a pH gradient is established instead of a density gradient, proteins become concentrated in bands where they are stationary at a particular pH (Fig. 2–32). This is known as isoelectric focusing (Svensson, 1961). More recently electrophoresis has been applied to the study of *whole* cells. A suspension of cells is placed in a tube and an electric

Fig. 2-31 (right) Gel-electrophoresis, showing buffer chambers containing anode and cathode. The column is packed with gel. The sample is placed at the top of the gel and moves downwards. The fractions are collected at the elution outlet

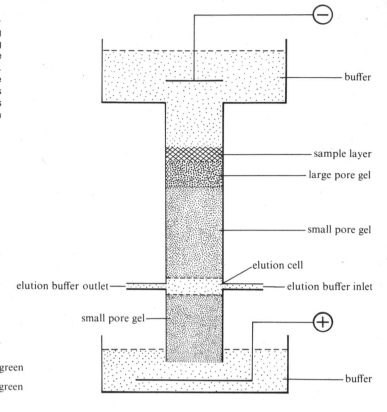

Fig. 2-32 The use of a pH gradient combined with electrophoresis. Protein molecules are stationary at their isoelectric points

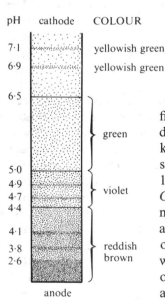

field is applied. The velocity of the cells can be measured by direct observation using a graticule. Measurements of this kind have helped in an understanding of the structure of the surface of the cell plasma membrane (James, 1956; Forrester, 1962; Ruhenstroth, 1962).

Gel filtration. A remarkable new development in separating macromolecules has been the use of 'molecular sieves'. These are gel-like materials, usually cross-linked dextrans (celluloses), of trade name Sephadex. They possess a molecular structure which allows some molecules to pass through, but retains others. A column of the material is prepared in a vertical tube and the solution containing the mixture of proteins or other macromolecules is placed in the top. The smaller molecules pass into the structure of the gel more quickly than the larger molecules (Fig. 2-33). The smaller molecules therefore pass out of the base of the column more slowly than the large molecules. The liquid flowing out of the column is collected in a *fraction collector*. This consists of a horizontal disk which contains small collecting tubes on its circumference. An electrical control rotates the disk about a vertical axis. The number of drops leaving the column and falling into a tube is counted automatically. The disk then rotates to bring the next tube under the column and so on until all the solution has been collected in separated fractions.

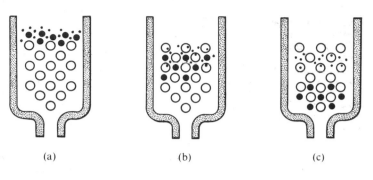

Fig. 2-33 The use of a molecular sieve for the fractionation of substances of high and low molecular weight: large dots—high molecular weight molecules; small dots—low molecular weight molecules; large circles—particles of Sephadex gel. **(a)** Solution placed at top of column. **(b)** Small molecules penetrate into Sephadex particles and are trapped. **(c)** Large particles pass through column

(a)   (b)   (c)

Instead of molecular sieves, ion-exchange materials are sometimes used. These depend on the presence of electrically charged groups, which produce a selective adsorption of protein or other molecules on their surface. Ion exchangers derived from Sephadex have a combined property of molecular sieving and selective adsorption, and are used in the isolation of mucopolysaccharides as well as proteins (Berman, 1962).

Physical analysis. *Size and shape of macromolecules.* The first important measurement to be made on a protein or other macromolecule is its molecular weight. The molecular weights of biological macromolecules vary widely from 4,000 for insulin to 10,000,000 or more for nucleic acids. Early measurements were made by Adair and Adair in the 1930s, from simple osmotic pressure measurements. The osmotic pressure, P_0, of a solution separated by a semi-permeable membrane from another solution is given by $P_0 V = RT$ and a solution containing one gramme-molecule of non-ionizable solute in 22·4 litres exerts an osmotic pressure of 1 atmosphere at 0°C. The use of a delicate manometer by Adair and Adair enabled them to determine the molecular weight of proteins in solution by comparing their osmotic properties with those of known solutes. The analytical ultracentrifuge is also widely used (Svedberg, 1925). This works on the same principle as the preparative centrifuges described above, except that a sharp boundary is formed between the protein solution and the buffer. This is viewed with a special optical system (known as Schlieren optics) while the centrifuge is rotating. The proteins are separated in several bands, depending on their shape and molecular weight.

Viscosity measurements are also important since viscosity depends both on the size and shape of molecules. Very elongated molecules like nucleic acids have a high viscosity because the long chains make contact in a network even at very low concentrations. Compact protein molecules, such as enzymes, on the other hand, have a low viscosity. A combination of ultracentrifuge and viscosity measurements therefore gives useful information about both the size and shape of the molecules. One type of viscometer in common use is the

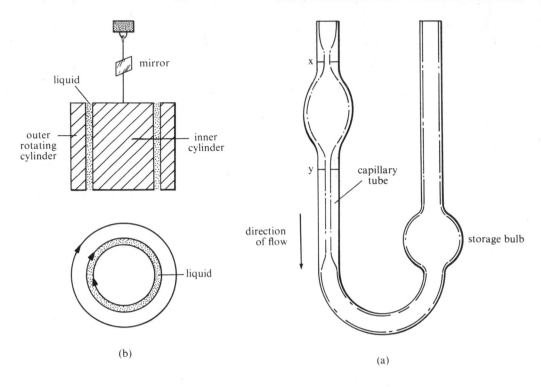

(b)

(a)

Fig. 2-34   The measurement of viscosity. **(a)** Capillary viscometer; the time taken to flow between x and y is recorded. **(b)** Couette viscometer; upper diagram is vertical cross-section and lower is horizontal; the outer cylinder rotates. The angular displacement of the inner cylinder about the torsion wire is recorded by a light beam reflected from a mirror

capillary viscometer shown in Fig. 2–34a. A highly viscous liquid flows very slowly through the viscometer while a low-viscosity liquid flows very rapidly. Relative viscosities are therefore measured from rates of flow. The Couette viscometer consists of concentric cylinders. The outer cylinder rotates and the force required to rotate the inner cylinder is measured (Fig. 2–34b).

Structure of proteins and nucleic acids.   Most of what is known about the structure of proteins and nucleic acids has emerged from a study of the materials in their solid form, either as fibres (silk, hair keratin, muscle proteins, nucleic acids) or crystals as (haemoglobin, myoglobin, ribonuclease, insulin).

The most fruitful method of analysis has been X-ray crystallography (Bragg, 1913; Bernal, 1934; Hodgkin, 1930s onwards; Wilkins and Franklin, 1953). This method depends on the fact that crystals diffract X-rays rather as a diffraction grating diffracts light waves. Since the wavelength of X-rays is much shorter than that of light waves, diffraction phenomena are observed between planes of atoms lying within the crystal (Fig. 2–35). In addition, the intensity of reflections depends on

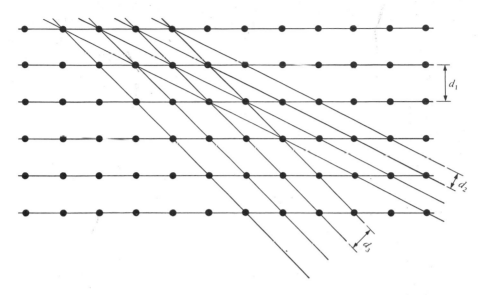

Fig. 2–35 X-ray diffraction. Various regular spaces, d_1, d_2, and d_3, occurring in a simple crystal lattice which can give rise to X-ray diffraction

the mass and position of atoms in relation to the reflecting planes (Figs 2–36 and 2 37). In the early days of X-ray crystallography of proteins, the problem was that it was impossible to decide whether particular atoms were on one side or other of a given reference point; this is the so-called 'phase problem'. The discovery by Perutz and coworkers (1954) that it was possible to introduce heavy atoms such as mercury into the protein permitted the determination of phases. The heavy atoms scatter X-rays particularly strongly and this can be interpreted in the final analysis. X-ray crystallography has proved to be a tremendously powerful tool. The structures of deoxyribonucleic acid (Chapter 3) and of several crystalline proteins (Chapter 4) have been worked out in detail by this method.

Infra-red spectroscopy (Ambrose and Elliott, 1950) makes use of another property. The vibrations of chemical

Fig. 2–36 Equipment for studying X-ray diffraction: *left*, the X-ray tube and collimating system; *right*, the crystal, which rotates about a vertical axis. Successive planes as shown in Fig. 2–35 reflect spots onto the cylindrically arranged photographic film. The camera is placed on the right

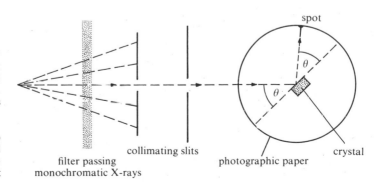

filter passing monochromatic X-rays

collimating slits

photographic paper

spot

crystal

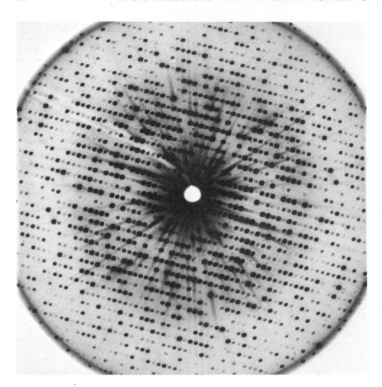

Fig. 2-37 The X-ray diffraction pattern produced by a single crystal of the protein myoglobin. (With permission from J. C. Kendrew.)

bonds in molecules give rise to absorption bands in the infrared spectrum (3–15 μ wavelength). The peptide group

$$
\begin{array}{c}
\text{H} \\
| \\
\diagdown\text{N}\diagdown \\
\diagdown\text{C}\diagup \\
|| \\
\text{O}
\end{array}
$$

of proteins has characteristic absorption bands in the infra-red region of the spectrum. If the radiation is polarized, this absorption only occurs when the wave is vibrating in the same direction as the chemical bond (Fig. 2–38). By this method the direction of various chemical bonds in complex molecules can be determined. This has led to the finding that the $\diagdown$NH and $\diagdown$C=O bonds in folded protein molecules lie

Fig. 2-38 The absorption of polarized infra-red radiation. Radiation polarized in the vertical plane is absorbed by the N—H bond when the atomic nuclei vibrate along the length of the bond (stretching frequency). No absorption occurs when the radiation is polarized in a horizontal plane

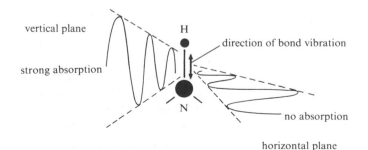

vertical plane

strong absorption

H

direction of bond vibration

N

no absorption

horizontal plane

along the long axis of the folded molecule, from which followed elucidation of the α-helical structure of proteins.

Another type of spectroscopy which is widely used, particularly for the nucleic acids, is ultraviolet spectroscopy. Nucleic acids and proteins have characteristic absorption bands in the ultraviolet region (2,000 Å–3,300 Å). These can be used to follow the isolation of nucleic acids (Fig. 2–39) as in fractions leaving a column, or for direct microscopic observations on living cells (Caspersson, 1936, and others).

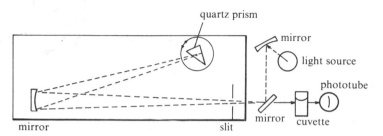

Fig. 2–39 The ultraviolet spectrometer. The incident beam is reflected in the quartz prism after dispersion within the prism. The intensity of transmission through the cuvette is recorded photoelectrically

Free energy changes. There are many experimental difficulties in the determination of free energy changes. Three methods are outlined below, the details of which may be found in standard physical chemistry text-books.

One method is to determine the equilibrium constant of a reversible chemical reaction in a homogeneous system at constant temperature and pressure, and to calculate the ΔF^0 from the equilibrium constant (see Section 2–7). The concentrations of reactants and products are usually assumed to be equal to their respective activities. Considerable errors in the determination of the equilibrium constant occur when the reaction lies very far in either direction; the accuracy may then be improved by labelling the reactants with isotopes, which may be estimated with considerable accuracy even when present in very small quantities (see below).

The second method is the use of thermal measurements The standard free energy of formation of a compound may be calculated from measurements of the heat of formation by calorimetry. Data are available on the standard free energies of formation of many substances of biochemical importance, such as carbon dioxide, acetic acid, glucose, glutamic acid, and so forth. The difference between the free energies of formation of the products and the free energies of formation of the initial reactants gives the free energy change in various reactions.

A third method depends upon the relation between the free energy change of a reaction and the difference in oxidation–reduction potentials (ΔE_0) of the reactants. For instance, if two oxidation–reduction systems

$$A(\text{reduced}) \rightleftharpoons A(\text{oxidized}) \quad \text{and} \quad B(\text{reduced}) \rightleftharpoons B(\text{oxidized})$$

are placed together and allowed to reach equilibrium, then

$$A(\text{reduced}) + B(\text{oxidized}) \rightleftharpoons A(\text{oxidized}) + B(\text{reduced})$$

The change in redox potential is equal to the difference between the potential of the system $B(\text{reduced}) \rightleftharpoons B(\text{oxidized})$ and that of the system $A(\text{reduced}) \rightleftharpoons A(\text{oxidized})$.

Thermodynamically, $\Delta F = -n\mathscr{F}\Delta E$ at a particular pH, where n is the number of electrons transferred in an oxidation–reduction reaction, and $\mathscr{F}$ is a constant known as Faraday's constant. This relationship provides another experimental technique for the determination of free energy change in certain reversible reactions.

Use of isotopes.   Radioactive isotopes of various elements are now widely used in biochemistry, molecular biology, and cytology. The most usual method is to obtain a simple molecule (metabolite) which has been labelled with an isotope of carbon, such as ^{14}C which has a half life of more than 5,000 years, or with an isotope of hydrogen, ^3H, which has a half life of about 12 years. The metabolite is introduced into a cell culture or a homogenate. If a nucleotide, it becomes incorporated into nucleic acids, and if an amino acid, into proteins. The amount incorporated in various fractions leaving a column, or separated on a paper strip in electrophoresis, may be measured with a Geiger counter (Fig. 2–40). A very sensitive method is to use a scintillation counter. Here, the radioactive material in solution emits energy which is absorbed by substances known as scintillants added to the fluid. The scintillants emit visible light, which is detected and measured by a photoelectric cell device.

Fig. 2–40 The Geiger-Müller counter for the detection of ionizing radiation: A source of potential; B Geiger–Müller tube containing ionizable gas; C lead shield; D sample pan holder; E shield for cathode; F anode wire; G mica window to admit radiation; H recorder of signals (scaler)

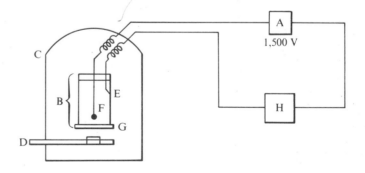

With cells or sections of tissues, auto-radiography is generally used (Howard and Pelc, 1951; J. H. Taylor, 1956, and others). A thin strip of photographic film is deposited on the surface of the slide, which is then kept in the dark for about one month. Regions of radioactivity produce silver grains in the photographic film and these can be observed at the same time as microscopic examination of the cells through the film is carried out. Using this method it is possible to observe labelling of cell nuclei, or cytoplasm; with the best resolution, the labelling of individual chromosomes can be distinguished.

Measuring rates of respiration. Living cells require a continual supply of energy which they obtain by the oxidation of glucose. This process is accompanied by the consumption of oxygen and the liberation of carbon dioxide. Many other biochemical reactions involve the formation or utilization of carbon dioxide or oxygen. An apparatus originally designed by Warburg can be used to measure respiration rates (Fig. 2–41). The tissue slice (or cell suspension) is placed in a small

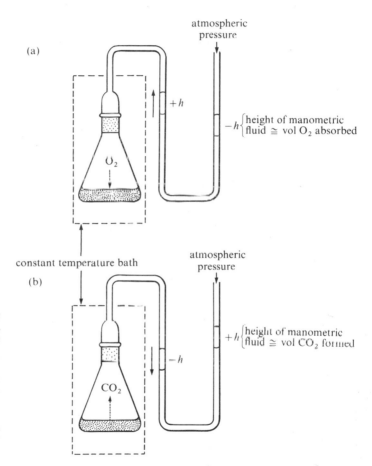

Fig. 2–41 The principle of the Warburg manometer for measuring the respiration of biological tissues (much simplified). **(a)** Absorption of O_2 during respiration causes a reduction of pressure, which is then measured. **(b)** Release of CO_2 from the solution is measured as an increase of pressure by the manometer

sintered glass vessel inside a conical flask. The rate of O_2 consumption is measured by the rate of change of pressure recorded by the U-tube manometer, the CO_2 evolved being absorbed by alkali. In a separate experiment the net change of pressure due to O_2 loss and CO_2 increase is then measured, and from this the amount of CO_2 liberated may also be determined. The principle is illustrated in Fig. 2–41. Micromethods have been worked out for measuring respiration of single cells.

pH measurements. The pH of a solution is generally measured with a glass electrode (Fig 2–42). When a thin glass membrane of high electrical conductivity is placed between two solutions of different pH, a potential difference related to this pH difference is obtained. The glass electrode can be used in almost any kind of solution, except those which are very acid or very alkaline. A typical arrangement is shown in Fig. 2–42. The electrode contains a platinum wire coated with silver/silver chloride in 0·1 M hydrochloric acid. The reference electrode is usually a calomel one containing mercury coated with mercurous chloride in saturated potassium chloride solution. The two electrodes are placed in the solution and the potential difference between them is measured with an accurate potentiometer. (Nowadays this is generally read directly with a pH meter.)

The difference E_g (glass electrode)$-E_{ref}$ (reference electrode) is related to the pH as follows:

$$pH = \frac{E_g - E_{ref}}{0·0591} \text{ at } 25°C.$$

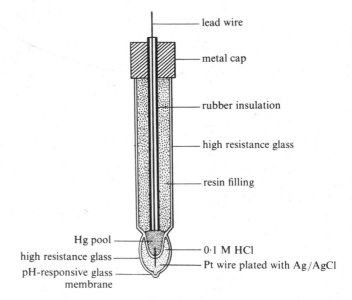

Fig. 2–42 A glass electrode for the measurement of pH. The bulb contains a solution of 0·1M HC1. The thin glass membrane at the base is conducting, and makes electrical connection with the test solution

Labels: lead wire; metal cap; rubber insulation; high resistance glass; resin filling; Hg pool; high resistance glass; pH-responsive glass membrane; 0·1 M HCl; Pt wire plated with Ag/AgCl

Bibliography

Brown, G. I., *Introduction to Physical Chemistry*. Longmans Green, London, 1968.

Conn, E. C. and P. K. Stumpf, *Outlines of Biochemistry*. John Wiley, New York, 1964.

Pollister, A. W. (Ed.), *Physical Techniques in Biological Research*, 2nd edn. Academic Press, New York, 1969.

Tedder, J. M. and A. Nechvatal, *Basic Organic Chemistry*. John Wiley, New York, 1966.

Wilson, J. F. and A. B. Newall, *General and Inorganic Chemistry*. Cambridge University Press, London, 1968.

PART 2   Structure, function, and synthesis of cellular components

We now come to consider the structure, function, and synthesis of the various cellular components. Which region of the cell should we study first? The problem is that no particular cellular component can be studied in isolation. The cell is a highly integrated system consisting of many interdependent processes, and to understand any one of these fully we must consider it in relation to all the others. In other words, to understand a particular part in some depth it is necessary first to appreciate the functioning of the cell as a whole. However, a start has to be made somewhere, and it seems logical to describe first the structure and function of the nucleus, because it plays a central role in the life of the cell. The replication of the deoxyribonucleic acid (DNA) molecule and the synthesis of ribonucleic acid (RNA) are described first, both in Chapter 3. Chapter 4 then deals with protein synthesis. Subsequent chapters in this part of the book describe the part played by various other cellular components and by the lipid membrane structures which enclose and protect intracellular organelles as well as forming a protective barrier surrounding the entire cell.

Throughout these chapters a certain amount of biochemistry will be introduced at various stages, particularly in the study of the nucleus, of protein synthesis, and of the energy requirements of cells. Detailed biochemistry concerned with the synthesis of the various cellular building units of nucleotides, proteins, carbohydrates, and lipids is postponed until Chapter 7, which also deals with the function of enzymes.

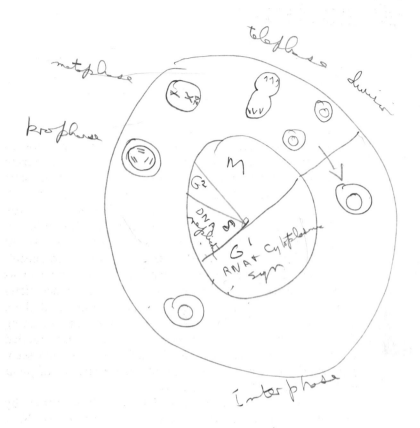

telophase
metaphase
prophase
dividing

M
G2
DNA replication
S
G1
RNA + Cytoplasme syn

interphase

3 The interphase nucleus

The *interphase* nucleus is the nucleus as it functions between mitoses in growing and dividing cells (Fig. 9–1). It is during interphase that the synthesis of RNA and various cytoplasmic components takes place; this is mainly in the G_1 phase, and, to a lesser extent, during the synthesis of DNA in the S phase, before mitosis. During mitosis and during the formation of the germ cells in meiosis the nucleus performs very different functions from those of interphase, being concerned specifically with the transfer of genetic material. In fact the nucleus no longer exists as the same entity and its appearance alters considerably; the nuclear membrane disappears and there are profound changes in the internal organization, including the disappearance of the *nucleoli*. These are irregular rounded or oval structures, one or more of which may be clearly seen, particularly after staining, in the interphase nucleus of plant and animal cells (Figs 1–3b and 1–3c).

Control of synthesis, particularly protein synthesis, by nucleic acids occurs during the early part of interphase. Now and in the next chapter we consider the functioning of the nucleus during this phase, and also during late interphase, when DNA is synthesized. The role of the nucleus in differentiation and development will be dealt with in Chapter 13.

In 1700, van Leeuwenhoek detected refractile bodies in the centre of the blood corpuscles of salmon blood. These structures, which must have been nuclei, were seen with the simple lenses which were ground by the Dutch draper in his spare time. Fontana in 1781 observed similar oval bodies inside the skin cells of an eel, and the Scottish botanist Robert Brown described in 1833 how nuclei can be seen quite generally within the cells from several tissues in flowering plants.

Evidence that the nucleus is not homogeneous, but contains thread-like structures, came from studies of dividing cells. Hofmeister, in the 1840s and 1850s, observed that when cells divide, the nuclei also divide to produce two daughter nuclei. He also recorded that small bodies become visible as the nucleus is about to divide. It was discovered some years later that these structures could be stained with aniline dyes, for which reason they became known as chromosomes (Greek: *chrōma*, colour, and *sōma*, body). Sutton in 1903 first suggested that hereditary factors or genes are carried on chromosomes. A few years later, Morgan demonstrated that certain genetic traits of the fruit fly *Drosophila* could in fact be traced to individual chromosomes (see Chapter 10).

Meanwhile much had been discovered about the chemistry of cell nuclei. In 1869, Miescher in Switzerland was able to isolate nuclear material, which he called nuclein, from the pus cells on hospital bandages. He later worked with salmon sperm cells because he observed that these consist almost entirely of nuclei. He showed that the material he extracted contained the elements phosphorus, carbon, oxygen, hydrogen, and nitrogen. When nuclein was found to be acidic in solution its name was changed to nucleic acid.

The great German chemist, Fischer, identified in the 1880s two heterocyclic structures related to uric acid. These were pyrimidines and purines (see Fig. 2–10):

pyrimidine structure   purine structure

Kossel was able to demonstrate the presence in nucleic acid of two pyrimidines (cytosine and thymine) and two purines (adenine and guanine). For this work he was awarded the Nobel Prize in 1910. Levene, a Russian-born biochemist working in New York, identified a five-carbon sugar, ribose (see Fig. 2–12), in nucleic acid in 1910. He later discovered that there was another type of nucleic acid which contains deoxyribose (Fig. 2–12). He also suggested that phosphorus is present in nucleic acids as a phosphate group, and that this group acts as a link between building units consisting of a sugar molecule and a basic purine or pyrimidine. In the early 1950s, Todd in Britain synthesized structures of this kind and found that they were identical with the material obtained from nucleic acids. He was awarded the Nobel Prize for his work in 1957.

The bringing together of the work done by cytologists and geneticists on chromosomes, and by chemists on the composition of nucleic acids, was accomplished in 1924 when Feulgen developed a staining technique which clearly showed that chromosomes contain the deoxyribose form of nucleic acid. In 1928, Griffiths, a bacteriologist, accidentally discovered that a factor present in heat-killed bacteria of one strain could confer characteristics of this strain on another type of bacterium. That this was due to the transfer of deoxyribonucleic acid was demonstrated later by Avery, Macleod, and McCarty in 1944, and they thus provided the first evidence that deoxyribonucleic acid is directly involved in hereditary mechanisms.

**3–1
Structure and
replication of
DNA**

*here there is nuclear
membrane, also
protein molecules
are in chromosomes!*

Studies since the 1940s have indicated that DNA is the universal genetic material of all forms of life except certain viruses, which have RNA as their genetic material. In myco-plasmas and bacteria the DNA does not appear to be asso-ciated with other molecules, but in the nuclei of plant and animal cells it is present in the chromosomes in association with protein molecules.

The building units of DNA are shown in Fig. 3–1. There are four different bases—the two purines, adenine and guanine, and the two pyrimidines, thymine and cytosine. The combination of one of these bases with deoxyribose is known as a *nucleoside* (Fig. 2–12). The carbon-1 atom of the sugar is linked to the 9-nitrogen in a purine base, and to the 1-nitro-gen in a pyrimidine (see Section 2–4). The phosphate ester derivative of a nucleoside is known as a *nucleotide* (Fig. 3–1), and it is from nucleotides that the DNA molecule is built. The units are joined together to form a long *polynucleotide* chain. All of this was known in the 1940s, when it was also thought that the units were joined by phosphate linkages. However, the exact spatial structure of DNA had not yet been determined, nor was it understood how such a molecule could carry the vast store of genetic information which would be needed to specify the many different types of cell in an organism.

A chemical study of DNA was made by Chargaff in 1947. He showed that DNA contains equal proportions of

Fig. 3–1 The basic building units of deoxyribonucleic acid. *Left*, the four bases, adenine, guanine, thymine and cytosine. *Right*, a pyrimidine nucleotide, formed from thymine, deoxyribose, and phosphoric acid

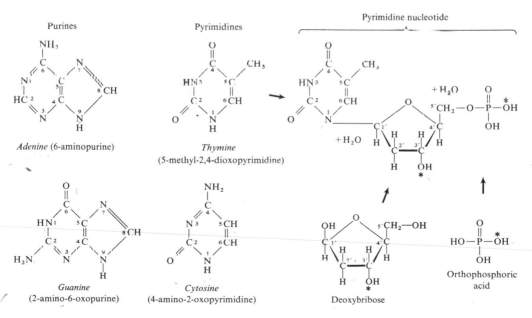

*Groups which condense to form the nucleic acid chain.

Purine   Pyrimidine
Adenine + Thymine
Cytosine + Guanine

what about extra protein molecules??

Handedness?

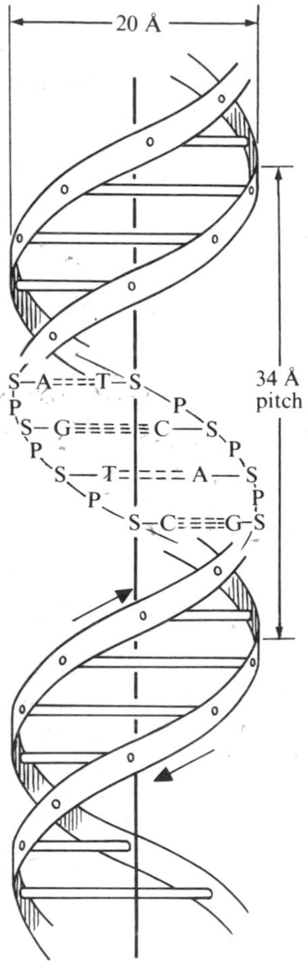

Fig. 3–2 The double helix of DNA (Watson and Crick model). The two chains are held together by hydrogen bonds between the bases. A adenine; T thymine; G guanine; C cytosine; P phosphate; S deoxyribose sugar

the large purine bases and the smaller pyrimidines, and—even more interesting—that adenine and thymine are present in equimolecular proportions and so are cytosine and guanine. The significance of this observation became clear when the structure of DNA was finally elucidated a few years later.

In 1950, Wilkins and his coworkers started work on the X-ray diffraction patterns of purified fibres of DNA prepared from animal cells, comparing their experimental observations with those to be expected from molecular models of various theoretical arrangements. They found that the purine and pyrimidine bases were placed regularly along the molecule at a distance of 3·4 Å. They also found that another regularity in structure occurred at a repeating distance of 34 Å (i.e., every 10 nucleotide units), and therefore suggested that the molecule was not linear, but twisted into a helix, one complete turn occurring every 34 Å. Density measurements indicated that the molecule did not consist simply of one nucleotide chain.

Pauling and Corey, in 1951, worked out the α-helical structure of fibrous proteins (Section 4–13), and later extended these ideas to the DNA molecule. They suggested that DNA consisted of three nucleotide chains coiled to form a helix, with the phosphate groups orientated inside the structure and the purine and pyrimidine groups projecting outwards. However, biochemists working on the problem of DNA structure had suggested that hydrogen bonding between the bases probably played a part in stabilizing the molecular structure. This could only occur if, in contrast to Pauling and Corey's suggestion, the bases on the polynucleotide chains pointed inwards towards each other.

It was in 1953 that Watson and Crick constructed their famous *double helix* model of DNA, which explained all the evidence then available, and for which they and Wilkins were later awarded the Nobel Prize. The double helix model is illustrated in Fig. 3–2. It consists of two twisted polynucleotide chains, in each of which the deoxyribose sugar units on adjacent nucleotides are linked by phosphate groups to form an outer sugar–phosphate backbone. Figure 3–1 shows that the only OH groups available for ester linkages on the deoxyribose of a nucleoside unit are those on the 3′- and 5′-carbon atoms. In fact each phosphate links the 3′-carbon atom on one sugar to the 5′-carbon on the sugar of the next nucleotide (Fig. 3–3).

The purine and pyrimidine bases of the nucleotide units are turned inwards and are linked by hydrogen bonds, each base on one chain being paired with a base on the other chain. This, however, is a very specific form of base pairing, occurring only between adenine and thymine and between cytosine and guanine (no other associations would give the necessary spatial fit for a regular helix to be possible, and for hydrogen bonding between the bases to occur). The very exact nature of

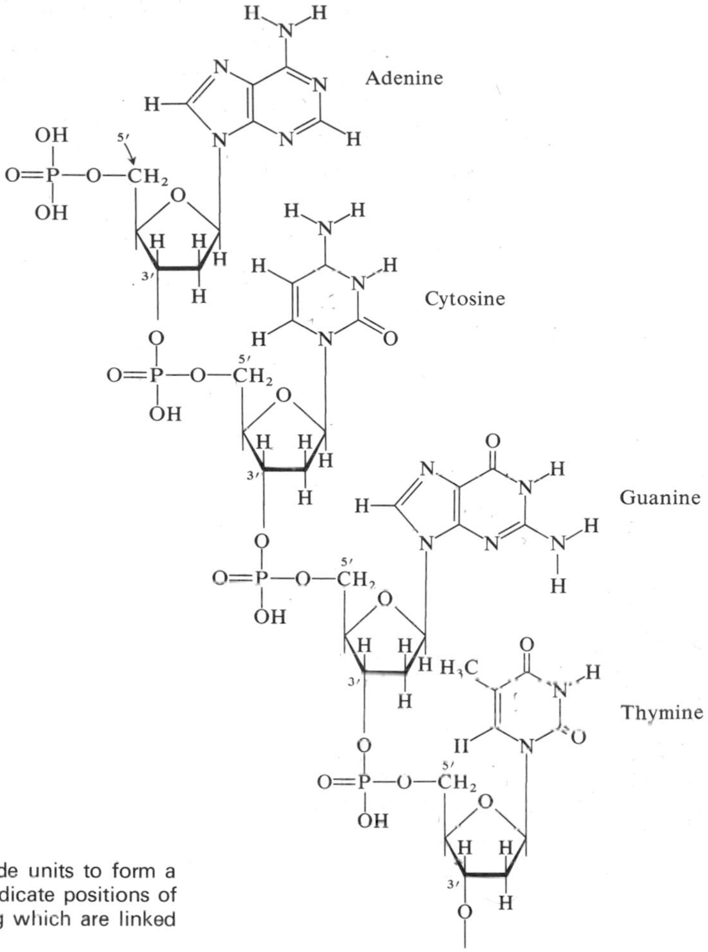

Fig. 3-3 Linking of nucleotide units to form a single DNA chain. 3′ and 5′ indicate positions of carbon atoms in the sugar ring which are linked by phosphate groups

the fit is illustrated in Fig. 3–4; two hydrogen bonds can form between adenine and thymine, and three between cytosine and guanine. Because of base pairing the model satisfies Chargaff's chemical observations that the DNA molecule contains equal numbers of adenine and thymine bases, and of cytosine and guanine bases. It also fulfils all of the requirements of Wilkins' X-ray diffraction observations. As shown in Fig. 3–2, the diameter of the helix is 20 Å, and the two twisted chains form a molecule with alternate wide and narrow

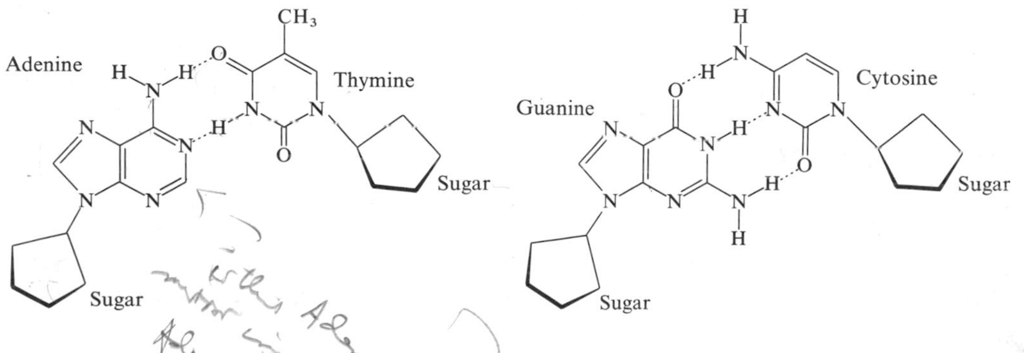

Fig. 3–4 (below) Hydrogen bonding between adenine and thymine, and between guanine and cytosine

grooves. A complete turn of the chain (pitch) occurs every 34 Å, and 10 nucleotide units are present in this chain length.

The two polynucleotide chains are said to be *complementary*, in that there is a complementary relationship between their sequences of bases. Thus if one chain has a region which goes —adenine—guanine—cytosine—guanine— , then the corresponding region of the complementary chain will go —thymine—cytosine—guanine—cytosine— . Base pairing means that the phosphate–sugar linkages run in opposite directions on the two chains. In Fig. 3–3 it may be seen that, running from the top to the bottom of the page, the phosphate–sugar links go from a 3'-carbon to a 5'-carbon. In the complementary chain, however, with the bases facing to the inside in order to pair with the first sequence of bases, the direction of the phosphate–sugar linkages (again from the top to the bottom of the page) must go from a 5'-carbon to a 3'-carbon. This may be more clearly seen by rotating Fig. 3–3 through 180°, which gives the steric arrangement of the complementary sugar–phosphate backbone.

The formation of hydrogen bonds between the poly-nucleotide bases stabilizes the double helix, and yet the bonds are sufficiently weak to be capable of breaking and re-forming at room temperature with comparative ease.

As a result of magnificent work by molecular biologists, the way in which genetic information is carried in the DNA molecule is now largely understood (Chapter 4). It depends on the order or sequence in which the four bases adenine, thymine, cytosine, and guanine are arranged along the DNA chain. DNA molecules may be very large; for instance in the bacterium *Escherichia coli*, in which DNA is in the form of a closed molecule (Fig. 3–8), it is about 1,000 μ in length. Using the value of a 3·4 Å distance between bases on the polynucleo-tide chain, there must be about 3×10^6 nucleotide pairs present in the DNA of *E. coli*. The molecular weight of a nucleotide pair is about 600, so that the molecular weight of *E. coli* DNA must be of the order of $600 \times 3 \times 10^6$, or $1·8 \times 10^9$. Plant and animal cells may possess DNA molecules which are even longer. Since DNA molecules can be so large and contain so many nucleotide units, the order of the bases along the molecule can vary greatly. To ensure that the exact sequence of nucleotide bases in the DNA of an organism is transmitted unchanged from one generation to the next, which is essential for continuity of genetic character, DNA must be replicated very exactly during the life cycle of a cell before division (see Fig. 9–1).

Watson and Crick stated in 1953, 'it has not escaped our notice' that the double helix structure itself suggests the manner in which DNA molecules are replicated. If the two strands were to begin to unwind and separate in the presence of a supply of the building units which make polynucleotide

Fig. 3–5 Replication of the DNA double helix. The strands unwind and a new polynucleotide chain is built up from nucleoside triphosphate building units on each of the old chains. This results in two identical double helices, each consisting of one old and one new chain. It is presumed that a specific enzyme, DNA polymerase, is involved

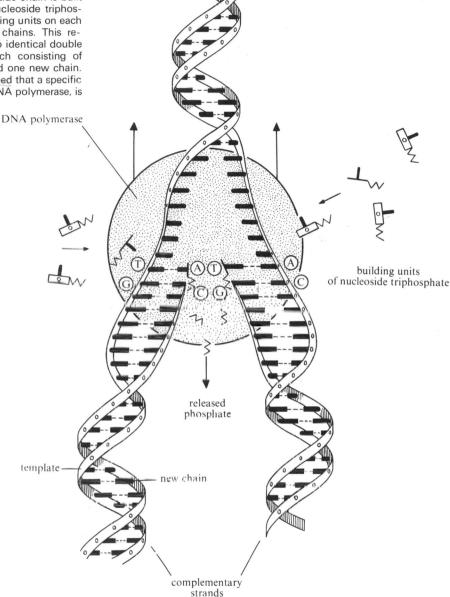

complementary strands

DNA polymerase

building units of nucleoside triphosphate

released phosphate

template

new chain

complementary strands

chains, then one would expect these units to attach by their bases to corresponding bases on the single chains (Fig. 3–5). In each case, attachment by hydrogen bonding would only be possible for the correct complementary base, and no other. In this way a completely complementary set of nucleotide units would progressively be built up by base pairing on each of the two original strands acting as templates. The final stage would then involve the linking up of the complementary sets of units to form *two* DNA molecules, each of them exact copies of the original molecule. This type of DNA synthesis is described as *semi-conservative replication*, since each of the daughter DNA molecules consists of one 'old' polynucleotide strand from the parent molecule, and one newly synthesized strand.

Experiment has borne out Watson and Crick's proposal that the DNA molecule unwinds so that each strand may serve as a template for a new complementary strand. In 1958, Meselson and Stahl demonstrated by some elegant experiments with *E. coli* bacteria that replication is a semi-conservative process. The genetic material of *E. coli* is known to be a long, continuous molecule of DNA, whereas the chromosomes of plant and animal cells are much more complex in structure and organization. Meselson and Stahl first grew the bacteria in a medium containing a heavy isotope of nitrogen, ^{15}N. The DNA of bacteria grown in this medium is more dense than that of bacteria grown in a medium containing ordinary nitrogen compounds. By carrying out high-speed ultra-centrifugation of solutions of the two types of DNA, one extracted from ordinary cells, and the other from isotopically labelled cells, they found that the two preparations sedimented at different distances from the axis of rotation, depending upon their density (Fig. 3–6). Meselson and Stahl took bacteria which had been grown for several generations in ^{15}N and contained heavy DNA, and allowed them to reproduce

Fig. 3–6 Meselson and Stahl's experiment with *E. coli* bacteria demonstrating that DNA replication is a semi-conservative process

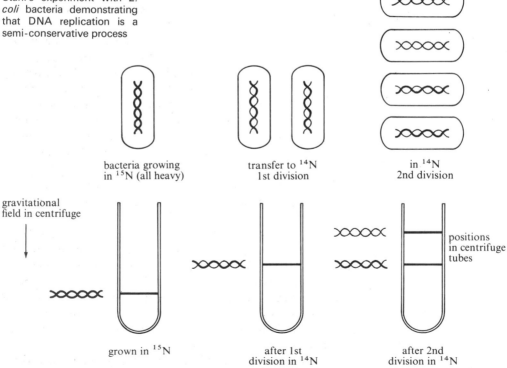

bacteria growing
in ^{15}N (all heavy)

transfer to ^{14}N
1st division

in ^{14}N
2nd division

gravitational
field in centrifuge

positions
in centrifuge
tubes

grown in ^{15}N

after 1st
division in ^{14}N

after 2nd
division in ^{14}N

in ordinary medium. They reasoned that if the Watson–Crick model were correct, the DNA of all the bacteria of the next generation would consist of an intermediate form, consisting of one heavy and one normal strand of DNA, and that this would sediment somewhere between the heavy and the normal DNA. This in fact happened (Fig. 3–6). In the next generation, grown in normal medium, half of the DNA would be expected to be the normal form (synthesized on the normal strand) and the other half would be the intermediate form (synthesized on the heavy strand). This too was found to be the case and their results conclusively proved that DNA replication takes place by a semi-conservative process.

Kornberg showed in 1956 that in an *in vitro* system a specific enzyme is needed for DNA synthesis. When each of the new nucleotide units has joined on to the template strand, then an enzyme, *DNA polymerase*, is required to link them all together and form a continuous complementary strand. Working with material from *E. coli*, Kornberg showed that DNA polymerase can bring about the synthesis of DNA *in vitro* in a cell-free system, although it is not certain if it plays the same role *in vivo*. Kornberg's system consisted of the four DNA nucleotides, the enzyme DNA polymerase, some ATP (to supply energy for the synthetic process), and some DNA to act as a 'primer'. In the absence of DNA no synthesis occurred, which is further evidence in favour of the view that new DNA molecules are always synthesized on existing DNA. By using radioactive labelling of one of the nucleotide units, Kornberg was able to show that synthesis of DNA had taken place.

He later used the nucleoside triphosphates (nucleoside —(P)—(P)—(P), see Section 2–4) instead of the nucleotides and ATP. The two extra phosphate groups split off, freeing the nucleotides to attach and link up into a continuous chain, and also supplying the energy required for synthesis. Kornberg found that whatever the relative proportions of nucleotides in his mixture, the synthesized DNA was always an exact replica of the original DNA, thus providing further evidence for the proposed mechanism of synthesis.

An apparent discrepancy came to light when it was found that DNA polymerase works on the strand which it is copying in one direction only—from the 3′ end to the 5′ end. (When two strands of DNA are twisted into a double helix, they run in opposite directions, the 5′ end of one molecule lying opposite the 3′ end of the other.) So when the double helix unwinds and replication begins, it is impossible for a continuous complementary strand to be built up on each of the old strands in the direction of unwinding. This is illustrated in Fig. 3–7. It is only on the left-hand strand, in the 3′ to 5′ direction, that continuous synthesis can take place.

To overcome this difficulty, Kornberg and his coworkers

Only left hand DNA copied because of poler traveling of enzyme like duch??

suggested in 1967 that synthesis takes place 'backwards' on the other (right-hand in Fig. 3–7) strand—in the opposite direction to that in which unwinding occurs, so that the enzyme is still operating in the 3' to 5' direction. This means that synthesis cannot be continuous, but takes place in short stretches, the enzyme coming off the template strand after each stretch and returning higher up the molecule to start the next stretch (Fig. 3–7). There must then be a mechanism for

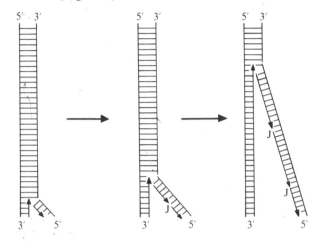

Fig. 3–7 DNA replication as suggested by Kornberg. The replicating enzyme travels only from the 3' to the 5' end of the DNA template, so one parent strand is copied 'backwards' in short sections. These sections are subsequently linked by a joining enzyme or ligase. Arrows indicate direction of movement of DNA polymerase; J indicates point of action of ligase

linking these disconnected segments to form a continuous chain. In fact, a new enzyme was discovered in 1967 which is capable of linking up short lengths of DNA. The enzyme, discovered by Khorana and others, is known as *DNA-joining enzyme* or *DNA ligase*. By incorporating this in his cell-free systems for DNA synthesis, Kornberg (1968) succeeded in synthesizing biologically active DNA. This he had failed to do in earlier experiments, presumably because the new DNA on one strand was in disconnected sections. Okazaki (1968) obtained further proof of <u>discontinuous DNA synthesis</u>. With a special radioactive labelling technique known as short pulse labelling, he found that nearly all of the label was in short sections of DNA of about 1,000 to 2,000 nucleotides in length.

How does it get linked in order > J J J etc

   The existence of DNA-joining enzyme is of importance in DNA repair mechanisms. Many bacteria and at least some mammalian cells can repair damage to their DNA caused by ultraviolet and ionizing radiations by removing the damaged regions and replacing them with newly synthesized sections of nucleotide chain. A joining enzyme is then necessary to link each end of the new sections to the existing polynucleotide strand. Kornberg has obtained evidence (1969) that DNA polymerase is involved in DNA repair mechanisms. It is possible that this may be its chief, or even its only, function in living cells, and that perhaps another enzyme is necessary for DNA replication *in vivo*.

   Another interesting aspect of the DNA replication model concerns the process of unwinding of the parent molecule. It seems unlikely, as Crick has pointed out, that the molecule has to unwind completely before any polynucleotide synthesis can start; the length of the DNA molecule, and the considerable amount of untwisting involved, make this seem improbable. Crick has suggested instead that the two processes are simultaneous, and that as soon as unwinding starts, the formation of the two new chains begins.

Evidence that this is so was obtained by Cairns, who carried out autoradiographic studies on labelled and carefully isolated DNA from *E. coli*. If synthesis and replication are taking place simultaneously, then it should be possible to observe a Y-shaped region of the autoradiograph, the vertical part of the Y corresponding to the double-stranded parent DNA molecule which has not yet unwound, and the V-shaped part of the Y corresponding to the two separated parent strands on which synthesis is taking place. An actual autoradiograph is reproduced in Fig. 3–8. The DNA was labelled

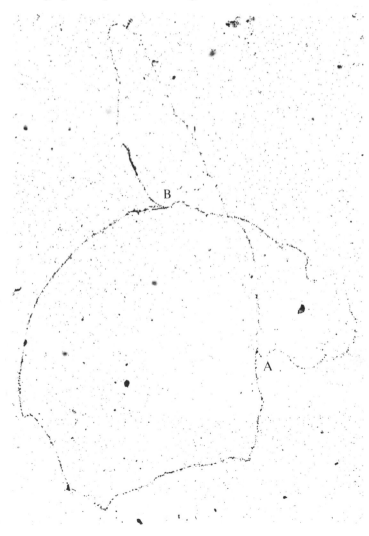

Fig. 3–8 Autoradiograph showing the Y-shaped growing point of *E. coli* DNA. The circular molecule has been two-thirds duplicated. The symbol A marks what is probably the growing point, and B the starting and finishing point. Magnification× 270. (With permission from J. Cairns.)

with [³H]thymidine for two generations, so that the whole of the ring-shaped molecule is visible. At the stage shown, the molecule is about two-thirds replicated. It is possible to observe two Y-shaped forks, one of them at the point where replication started (and must finish, since the molecule is continuous) and the other at the growing point. It is probable from the amount of labelling that B is the starting-point in Fig. 3–8, and the two long loops (left and middle) from B to A are the two daughter DNA molecules. A is then the growing point, and the parental DNA molecule has not yet unwound in the short loop from A to B on the right. The autoradiograph makes it clear that unwinding and synthesis are taking place simultaneously.

How can a closed molecule, consisting of two twisted strands, unwind completely? The replicating molecule may be compared to a ring of two-ply wool in which unwinding of the two strands has been started at a particular point (B in Fig. 3–8). As unwinding proceeds, torsion in the unwound part (A to B) steadily builds up, and, as the final point is approached, separation becomes increasingly difficult. It was therefore suggested by Cairns (1963) that in replicating bacteria there is a *swivel* at the point where the process begins, and that at this structure the unwound part of the molecule rotates with respect to the rest, thus preventing the build-up of torsion, and facilitating strand separation. However, in *E. coli* and *Bacillus subtilis* the DNA molecule in the interphase stage is closely folded upon itself (see Fig. 14–14), and it is difficult to see how it can possibly rotate along its entire length. This is even more of a problem in plant and animal chromosomes, where the DNA molecule, as well as being folded and twisted, is associated with proteins. It appears necessary, therefore, to postulate a number of swivel points, or alternatively points of breakage and reunion, which would allow parts of the DNA molecule to rotate, uncoil, and replicate more easily.

3–2
Appearance of
the interphase
nucleus

The nuclei of dividing cells are generally spherical, but interphase nuclei of plant and animal cells may vary considerably in shape, as well as in size relative to total cell size. Figure 3–9a shows a human heterophil leucocyte; the nucleus is polymorphous, consisting of several (usually three) irregular oval lobes, connected by thin threads of DNA-containing material. In contrast, the nuclei of the smooth-muscle cells of human intestine (Fig. 3–9b) are elongated in the long axis of the cell, and have a cylindrical or oval form with pointed or rounded ends. During contraction of the cell in peristalsis the nucleus becomes folded or twisted.

Usually little or no structure can be seen in the interphase nucleus, apart from the nucleolus (or nucleoli) (see Fig. 5–11) either by light or electron microscopy. Even with staining

|← ca 8 μ →|

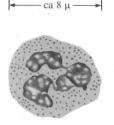

(a)

techniques it is extremely difficult to see chromosomes—only a network or reticulum of chromosomal material (known as *chromatin*) is visible. The reason is that during most of the cell cycle the chromosomes are in a highly hydrated, and therefore swollen, state. In this swollen gel form the refractive index of the chromosomes is about the same as that of the surrounding fluid, or *nuclear sap*. On the basis of staining properties, chromatin was classified as *heterochromatin*, which stains deeply, or *euchromatin* which is much less deeply stained. The difference between the two regions is now thought to be a

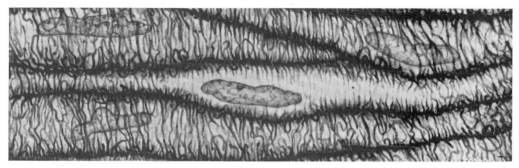

(b)

Fig. 3–9 (a) Photograph of human heterophil leucocyte (Wright's stain). (b) Drawing of longitudinal section through smooth muscle of human intestine, stained with hematoxylin and the Bielschowsky silver method for reticular fibres (Magnification × 2,000). Average length of cells is ca 200μ. (Drawn by Miss E. Bohlman and reproduced with permission from A. A. Maximow and W. Bloom, *Textbook of Histology*, 5th edn. W. B. Saunders, Philadelphia, 1948.)

Fig. 3–10 Chromosomes seen in the resting nucleus of a somatic cell (neuroblast of locust). (With permission from E. J. Ambrose and A. R. Gopal-Ayengar.)

functional one; the heterochromatic regions appear to be parts of the chromosome which can replicate but lack the ability to direct the synthesis of other molecules. At mitosis and meiosis, the chromosomes become clearly visible as threads, because of the increase in refractability caused by condensation and coiling of their already existing structures.

With a very sensitive type of microscope it is in fact possible to see the chromosomes in certain interphase nuclei, as in Fig. 3–10, taken using interference contrast. Ris and Mirsky showed in 1949 that by careful adjustment of the salt concentration of the medium they could make the interphase chromosomes of grasshopper cells stand out sharply in living nuclei.

There are one or two unusual cases in which interphase chromosomes are visible. In particular, the giant chromosomes of the salivary gland cells of dipteran larvae, first observed in the late 19th century by Balbiani, have proved extremely useful for the study of chromosome structure and behaviour. The peculiarity of these cells is that instead of growing to a certain size and then dividing, they continue to grow and become very large without going into mitosis. The chromosomes grow to an enormous size, because the chromosomal constituents are duplicated many times over, with the newly duplicated regions remaining attached to the original chromosome. This gives rise to structures which are very much magnified both in length and width; these are known as *polytene* chromosomes, and are readily visible under the microscope and easily studied. In *Drosophila*, the polytene chromosomes of the late larval stage are approximately 100 times as long as the chromosomes

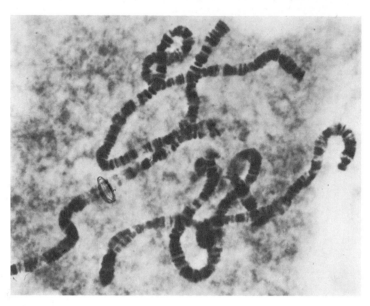

Fig. 3–11 Salivary gland chromosomes in *Sciara coprophila*. (With permission from H. Crouse.)

of the somatic cells (i.e., any tissue cells as distinct from germ cells) observed in metaphase. Metamorphosis of the larva to the pupa stage involves the complete breakdown of the cells containing giant chromosomes. All of the cellular biochemical materials are thrown into the melting pot, and a new process of differentiation commences, leading eventually to the development of the adult fly.

A photograph of one of these giant chromosomes is reproduced in Fig. 3–11. It is a structure containing many dense lateral bands. After Feulgen staining, these bands are stained very intensely, but the interband regions hardly at all, indicating that the DNA is concentrated almost entirely in the bands. By u.v. absorption methods, treatment with proteases (enzymes which break protein molecules down into peptides), and other techniques, protein has been found in the bands and in the interband regions, which are composed largely of protein. Whether DNA runs continuously along the length of the giant chromosome is not yet clear, but the evidence indicates that it does. Further evidence that DNA forms a continuous strand running longitudinally along chromosomes has been obtained from studies on another type of very large chromosome found in amphibian eggs (see 'lampbrush' chromosomes below). Experiments on the stretching of giant chromosomes using a micromanipulator fitted with microdissection needles indicate that they are very extensible. They can recover completely after being stretched up to 10 times their normal length. This implies that their basic structure is probably a long fibril, which in the chromosome's normal state is considerably folded and coiled. This is certainly in agreement with present ideas about the nature of somatic cell chromosomes; it is thought that the basis of chromosome structure is a fibril consisting of nucleic acid associated with protein.

Since it was realized that certain gene loci (see Chapter 10) can be identified with particular bands in polytene chromosomes, geneticists have made considerable use of such chromosomes in investigating the relationship between the gene and the chromosome. It was observed as early as 1881 that the structure of giant chromosomes is modified in some regions. These are caused by a band becoming enlarged and in extreme cases puffing out to form a large ring around the chromosome (Fig. 3-12a). Such structures are known as *chromosome puffs* or *Balbiani rings*. It was found by staining methods that, in addition to DNA and protein, the puffs contain large amounts of the second type of nucleic acid, ribonucleic acid (RNA). By autoradiography, using a labelled compound required for the synthesis of RNA, it was found that puffs are in fact regions where RNA is being actively synthesized, and that the rate of synthesis is correlated with the degree of enlargement. Puffing is thought to be due to the unwinding or uncoiling of chromosome fibres which are usually closely folded or coiled in the dense band regions; these then project in the form of loops (Fig. 3-12b). Giant chromosomes have been observed in other tissues of dipterans besides the salivary glands, and puffs have been found to occur on different regions of the chromosomes in different tissues. This implies differences in genetic activity with respect to RNA synthesis in different tissues.

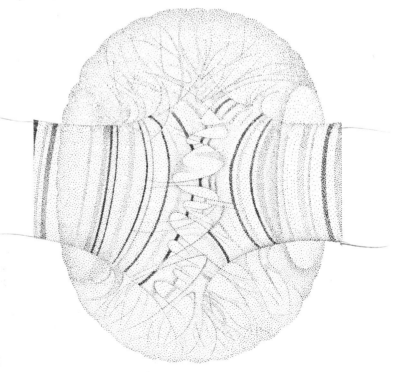

Fig. 3-12 (a) Diagrammatic representation of the structure of a large chromosome puff as seen in the light microscope. Some of the fibrils that make it up are visible. (From W. Beermann and U. Clever, Chromosome puffs Copyright © April 1964 by Scientific American, Inc. All rights reserved.)

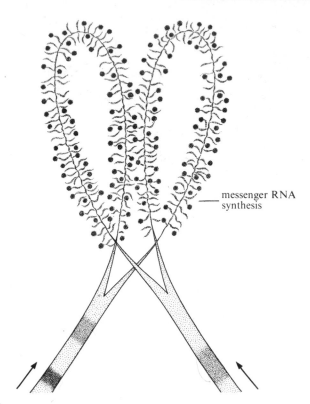

messenger RNA
synthesis

Fig. 3–12   (b) Suggested form of the individual loops in chromosome puffs. (From W. Beermann and U. Clever, Chromosome puffs. Copyright © April 1964 by Scientific American, Inc. All rights reserved.)

There is another type of giant chromosome which shows considerable RNA activity. This is the 'lampbrush' chromosome (so called because it resembles the brushes which were used for cleaning the glass chimneys of old-fashioned paraffin or kerosene lamps), which occurs in the developing eggs of amphibia, and probably other species. Lampbrush chromosomes take the form of long strands, about 200 µ in length, with many symmetrical lateral loops (Fig. 3–13), which are clearly visible in the light microscope, and which arise from dense regions on the central strand. These loops have a fluffy appearance, which Callan has shown disappears after treatment with RNAase (an enzyme which attacks RNA) whereas the loop itself remains intact. Staining techniques also indicate the presence of RNA. Presumably these also are regions where RNA is being actively produced on loops of DNA. Enzyme treatment and electron microscopic studies indicate that lampbrush chromosomes have a continuous longitudinal strand of DNA; like dipteran polytene chromosomes they are highly extensible (Figs 3–13b and 3–13c).

There is another region of the interphase nucleus which contains large amounts of RNA, and this is the nucleolus, the structure and function of which will be discussed later.

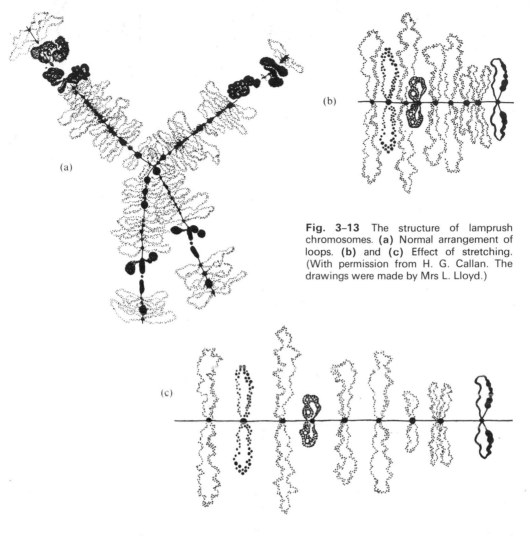

(a)

(b)

(c)

Fig. 3-13 The structure of lamprush chromosomes. **(a)** Normal arrangement of loops. **(b)** and **(c)** Effect of stretching. (With permission from H. G. Callan. The drawings were made by Mrs L. Lloyd.)

**3–3
Structure and
synthesis of
ribonucleic acid**

Like DNA, RNA is a long-chain molecule built up of repeating nucleotide units linked by 3′ to 5′ phosphate diester bonds. There are two differences in the building units of RNA and DNA; first, the sugar component of RNA is not deoxyribose but ribose (Fig. 3–14), and second, although three out of four of the bases, adenine, guanine, and cytosine, are the same in RNA as in DNA, the fourth base, thymine, is replaced in RNA by *uracil* which has one methyl group less (Fig. 3–14). Other bases occasionally occur in one of the three types of RNA, as will be mentioned later.

Work carried out in the 1950s and 1960s indicates that RNA is usually synthesized by a process analogous to the replication of DNA. The RNA molecule is built up from nucleoside triphosphate units by a copying mechanism which works off one strand of the DNA molecule. As for DNA replication, the DNA strands must unwind for RNA to be copied off them. A specific enzyme, *RNA polymerase*, is

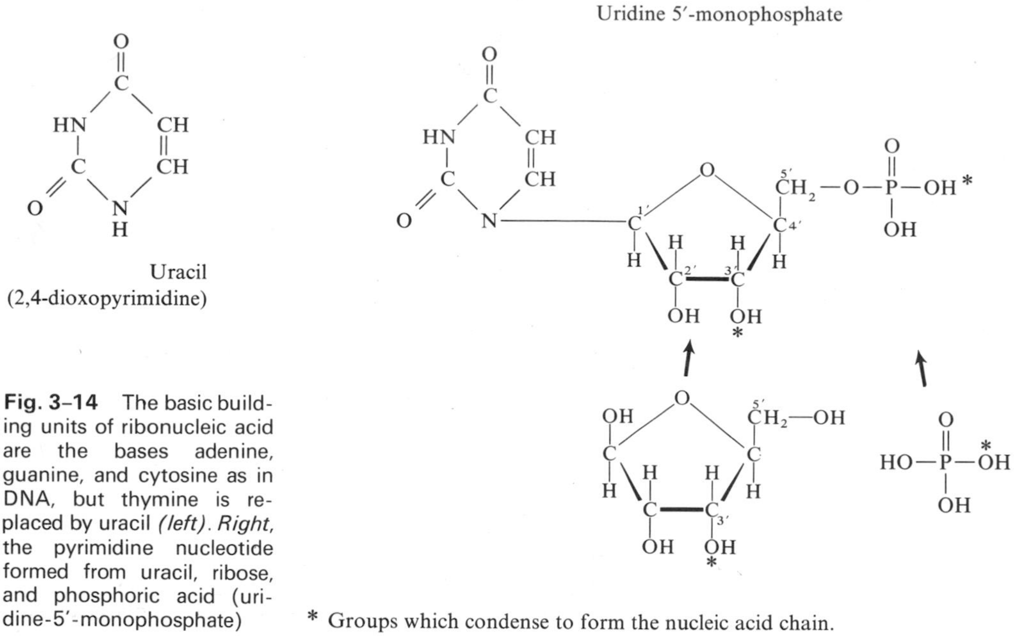

Uridine 5′-monophosphate

Uracil
(2,4-dioxopyrimidine)

Fig. 3–14 The basic building units of ribonucleic acid
are the bases adenine,
guanine, and cytosine as in
DNA, but thymine is re-
placed by uracil *(left). Right,*
the pyrimidine nucleotide
formed from uracil, ribose,
and phosphoric acid (uri-
dine-5′-monophosphate)

* Groups which condense to form the nucleic acid chain.

required (Fig. 3–15) to link together the ribonucleotide units
by means of ester linkages. This method is the one by which
RNA appears to be synthesized in a variety of organisms.
The synthesis has been carried out *in vitro*, using the appropriate building units and RNA polymerase; as in DNA
replication, some DNA must be present to act as a template.

Chemical analysis by Bautz and Hall (1962) showed that
cytosine and guanine, which base-pair in DNA, are present
in differing proportions in the RNA synthesized in bacterial
cells after viral infection. This implies that RNA chains do
not pair in a complementary way along their length, as DNA
chains do, and that RNA exists as a single-stranded molecule.
This agrees with the hypothesis that RNA is copied as a single
strand from one strand only of the DNA molecule. So the
newly formed RNA would be expected to be complementary
with one strand of the template DNA. That this is so has been
demonstrated by some elegant experiments; when double-
stranded DNA is heated to near 100°C, the hydrogen bonds
between the two chains break and the strands separate. When
it cools under appropriate conditions the bonds can re-form.
Hall and Spiegelman, working with materials from *E. coli,*
added RNA to the separated strands of DNA. They found
that the RNA was able to form a close association with a
particular region of the DNA (Fig. 3–16). This process is
called *hybridization*; it is highly specific, and its effectiveness
depends on having complementary base sequences on the two
molecules. The technique of hybridization has proved to be an
extremely powerful tool for the study of base sequences in
DNA and RNA molecules.

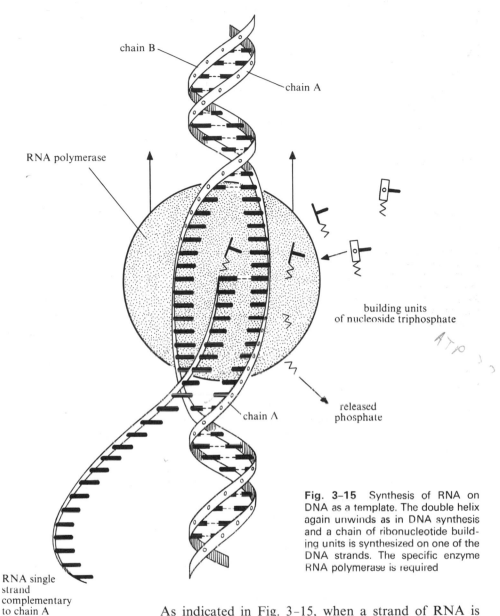

chain B

chain A

RNA polymerase

building units
of nucleoside triphosphate

chain A

released
phosphate

Fig. 3-15 Synthesis of RNA on DNA as a template. The double helix again unwinds as in DNA synthesis and a chain of ribonucleotide building units is synthesized on one of the DNA strands. The specific enzyme RNA polymerase is required

RNA single
strand
complementary
to chain A

As indicated in Fig. 3–15, when a strand of RNA is synthesized on a DNA template, it appears that the RNA molecule rapidly detaches itself, and in the region where the DNA helix has unwound the two individual strands again come together and form hydrogen bonds. The explanation seems to be that the DNA–DNA double helix is energetically more stable than the DNA–RNA hybrid structure, and so the RNA strand is rapidly replaced by the original DNA strand after the polymerase enzyme has moved over the region of synthesis.

It is now known in what direction along the DNA strand the synthesis of RNA proceeds. A study of growing chains has indicated that newly attached ribonucleotides are

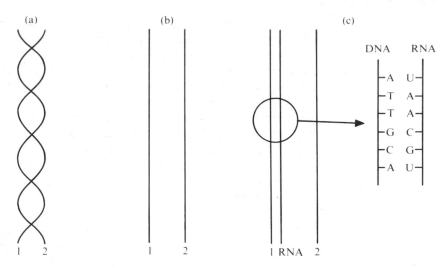

(a)   (b)   (c)

DNA	RNA
A	U
T	A
T	A
G	C
C	G
A	U

1   2   1   2   1 RNA 2

Fig. 3-16 Hybridization of DNA and RNA. **(a)** Double-stranded DNA. **(b)** Denatured DNA (strands separated). **(c)** Hybridization of RNA with one strand of DNA. Base sequences must be complementary

found at the 3′ end of the molecule, and that the nucleotide at the beginning has an attached triphosphate group (which must be on the 5′ carbon atom—see Fig. 3-14). So RNA chains must grow in the 5′ to 3′ direction, as do DNA chains in replication (Fig. 3-7).

A number of RNA molecules, all considerably smaller than the ordinary DNA template, can be synthesized on one DNA molecule. There must therefore be a number of places where RNA polymerase can attach to DNA and synthesis can begin, and equally a number of places where the signal to stop is given. Work by Burgess and Travers (1969) and by Dunn and Bautz (1969) on *E. coli* bacteria suggests that it is a component of the RNA polymerase molecule which in some way specifies where synthesis of RNA should begin on the DNA template. This component has been called the sigma factor.

There are one or two exceptions to the rule that RNA is formed on a DNA template; for instance, in certain viruses which contain RNA but no DNA, it appears that the RNA molecule can itself act as a template for its own replication (Section 14–2).

Some device must exist for ensuring that only one DNA strand is copied in RNA synthesis. As already mentioned, the sequence of bases in DNA molecules determines the sequence of amino acids in the proteins synthesized in cells. Only one RNA molecule, copied on one of the DNA strands, is required to carry this message from the nucleus to the cytoplasm where protein synthesis takes place. If two RNA molecules were synthesized simultaneously, one on each DNA strand, then they would each carry a different message, which would lead to confusion. The mechanism by which one chain is preferentially chosen is not yet clearly understood.

Cells which synthesize a lot of protein contain large amounts of RNA. By the same token, little RNA can be detected in cells which do not manufacture much protein. Three types of RNA are involved in protein synthesis in the cytoplasm—*messenger, transfer,* and *ribosomal* RNA (the functions of these three RNAs will be explained in Chapter 4). All three types of RNA are synthesized on the DNA template. Beermann and coworkers have obtained evidence that the RNA present in the puffs of diplotene chromosomes is in fact messenger RNA and, since puffs occur on different regions of the chromosome in different tissues, the inference is that there are particular regions of genetic activity for messenger RNA synthesis in different cell types. Hence the base composition of different messenger RNAs must vary considerably, as well as the molecular size. Transfer RNA molecules also possess a range of structures, as will be discussed in Chapter 4. Ribosomal RNA (present in the ribosomes, the sites of protein synthesis in the cytoplasm) varies little in overall base composition whether it comes from bacterial, plant, or animal cells. There was long thought to be a connection between ribosomal RNA and the nucleolus, which is known to contain large amounts of RNA. There is now evidence to indicate that the nucleolus is the site of synthesis of ribosomal RNA.

The nucleolus (Figs 3-17 and 5-11) is an irregularly shaped, often dense and compact, body consisting largely of

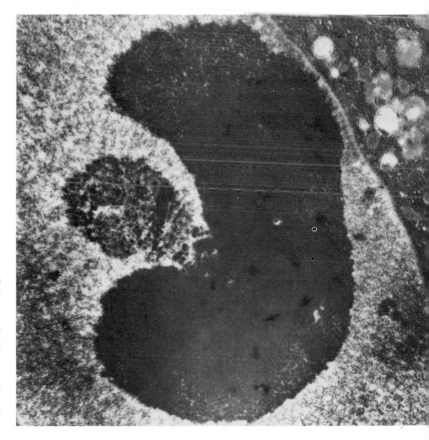

Fig. 3-17 Electron microscope autoradiograph of a cell of *Smittia*. The section is cut across a chromosome in the region of the nucleolus. The chromosome is to the left of the centre, and the large central irregularly shaped object is the nucleolus. Dense regions due to the silver staining are those of active RNA synthesis. (With permission from J. Jacob.)

RNA and protein, the main component (up to 80 per cent of the dry weight) being protein. There may be one or more nucleoli per nucleus, depending on the type of organism, and the ratio of RNA to protein probably varies considerably in different cell types. The nucleolus is associated with a specific site, known as the *nucleolar organizer*, on a specific chromosome and probably remains attached to this region during interphase. It is therefore in close contact with the DNA of the organizer. Electron microscopy has shown the nucleolus consists of two regions—one being composed of fibres and the other of particles; both regions are formed of ribonucleic acid associated with protein. The background consists of a network of protein. There is no limiting membrane surrounding the nucleolus and separating it from the rest of the nucleus. Nucleolar activity, as judged by size, is much more pronounced in cells which are making large quantities of protein than in other cells. Certain regions of the chromosome contained within the nucleolus of actively metabolizing cells are in the form of loops very similar to those of lampbrush chromosomes, and are presumably in a state of active RNA synthesis. It now seems that one of the chief metabolic functions of the nucleolus is the synthesis of ribosomal RNA, and possibly even the formation of complete ribosomal particles (see Section 4–4) from RNA and ribosomal proteins.

The nucleolus is thought to perform another important function—that of information transfer. It has been observed that genetic information is not transferred from the nucleus to the cytoplasm unless the cell has a functional nucleolus (H. Harris, 1967). It has therefore been suggested that a component from the nucleolus may play a protective role in shielding messenger RNA (and possibly other types of RNA required for export to the cytoplasm) from enzyme attack and degradation by RNAases in the nucleus. Ribosome-like particles have been isolated from nucleoli (Birnstiel, 1963); but very few complete ribosomes are observed in the nucleus, so it seems unlikely that the ribosomes are finally assembled until the components have crossed the nuclear membrane on their passage to the cytoplasm.

3–4
Proteins of the cell nucleus

Apart from nucleic acids, the other important constituent of the cell nucleus is the protein component. The chromosomes of plant and animal cells, unlike the genetic material of bacteria, consist of DNA associated with protein. In giant polytene chromosomes protein has been found in the band and interband regions, and in the puff regions of gene activity. Protein is associated also with the extended loops of lampbrush chromosomes. Nucleoli contain a high proportion of protein, and there are of course proteins present in the nuclear sap or nucleoplasm, and, in association with lipids, in the nuclear membrane (Section 5–3).

Improved methods of isolation and fractionation of chromosomes from dividing cells and chromatin from interphase cells have permitted chemical analysis of chromosome composition. Considerable amounts of two types of protein are associated with DNA and RNA in chromosomes; these are basic proteins known as *histones*, and acidic proteins. Small quantities of other proteins are also present. The detailed structure of proteins will be described in Chapter 4, and here we only mention that histones contain a high proportion of the basic amino acids lysine and arginine (Fig. 4–1a), and that the acidic proteins linked to DNA probably have a high content of dicarboxylic amino acids, although much less is known about the acidic proteins than about histones. These two types of protein form an association with DNA known as *nucleoprotein*. The actual form of this association is still far from clear, but the fact that it involves an interaction between nucleic acid and protein is indicated by its physical properties. DNA forms a highly viscous solution in water; nucleoprotein, on the other hand, has strong gel-forming properties. The behaviour of nucleoprotein in water depends on its concentration, which indicates that the macromolecules interact strongly with each other. Gel formation is probably brought about by the cross-linking of DNA strands by protein molecules to form an interlocking structure. A likely explanation is that the basic groups of the histones form linkages with the acidic phosphate groups of the DNA, and that the acidic proteins similarly form bonds with the poly-nucleotide bases.

Many of the studies on nuclear proteins have been confined to histones. These have been fractionated and classified according to the relative proportions of basic amino acids which they contain (Butler, Johns, Phillips, and others, 1960 onwards). The biological function of histones has intrigued research workers for a long time. One of the functions ascribed to them is that of acting as a chromosomal 'glue', binding together the genetic units of DNA. It is known also that DNA–histone complexes are more resistant to denaturation by heat than DNA alone, and that proteins (mainly histones) in the nucleoprotein complex can partially protect DNA from radiation damage. Evidence now indicates, however, that the really important role played by histones lies in repressing the genetic activity of cells.

It was suggested by Stedman and Stedman in 1950 that histones interact with DNA in a specific way, preventing it from acting as a template for RNA synthesis and thus preventing the transfer of genetic information to the cytoplasm. A great deal of evidence has now been obtained in support of the 'masking' theory (Barr and Butler, 1963; Bonner, Huang, and Gilden, 1964). *In vitro* experiments have confirmed that, when DNA–histone complexes are formed by gradually adding histones to DNA, the decrease in RNA synthesis on the DNA template is directly proportional to the amount of added histone. Since different regions of DNA can be shut down in this way, the implication is that there must be many different kinds of histone, each one specific for a particular region of DNA. However, this is not so; histones are remarkably similar in composition not only in different tissues, but in species as different as cows, chickens, peas, and sea urchins (Bonner and others, 1966 onwards). This suggests that the structure is very critical and that even a slight change would make it non-functional. Bonner suggests that histones are the 'heavy switch-gear' involved in repressing DNA activity, and that other molecules act as 'trip switches' and convey site specificity to the process.

Balis and colleagues suggested in 1964 that acidic nuclear proteins may act as derepressor molecules. There is evidence that this is so; Paul and Gilmour (1969) carried out experiments on the synthesis of RNA *in vitro* on DNA, DNA plus histone, and DNA plus histone plus nuclear acidic protein—that is, reconstituted nucleoprotein. They found that the acidic proteins do make specific regions of DNA available for RNA synthesis. The processes involved are not yet clearly understood, but there may be a region of specific composition on the histone molecule which interacts with a specific controlling acidic nucleoprotein.

In most cells which have been studied it appears that only a small part of the DNA is in a derepressed state and capable of RNA synthesis. This means that the considerable amounts of protein associated with DNA must be duplicated before mitosis, and then associated with the DNA in the daughter cells as soon as mitosis is completed. It has been suggested that the actual site of histone biosynthesis is the nucleolus (Birnstiel, 1964). Certainly intensive protein synthesis has been observed in nucleoli, and ribosome-like particles have been isolated from them.

It is interesting that in sperm the histones of somatic cells are partly or completely replaced by *protamines*, which are more basic and structurally more simple proteins. They have been shown *in vitro* to be less effective repressors of DNA template activity than histones. This would appear important for easier derepression of the genetic material of the sperm during fertilization, enabling it to open up rapidly as development commences (see Section 13-4). Another advantage is that protamines are smaller than histones, which means that the chromosomes and nuclei can be packed into a smaller space, allowing greater mobility of the sperm.

3-5 Chromosome structure

Few signs of structure can be seen in interphase nuclei by means of either light or electron microscopy. Studies mostly carried out on cells in mitosis and meiosis indicate that

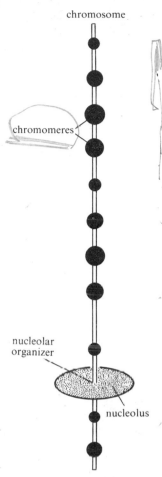

chromosome

chromomeres

nucleolar
organizer

nucleolus

Fig. 3-18 Diagrammatic representation of the interphase chromosome

chromosome structure in interphase may be shown diagrammatically as in Fig. 3-18. The chromosome is thought to be a long fibrous structure, with dense regions of unequal size spaced at unequal intervals along its length. These regions, known as *chromomeres*, may be clearly seen in the early stages of meiosis. They are considered to correspond, when many times multiplied, to the dense bands of giant polytene chromosomes, and to the regions on lampbrush chromosomes from which the lateral loops emerge. The nucleolus is associated with a particular site on the chromosome, known as the nucleolar organizer. Both animal and plant cells vary considerably in chromosome size and DNA content, even in related organisms.

Considerable interest has centred round the question as to whether the DNA in chromosomes is present as a single strand consisting of one DNA double helix, or as a multi-stranded structure built like a cable from a number of double helices. Geneticists have tended to favour the former because of the difficulty that would otherwise arise of explaining mutation and recombination, while many cytologists have claimed to observe multi-stranded chromosomes. It would be very difficult, however, to distinguish by means of light microscopy between a multi-stranded structure and a chromosome in which a single thread of DNA is folded backwards and forwards upon itself many times to give a multi-stranded appearance. Studies on the stretching of lampbrush chromosomes and on the effects of various enzymes on their structure (Gall, 1963; Miller, 1964), on labelling experiments on duplicating chromosomes (Peacock, 1963; Taylor, 1965; and others), and on damage to chromosomes caused by X-rays (Evans, 1967) all provide strong arguments in favour of single-strandedness. The simplest model therefore is that in which each chromosome contains one very long molecule of DNA which, however, may be considerably folded or coiled.

Chromosomes are known to be composed of DNA, RNA, histones, and acidic proteins, but the exact way in which the nucleoprotein complex is assembled is still not clear. It is thought that RNA is localized at certain regions of the chromosome, but proteins are associated with the DNA molecule along its entire length. Clearly, the electrical properties of nucleoprotein largely determine the stability of chromosomes. As mentioned above, histones partly neutralize the negatively charged phosphate groups on DNA. Further neutralization is brought about by calcium bridges formed by the Ca^{2+} ion between phosphate groups. Any factors of this type which reduce the negative charges on nucleoprotein will tend to lead to the formation of a more compact structure, and conversely any factors which increase the negative charges will cause repulsive forces to operate, and thus expand the chromosome. This has been demonstrated by Ambrose and

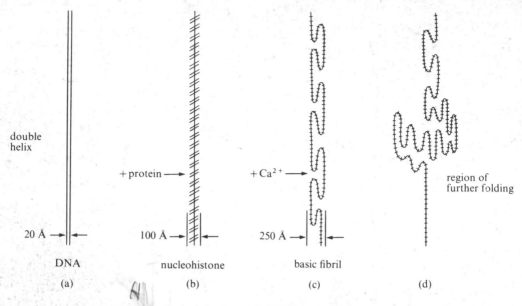

double
helix

+ protein →   + Ca²⁺ →

region of
further folding

20 Å →|←   100 Å →|←   250 Å →|←

DNA   nucleohistone   basic fibril

(a)   (b)   (c)   (d)

Fig. 3–19 Model for chromosome fibres, after Ris. **(a)** DNA double helix, 20 Å diameter. **(b)** DNA and associated histone, ca 100 Å diameter. **(c)** Folding due to calcium bridges, giving a fibre ca 250 Å diameter. **(d)** Regions of further folding postulated by Ris

Gopal-Ayengar (1952) in experiments using the giant chromosomes of *Chironomus* larvae. The chromosomes were made to expand and contract reversibly many times simply by changing the salt concentration in the medium surrounding the cells, thus changing the effective electrical charge on the molecules of nucleoprotein.

There have been various ideas about the precise way in which the protein molecules are arranged around the DNA. Wilkins suggested in 1957 that in nucleohistone the protein chains might be coiled round the DNA, lying in the grooves of the double helix (Fig. 3–2), to give a nucleohistone thread of about 30 Å diameter. From X-ray diffraction studies, Wilkins and coworkers suggested in 1967 that the nucleohistone is itself supercoiled to give a helix of about 100–130 Å in diameter. Ris has suggested (1967) that the histone is associated with the DNA in some regular but unspecified way, and that some type of coiling or folding again takes place to give a nucleohistone fibre of about 100 Å (Fig. 3–19). Further folding takes place because of calcium bridges to give a basic fibril of about 200–250 Å. Ris has obtained electron microscope pictures of both 100 Å and 250 Å fibrils (Fig. 3–20) by spreading chromosomal material at an air–water interface. He postulates regions of still further folding. DuPraw and others have also observed a 200–250 Å fibre thickness in the

Fig. 3–20 (opposite) (Top) Fibres of 250 Å diameter from erythrocyte nuclei of *Triturus viridescens*, spread on water, fixed in ethanol, and shadowed with carbon–platinum. The fibres contain DNA and histones. Calcium bridges are present. (Magnification×56,000.) (Bottom) 100 Å fibres from the same material as above, spread on sodium citrate solution, which breaks the calcium bridges, fixed in ethanol, and shadowed with carbon–platinum. (Magnification×56,000.) (With permission from H. Ris in V. V. Koningsberger and L. Bosch (eds), *Regulation of Nucleic Acid and Protein Synthesis*. Elsevier, Amsterdam, 1967.)

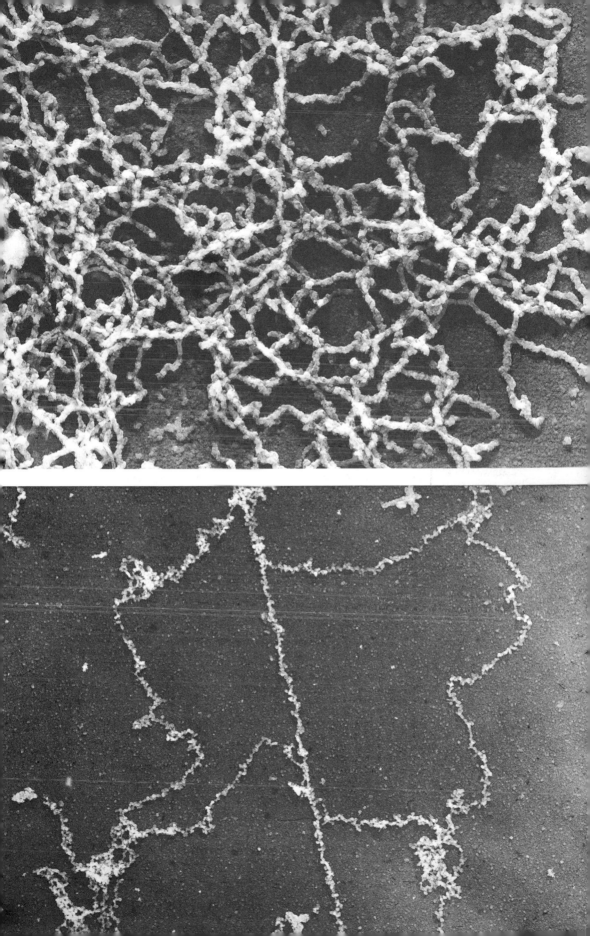

electron microscope. DuPraw suggests that the final process of condensation of the nucleoprotein thread is one of helical coiling and not of folding.

The model of chromosome structure that is emerging appears to be a long strand of nucleoprotein which is considerably coiled and folded, and has extremely dense regions because of further aggregation. These dense compact regions are assumed to correspond to the chromosomes.

It was suggested by Belling in 1928 that chromomeres might be the separate units of functional genetic activity, or *genes* (see Section 10-9), containing the information relating to a specific cell function. The suggestion received little support at the time because chromomeres seemed much too large for such a function. Recent work indicates, however, that the chromomere may be the basic unit of organization of the chromosome. Callan and Lloyd (1960) suggested from morphological, labelling, and inheritance studies on lampbrush chromosomes that each loop pair and thus each chromomere is associated with the activity of one specific gene. Further support for the idea that a chromomere corresponds to a gene was obtained by the work of Beermann (1967) on mutations in giant polytene chromosomes; in this case the dense transverse bands, as mentioned above, are the counterparts of chromomeres in ordinary chromosomes.

To explain the large size of the chromomere and of the lampbrush loop, Callan and Lloyd postulated that each loop consists not of one gene, but of a number of duplicate copies, linearly arranged, of one gene. There is a 'master' copy at each chromomere and information is transferred from this to each of the 'slave' copies which are matched against it to ensure that they are all identical to the master. The master and slave hypothesis is illustrated in Fig. 3-21. Eight genes are shown, each loop corresponding to a chromomere. The double trans-

Fig. 3-21 Model for chromosome structure based on studies of lampbrush chromosomes by Callan et al. (With permission from H. L. K. Whitehouse, *Towards an Understanding of the Mechanism of Heredity*, 2nd edn. Edward Arnold, London, 1969.)

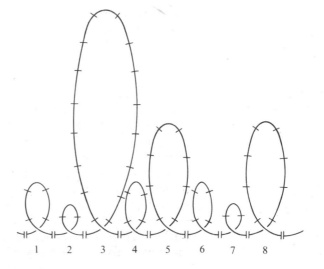

1   2   3   4   5   6   7   8

verse lines indicate the point at which one gene type ends and another starts, and the single transverse lines indicate the ends of the duplicate copies. Gall and Callan had observed that the lateral lampbrush loops always have one thin and one much thicker end at the point of insertion in the chromomere (Fig. 3–13c). They suggested that the loops are spun out from the chromomere at the thin end and returned to it at the thick end, which explains the greater accumulation of RNA, and thus the greater thickness, on one side of the loop. Callan suggested that the master copy of a gene does not take part in RNA synthesis, but only the slave copies in the loops. The advantage of having a number of duplicate copies of a gene is that a higher rate of RNA synthesis is possible. The addition of histones to lampbrush chromosomes, as in the experiments when histone is added to DNA, stops the synthesis of RNA, and causes the loops to retract into the chromosomes.

The RNA associated with the puffs of polytene chromosomes has a characteristic base composition for each puff, again implying specific genetic activity. According to Callan's hypothesis the puff must correspond to a set of copies of one gene in each of the many strands which make up the giant chromosome.

There is evidence that the nucleolar organizer resembles other chromomeres and contains a large number of duplicate gene copies on which ribosomal RNA is synthesized. It seems probable too that there are duplicate copies of the genetic material on which transfer RNA is synthesized.

Further experiments indicate that chromomeres may also be the units of replication (*replicons*) of chromosomes (see Chapter 9).

Britten and coworkers have also obtained evidence suggesting that cells in higher organisms contain a considerable proportion of genetic material in single copies linearly arranged along the molecule, but that the remaining DNA is present in all degrees of repetition. The suggested role of these reiterated DNA sequences as extremely accurate controllers of genetic activity will be discussed in Chapter 13.

3–6
Nucleus as a
functional unit

In summary, the functions of the nucleus are: to preserve the genetic material intact and hand it on unchanged from one generation to the next; and to direct the synthesis of RNA, which is required for the synthesis of proteins and of other cellular components. The nucleus is a structural unit, but it combines structure with plasticity. The chromosomes are elastic bodies and the nuclear sap is quite fluid. As will be explained in Chapters 9 and 10, the nucleus is radically reorganized in mitosis and meiosis. Considerable movement is also involved, however, in the functioning of the interphase nucleus. There is, of course, an increase in nuclear size during interphase, as the synthesis of various cellular components proceeds. Materials such as messenger RNA molecules and ribosomal precursors move outwards to the periphery of the nucleus, and through the nuclear membrane into the cytoplasm. There is not a one-way traffic, however; it is known that signals from the cytoplasm are passed back to the nucleus. Changes in the external environment of the cell result in changes within the cell, which adapts to the altered conditions. The advantages of there being a turnover of messenger RNA will be discussed in Section 9–5.

Bibliography   Allfrey, V. G. and A. E. Mirsky, How cells make molecules. *Scientific American*, September 1961.

Callan, H. G., The nature of lampbrush chromosomes. *Int. Rev. Cytol.,* **15**, 1, 1963.

Darlington, C. D. and K. R. Lewis (eds), *Chromosomes Today*, vol. I. Oliver and Boyd, Edinburgh, 1966.

Haggis, G. H. (ed.), *Introduction to Molecular Biology*. Longmans Green, London, 1964.

Harris, H., *Nucleus and Cytoplasm*. Oxford University Press, London, 1968.

Hnilica, L. S., Proteins of the cell nucleus. *Prog. Nucl. Acid Res. Mol. Biol.,* **7**, 25, 1967.

The Nucleolus, its Structure and Function. Nat. Cancer Institute Monograph No. 23: International Symposium, December 1966 (U.S. Dept of Health, Education and Welfare).

Watson, J. D., *Molecular Biology of the Gene*. W. A. Benjamin, New York, 1965.

Watson, J. D. and F. H. C. Crick, Molecular structure of nucleic acids: a structure for deoxypentose nucleic acids. *Nature*, **171**, 737, 1953.

4 The genetic code and protein synthesis

Proteins are very widely distributed biological molecules which have many specialized functions. For instance, all known enzymes are proteins; proteins occur in cell membranes and walls; the fibrous structures of hair, collagen, and muscle are all proteins; haemoglobin, present in red blood cells, is a protein and so is insulin, the hormone which helps to control the level of blood sugar.

Proteins (from the Greek word for 'first place') were so named by Mulder in 1838, before the role of nucleic acids in protein synthesis was understood and when it was thought that proteins gave rise to the first life. (Whether proteins or nucleic acids came first in the evolution of biological molecules will be discussed in Chapter 15.) In 1897, the Buchner brothers isolated from yeast cells a cell-free extract which undoubtedly contained protein. They demonstrated that the extract encouraged the fermentation of sugar into alcohol; in other words, that proteins can behave as biological catalysts or enzymes.

It was thought in the 19th century that proteins lacked structure. Haeckel (1868) referred to proteins as homogeneous masses, structureless and amorphous. Subsequent physical and biochemical studies have caused this view to be revised. Proteins are in fact highly organized and complex structures. The units from which they are made are the amino acids. In proteins, amino acids are linked together by peptide bonds to form long chains known as polypeptides. A peptide bond is formed when the carboxyl group of one amino acid combines with the amino group of another, with the elimination of a molecule of water. For instance:

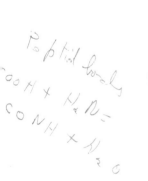

where R^1 and R^2 represent the side-chains of two amino acids.

123

Glycine (Gly) L-Alanine (Ala) L-Valine (Val) L-Isoleucine (Ile)

L-Leucine (Leu) L-Serine (Ser) L-Threonine (Thr) L-Proline (Pro)

L-Aspartic acid (Asp) L-Glutamic acid (Glu) L-Lysine (Lys) L-Arginine (Arg)

L-Asparagine (Asn) L-Glutamine (Gln) L-Cysteine (Cys) L-Methionine (Met)

L-Tryptophan (Trp) L-Phenylalanine (Phe) L-Tyrosine (Tyr) L-Histidine (His)

(a)

There are 20 common amino acids (Fig. 4–1a) from which the giant protein molecules are constructed.

Because of the spatial arrangement of the various groups, the polypeptide chain has a zig-zag form. The polypeptide chain in Fig. 4–1b, reading from left to right, is formed from the amino acids glycine, lysine, alanine, serine, glutamic acid, and phenylalanine. The chain in Fig. 4–1c is made of the same amino acids, but arranged in a different order—that is, serine,

Fig. 4–1 (a) (left) The twenty common amino acids which occur in proteins. **(b)** Polypeptide chain formed from (left to right) the amino acids glycine, lysine, alanine, serine, glutamic acid, and phenylalanine. **(c)** Polypeptide chain formed from the same amino acids as in **(b)** but arranged in a different sequence. From left to right, serine, alanine, glutamic acid, lysine, glycine, and phenylalanine

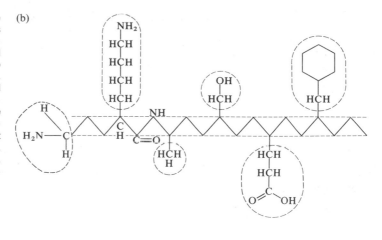

(b)

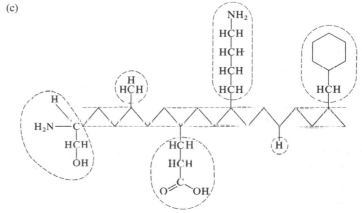

(c)

alanine, glutamic acid, lysine, glycine, and phenylalanine. The two chains differ greatly in shape, and possess different properties because they consist of different sequences of the same amino acids. All of the diverse properties of the many kinds of protein molecules arise simply from the fact that each possesses a characteristic sequence of amino acids in its polypeptide chain. This sequence determines the way in which a polypeptide chain coils up to form a compact protein molecule.

Aristotle and Theophrastus in the 4th century B.C. believed that all animals and plants, however complicated, are constituted from a few basic elements which are repeated in each of them. Studies in cell biology and molecular biology have borne out the truth of this statement at the molecular level for protein structure.

Various methods have been used for studying proteins. In 1926, Svedberg developed the high-speed (ultra) centrifuge which he used to determine the molecular weights of proteins. He found that these lay in the range of about 20,000 to 200,000. In 1928, Meyer and Mark found that they could obtain X-ray diffraction patterns from silk fibres, and Astbury used the same method to study the fibres of hair and wool. X-ray crystallo-graphy techniques were applied to non-fibrous, or globular, proteins in the 1930s when it became possible to obtain these in crystalline form. An ultraviolet absorption method was developed by Caspersson in the 1930s to measure the DNA, RNA, and protein content of cells. He found that the amount of RNA is greater in cells which synthesize large amounts of protein and suggested that RNA might control protein synthesis. Similar views were put forward independently by Brachet, based on studies of sea urchin eggs. But almost all of the detailed knowledge of the processes involved in protein synthesis has come from work of microbiologists. Some bacteria can divide once every 20 minutes, and experiments which would take weeks with other cells can be carried out in a few hours with bacteria. The picture of protein synthesis we give below has been derived from bacterial studies, and the situation in the more complex cells of plants and animals may turn out to differ in certain details.

4–1
Protein synthesis

The sequence of nucleotide bases in the DNA molecule determines the structure of proteins. The relationship between the nucleotide sequence in DNA and the amino acid sequence in proteins is called the *genetic code*. Information in the sequence of the DNA molecule is copied on to strands of messenger RNA (mRNA). The mRNA passes into the cytoplasm, where the information it carries is used to synthe-size particular protein molecules.

Two other kinds of ribonucleic acid are important in protein synthesis: transfer RNA (tRNA) and ribosomal RNA (rRNA). The part played by tRNA is the selection of particular amino acids from the cytoplasm and the transfer of each amino acid as required to the site of protein synthesis. The actual site of protein synthesis is the ribosome, and it is here that rRNA plays a part.

The first step in protein synthesis is known as *tran-scription*, because it involves copying on the mRNA the same base coding that is present on the DNA (except that where DNA has thymine, RNA has uracil, as explained in Chapter 3). The second step is called *translation*, because it involves a change from the nucleotide language of mRNA to the amino acid language of proteins. The actual steps are quite complex; energy and specific enzymes are needed for the various highly specific reactions involved.

4–2
The genetic code

The four bases of mRNA—adenine, cytosine, guanine, and uracil—are usually symbolized as A, C, G, and U (Fig. 4–2a). These must code in some way for 20 different amino acids. How many bases are needed to specify one amino acid? If one nucleotide base coded for one amino acid, only four kinds of amino acid could be specified. If a combination of two bases

Fig. 4–2 (a) The four bases of mRNA. (b) The code-words for the peptide chain shown in Fig. 4–1 (b). (c) The code-words for the peptide chain shown in Fig. 4–1 (c)

(a)		A, C, G, U				
(b)	GGU	AAA	GCU	UCU	GAA	UUU
	Gly	Lys	Ala	Ser	Glu	Phe
(c)	UCU	GCU	GAA	AAA	GGU	UUU
	Ser	Ala	Glu	Lys	Gly	Phe

were used, there would be only $4 \times 4 = 16$ different arrangements. Only if a combination of three bases codes for one amino acid are there enough possibilities, since such a triplet combination can be arranged in $4 \times 4 \times 4 = 64$ different ways.

Direct evidence that the code is in fact a triplet was provided by the genetic studies of Crick et al. (1961), who carried out experiments on a virus, T4 bacteriophage. They found that treatment with a chemical called proflavine damaged the virus by either adding or removing a base in its DNA molecule, thus producing an altered, or mutant, form of the virus (Figs 4–3 a and b). When the addition of a base into the DNA was followed by a deletion not too far away, a normal virus was produced. This implied that the normal sequence of bases in the DNA molecule had been restored by this second change (Fig. 4–3c). Crick further showed that three alterations of the same type (either all of them additions or all deletions) and again in close proximity to each other, also resulted in the production of a normal virus, and presumably therefore of restoration of the normal base sequence in the DNA (Fig. 4–3d). This, then, was evidence that the code operates in sequences of three bases along the DNA molecule, and that these sequences do not overlap. Each triplet of bases coding for a particular amino acid is called a codon.

Fig. 4–3 Effect of additions and deletions of bases in the genetic code: (a) normal triplet code; (b) addition of base X* changes the triplet code to CAB; (c) further deletion of base A‡ restores the code to ABC; (d) three close additions of the bases, X°, Y°, and Z° mean that the normal sequence is restored

(a)	/ ABC / ABC / ABC / ABC /
(b)	/ ABC / AX*B / CA‡B / CAB /
(c)	/ ABC / AXB / CBC / ABC /
(d)	/ ABC / AX°B / CAY° / BZ°C / ABC /

What is the 'spelling' of a codon for any particular amino acid? Nirenberg and his coworkers in 1961 went a long way towards answering this when they synthesized a protein in vitro. To start with they synthesized mRNA in vitro from only one base, uracil. This synthetic mRNA was then added to bacterial extracts containing ribosomes, tRNA, amino acids, and the various enzymes required to carry out protein synthesis. Protein was rapidly synthesized by the system when the uracil polymer was present, but not in its absence. By labelling the various amino acids with a radioisotope as a marker, and using them in different combinations, Nirenberg discovered that protein was only synthesized from the amino acid phenylalanine, the product being a polypeptide, poly-phenylalanine. He thus showed that uracil codes for phenyl-alanine or, assuming a triplet code, that UUU is the codon for phenylalanine.

Nirenberg, and Ochoa independently, synthesized further artificial mRNA molecules from the four bases. Comparing the amino acids incorporated in a particular polypeptide chain with the calculated triplet base composition of the synthetic mRNA, they determined which amino acid was coded for by each triplet of bases. These experiments gave the codon for each amino acid, but not the sequence of bases in it.

The first spelling of a codon was determined by Nirenberg and Leder in 1964. They prepared an artificial mRNA from the bases uracil and guanine, in the proportion of two of uracil to one of guanine. They then broke the mRNA into pieces consisting of 3-base units. From these they separated the three combinations GUU, UGU, and UUG. Again using a system containing tRNA, and radioactively labelled amino acids, they were able to show that GUU codes for the amino acid valine.

Soon after, the spellings of further codons were deter-mined by Nirenberg and by Khorana and his coworkers. Khorana prepared synthetic mRNA molecules in which the exact sequence of bases was known. By analysing the poly-peptide chains which were produced by the synthetic mRNAs, he obtained further evidence that the code is founded on triplets of nucleotides and that the triplets do not overlap each other. For instance, a synthetic mRNA composed of the bases uracil and cytosine alternately, UCUCUCUC . . . , led to the alternate incorporation of the two amino acids serine and leucine. A triplet code would allow the alternate readings UCU and CUC, corresponding to two different amino acids, from this synthetic mRNA.

This and other evidence proved that the genetic code consists of triplets of bases which are read in sequence along the DNA molecule and do not overlap.

The codons for the two peptide chains in Figs 4–1b and 4–1c are given in Figs 4–2b and 4–2c.

4–3
Role of
messenger RNA

Jacob and Monod in 1961 postulated a 'messenger' which they believed was necessary to convey the message of the genetic code from the DNA of the nucleus to the cytoplasm. This messenger, now known to be mRNA, acts as a template for the translation of the DNA code into a specific protein.

Messenger RNA molecules necessarily exist in a large variety of lengths, depending on the length of the polypeptide chain for which they code. There must therefore be consider-able variation in molecular weight, and in sedimentation values in the ultracentrifuge, of different mRNA molecules.

(*s*, the sedimentation coefficient, usually given in terms of Svedberg units, S, depends on the size and density of a particle, and is constant for a given particle in a given medium.) There are very few polypeptide chains which contain less than 100 amino acids, and so very few mRNA molecules which contain less than 3×100 nucleotide units. Many mRNA molecules are, of course, considerably larger than this.

The length of mRNA which carries the information necessary to determine the complete polypeptide chain of a protein molecule is called a *cistron*. The concept of a cistron was put forward by Benzer, who found in his genetic studies that he could conveniently sub-divide the genetic material of the virus he was studying into functional regions which he called cistrons, in which an alteration leading to a mutant form of the virus had occurred (see Section 10–9). Some mRNA molecules carry the information for the synthesis of more than one protein molecule; such mRNAs are said to be *polycistronic*.

Although the existence of mRNA has been established using a bacterial system, attempts to isolate and characterize it have largely proved unsuccessful. It comprises a very small fraction, only 3 to 5 per cent, of the total cellular RNA, and it has no specific characteristic enabling us to identify it.

The average lifetime of mRNA molecules can vary greatly. In some bacteria mRNA molecules are metabolically unstable, the average lifetime of the molecules being only a few minutes at 37°C, after which they are broken down by enzymes. The free nucleotides released are used to synthesize new mRNA molecules. This situation is an advantage to the bacterial cell, because it helps it to adapt rapidly to changes in the environment. On the other hand, mammalian reticulocytes (immature red blood cells which no longer have a nucleus) continue to synthesize haemoglobin for many days. Since these cells cannot synthesize mRNA, the mRNA already present must be long-lived. There is, then, considerable variation in the average lifetimes of mRNA molecules, the longest lived being those occurring in cells with a long lifetime or cells which have lost their nuclear function. It is possible that the lifetime is determined genetically by the nucleotide sequence in the molecule.

Studies using cells incubated with radioactive building units of RNA show that the labelled material is first incorporated into the nuclear RNA and appears later in the cytoplasmic RNA. This indicates that, when the code has been transcribed from DNA on to mRNA, the latter leaves the nucleus, passing through the nuclear membrane into the cytoplasm. There it moves to the site of protein synthesis, the ribosome. The mRNA molecule attaches reversibly to the surface of the ribosome, always binding to the smaller sub-unit (the 30S, see Section 4–4). The magnesium cation, Mg^{2+}, is involved in the formation of a complex between mRNA and the ribosome.

4–4
Ribosomes

Ribosomes are composed of RNA and protein. In *E. coli* the ratio of RNA to protein in ribosomes is 2 to 1, but in many other organisms it is approximately 1 to 1. Most of the protein of ribosomes probably has a structural role. In *E. coli*, ribosomes comprise about a quarter of the total cell mass, which reflects the importance of protein synthesis in cellular activity. Electron microscopy indicates that ribosomes are composed of two rounded sub-units fitted together to give a complete unit about 200 Å in diameter (Fig. 4–4). One sub-unit is approximately twice the size of the other. However, studies with rapidly frozen material, which is later 'shadowed' with metal atoms, indicate that the larger sub-unit may have a regular icosahedral structure, the usual rounded appearance being due to collapse of structure during fixation (Nanninga, 1968). The advantages of an icosahedral structure in certain simple viruses are discussed in Chapter 14.

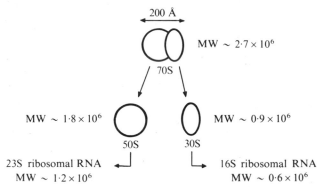

Fig. 4–4 Physical properties of intact ribosomes and of ribosomal sub-units

The sedimentation coefficient in the ultracentrifuge of intact ribosomes is 70S, while the sub-units have values of 50S (large sub-unit) and 30S (small). The aggregation of the two sub-units depends reversibly upon the concentration of magnesium ions; at low concentrations of Mg^{2+} the sub-units dissociate. Two sizes of RNA are found in ribosomes. The smaller one, with a sedimentation value of 16S, is found in the smaller sub-unit, and the larger one, with a 23S value, occurs in the larger sub-unit. The part that ribosomal RNA plays in protein synthesis will be discussed below.

Nomura (1968) has reconstituted *in vitro* a unit, similar to the 30S ribosomal sub-unit and active in *in vitro* polypeptide synthesis, from 16S ribosomal RNA and proteins from a variety of bacteria. He suggests that only certain small regions of the RNA have 'conserved' during evolution the ability to interact with 30S ribosomal proteins from a wide variety of bacteria. The 'non-conserved' regions, specific for the bacterium of origin, may have important functions other than direct interaction with protein in the assembly of ribosomal particles.

4–5
Role of transfer RNA

The existence of transfer RNA was postulated by Crick and demonstrated by Hoagland and his coworkers in 1955. The role of tRNA (also sometimes called soluble RNA, or sRNA) is to transfer amino acids present in the cytoplasm to the site of protein synthesis on the ribosome. Transfer RNA thus acts as an intermediate or adaptor molecule between a particular amino acid and the triplet that codes for it on the mRNA molecule. There is a different, specific, tRNA molecule for each amino acid.

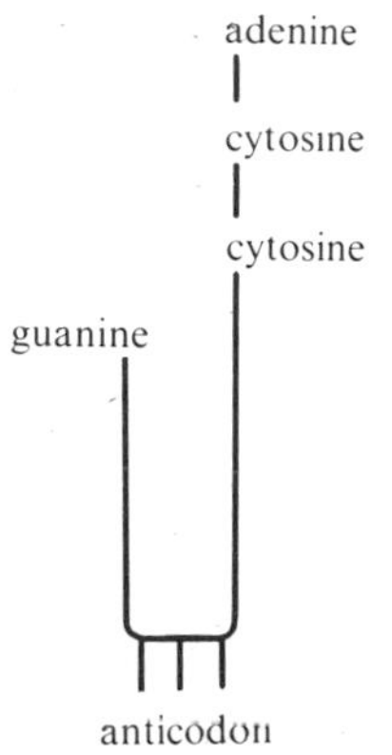

The part of the tRNA molecule which becomes attached to its particular amino acid appears to be the same for all tRNAs. This is the 3′ end of the molecule, terminating in the sequence of three bases –cytosine–cytosine–adenine (–C–C–A). Without this sequence no attachment of an amino acid takes place. The 5′ end of the molecule terminates in the base guanine (Fig. 4–5). A specific adaptor enzyme (an amino-acyl–tRNA synthetase) is required to couple each amino acid to its specific tRNA. There must therefore be one particular site on the tRNA molecule concerned with the enzyme–amino acid binding, and another specific site concerned with the appropriate sequence of three bases on the mRNA. So tRNA carries one word of the nucleotide language, and one word of the amino acid language.

The process by which the amino acid becomes attached to the tRNA is quite complicated. ATP is required since the two molecules become bonded in such a way that the bond energy can be used later in the formation of the lower-energy

adenine
|
cytosine
|
cytosine

guanine

anticodon

Fig. 4–5 Structures known to be present in tRNA molecules

Fig. 4–6 The attachment of an amino acid to its specific tRNA; * represents the specific site on the tRNA molecule which combines with the appropriate amino-acyl-tRNA synthetase

Transfer RNA

Amino acid – transfer RNA complex

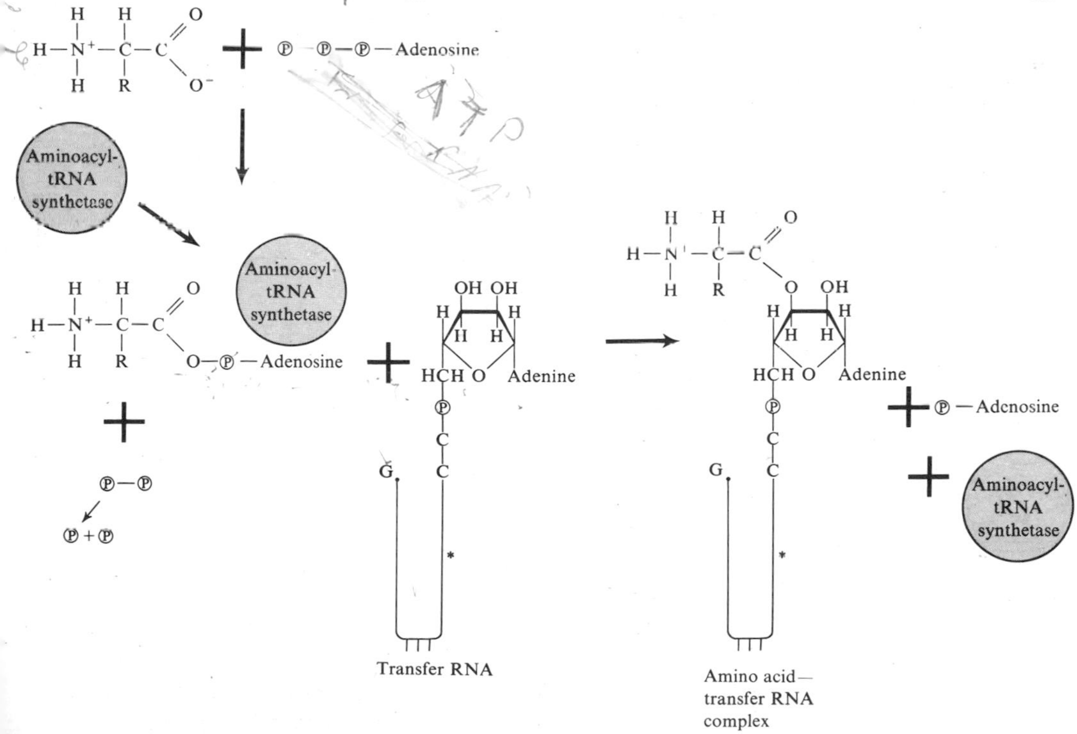

peptide bond of the protein molecule. The amino acid is first activated by ATP, the specific enzyme aminoacyl–tRNA synthetase being necessary for this (Fig. 4–6). Some site on the enzyme must recognize not only a specific amino acid, but also a particular sequence of bases on the correct tRNA. The link formed between the amino acid and tRNA is a covalent bond between the carboxyl group of the amino acid and the ribose of the terminal adenine nucleotide of the tRNA (Fig. 4–6). Thus there are three components necessary for this reaction —a specific amino acid, a specific enzyme, and the specific tRNA. When protein synthesis is taking place, these amino acid–tRNA complexes bind to the ribosome.

4–6 Attachment of tRNA to mRNA

The specific site on the tRNA molecule which attaches to the mRNA is a sequence of three bases, complementary to the codon on the mRNA, and known as the *anticodon*. Experiments with synthetic mRNAs have shown that although the first two letters of the codon for any particular amino acid are nearly always unchanging (see Fig. 4–7) the third letter is sometimes different. For instance, serine (Ser) is coded for not only by AGU but also by AGC. This implies that the first two bases alone usually determine the codon site for the tRNA molecule; that is, the fit of the third base of the codon to that of the anticodon is not nearly so specific as the fit of the first two bases. Because of this, Crick in 1966 put forward his 'wobble' hypothesis, the wobble being the rather loose fit between the third pair of bases compared with the exact requirements of the first two pairs of bases. Since alternative codings exist for the same amino acid, the code is said to be *degenerate*.

First position	Second position				Third position
	U	C	A	G	
U	Phe	Ser	Tyr	Cys	U
	Phe	Ser	Tyr	Cys	C
	Leu	Ser	non.	non.	A
	Leu	Ser	non.	Trp	G
C	Leu	Pro	His	Arg	U
	Leu	Pro	His	Arg	C
	Leu	Pro	Gln	Arg	A
	Leu	Pro	Gln	Arg	G
A	Ile	Thr	Asn	Ser	U
	Ile	Thr	Asn	Ser	C
	Ile	Thr	Lys	Arg	A
	Met ('capital letter')	Thr	Lys	Arg	G
G	Val	Ala	Asp	Gly	U
	Val	Ala	Asp	Gly	C
	Val	Ala	Glu	Gly	A
	Val	Ala	Glu	Gly	G

Fig. 4–7 The genetic code table. The first letter of a codon is given in the vertical column on the left of the table, the second letter in the horizontal row at the top, and the third letter in the vertical column on the right: 'non' represents nonsense or release codons, 'capital letter' indicates the codon for chain initiation

Mainly through the work of Nirenberg and Khorana in assigning codings to specific amino acids, 61 of the 64 possible triplets have been found to code for the 20 common amino acids. Consequently there must be several alternative codings for many amino acids. The three remaining codes appear to play a role in terminating the polypeptide chain.

4–7 Structure of tRNA

Each tRNA contains approximately 80 nucleotides. The 3' end of the molecule always ends in the sequence —CCA, which is believed to be added to the RNA by a special system in the cytoplasm, and the 5' end is always guanine. The first complete nucleotide sequence of a tRNA was established by Holley in 1964 for a tRNA from yeast. It contained, like most tRNA molecules, a number of unusual bases, such as inosine (I) and pseudouracil (ψ) (Fig. 4–8a), and certain methylated forms of normal bases. These unusual nucleotides probably arise by enzymic modifications of the already formed poly-nucleotide chain. There are known to be double helical regions in tRNA molecules, where complementary base pairs form hydrogen bonds. These unusual bases, some of which cannot form conventional base pairs, may prevent pairing at various points, thus exposing certain groups. These groups are then left free to form secondary bonds with mRNA, or with the specific enzyme required to attach to a certain amino acid, or with a ribosome.

Fig. 4–8 The structure of tRNA. **(a)** The unusual bases inosine (symbol I) and pseu-douracil (symbol ψ).* Represents the point of attachment to ribose. **(b)** The clover leaf model of tRNA proposed by Holley. **(c)** A suggested three-dimensional arrange-ment of the clover leaf model (after Ninio et al., 1969)

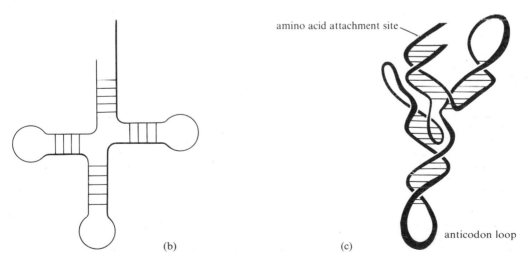

(a) Inosine

Pseudouracil

amino acid attachment site

anticodon loop

(b)   (c)

Although the sequence of bases in certain tRNAs is known, the actual three-dimensional structure has not yet been established, because of the problems of obtaining clear X-ray diffraction pictures of an irregular molecule. On the basis of the nucleotide analysis of yeast tRNA, Holley suggested a clover-leaf model with four regions of hydrogen bonding (Fig. 4–8b). The problem is to determine how the arms of the clover leaf are arranged in space. Various models have been suggested, one of which is shown in Fig. 4–8c. The arm bearing the anticodon is well away from the other three, allowing the rest of the molecule to bind to the ribosome surface. The amino acid recognition site is held by the other two arms in a specific steric arrangement which is probably slightly different for each type of tRNA, since the shape of the whole molecule probably determines the recognition of a specific amino acid.

4–8
Ribosomal RNA

Ribosomal RNA forms an integral part of ribosomal structure, and collapse occurs when it is removed. The two different size classes of rRNA mentioned earlier have molecular weights of about 6×10^5 and 1×10^6. Both contain double helical regions due to base pairing, as in tRNA. Once again the irregular shape of the molecule makes X-ray diffraction pictures difficult to obtain.

The function of rRNA in protein synthesis is not yet completely clear. The unpaired bases in the molecule may bind mRNA and tRNA to ribosomes, possibly by Mg^{2+} linkages between phosphate groups on the two molecules. There is also some evidence that rRNA may serve as a template, possibly in the synthesis of ribosomal proteins; then, following methylation of some bases (known to take place, as in tRNA), it becomes part of the ribosome structure and can no longer function as a template.

The actual function of the ribosome seems to be to ensure that the attached RNA molecules are orientated correctly, so that the codons on the mRNA are lined up next to the appropriate sites on the tRNAs with their corresponding amino acids, and also so that it is stereochemically possible for the polypeptide chain to be built up from its constituent amino acids.

In fact, a strand of mRNA does not attach just to one ribosomal unit, but several of these units become attached along the mRNA molecule, to form a complex known as a polyribosome. This greatly increases the efficiency of the process of protein synthesis, as one messenger RNA can be read simultaneously by several ribosomes. The whole process is rather like the assembly line of a mass production factory.

Why does the ribosome consist of two sub-units? The work of Nomura and Lowry (1967) indicates that mRNA is bound to the smaller sub-unit only in the absence of the larger sub-unit; the presence of the latter inhibits the binding. However, when the first tRNA has attached (see Section 4–10) the larger ribosomal sub-unit binds to form a complete ribosome. This is possibly a safeguard against protein synthesis starting in the middle of a mRNA molecule instead of at the correct starting point.

The implication of this work is that the two ribosomal sub-units should be able to change partners continually, and later experimental studies indicate that this is so. Models have been devised on this basis by Bretscher (1968) and by Spirin (1968) in which the tRNA carrying the growing polypeptide chain may go through the necessary shifting process with minimum risk of detachment. Studies on the molecular structure of ribosomes indicates that they possess protected channels or grooves, and Cox (1968) has suggested a model in which the mRNA lies along a groove between the two sub-units, and the growing polypeptide chain emerges from a hole in the centre of the large sub-unit.

4–9
Chain growth

A tRNA molecule, bound to its specific amino acid, attaches to a region on the ribosome near to its specific codon on the mRNA (Fig. 4–9). Then another tRNA–amino acid complex, specific for the adjacent codon on the mRNA, attaches in the same way to the ribosome. The covalent bond between the first tRNA and its amino acid breaks, and the carboxyl group of the amino acid thus released forms a peptide bond with the

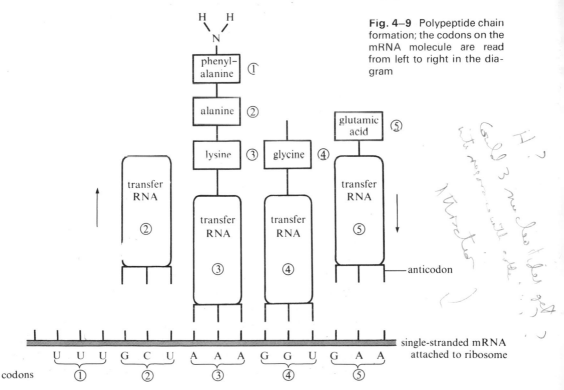

Fig. 4–9 Polypeptide chain formation; the codons on the mRNA molecule are read from left to right in the diagram

anticodon

single-stranded mRNA attached to ribosome

codons

Biology : hobby ??
Do ribosomes carrying growing peptide chain move
along the mRNA. How & why?

How does ribosome move along mRNA

terminal amino group of the second amino acid, which is still attached by means of its tRNA to the ribosome. The first tRNA then moves away from the site of synthesis on the ribosome. The repetition of this process, with the addition of one amino acid each time, results in the growth of a polypeptide chain. The complete process is represented in Fig. 4–10.

The formation of the peptide bond between amino acid units requires an enzyme, peptide synthetase, which seems to be the same for all amino acids; and although some energy is obtained from the breaking of the covalent bond between an amino acid and its tRNA, a further supply is required from ATP. The beginning of the chain always has an —NH$_2$ or an NHR terminal group, and at the other end of the chain the carboxyl group of the most recently added amino acid is still bound by its tRNA to the ribosome.

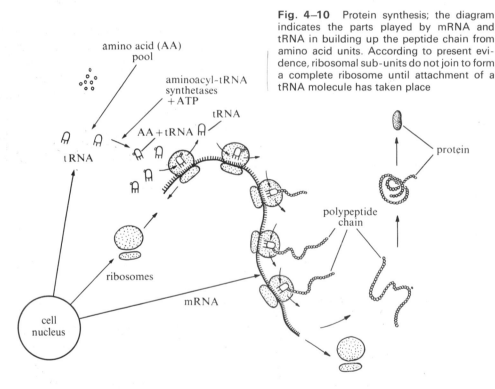

Fig. 4–10 Protein synthesis; the diagram indicates the parts played by mRNA and tRNA in building up the peptide chain from amino acid units. According to present evidence, ribosomal sub-units do not join to form a complete ribosome until attachment of a tRNA molecule has taken place

amino acid (AA) pool

aminoacyl-tRNA synthetases + ATP

tRNA

AA + tRNA

tRNA

protein

polypeptide chain

ribosomes

mRNA

cell nucleus

**4–10
Beginning and
ending of
the message**

It was suggested by Marcker in 1965 that the actual beginning of many polypeptide chains in bacteria is a specially adapted type of the amino acid methionine, formylmethionine, which would correspond to a 'capital letter' codon on the mRNA. Marcker and Sanger showed that formylation of methionine can occur after it has attached to its specific tRNA. Adams and Capecchi in 1966 confirmed that formylmethionyl-tRNA can function as a polypeptide initiator. This means that where a completed polypeptide chain does not have formylmethionine at one end there must be some method, probably enzymic, of removing it from the chain after synthesis is completed. Studies have indicated that an AUG triplet (or sometimes a GUG triplet, due, apparently, to a wobble at the third position in the anticodon) codes for formylmethionine.

The work of Steitz and of Hindley and Staples (1969) on the RNA bacteriophage viruses R17 and Qβ has thrown light on how the triplet (only AUG for these viruses) may be selected at the beginning of a cistron, while the same sequence in the middle of a message is not selected. The deciding factor is not simply proximity to the 5′ end of the molecule. Termination of one message on a polycistronic mRNA is not followed immediately by the beginning of another message, and cistrons may be quite widely separated from one another. There may be specific regions of nucleotides on either side of the initiator codons, and folding of the polynucleotide chain may give a looped structure with the codon prominently placed at the turn of the loop.

Just as there is a special mechanism for the initiation of protein synthesis, so there must also be one for chain termination. Hence it was suggested that mRNA might have a 'release' or 'nonsense' codon—that is, a group of three bases which does not code for any amino acid at the end of a message,

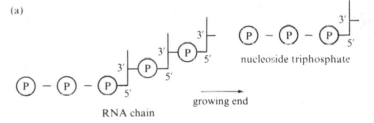

Fig. 4–11 (a) The synthesis of RNA on the DNA template is in the direction 5′ to 3′. **(b)** The direction of reading of mRNA is also 5′ to 3′

(a)

nucleoside triphosphate

growing end

RNA chain

(b)

growing polypeptide chain

5′   3′

and that this provides a release mechanism. Brenner and his coworkers, and independently Weigert and Garen, found such codons in 1965. The triplets UAA, UAG, and UGA function as release codons, but their operation is not yet clear.

Synthesis of RNA on the DNA molecule begins at the 5′ end of the molecule and finishes at the 3′ end (see Fig. 4–11, and Section 3–3). Also the direction of reading of the code on mRNA is continuous from the 5′ end of the codon to the 3′ end (Fig. 4–11). So in bacterial cells, which have no nuclear membrane, ribosome attachment and protein synthesis can begin at one end of a mRNA molecule while the other end is still being synthesized on the DNA. Ribosomes may facilitate the release of mRNA from the DNA molecule. With long-lived mRNA in nucleated cells the messenger must be released completely into the cytoplasm. When all of the codings on a length of mRNA have been read, the mRNA is released from the ribosome and may be re-used. Transfer RNA is also released from the ribosome after its amino acid has formed a peptide bond, and it may be used again to attach to another amino acid. Consequently neither tRNA nor mRNA is used up during protein synthesis.

4–11 Universality of protein synthesis

Remarkably, it seems that protein synthesis proceeds in the same way for all living things on this planet. The same molecules are used in all plant and animal cells and in the smallest bacterial and mycoplasmic cells. Viruses, which can infect and control the cellular synthesis of host cells, also contain the same types of molecules. All cells contain ribosomes and all cells contain mRNA, carrying the genetic code, and tRNA. The ribosomes are larger, however, in higher organisms, and tRNA molecules differ from one species to another.

Errors can occur during protein synthesis. For instance, they may occur in the replication of the DNA, or in the synthesis of the mRNA on the DNA template, or in the formation of the complexes between the various tRNAs and their amino acids, or in the interaction between these complexes and the mRNA on the ribosome.

4–12 Regulation of protein synthesis

Some proteins always seem to be synthesized at a steady rate within the cell, and their amounts do not seem to alter when the cellular environment varies; these are known as *constitutive proteins*. In general, however, the production of enzymes and other proteins is geared to the changing needs of the cell.

For instance, if lactose is added to a culture of *E. coli*, then these bacteria begin to synthesize β-galactosidase, an enzyme which attacks and breaks down the sugar. When all the lactose has been broken down, production of β-galactosidase ceases. This is an example of an *inducible enzyme system*, in which the cell is induced to produce a particular enzyme as a means of adjusting to a change in the extracellular environment.

Accumulation of the end product of a synthetic process in some instances helps to suppress the process itself. For instance, Monod and Cohen-Bazire, studying *E. coli*, showed that a high concentration of tryptophan in the medium suppresses the production of trytophan synthetase, the enzyme used by the cell to synthesize tryptophan. This is an example of a *repressible enzyme system*.

Since cells can make some proteins to order, protein synthesis must be controlled within the cell. How does this control system work? Little is known about the control system in mammalian cells, but Jacob and Monod in 1961 cast considerable light on how it works in bacterial cells, by studying the production of β-galactosidase by E. coli.

Part of the genetic material in E. coli is concerned with specifying the amino acid structure of the enzyme; they termed this part a *structural gene*. They also found that there must be another gene which regulates the production of the enzyme, because a mutant form of the bacterium lacking this gene produced the enzyme continuously, regardless of the concentration of lactose. Only when this gene is functional can the production of the enzyme be controlled. Jacob and Monod called this second gene a *regulator gene*. In addition, they found that a third gene is necessary for the switching-off process even when the regulator gene is functional. This third type of gene is known as an *operator gene*. By genetic analysis they showed that the operator is very close to the structural gene (or, usually, group of genes) which it controls, the whole group being termed an *operon*. The regulator gene in this case also lies close to the operon.

An operon was defined by Jacob as a group of several closely linked genes which appear to affect different steps in a single biosynthetic pathway. These genes have been found to be transcribed by a single mRNA molecule, so an operon may be looked upon as a unit of transcription. In the system responsible for metabolizing lactose in E. coli there are three structural genes, concerned with the production of three different enzymes. The complete system is known as the *lac* operon. This has been widely studied, and Beckwith and co-workers in 1969 managed to isolate most of the *lac* operon in the form of a short length of double-helical DNA.

They exploited the fact that when E. coli is infected by certain bacteriophages the viral DNA can become integrated into the bacterial DNA next to the *lac* operon, and the DNA of the *lac* operon is sometimes carried by the virus into the next host. Beckwith obtained phages with the *lac* operon on complementary strands of the DNA. These regions therefore hybridized when the two extracted DNA molecules were mixed, while the rest of the DNA, which was not complementary, was removed by an enzyme which digests single-stranded DNA only. This achievement opens up possibilities for studying the control mechanisms of a single operon in a test tube.

The control of protein synthesis, as postulated by Jacob and Monod for an inducible enzyme system, is illustrated in Fig. 4–12a. Two structural genes, *A* and *B*, producing two mRNA cistrons, are controlled by the operator gene, *O*. When the operator is switched off, no mRNA is formed, and protein synthesis is inhibited. When the operator is switched on, RNA is transcribed on the DNA, and protein synthesis begins. This switching on and off of the operator is controlled by the

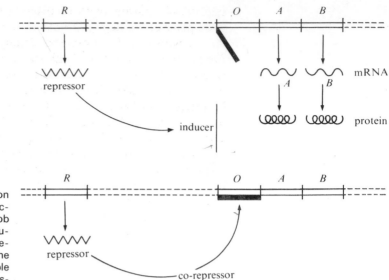

Fig. 4–12 The regulation of protein synthesis in bacteria, as postulated by Jacob and Monod. **(a)** An inducible enzyme system; the repressor is inactivated by the inducer. **(b)** A repressible enzyme system; the repressor is activated by the co-repressor

regulator, *R*. The signal from the regulator to the operator is made by means of a substance, the *repressor*, which is produced by the regulator. The repressor combines directly with the operator and inactivates it, thereby placing the system in the switched-off position. However, when a certain metabolite, the *inducer*, is present, it combines with the repressor, preventing it from reaching and inactivating the operator; the system then returns to the switched-on position and protein synthesis can proceed.

    The control of a repressible enzyme system is illustrated in Fig. 4–12b. Here the repressor molecule is not itself sufficient to inactivate the operator. It must first combine with a metabolite, the *co-repressor* (one of the end products of the process) which activates the repressor so that it can combine with the operator and thus switch off protein synthesis.

    The small molecules which act as inducers and co-repressors probably do not form covalent bonds with the repressor, but only weak bonds which may be easily made and easily broken. This means that the state of activation of a particular repressor may be quickly adjusted to the physiological needs of the cell.

    What is the repressor substance? It must be specific for a particular region (the operator gene) of the DNA molecule. As its effectiveness is determined by interactions with certain small molecules, one would suppose that it might be a protein, since a protein molecule would be considerably altered in structure and properties by such interactions. Evidence was obtained by Gilbert and Müller-Hill in 1966 that the repressor molecule of the *lac* operon is in fact a protein.

    The Jacob–Monod theory of the control of protein

synthesis in certain types of bacteria shows that regulation takes place at the stage of transcription, and suggests that the rate of synthesis of a protein is directly related to the rate of formation of its mRNA. Plant and animal chromosomes are considerably more complex than the genetic material of bacteria and here different mechanisms control protein synthesis. How histones act as repressors of mRNA synthesis in mammalian cells has been discussed in Chapter 3, and the action of hormones in controlling protein synthesis will be discussed in Chapter 9. Regulation cannot always occur at the transcriptional stage, because the various proteins coded for by a polycistronic mRNA are sometimes produced at different rates. Hence there must be other control mechanisms, probably operating at the translational stage. How repetitious sequences of DNA may control gene function will be discussed in Chapter 13.

4–13 Structure of proteins

In Fig. 4–10 the newly synthesized polypeptide chains which leave the ribosomal production line are shown coiling up to form highly compact molecules. The amino acid sequence of a protein molecule is known as the *primary structure*. This determines how interactions between the polypeptide chains give a regular arrangement known as the *secondary structure*, and also how further folding or coiling gives the highly specific, complex *tertiary structure*. The elucidation of the structure of proteins has been one of the major triumphs of molecular biology.

Fig. 4–13 The primary structure of insulin. There are slight variations in different species in the amino acid sequence at a fold of one of the polypeptide chains

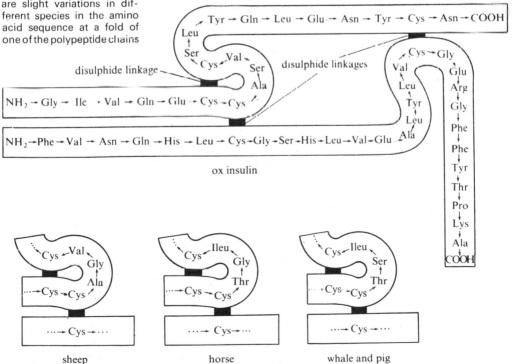

ox insulin

sheep   horse   whale and pig

Amino acid sequence. Detailed chemical analysis has determined the amino acid composition and the actual sequence of amino acids in many polypeptide chains. The first complete amino acid sequence of a protein, that of ox insulin, was determined by Sanger in 1953 (Fig. 4–13). This was a considerable achievement, because although insulin is a relatively small protein molecule it contains 51 amino acids. There is a slight species variation in the amino acid sequence of one of the looped regions of insulin, as shown in Fig. 4–13. Since 1953, using newer techniques, the amino acid sequence of further proteins has been worked out. Coincidentally the first protein molecule to be synthesized *in vitro* was also insulin, this great achievement being carried out by a team of Chinese scientists in 1965. Ribonuclease, which consists of 124 amino acid residues compared to 51 in insulin, was synthesized by two independent American teams in 1969 (Denkewalter and Hirschmann; and Merrifield and Gutte).

Secondary structure. A major step towards elucidating the structure of protein molecules was the discovery that fibrous proteins give X-ray diffraction patterns. Meyer and Mark (1928), applying this technique to a study of silk threads, were the first to show that a protein molecule can exist in the form of an extended chain. It had been thought previously that all protein molecules were spherical or nearly spherical in shape.

Astbury, in the early 1930s, grouped proteins into two classes—fibrous proteins such as silk (fibroin), wool and hair (keratin), and muscle (actomyosin); and globular proteins such as haemoglobin and myoglobin. He suggested that the two groups are basically similar, being built up in the same way from primary chains, which under certain conditions can change their shape and fold up.

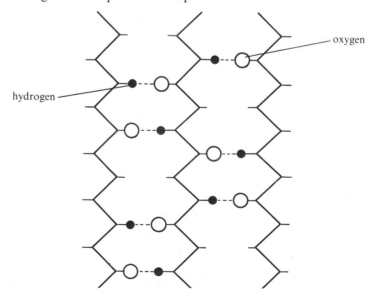

Fig. 4–14 The β-configuration of proteins

In 1933, Astbury and Street found that wool or hair fibres when placed in steam can be stretched to twice their original length, and that the stretched fibres yield X-ray diffraction patterns similar to those for silk fibres, and quite different from those for unstretched wool fibres. They called the unstretched keratin the α-form, and the stretched type the β-form. The structure of the β-form, as later modified by Pauling and Corey, is shown in Fig. 4–14. Parallel zig-zag

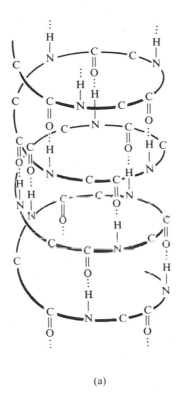

(a)

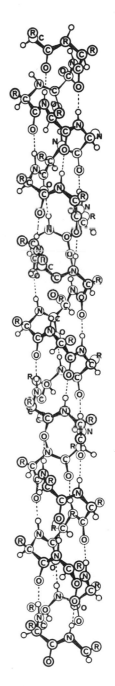

Fig. 4–15 The α-helix model of protein coiling. **(a)** Intra-chain hydrogen bonding gives stability to the molecule. (With permission from J. C. Dearden. *New Scientist*,**37**,589,629,1968.) **(b)** Molecular model of the Pauling–Corey α-helix. (With permission from J. A. Ramsay, *The Experimental Basis of Modern Biology*. Cambridge University Press, London, 1965.)

(b)

polypeptide chains form a pleated sheet-like structure, stabilized by the formation of hydrogen bonds between the $>$NH and $>$C=O groups on neighbouring chains. Side-chains attached to the amino acid residues lie above and below the hydrogen-bonded sheets. This structure is a stable arrangement where these side-chains are small, as they are in fibroin, the protein of silk, and do not cause distortion of the pleated structure. In fibroin, the chains run anti-parallel to each other, that is, the free amino (or carboxyl) groups are at opposite ends of neighbouring chains.

Astbury realized that in the α-form of keratin the chains must be folded up in some way, but it was not until polypeptide chains of high molecular weight were synthesized *in vitro* that the structure could be resolved. These chains were examined in the polarizing infra-red spectrometer by Ambrose, Elliott, and others (1950). Their work indicated that in the α-form the hydrogen bonds between the $>$NH and $>$C=O groups lie parallel to the long axis of the chains, not at right angles as they do in the β-structure (Fig. 4–15a). They proposed a folding of the polypeptide chain by the formation of hydrogen bonds within the chain. Later crystallographic work by Pauling and Corey (1951) led to the α-helix model shown in Fig. 4–15b. The polypeptide backbone is here twisted into a helix, and intra-chain hydrogen bonding results in the formation of internal rings. Perutz finally proved that this model is correct by showing that the predicted and measured distances between the side-chains are 1·5 Å.

The α-helix is the most important regular arrangement of the polypeptide chain. Large sections of the backbone of myoglobin exist as α-helices, and evidence suggests that there are helices in a great variety of proteins.

Tertiary structure. Very few protein molecules exist as a simple α-helix. Further degrees of folding or coiling of the polypeptide chains give a complex three-dimensional structure, often containing helical and non-helical regions. For instance, in keratin further twisting of the α-helix gives a coiled coil structure, and in haemoglobin and myoglobin (Fig. 4–16) considerable folding and twisting of the polypeptide chain produces a globular or corpuscular molecule.

Why do these irregular structures arise? There are various reasons, the chief one being the great variety of amino acid side-chains. The side-chains tend to take up positions where they can interact with each other to form energetically stable structures. The interactions may be in the form of hydrogen bonds—for example, the free hydroxyl group on serine tends to take up a position where it can hydrogen-bond with another group. Also, polar side-chains tend to react with

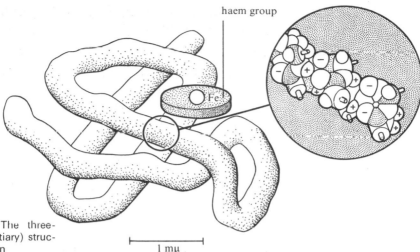

haem group

1 mμ

Fig. 4–16   The three-dimensional (tertiary) structure of myoglobin

groups of opposite charge—for instance, the carboxyl group of glutamic acid will react with the amino group of lysine. Disulphide bonds between cysteine residues are also formed, and van der Waals attractive forces may operate between non-polar groups.

Folding of the α-helix occurs where the amino acid proline is present. Proline contains an imino (=NH) group instead of an amino group and so where it is present in a polypeptide chain there is an interruption in the regular internal hydrogen bonding stabilizing the α-helix.

Factors such as these help to determine the most stable arrangement of a particular protein according to its own unique amino acid composition.

Work with the ultracentrifuge showed that globular or corpuscular molecules in solution are generally ellipsoid, but the final analysis of structure came from X-ray crystallography studies. In 1933, Bernal and Crowfoot found that globular proteins can be crystallized and that the crystals give X-ray diffraction patterns. The presence of peptide bonds and hydrogen bonds as in fibrous proteins was clearly shown by the infra-red studies of Ambrose and Elliott (1950). The first complete structures were elucidated after many years of magnificent work by X-ray crystallographers, particularly Kendrew and Perutz in association with Sir Lawrence Bragg. The structure of myoglobin as worked out by Kendrew (1958) is shown diagrammatically in Fig. 4–16. Myoglobin is composed of 153 amino acid residues, and contains only one polypeptide chain. The α-helix undergoes secondary folding to give a highly complex, unsymmetrical molecule. Close inspection shows that the chain is folded so as to bring together side-chain groupings which can interact. Most of the straight sections of the molecule consist of α-helices, and the proline molecules which it contains occur at angles in the amino acid chain as expected. The insoluble side-chains are arranged in the interior of the molecule, with soluble ones on the outside, which must clearly have a marked effect on the solubility of myoglobin.

The structure of haemoglobin is rather similar to that of myoglobin, and was determined by Perutz and coworkers (1960). It is an example of a protein which shows still further organization, being made up of four polypeptide chains; this further organization is known as the quaternary structure. The chains all undergo secondary folding, two of the structures consisting of what is called the α-chain and two of another β-chain. All four chains fit together in a compact tetrahedral arrangement to form the complete haemoglobin molecule.

The structure of various other globular proteins has now been worked out, including that of insulin (Hodgkin and coworkers, 1969). As shown in Fig. 4–13, the insulin molecule consists of two polypeptide chains joined by disulphide linkages; insulin exists in the pancreas in the form of a hexamer.

4–14
The α-helix and life

An enormous number of different proteins is found in nature. Probably all enzyme molecules are globular proteins. Since these structures arise from a further degree of organization imposed upon the α-helix, it is likely that, as newly synthesized polypeptide chains grow at the ribosomal assembly line, they rapidly take up the α-helical form. Then, as the complete protein molecule is synthesized, further folding and twisting occurs to give a unique structure for each protein.

That the three-dimensional structure of a protein is the most stable for its own composition has been indicated by some interesting experiments. If a protein is exposed to heat or some other abnormal condition, its complex structure breaks down to give single polypeptide chains which are randomly oriented and inactive biologically. This process, known as denaturation, may be irreversible, as in the death of the cell by chemical fixation. If, however, the separated chains are carefully returned to their normal environment, it has been found that some of them can again take up their original structure, and are active biologically.

The β-form of fibrous protein structure, as in silk fibroin, is strongly bonded and quite insoluble. If the newly synthesized polypeptide chains within the cell were to assume this structure, they would form a dense precipitate and the cell would die very rapidly. The α-form is, however, soluble—it does not form bonds with other chains because its hydrogen bonds are satisfied within the chain. Were it not for the α-form of the polypeptide chain, the dynamic state combined with stability which is so necessary for life almost certainly would not exist.

Bibliography   Crick, F. H. C., On protein synthesis. *Symp. Soc. Exptl. Biol.,* **12,** 138–163, 1958.

Haggis, G. H. (Ed.), *Introduction to Molecular Biology.* Longmans Green, London, 1964.

Jacob, F. and J. Monod, Genetic regulatory mechanisms in the synthesis of proteins. *J. Mol. Biol.,* **3,** 318–356, 1961.

Kendrew, J. C., The three-dimensional structure of a protein molecule. *Scientific American,* December 1961.

Loewy, A. G. and P. Siekevitz, *Cell Structure and Function,* 2nd edn. Holt, Rinehart, and Winston, New York, 1969.

Nirenberg, M. W., The genetic code. *Scientific American,* March 1963.

Rich, A., Polyribosomes. *Scientific American,* December 1963.

Warner, J. R. and R. Soeiro, The involvement of RNA in protein synthesis. *New Eng. J. Med.,* **276,** 563, 613, and 675, 1967.

Watson, J. D., *Molecular Biology of the Gene.* W. A. Benjamin, New York, 1965.

5 Cytoplasmic organelles

When cells were first studied with the light microscope it was observed that, although their individual shapes varied considerably, most were divided into two main compartments—an inner nucleus, and an outer region known as the cytoplasm. It was recognized that the complete cell must be enclosed by some limiting boundary or membrane, which served as a protective layer between the cell and its environment, and in mammalian cells also preserved the shape of the cell and restricted the passage of molecules into and out of it. This membrane was far too thin to be observable by light microscopy and it was not until the advent of the electron microscope in the 1940s that it could actually be seen and its approximate thickness estimated.

Before the electron microscope came into use, it was impossible to detect all of the various structures or organelles which now appear to be well-defined compartments within the cytoplasm (Fig. 5–1). Indications of the existence of some forms of cytoplasmic structure had, however, been obtained. As early as 1898, the Italian cytologist, Golgi, observed by means of staining methods that there is a reticular structure in the cytoplasm of certain nerve cells. His original observations have been proved substantially correct, and the *Golgi apparatus*, or Golgi body, has been revealed by electron microscopy as a complex of interconnecting membranes.

In 1910, Gadukov, using the ultra microscope, which produces a dark field in which objects are illuminated by scattered light, detected in the cytoplasm of cells particles which were invisible under the light microscope. These particles were later identified as mitochondria. Bayliss, in 1920, using a similar method, detected particles undergoing Brownian movement in the plasma sol of amoebae.

The various subcellular organelles shown in Fig. 5–1 possess limiting boundaries or membranes, similar to the cell membrane. A combination of chemical, physical, and electron microscopic studies has shown that all of these boundaries, including the cell membrane, consist of a *lipid membrane*. Lipid molecules, particularly those which contain polar groups at one end, can become orientated at air–water and lipid–water interfaces to produce monolayers, and this largely

148

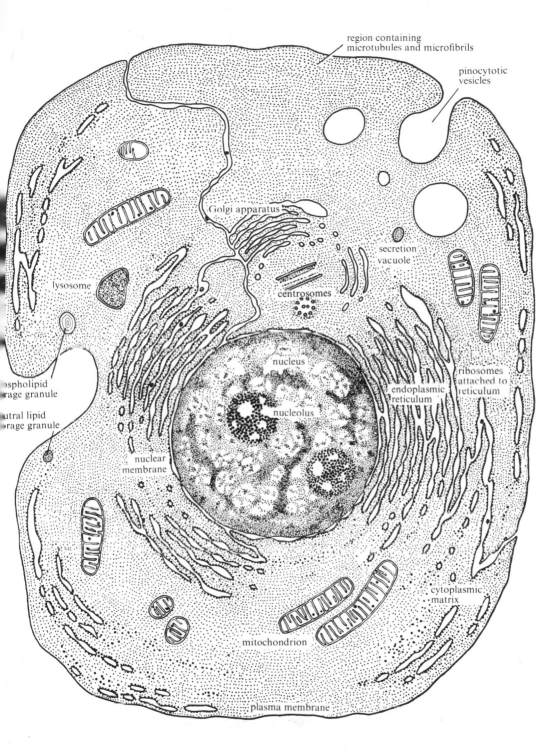

Fig. 5–1 Generalized diagram of an animal cell, showing the nucleus, membrane systems, and main cytoplasmic inclusions. (From J. Brachet (ed.), *The Living Cell.* Copyright © September 1961 by Scientific American, Inc. All rights reserved.)

determines the formation of lipid membranes (Chapter 2).

We begin, then, by dealing with the structure and properties of lipid membranes, before considering the individual cytoplasmic organelles in detail. The mitochondria of animal and plant cells, and the chloroplasts of cells of green plants and some micro-organisms, will be discussed separately in Chapter 6. We consider the plasma membrane in greater detail in Chapter 8; because it is the only means of communication between the cell and the extracellular environment it has a rather specialized structure.

5–1
Historical survey

Early studies on the nature of cellular membranes were confined to work on the outer surface of the cell—the cell plasma membrane itself. Pfeffer in 1877 observed that, on crushing the root hairs of the plant *Hydrocharis*, small droplets were released from the cells, and that these droplets behaved as individual cells so far as osmotic effects were concerned. This observation is very much in line with present ideas, according to which the plasma membrane is not unchanging, but labile. If the membrane is ruptured the fragments tends to reorganize to form smaller aggregates with similar properties.

Overton in 1895 obtained evidence suggesting that the cell membrane contains lipid. He studied the permeability properties of various unfertilized egg cells and found that fat-soluble substances easily penetrate the cell membrane, whereas substances insoluble in fats penetrate very slowly. He therefore suggested that a layer of lipid must be present on the cell surface, since ionic substances, which are insoluble in lipids, enter a lipid layer only with great difficulty.

A quite different line of approach, which also indicated that the cell surface contains lipid, was to measure the electrical impedance of the cell. The studies of Höber (1910), Fricke (1925), and later workers showed that the cell contents have a comparatively high conductivity (which would be expected because the cytoplasmic fluid consists of solutions of salts and proteins) while the intact cell has a very low conductivity. These results indicated that a thin lipid layer existed at the cell surface, since this would have the necessary high electrical impedance, whereas a thin layer of protein would not. Harvey (1912) and Collander (1927) studied the ease with which various substances can penetrate artificially prepared protein membranes. They found that these membranes were not selectively permeable to the lipid-soluble molecules which penetrate cells preferentially, supporting the view that the cell membrane is not composed primarily of protein.

Wetting cells with oil confirmed the view that there is a lipid layer on the cell surface (Dawson and Belkin, 1929; Mudd and Mudd, 1931; and others). For instance, if a small droplet of oil is placed in contact with the surface of *Amoeba dubia* it adheres to and spreads over the cell surface. This wetting by oil is to be expected if lipid is present on the surface of the cell plasma membrane.

5–2
Basic membrane
structure

Further evidence for a lipid membrane came from Gorter and Grendel in 1925. Their work constituted a major advance in the field because they were the first to suggest a possible membrane structure. The cells they studied were erythrocytes (red blood cells), which contain no obvious organelles in the cytoplasm, so that the only membrane structure to be considered is almost entirely that of the cell surface. Gorter and Grendel extracted the total lipid of erythrocytes and found that when this is spread as a monolayer at an air–water interface the film formed covers an area twice the total surface area of the original cells. From this they suggested that the surface membrane of the red cell is composed of a double layer of lipid molecules, with the polar groups situated on the outside of the layers. This bimolecular structure (Fig. 5–2a) is now considered substantially correct, and is basically similar to the lipoprotein model later proposed by Danielli (Fig. 5–2b).

Studies of the surface tension of cells by Harvey and Cole (1931) revealed that this is only about 0·2 dyne/cm. This is much lower than would be expected for a simple oil–water interface (about 35 dyne/cm). Extraction of the oil from fish egg cells by Danielli and Harvey (1935) and further surface tension measurements showed that the tension at an oil–seawater interface is almost 10 times that at the interface between seawater and oil in contact with egg extracts. They further showed that in this case the substance responsible for the

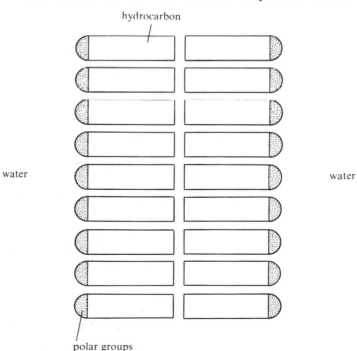

hydrocarbon

water   water

Fig. 5–2   (a) The structure of the red cell plasma membrane, after Gorter and Grendel

polar groups

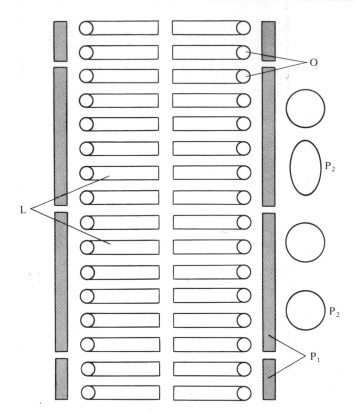

Fig. 5–2 (b) The lipoprotein structure of the cell plasma membrane according to Danielli. L   chains of lipid molecules; O   polar groups of lipids; P_1 tangentially arranged protein chains; P_2 globular proteins on outer surface

decrease in tension is a globulin-like protein. This led Danielli to propose that the lipid layer of the cell membrane is coated with a thin sheet of protein, which would reduce the tension at the cell surface. From these ideas gradually emerged the model of the cell membrane in which a bimolecular lipid layer, 30–50 Å thick, is sandwiched between two protein layers (Danielli, 1938) (Fig. 5–2b). The lipid part of the structure lies within the membrane, with the hydrocarbon chains perpendicular to the surface, while the polar regions of the lipid lie towards the outside and are associated with the protein molecules. More recent studies have slightly modified and extended the basic Danielli model.

From 1950 onwards many attempts were made to determine the composition of cell membranes by direct chemical analysis. There are immense difficulties in isolating membranes in a pure state—even with the erythrocyte, which contains little besides haemoglobin in its cytoplasm. We now know that not only do the membranes of different species and different types of cell differ in their composition, but also that the same cell type can change its composition, depending upon its environment. The cell membrane can even alter in response to changes in diet of the whole animal. Furthermore, different methods of preparation can give different results with the same cells. It has now been established, however, that the membranes of animal cells consist largely of protein and lipid together with

Fig. 5–3 The chemical composition of plasma and mitochondrial membranes The figures are expressed as a percentage of the total dry weight

Cell type	Phospholipids					Cerebrosides (cerasine, etc.)	Cholesterol	Other lipids	Protein	Carbohydrate (or carbohydrate component of glycoprotein)
	Phosphatidic acid	Phosphatidyl-ethanolamine	Phosphatidyl-choline	Phosphatidyl-serine	Others (sphingomyelin, etc.)					
Bovine red cells [1, 6]	less than 1%	~2%	~3%	~3%	10 to 15%	Very little	~10%	less than 3%	60 to 70%[a]	Carriers of blood group markers, in human red cells, at the outer surface of the membrane.
Bacterial protoplasts [2] (*Staphylococcus aureus*)	Total lipids ~20% (very little cholesterol)								~40%[b]	Up to 40%
Estimate for myelin sheath of nerve [3, 4]	—	~10%	5 to 10%	~5%	—	10 to 15%	15 to 20%	—	20 to 40%	Histochemical staining shows carbohydrate present.
Mitochondrial membranes [5]*	—	5 to 20%	10 to 25%	less than 5%	—	—	~1%	—	50 to 60%[c]	

(Plasma membranes: Bovine red cells, Bacterial protoplasts, Estimate for myelin sheath of nerve)

1. PAPPART, A. K. and R. BALLENTINE (1952) in *Modern Trends in Biochemistry*, ed. E. S. C. Barron, Academic Press, New York.
2. MITCHELL, P. (1959) *Biochem. Soc. Symposia*, **16**. 73.
3. ENGSTROM, A. and J. B. FINEAN (1958) *Biological Ultrastructure*, Academic Press, New York
4. DAWSON, R. M. C. (1957) *Biol. Revs.*, **32**, 188.
5. BALL, E. G. and C. D. JOEL (1962) *Int. Rev. Cytol.*, **13**, 99.
6. DAWSON, R. M. C., N. HEMINGTON and D. B. LINDSAY (1960) *Biochem. J.*, **77**, 226.

* Inner and outer membranes show some differences in structure and properties.
(a) Including enzymes of active transport.
(b) Including a wide variety of enzymes.
(c) Including enzymes of the electron transfer chain.

a small amount of carbohydrate (Fig. 5–3). Although the proportion of protein to lipid varies greatly in most cases a phospholipid, lecithin (phosphatidylcholine), and a steroid molecule, cholesterol, represent a large part of the lipid component. The structures of some of the lipid molecules listed in Fig. 5–3 are shown in Fig. 5–4.

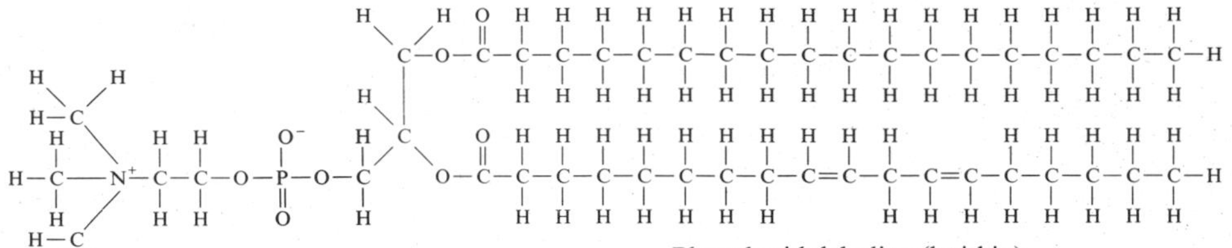

Phosphatidylcholine (lecithin)

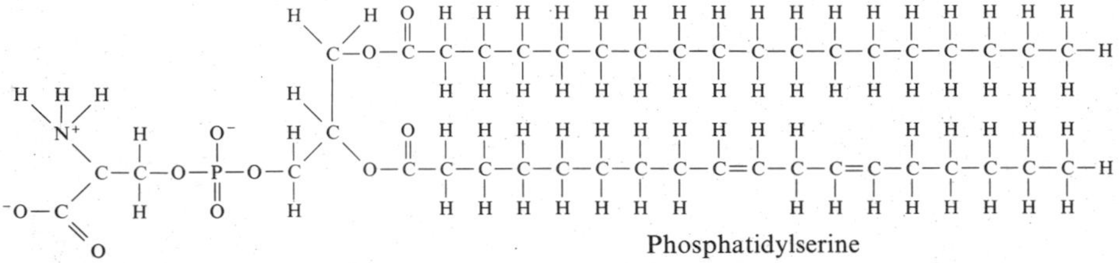

Phosphatidylserine

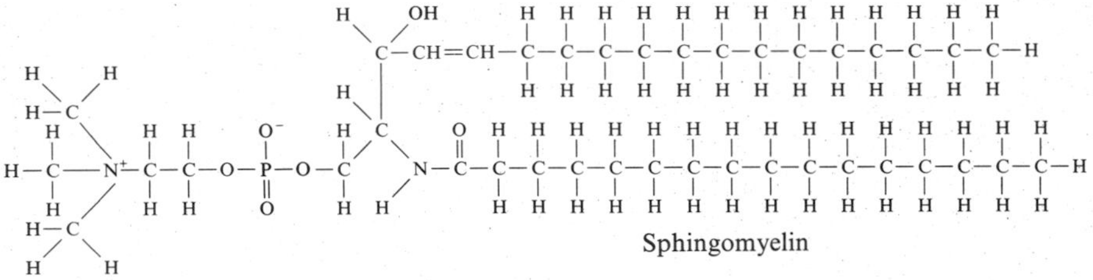

Sphingomyelin

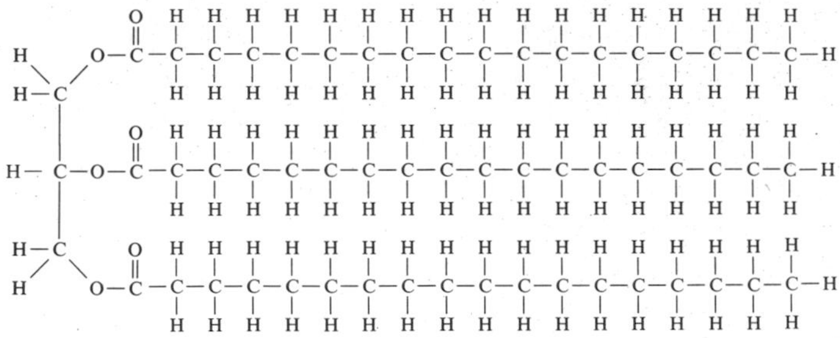

Triglyceride

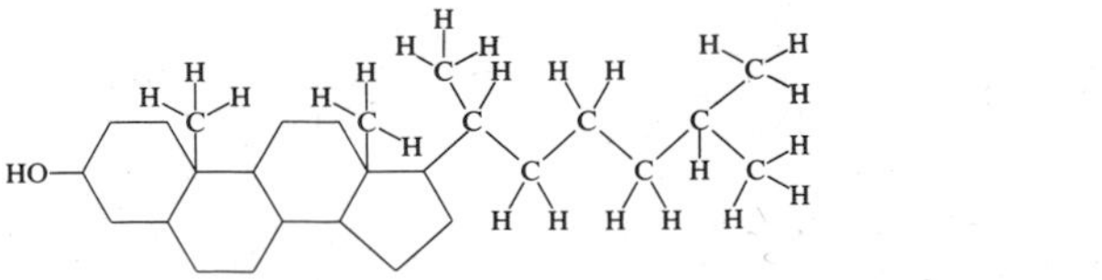

Fig. 5–4 The chemical structures of some of the chief lipid molecules found in cell membranes (approximately to scale). The lengths of the fatty acid side-chains in neutral lipids and phospholipids may vary, and also the degree of unsaturation

Cholesterol

Physical characteristics of lipid systems. Since phospholipids appear to be the chief lipid component of cell membranes, their physical properties must be of great importance in determining the molecular organization of membranes. Notable in this field has been the work of Luzzati and Husson (1962). They studied lecithin–water mixtures (see Chapter 2) by X-ray diffraction and light-scattering methods and found that in aqueous systems phospholipids tend to aggregate so that their hydrocarbon chains are isolated from the aqueous phase (Fig. 5–5). When the water content is very low, the lecithin molecules produce micelles in which their ionic groups are directed towards a central aqueous core. When the water content is increased, the arrangement of molecules takes a lamellar (or liquid crystalline) form. With even higher water contents, a micellar form again occurs, but this time with the ionic groups on the outside; this last form is obtained with concentrations of lecithin as low as 5×10^{-7} g/ml of water.

One would expect cholesterol in the cell membrane to modify the properties of a pure phospholipid layer. The result of varying proportions of lecithin and cholesterol in an aqueous medium has therefore been studied. The two types of lipid will form mixed monolayers at an air–water interface, and the area of the mixed film is less than that calculated for the two separate components. This has been shown by Bernhard (1958) to hold true over a wide range of molecular proportions (from 2 parts of lecithin: 1 part of cholesterol, down to 1 part of lecithin: 2 parts of cholesterol). This packing together of the two types of molecules indicates that they interact in some way and that the compact structure so formed is of increased

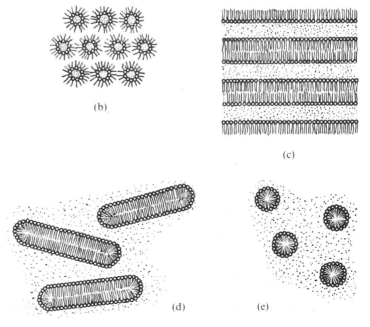

(a)   (b)   (c)   (d)   (e)

Fig. 5–5 Diagrammatic representation of the molecular arrangements occurring in phospholipid–water systems: **(a)** anhydrous phospholipid; **(b)** water in lipid phase, hexagonal packing; **(c)** lamellar or liquid crystalline phase; **(d)** and **(e)** lipid in water micellar arrangements. (With permission from J. B. Finean, The molecular organization of cell membranes. *Prog. Biophys. Molec. Biol.* **16**, 145, 1966.)

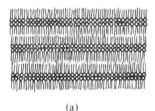

stability. Nothing is directly known about the type of association that occurs between the two molecules. As can be seen from Fig. 5–4, cholesterol has very little ionic character and is much shorter than the phospholipid molecules.

Finean (1953) suggested that the polar end of the phospholipid molecule is bent round rather like a walking-stick, and that this forms a bond with the cholesterol which lies along one of the hydrocarbon chains (Fig. 5–6a). In an alternative model proposed by Vandenheuvel (1963), there is less deformation of the phospholipid, both hydrocarbon chains are in close proximity to the cholesterol (Fig. 5–6b), and the chain, which is unsaturated at the ninth carbon atom, curls around the end of the cholesterol. It may be that different types of packing occurs in different situations, and of course the conditions in a bimolecular layer may differ from those in the monolayers which have been investigated.

Model systems have also been studied to determine how protein interacts with lipid layers. Matalon and Schulman (1949) found that the surface pressure of a layer of phospholipid at an air–water interface changes when soluble protein is

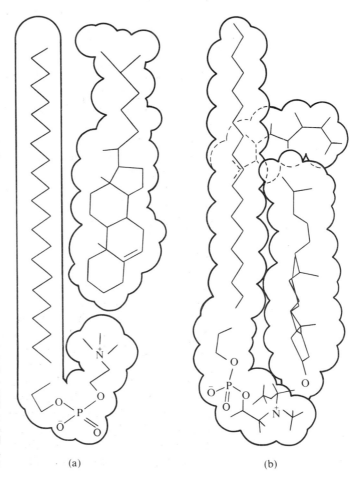

Fig. 5–6 Suggested arrangements of the phospholipid–cholesterol complex in cell membranes: **(a)** according to Finean (1953); **(b)** according to Vandenheuvel (1963). (With permission from J. B. Finean, The molecular organization of cell membranes. *Prog. Biophys. Molec. Biol.* **16**, 145, 1966.)

(a)   (b)

added below the monolayer. This indicates an interaction and partial penetration of the monolayer by some of the protein side-chains. It was also shown by electron microscopy (Stoeckenius, 1959) that, when protein is added to aqueous dispersions of phospholipid, the surface of the micelles thickens, again indicating some form of association between the protein and lipid, and incomplete penetration by the protein molecules.

In most of the investigations of lipid–protein interactions, water-soluble protein was used, presumably because of the charged groups on its surface. The resulting interaction was largely electrostatic, involving a neutralization of charge on the phospholipid. But Green and Fleischer (1964) have shown that a predominantly non-polar interaction can occur with other proteins, and that in mitochondria both electrostatic and non-polar interactions between proteins and phospholipid probably play a part in determining structure.

Other factors must influence the organization of membranes—for instance, the presence of Ca^{2+} ions during the formation of a phospholipid monolayer seems to result in a much more coherent and less permeable film—but the structure which has been built up using surface films of mixed lipid composition, with adsorbed protein on the surface, provides quite a satisfactory model for the cell membrane.

Electron microscopic studies. Studies with the electron microscope tell us little about the chemistry of the membrane, but are of great value in determining whether the structures observed are consistent with the models built up by physical and chemical studies. In fact, the model of membrane structure inferred from electron microscopy agrees remarkably well with the Danielli model.

The work of Sjöstrand and his coworkers from 1953 onwards and of Robertson from 1959 has established that the cell plasma membrane does in fact appear as a triple-layered structure (Fig. 5–7) consisting, after osmium fixation, of two dense bands (corresponding to the protein and polar layers in the Danielli model), separated by a central clear zone of lipid. Robertson suggested a unit membrane structure with a central lipid layer and two outer layers of protein, or perhaps carbohydrate. The precise dimensions depend on the way the tissue is prepared, but the total thickness of the triple layer is in the range 75–100 Å. The lipid layer is 25–30 Å thick, the outer protein layer is 25 Å thick, and the inner one is 25–35 Å thick, depending upon the procedures used.

One problem in interpreting the results of electron microscopy is the variation caused by different methods of fixation, embedding, and sectioning. Hence it is of great value to examine a structure where the findings of electron microscopy may be compared with those of some other technique.

Fig. 5–7 Electron micrograph of the human red cell plasma membrane; permanganate fixed (magnification × 280,000). (With permission from J. David Robertson.)

Such a structure is the myelin sheath of nerve cells, which is derived from the membrane of the Schwann cell. An extension of the cell, with a surface membrane on both sides, wraps itself around the nerve axon forming a tightly wound helix consisting of layers of membranes; this multilayer forms the myelin sheath. Schmidt examined this structure in the polarizing microscope and inferred that the lipid molecules are arranged with their chains at right angles to the membranes, and that the protein lies tangentially orientated between the turns of the lipid helix; this agrees very well with the lipid-bilayer Danielli model. In the electron microscope, the myelin sheath appears as a series of dark and light bands, with a repeating distance of 120 Å, which must represent two closely apposed plasma membranes. Since the myelin sheath is a repeating structure it can give rise to an X-ray diffraction pattern, and this has been extensively studied, using fresh and fixed material, by Finean from 1958 onwards. His studies of fixed material have confirmed a repeating distance of 120 Å. As a result of all this work the basic packing of lipids and proteins in nerve myelin is fairly well understood.

Stoeckenius has prepared bimolecular lipid layers, with protein adsorbed on the surface, and examined them under the electron microscope (Fig. 5–8). Using osmium fixation he observed two dark bands, of 25–50 Å thickness, with a clear zone, 20–25 Å wide, in between. This again is remarkably similar to the appearance of cell membranes in tissue sections.

Under the electron microscope the membranes surrounding many cytoplasmic structures also bear a marked resemblance to the plasma membrane, and the Danielli model is still

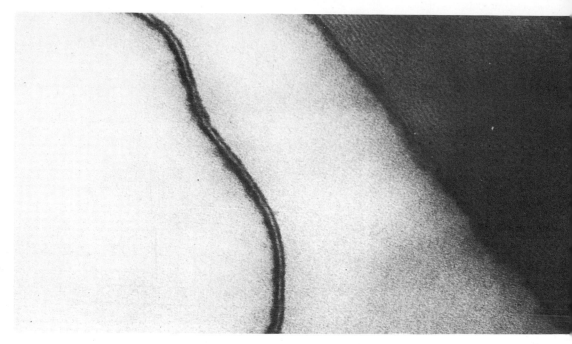

Fig. 5–8 Electron micrograph of a triple-layered structure formed in a protein lipid–water mixture. The two osmiophilic lines which are 25–50 Å wide, representing the polar ends of the lipid and the protein, are separated by the clear zone of the hydrocarbon chains of the lipid (20–25 Å wide). (With permission from W. Stoeckenius.)

considered to be the basic structure for all cell membranes. In its simplest form, however, it cannot account for all of the many different activities of the cell membrane. Interestingly, there is evidence for a micellar membrane structure in certain cells.

The hexagonal repeating structure observed by Dourmashkin *et al.* (1962) in cell plasma membranes treated with saponin was later shown to be due to a saponin–lipid complex, rather than to a basic structure of the cell membrane which had been revealed as a result of the removal of a component by saponin. Nonetheless, this work stimulated investigations of mixtures of lecithin and cholesterol in the presence of saponin (Bangham and Horne, 1964; Lucy and Glauert, 1964). It was realized that phospholipid mixed with other fatty substances can exist in many forms of lamellar, tubular, and helical structures, believed to be built of micelles 35–40 Å in diameter, and these can be readily interconverted (compare Fig. 5–5).

A conversion of even a small area of the cell membrane from a bimolecular layer into a micellar arrangement for only a short period of time could have a very marked effect on membrane functioning. There have been reports of globular micellar and hexagonal arrangements in various membranes, and Benedetti and Emmelot (1965) have observed a hexagonal pattern of repeating units in rat liver cell membranes (Fig. 5–9). This arrangement was obtained only after particular conditions of fixing and treating the cells. Possibly the micellar arrangement is a less stable form of the cell membrane than the bimolecular leaflet of the Danielli model, the membrane only assuming this micellar form under certain conditions.

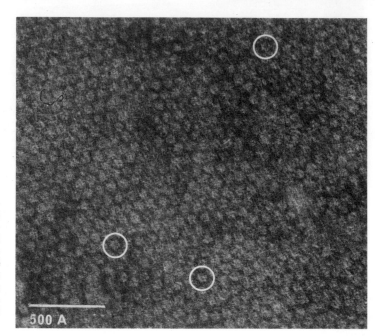

Fig. 5–9 High-magnification electron micrograph of isolated plasma membranes negatively stained at 37°C showing the hexagonal lattice in which some pentagonal units (circled) are inserted (line indicates magnification). (With permission from E. L. Benedetti and P. Emmelot.)

5–3 Cytoplasmic components

The various cytoplasmic components all play an important part in the functioning of the cell. The nuclear membrane protects the genetic material from the many different activities of the cytoplasm. The mitochondria in animal and plant cells, and chloroplasts in plant cells, control the production of the energy required for various metabolic processes, and are such an extensive and important subject that they will be dealt with separately in Chapter 6. The endoplasmic reticulum is a reticular network to which are attached the ribosomes, the sites of protein synthesis. The Golgi apparatus consists of a group of membranous sacs; it plays an important part in the formation of membranes, including those of storage and secretory vacuoles. Lysosomes are vacuoles of variable size and shape, containing enzymes which are of considerable value to the cell, but could nevertheless cause its death by autolysis if they were not enclosed in some way. It is clear that most of the structures seen in animal cells must also occur in plants, but there are certain structures which are characteristic of plant cells only, and these will be dealt with in a separate section.

All of these structures depend for their existence on some type of lipid membrane.

Nuclear membrane. A limiting membrane which separates the nucleus from the cytoplasm is present in all animal cells except bacteria and mycoplasmas, and in all plant cells except blue-green algae. It is not known why some cells lack a nuclear membrane, but it is thought that they represent a more primitive stage of evolution.

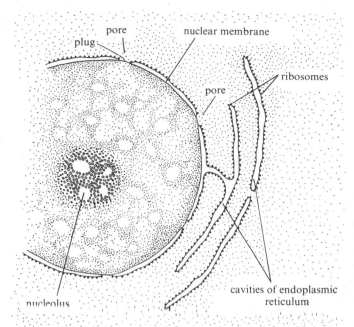

Fig. 5–10 Diagram of the nuclear membrane, showing how it may interconnect with the endoplasmic reticulum

The nuclear membrane (Fig. 5-10) consists of two of the unit membranes already described, the triple layers of which are somewhat narrower than those in the cell plasma membrane. Each membrane is 40–60 Å thick and the clear space between the membranes is of the order of 200 Å, although its width may vary considerably (Fig. 5-11).

There are large pores or holes over the whole surface of the nuclear membrane, formed by the fusion of the inner and outer units. These pores have a diameter of about 500 Å, and they occupy about 10 per cent of the total nuclear surface area of mammalian cells. Their function is not clear, but is presumably concerned with the transport of materials across the membrane; for instance, RNA molecules probably enter the cytoplasm from the nucleus by means of these pores. In some animal cells the pores may be partly or completely plugged with material which has a definite organized structure. This material appears to have a low electrical conductivity, and would present a considerable barrier to the passage of even small ions such as Na^+ or K^+. In plant cells, too, the nuclear pores are in some cases essentially closed, whereas in others—such as the nuclei of onion roots and yeast—they are open. It is likely that the pores play some sort of 'door-keeping' role.

During cell division in both plant and animal cells the nuclear membrane disappears; phase-contrast microscopy shows this clearly in living cells. During metaphase the membrane breaks up into fragments which then form rounded vesicles in the cytoplasm. At a later stage of cell division the vesicles move into both daughter cells, aggregate round the

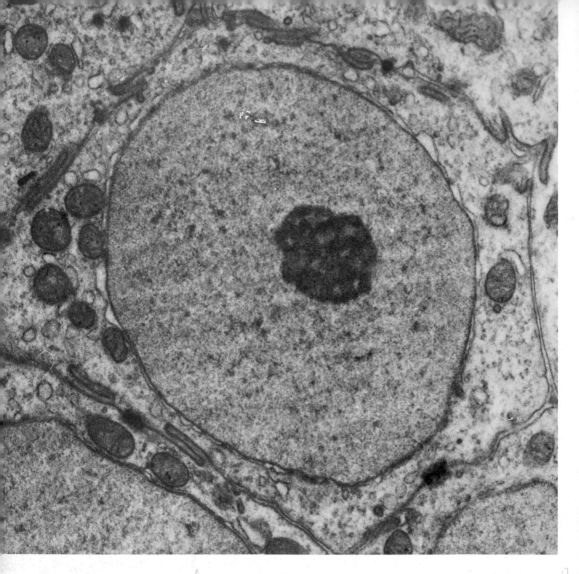

Fig. 5–11 Electron micrograph of mouse intestine showing variation in width of the space between the inner and outer layer of the nuclear membrane (magnification ×19,000). (With permission from M. S. C. Birbeck.)

chromosomal material of these cells, and then flatten out to form a new nuclear membrane.

In both plant and animal cells the outer nuclear membrane can form extensions into the cytoplasm which appear to be continuous with the endoplasmic reticulum (Fig. 5-10). These regions of continuity frequently appear in rapidly dividing embryonic or neoplastic cells, but rarely in the cells of differentiated tissues. This suggests that such continuity may be of short duration, and may disappear soon after cell division. The presence of ribosomes, frequently observed on the cytoplasmic surface of the nuclear membrane, also suggests that the membrane is at some time continuous with the endoplasmic reticulum. Since there is some evidence (see below) that certain areas of the endoplasmic reticulum are dynamically related to the plasma membrane of the cell, this suggests a possible mechanism of direct transfer of extracellular material to the nucleus of the cell (Fig. 5-1). It may be that the fat droplets absorbed by certain specialized cells and sometimes found in

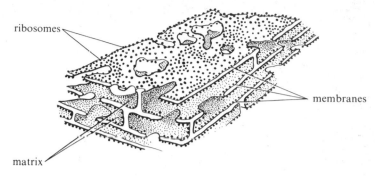

ribosomes

membranes

matrix

Fig. 5–12 Diagram showing the three-dimensional structure of the endoplasmic reticulum. (With permission from E. D. P. De Robertis, W. W. Nowinski, F. A. Saez, *Cell Biology*, 4th edn. W. B. Saunders, Philadelphia, 1965.)

Fig. 5–13 Electron micrograph of mouse liver showing regions of rough (R) and smooth (S) endoplasmic reticulum (magnification ×27,000). (With permission from M. S. C. Birbeck.)

the cavity of the nuclear membrane travel there by this mechanism. This may also explain how acridine dyes penetrate the nucleus without staining the cytoplasm.

Endoplasmic reticulum and Golgi region. A three-dimensional diagram of the structure of the endoplasmic reticulum is shown in Fig. 5–12. The name endoplasmic reticulum was first used to describe certain structures observed with the electron microscope by Porter and colleagues (1945) in the cytoplasm of thinly spread out fibroblast-like cells in cultures of chick embryonic tissues; it is now applied more generally to describe similar structures found in almost all cells. The characteristic appearance of the endoplasmic reticulum has

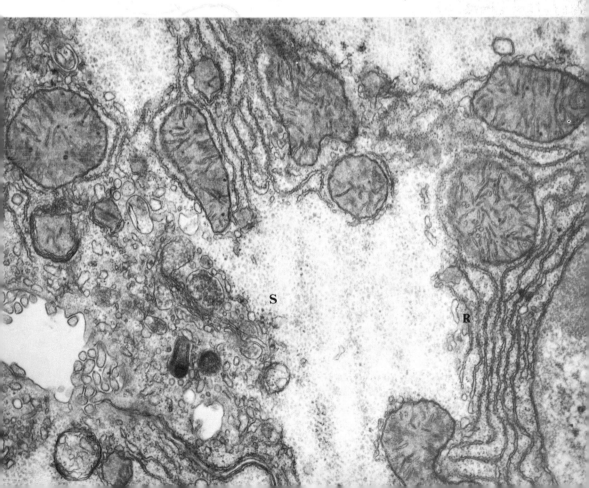

also been observed by light microscopy in ciné films of cultured cells. It consists of a complex branching network of membrane-bound cavities (or cisternae) and is more concentrated in the inner, endoplasmic region of the cell than in the peripheral or ectoplasmic region—hence its name. The form it takes varies considerably; it may show vesicles, tubules, and large flattened sacs or cisternae, and is frequently folded to form layers of membranes.

The branches of this network are intercommunicating and divide the cytoplasm into two main regions, the first being enclosed within the membrane system, the second forming the outer region or cytoplasmic matrix. The enclosing membrane appears in electron microscopy as parallel pairs of the single or unit membranes described above. The endoplasmic reticulum may be rough (granular) because of the attachment of ribosomes to its outer surface (Fig. 5–13) or smooth (agranular) in the absence of ribosomes. The actual amount of endoplasmic reticulum in a cell varies greatly with the age and function of the cell and with the extracellular conditions.

The Golgi region was first observed in certain nerve cells. It was later found to be a characteristic structure in the cytoplasm of all cells except bacteria and mycoplasmas and so must be regarded as necessary to cellular function in plants and animals. In animal cells it consists of an intricate and characteristic pattern of flattened, parallel, and smooth-surfaced double membranes and numerous smaller vesicles (Fig. 5–14), with some orientation with respect to the nucleus. The Golgi region is now thought to be an extension of the smooth-surfaced endoplasmic reticulum. Structures similar to those of the Golgi region have also been observed in plant cells. Botanists usually call them dictyosomes. The most striking difference between the arrangement of these structures in plant and animal cells is that dictyosomes are generally randomly distributed in the cytoplasm (Fig. 5–15).

The term 'microsomes' is often used when talking of cytoplasmic structures. This, however, is simply a convenient way of describing the type of material obtained in the lightest fraction when cells are disrupted and their contents separated by ultracentrifugation; they are not structures which exist in the cell itself. Microsomes are in fact small spherical vesicles formed from disrupted endoplasmic reticulum. The microsomal fraction of homogenized cells may also contain ribosomes.

Studies with the electron microscope have shown that the chief function of the endoplasmic reticulum and the Golgi region is probably the storage, segregation, and finally the transport of substances synthesized by the cell (particularly proteins) for extracellular use.

Palade, in the mid 1950s, concluded that there is a fairly close correlation between the amount of rough endoplasmic

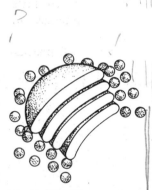

Fig. 5–14 Diagram showing the three-dimensional structure of the Golgi region. (With permission from *The Plant Cell* by William A. Jensen. © 1964 by Wadsworth Publishing Company, Inc., Belmont, California, 94002. Reprinted by permission of the publisher.)

reticulum in a particular cell and the quantity of protein synthesized for export from the cell. Examples of protein-secreting cells in animals are plasma cells which produce antibodies, fibroblasts which produce collagen, and pancreatic cells which produce digestive enzymes. The characteristic arrangement in these cells when they are actively producing protein is a highly developed rough endoplasmic reticulum consisting of large distended cisternae, containing varying amounts of material. In plants, too, those cells which appear to play a part in the nutrition of other cells also contain large amounts of endoplasmic reticulum. The proteins synthesized on the ribosomes somehow penetrate the cavities of the endoplasmic reticulum where they are stored and segregated from the rest of the cytoplasm. This has been clearly demonstrated for the γ-globulin of antibodies by using ferritin as an electron-dense label. In the plasma cells γ-globulin is localized exclusively in the cisternae of the endoplasmic reticulum, presumably being stored there for export.

On the other hand, when the proteins produced are used by or incorporated into the cell itself (as in reticulocytes—the precursors of red blood cells—which synthesize haemoglobin, or muscle cells which synthesize fibrous protein) there is little or no endoplasmic reticulum. The proteins are synthesized on free ribosomes and are stored in the cell matrix.

The exact mechanism of transport to the cell plasma

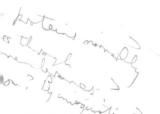

Fig. 5–15 Electron micrograph of the Golgi body (dictyosome) of a spinach plant cell which is actively producing wall material: GA Golgi region; Ve vesicles are produced at the margin of the Golgi region, migrate to the newly forming cell plate, and fuse with the wall; CW cell wall. (With permission from A. D. Greenwood.)

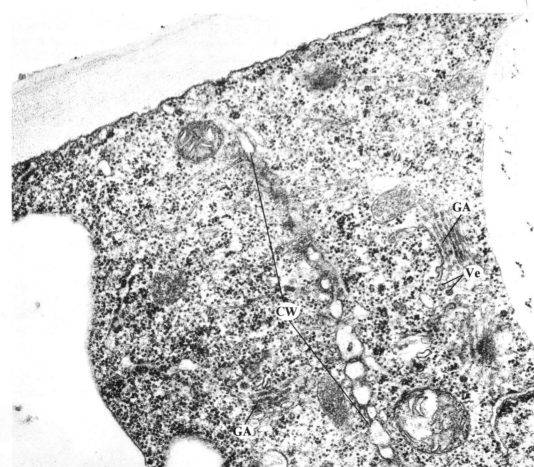

membrane of proteins for export has not yet been definitely established. Probably proteins synthesized on the rough-surfaced endoplasmic reticulum and stored in the cisternae become enclosed in vesicles formed by budding from the smooth endoplasmic reticulum, particularly in the Golgi region (Fig. 5–16). These vesicles then migrate to the cell surface where they release their contents. Pancreatic cells are of particular interest as they synthesize many well-characterized enzymes which play an important role in the breakdown of foodstuffs in the intestinal tract. Enzymes such as trypsin, chymotrysin, lipase, and amylase could have a disastrous effect on the protein, lipid, and carbohydrate content of the synthesizing cells themselves, so that after production and during transport they must be carefully segregated or inactivated. Trypsin and chymotrypsin are rendered harmless to the cell by being made and exported in an inactive form known as a zymogen, which is later converted into the active form in the duodenum. A careful study of the formation and transport of zymogen in the exocrine cells of the guinea pig pancreas has been made by Siekevitz and Palade (1960) and Caro and Palade (1964). They obtained excellent evidence that the inactive form of the enzyme secreted in the cisternae of the endoplasmic reticulum later accumulates in the Golgi region, where zymogen granules are formed. Portions of the Golgi region which contain granules then bud off and migrate to the cell plasma membrane. During the liberation of their contents outside the cell, the vesicular plasma membranes fuse (Fig. 5–16). This may apply to other protein-exporting cells.

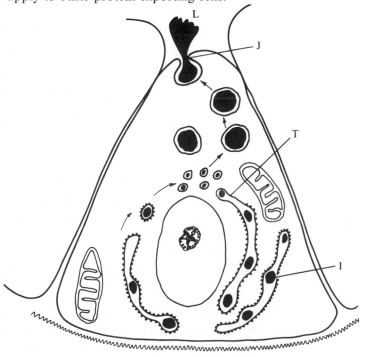

Fig. 5–16 Possible secretion mechanism in pancreatic cells. The vesicles fuse with the plasma membrane and liberate their contents into the lumen of the acinus: J plasma membrane; L lumen of the acinus; I small zymogen granules in the cavities of the endoplasmic reticulum; T transition from rough endoplasmic reticulum to smooth membranes bounding vesicles in the Golgi region. (With permission from G. H. Haggis (ed.), *Introduction to Molecular Biology*. Longmans Green, London, and John Wiley, New York, 1964.)

The situation has been studied in the parenchymal cells of liver. Some of the protein produced is for internal use and some is exported, the most important export product being serum albumin. The work of Bruni and Porter (1965) indicates that the two types of protein are somehow segregated in the Golgi region, but it is not yet known how the cell discriminates between the two so as to retain one type and export the other.

Smooth endoplasmic reticulum in plant cells develops at the surface where the cellulose wall of the cell is being formed, which implies a relationship with polysaccharide metabolism. In liver cells, too, the smooth form is present in regions of the cell which are rich in deposits of the polysaccharide glycogen. However, because biochemical studies have shown that it does not contain the necessary enzymes for glycogen synthesis, smooth endoplasmic reticulum is not now thought to be connected with this process. It is clearly important, however, for the metabolism of lipids and steroids, their storage in vacuoles in animal cells and plastids in plant cells (see below), and their export from the cell.

Since the endoplasmic reticulum forms a vacuolar system which divides the cytoplasm into two compartments, ionic gradients and electrical potentials can be set up across its membranes. This is particularly true of a specialized form of endoplasmic reticulum found in striated muscle fibres, and known as the sarcoplasmic reticulum. This is now regarded as an intracellular conducting system, by means of which impulses are transmitted from the surface membranes deep into the muscle cells.

The relationship between the endoplasmic reticulum and the cell membrane is not yet clear. Robertson (1962) has suggested that the two systems possess intercommunicating channels, and that both the endoplasmic reticulum and the nuclear membrane may be formed by invagination of the cell plasma membrane. The continuous membrane system so formed would provide a direct communication route between the cell nucleus and the extracellular environment (Fig. 5–1). Some workers have suggested that in plant cells also the endoplasmic reticulum is essentially an infolding of the plant cell membrane, although others consider it to be continuous with the membrane of the plant vacuole (see below). Electron microscope studies suggest that the endoplasmic reticulum in plants plays a special role in the interconnection of cells through the cytoplasmic strands called plasmodesmata.

There is evidence to suggest that the endoplasmic reticulum is the origin of the organelles known as primary lysosomes.

The Lysosome. The lysosome is basically a vacuole containing a high concentration of various enzymes which are used in digestive processes inside the cell; it is surrounded by a single outer unit membrane. The existence of lysosomal particles was determined partly by biochemistry and partly by electron microscopy.

In 1949, de Duve and his coworkers were studying enzyme levels in various fractions of homogenized animal cells after ultracentrifugation. They found that the level of the enzyme acid phosphatase in the mitochondrial fraction increased considerably after standing for several days, which indicated that this enzyme was being released from some source during ageing. It was known that cells containing acid phosphatase in large amounts could also contain many phosphate esters, apparently unattacked by the enzyme, so it seemed likely that the enzyme was normally segregated from the rest of the cell inside some type of organelle, and only when the membrane surrounding the organelle became permeable was the enzyme released. Because a rather gentle method of homogenization had been used in the biochemical studies, it was probable that intact particles containing acid phosphatase were still present in the initial homogenate, and that the enzyme was gradually released only on standing.

It was extremely difficult to demonstrate the existence of these particles by means of electron microscopy because they have no characteristic shape or internal structure. Depending upon their state of activity, their size, shape, and internal inclusions may vary widely. However, in 1955 Novikoff clearly identified lysosomes in rat liver cells with the electron microscope. A special method of electron staining has made their detection much simpler, and it is now thought that lysosomes may exist in all animal cells. Vacuoles very similar in appearance have also been observed in some plant cells, but their biochemistry has not been investigated systematically.

The particles identified by electron microscopy contain more than twelve enzymes, including proteases, nucleases, and glycosidases. These are digestive enzymes, which can break down cellular material such as proteins, nucleic acids, and polysaccharides. The particles are therefore known as lysosomes—that is, bodies which can digest or lyse substances. It is clearly essential for the digestive enzymes to be rigorously segregated from the rest of the cellular contents, to which they could cause irreparable damage. For this reason lysosomes are sometimes called 'suicide bags'. They are particularly numerous and large in cells such as macrophages which perform special digestive functions. Lysosomes are also believed to be involved in the fertilization of ova, morphogenesis, ageing, and some diseases; they may play a part in carcinogenesis (the changing of normal cells to cancer cells).

In cultures of cells *in vitro*, droplets of medium can be taken in by the cell and enter the cytoplasm. The way in which this happens is illustrated in Figs 5–1 and 11–11. The cell membrane first invaginates and then buds off internally to form

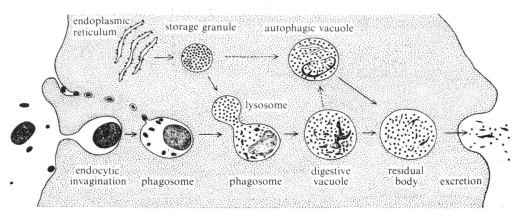

endoplasmic reticulum   storage granule   autophagic vacuole

lysosome

endocytic invagination   phagosome   phagosome   digestive vacuole   residual body   excretion

Fig. 5–17 Diagrammatic representation of the functions of lysosomes. (From Christian de Duve, The lysosome. Copyright © May 1963, by Scientific American, Inc. All rights reserved.)

a vesicle which encloses the droplet of medium in the cytoplasm. This process is known as *pinocytosis* and the vesicle so formed is called a *pinocytotic vesicle*. Some cells, such as Protozoa and certain cells of the reticulo-endothelial system, can take in solid particles in this way, the process being known as *phagocytosis*; the vacuole formed is called a food vacuole or *phagosome*.

Most of the foodstuffs needed by cells are absorbed through the cell membrane in the form of small molecules. Where the materials consist of large complex molecules, they must first be broken down or digested by the cell so that they are in a simpler form which may then be utilized. These large molecules are taken into the cell by the process of phagocytosis and are broken down by the lysosomal enzymes. Figure 5–17 shows how this is done. When the phagosome contacts a lysosome containing its store of enzymes, the membranes of the two fuse to form a digestive vacuole. Here large substances are broken down and the products diffuse through the membrane into the cell, where they may be used in various synthetic processes. Finally the vacuole, now known as a residual body, is left with indigestible material only inside it. This is ejected from some cells into the external environment by a process similar to phagocytosis in reverse. Lysosomes have now been recognized in various cells in all of these stages, from the storage granule to the digestive vacuole and finally the residual body.

The primary lysosomal storage vacuole may originate from the endoplasmic reticulum (Fig. 5–18). Novikoff (1965) has shown that when certain human cells cultured *in vitro* are injured, formation of whorls of endoplasmic reticulum, often associated with increased acid phosphatase levels, results. An increase in lysosomal activity would be expected in the living cells since these scavenge dead-cell debris in the culture. This certainly suggests that under these conditions the endoplasmic reticulum is involved in the formation of lysosomes and the transport of acid phosphatase to them.

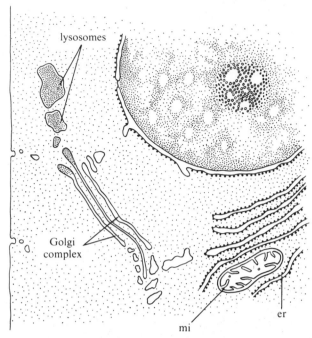

Fig. 5–18 Diagram showing the suggested formation of the primary lysosomal storage vacuole from the Golgi region: er endoplasmic reticulum; mi mitochondrion

Cell injury involves another aspect of lysosomal function, known as autodigestion or autophagy, in which some of the contents of the cell itself are engulfed by the lysosome and broken down. When injury is due to starvation, this process enables the cell to use some of its own materials for the formation of essential substances no longer available from outside sources, without causing irreparable damage.

Besides their clearly defined function in the general digestion of material, lysosomes perform other important biological functions. One of the most striking is concerned with the regression of the tadpole tail during the later stages of the tadpole's development. Weber has shown that regression is the result of lysosomal digestion and is connected with the dense secretion of lysosomes within the tail cells. This process illustrates the point that the death of cells is by no means always detrimental to the general well-being of an organism.

It has also been suggested that lysosomes play a part in the fertilization of ova, the spermatozoa releasing lysosomal enzymes which remove some of the structures surrounding the ovum.

In abnormal circumstances cells may be forced to ingest large quantities of material which they are unable to break down. This material will continue to accumulate in the phagosomes and will eventually harm the cell. This may happen in silicosis, an illness caused by inhaling silica dust. The accumulation of indigestible material also occurs in a certain disease involving disturbance of glycogen storage, where the enzyme necessary to digest this material is missing from the lysosomes.

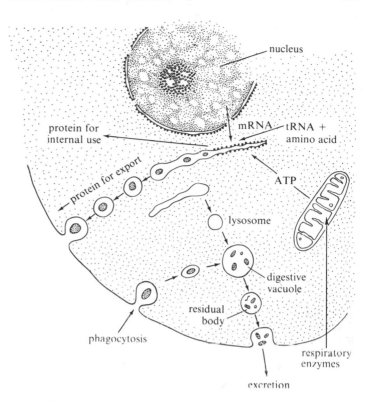

nucleus

protein for internal use

mRNA tRNA + amino acid

protein for export

ATP

lysosome

digestive vacuole

residual body

phagocytosis

respiratory enzymes

excretion

Fig. 5–19 Generalized diagram showing the functions of the various structures in the cytoplasm of animal cells.

Allison and Mallucci (1964) have shown that carcinogenic hydrocarbon compounds are taken up by cells cultured *in vitro* and selectively concentrated in the lysosomes, where they remain for a comparatively long time. Because three of the most effective co-carcinogens (substances which promote the effect of carcinogens) are croton oil, a certain class of detergent, and a high oxygen level, all of which are known to affect the permeability of lysosomal membranes, Allison and Mallucci suggested that lysosomes and the release of lysosomal enzymes may be involved in carcinogenesis.

Interestingly, lack of oxygen also affects lysosomes. It seems that, in cells suddenly deprived of oxygen, the lysosomes may rupture, resulting in the release of enzymes and the death of the cell by autodigestion. Fell and her colleagues (1961) have also shown that excess of vitamin A can cause the release of lysosomal enzymes in the cartilage and bones of animals, and that this probably accounts for the spontaneous fractures and other lesions which result from high doses of vitamin A. On the other hand, cortisone and hydrocortisone (cortisol) have a stabilizing influence on the lysosomal membrane. It may be possible in the future to utilize these two different groups of substances—the stabilizing substances to protect cells, and those which weaken lysosomal membranes to deal with unwanted and harmful cells.

Lysosomes, then, fulfil extremely important functions in cells, and it is now realized that, far from being suicide bags, they form a built-in mechanism for enabling the cell to adapt metabolically to rapidly changing conditions of food supply, oxygen concentration, and other environmental factors which occur in many living systems.

Summary.   Figure 5–19 summarizes the principal dynamic functions of the various structures described above.

5–4
Structures
characteristic of
plant cells

The vacuole.   Emphasis has been placed so far on the marked similarities between the structures found in animal cells and in plant cells (Fig. 5–20). There are, however, certain structures characteristic of most plants which do not occur in animal cells. The most striking of these is the large central vacuole which occurs in many plant cells (Fig. 5–21). The vacuole has a single-layer membrane, the *tonoplast*, which may be formed by expansion of the cisternae of the endoplasmic reticulum. It contains solutions of salts, sugars, and other substances.

In young plant cells there is no large central vacuole, but numerous small vacuoles are present which may not be

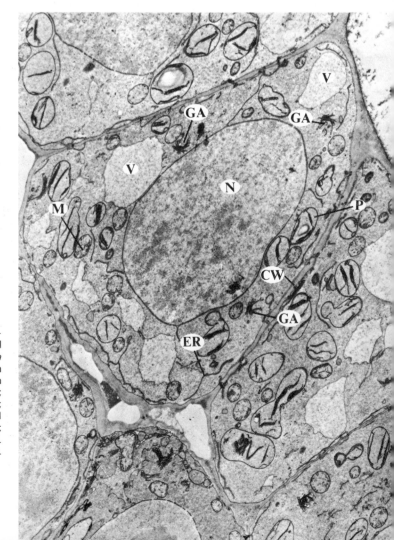

Fig. 5–20 Electron micrograph of a plant cell from a young leaf of *Elodea canadensis*, or Canadian pondweed (magnification ×5,500): CW cell wall; ER endoplasmic reticulum; GA Golgi apparatus; M mitochondria; N nucleus; P plastid; V vacuole. (With permission from R. Buvat.)

visible under the light microscope. As the cell grows, these vacuoles enlarge and coalesce until, in the mature cell, there is frequently only one vacuole which may occupy 90 per cent of the cell volume.

The vacuole carries out important physiological functions for the cell. In Chapter 1 we discussed the sizes and shapes of cells and the importance of the ratio of surface area to volume, since all of the materials required for cellular metabolism can only enter through the cell surface. Because plants have rigid cell walls they cannot solve this problem as animal cells do, by changing their shape, a manoeuvre which gives a large change in surface area for only a small change in volume. The large central vacuole in plants, however, helps to achieve the same result by forcing the cytoplasm to the outer edge of the cell (Fig. 5–21) where it forms a thin layer in which exchange of gases and nutrients readily occurs.

The membrane of the vacuole appears to have special permeability characteristics. The sugars, salts, and other substances which are stored in it are often present in large concentrations, and the high osmotic pressure developed helps to preserve the turgidity of the cell by inducing the movement of water into it. Various pigments may also be present in high concentration in the vacuole, giving it a characteristic colour; for example, the red colour of many flowers is due to a concentration of pigments in the vacuoles of the petals.

Crystals of various materials, in particular of calcium oxalate, are often found inside the vacuole, and the vacuole may serve as a depository for substances which the cell cannot use or no longer needs. Some Protozoa also contain a vacuole which is more complex than that of the plant because it is contractile. By this means the Protozoon probably gets rid of waste fluids or excess water by excreting them to the outside.

The Plastids.   Another group of organelles which is characteristic of plant cells is the plastids (Fig. 5–20). The most widely studied of these is the chloroplast, which is the site of photosynthesis (see Chapter 6), and which contains the green pigment chlorophyll. The chloroplast, like all plastids, is bounded by a double unit membrane and possesses a complex internal lamellar structure. There are various other types of plastid, which assume a wide variety of shapes (Fig. 5–22) and serve different functions in connection with storage and secretion, while retaining a common basic structure.

The second main category of plastid is the leucoplast, which as its name implies (Greek *leukos*, white) is colourless. It is distinguished from the chloroplast and the mitochondrion by the fact that it contains little in the way of internal lamellar and tubular structures and also by the nature of the substances which it elaborates. These substances are starch (when the leucoplast is known as an amyloplast) and lipids.

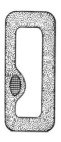

V

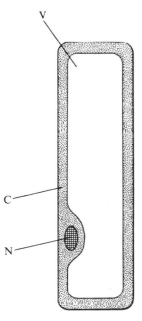

C

N

Fig. 5–21   Development of the large central vacuole in a growing plant cell: N   nucleus; V   vacuole; C   cytoplasm

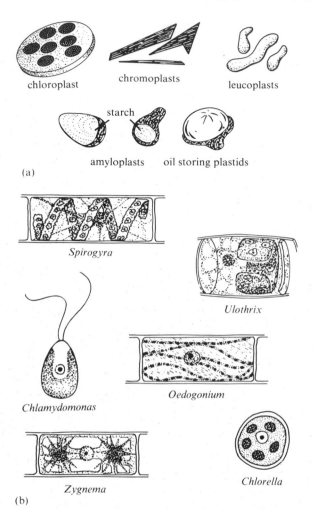

(a)

Spirogyra

Ulothrix

Chlamydomonas

Oedogonium

Zygnema

Chlorella

(b)

Fig. 5–22 Diagrammatic representation of **(a)** various types of plastids, and **(b)** types of chloroplasts found in algae. (With permission from *The Plant Cell* by William A. Jensen. © 1964 by Wadsworth Publishing Company, Inc., Belmont, California, 94002. Reprinted by permission of the publisher.)

Chromoplasts, the third main group, are coloured and contain the second class of pigments which are important in photosynthesis, the carotenoids.

Finally, there are smaller, spherical, structures, assumed to be young or immature plastids, which are known as proplastids. These are thought to be the common origin of all plastids (Fig. 5–23).

Considerable overlapping occurs in the products which plastids elaborate. For example, any of the three groups mentioned above could secrete starch, and examples are known of lipid accumulating in any of the three; chloroplasts and leucoplasts could equally well accumulate reserves of proteins. We assume, therefore, that plastids are all derived from the earlier, immature, proplastid, and that the way in which development takes place depends upon the requirements of the cell. Chromoplasts are probably not derived directly from proplastids, but by a process of secondary late differentiation from leucoplasts or chloroplasts.

Similarities in structure observed between mitochondria

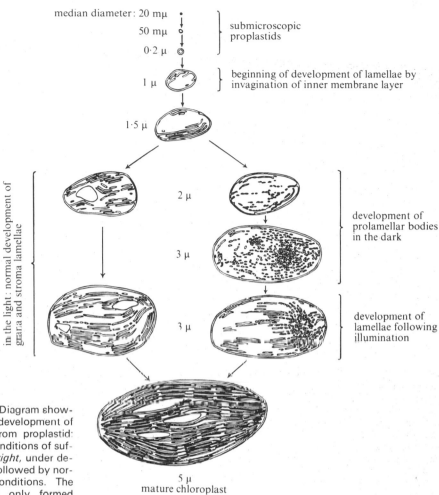

median diameter: 20 mμ

50 mμ } submicroscopic proplastids

0·2 μ

1 μ } beginning of development of lamellae by invagination of inner membrane layer

1·5 μ

in the light: normal development of grana and stroma lamellae

2 μ

3 μ } development of prolamellar bodies in the dark

3 μ } development of lamellae following illumination

5 μ
mature chloroplast

Fig. 5–23 Diagram showing possible development of chloroplast from proplastid: *left*, under conditions of sufficient light; *right*, under deficient light followed by normal light conditions. The lamellae are only formed under adequate light conditions. (Adapted from K. Mühlethaler and A. Frey-Wyssling With permission from R. Buvat.)

and plastids suggest that both are derived from a common primary structure. On the other hand, Strugger (1950), on the basis of his studies with the light microscope, suggested that proplastids contain a 'primary granum' endowed with genetic continuity. This would mean that proplastids must be quite independent of mitochondria. The relationship or otherwise of the two structures remains a controversial subject, and electron microscope studies have so far failed to resolve the problem.

Certain studies by Mühlethaler (1955) have suggested that spherosomes—the extremely small spherical particles rich in lipid and bounded by a single unit membrane, which occur in the cytoplasm of plant cells—closely resemble proplastids, so that the question of their origin must be linked to that of the origin of plastids. Other workers, however, think that spherosomes originate from the endoplasmic reticulum by the pinching off of a very small vesicle. This then enlarges to form a spherosome and frequently a fat body develops, which indicates that spherosomes may be involved in fat production.

However, their true function as well as their origin is uncertain.

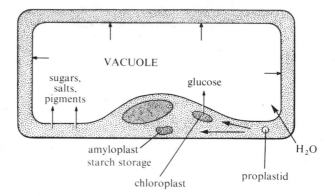

Fig. 5–24 Diagrammatic representation of the principal dynamic functions of the structures found only in plant cells

Summary. The principal dynamic functions of structures found only in plant cells are summarized in Fig. 5–24. For simplicity, structures common to both plant and animal cells are omitted.

Bibliography   Buvat, R., Electron microscopy of plant protoplasm. *Int. Rev. Cytol.*, **14**, 41, 1963.

Davson, H. and J. F. Danielli, *The Permeability of Natural Membranes*, 2nd edn. Cambridge University Press, London, 1952.

de Duve, C., The lysosome. *Scientific American*, May 1963.

Finean, J. B., The molecular organization of cell membranes. *Prog. Biophys. Mol. Biol.*, **16**, 145, 1966.

Haggis, G. H. (Ed.), *Introduction to Molecular Biology*. Longmans Green, London, 1964.

Jensen, W. A., *The Plant Cell*. Macmillan, London, 1964.

Maddy, A. H., The chemical organization of the plasma membrane. *Int. Rev. Cytol.*, **20**, 1, 1966.

Novikoff, Alex B., Mitochondria (chondriosomes), lysosomes, and related particles in J. Brachet and A. E. Mirsky (eds), *The Cell*, vol. 2. Academic Press, New York, 1961.

Palade, G. E., The endoplasmic reticulum. *J. Biophys. Biochem. Cytol.*, Suppl. 2, 85, 1956.

Robertson, J. D., The molecular structure and contact relationships of cell membranes. *Prog. Biophys. Biophys. Chem.*, **10**, 343, 1960.

Robertson, J. D., The membrane of the living cell. *Scientific American*, April 1962.

6 Sources of energy for the cell

Growing cells are continuously synthesizing nucleic acids, proteins, and other cellular components; muscle cells carry out mechanical work; plant cells pump water against the force of gravity. All these and other activities of living cells require a continuous source of energy. Although the ultimate source of energy is sunlight for both the plant and animal kingdoms, energy for immediate use within the cell is provided by the conversion of adenosine triphosphate (ATP) into adenosine diphosphate (ADP). Energy from sunlight is harnessed by photosynthesis and stored in carbohydrate molecules which can be utilized as a source of energy to produce ATP. This work is performed by two types of cellular organelle—chloroplasts, which are found in green plants, and mitochondria (Fig. 6-1), which occur both in animal and plant cells. The way in which these organelles function is the subject of this chapter.

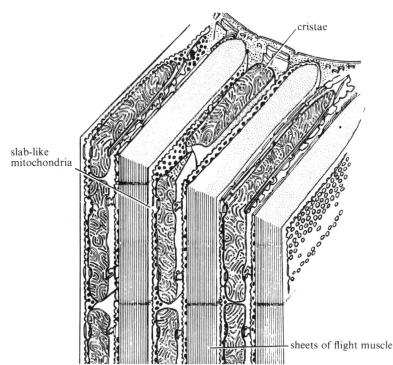

Fig. 6–1 Mitochondria of the flight muscle of a dragon fly, showing profuse cristae. (With permission from D. S. Smith. *J. Biophys. Biochem. Cytol.*, **11**, 119, 1961.)

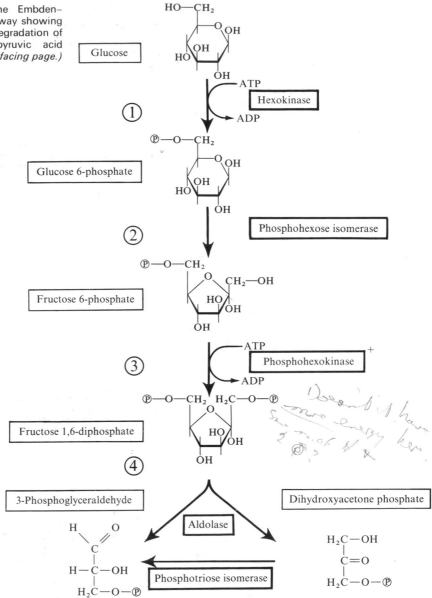

Fig. 6–2 The Embden–Meyerhof pathway showing the stepwise degradation of glucose to pyruvic acid *(Continued on facing page.)*

Glucose

① Hexokinase ATP → ADP

Glucose 6-phosphate

② Phosphohexose isomerase

Fructose 6-phosphate

③ Phosphohexokinase ATP → ADP

Fructose 1,6-diphosphate

④ Aldolase

3-Phosphoglyceraldehyde

Dihydroxyacetone phosphate

Phosphotriose isomerase

Chloroplasts contain a green pigment, chlorophyll, which absorbs quanta of red light. This energy is utilized in the step-by-step synthesis of glucose and is then stored in the form of starch molecules. For the immediate energy requirements of both plant and animal cells, enzymes are available in the cytoplasm and in mitochondria which are able to use glucose molecules derived ultimately from photosynthesis. The steps which occur in the cytoplasm are known as *glycolysis*; those occurring in the mitochondria comprise the Krebs or *citric acid* cycle (in the outer part) and *oxidative phosphorylation* (in the inner part). They lead to an efficient production of ATP.

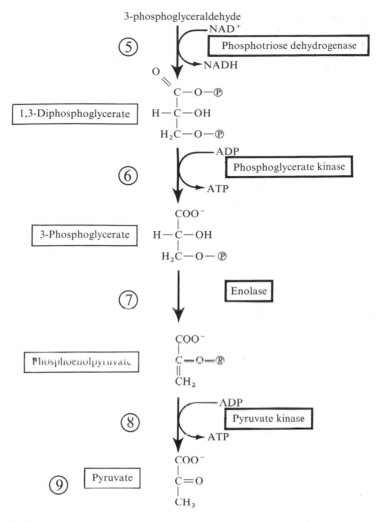

Glycolysis—the breakdown of glucose—occurs in stages during which the molecule is robbed of its energy and loses hydrogen atoms, until it is eventually broken down to carbon dioxide and water. The first stages in the process always take place within the cytoplasm in the cell sap and not within the mitochondria, the enzyme molecules involved being present as soluble molecules.

Embden–Meyerhof pathway. Named after the German biochemists who first worked out the stages, the Embden–Meyerhof pathway is the first example in this book of a *biochemical pathway*. Such a pathway involves a series of stages, each catalysed by a specific enzyme molecule. The products of the reaction of the first stage are passed on to the enzyme molecule of the second stage, and so on.

The individual steps and the names of the enzymes involved in the Embden-Meyerhof pathway are shown in Fig. 6–2. Several of the stages shown in the figure involve

Increase in Free Energy ↑?
means decrease in
Neg. Energy have less stable & formation using up energy ↑

Oxidized
Increase in Free
Energy
[Less stable]
formation absorbs energy?)

Reduced

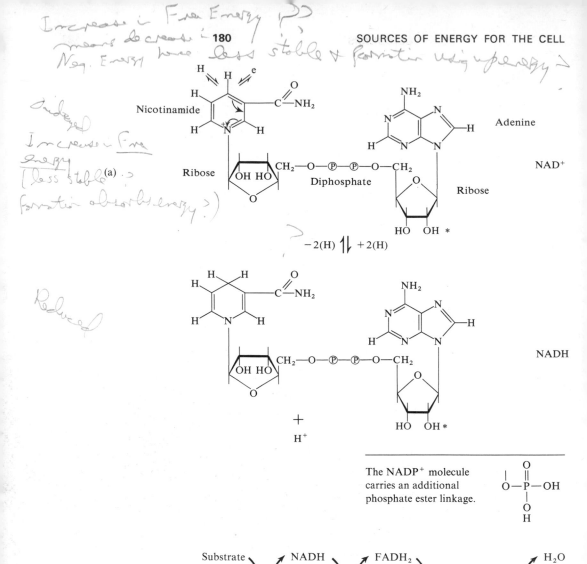

The NADP⁺ molecule carries an additional phosphate ester linkage.

Fig. 6–3 **(a)** The oxidized and reduced forms of nicotinamide–adenine dinucleotide. NAD⁺, the oxidized form, accepts H atoms to give NADH (the reduced form, a donor of H atoms) with an accompanying increase of free energy. (In NADP⁺ and NADPH the additional phosphate is attached to the —OH*. **(b)** The transfer of a pair of H atoms from one coenzyme to another, in this case from NAD to FAD. The final hydrogen acceptor in this case is oxygen. **(c)** Oxidized flavin mononucleotide FMN *Facing page:* **(d)** Oxidized flavin–adenine dinucleotide FAD. **(e)** The hydrogen acceptor part of FAD in the oxidized form. **(f)** The reduced form

(c) Oxidized flavin mononucleotide (FMN)

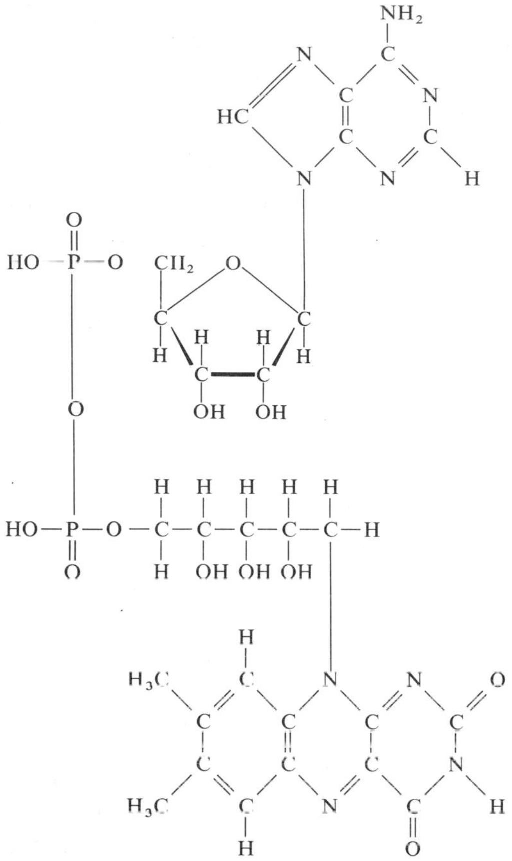

(d) Oxidized flavin–adenine dinucleotide (FAD)

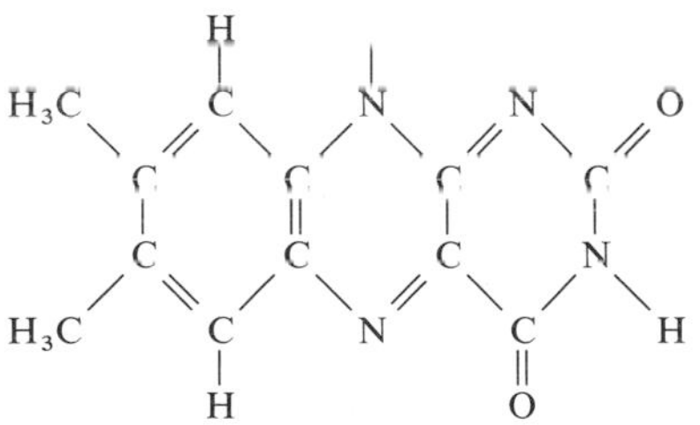

(e) oxidized form

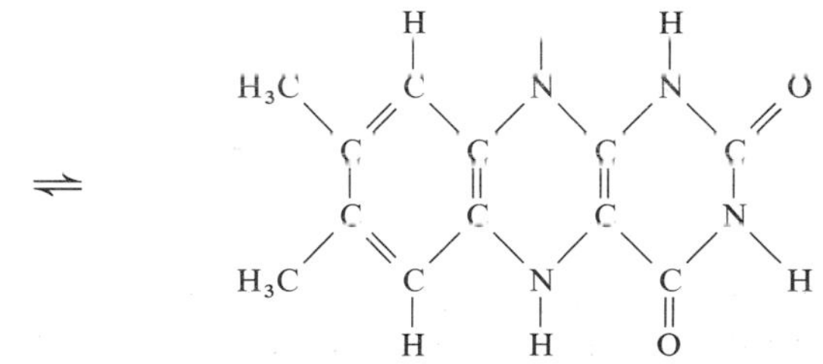

(f) reduced form

oxidation–reduction reactions of the type described in Chapter 2; their efficiency at the surface of the enzyme catalyst is greatly enhanced in some cases by essential cofactors. These are smaller molecules which are unchanged at the end of the reaction and so may be regarded as part of the catalytic process. They can be freely diffusing small molecules or they may be attached directly to the enzyme catalyst. Their function is to combine with an atom or residue which is then passed on to the other main reacting molecule. The cofactor which takes

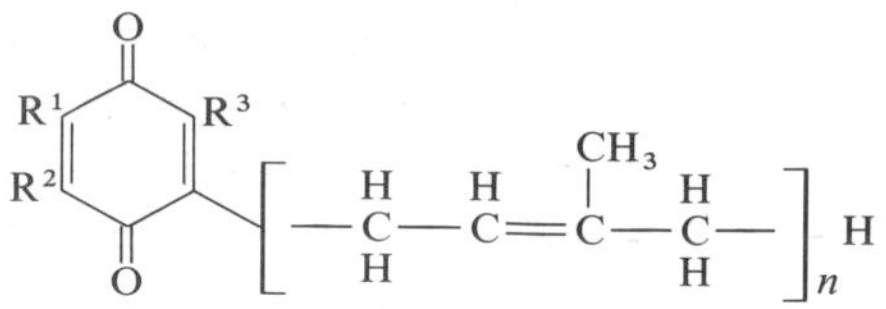

	n	R^1	R^2	R^3
Coenzyme Q_{6-10}	6–10	$-O-CH_3$	$-O-CH_3$	$-CH_3$
Plastoquinone	9	$-CH_3$	$-CH_3$	H
Vitamin K_2	4–7	CH———CH		CH_3

Fig. 6–4 The structure of the coenzyme Q compounds, plastoquinone (chloroplasts), and vitamin K. These compounds can undergo reversible reduction

part in the Embden–Meyerhof pathway is nicotinamide-adenine dinucleotide (NAD^+) (see Fig. 6–3a). The working part, NAD^+, insofar as oxidation–reduction reactions are concerned is the heterocyclic pyridine ring of nicotinamide. The NAD^+ can gain the equivalent of two *electrons* and so be reduced. One electron remains associated with a hydrogen atom attached to the ring while another hydrogen enters the solution as a hydrogen ion, H^+:

$$NAD^+ + 2(H) \rightleftharpoons NADH + H^+$$

As an extremely potent carrier of electrons, NADH takes part in later reactions. The point at which NAD^+ is involved in the Embden–Meyerhof pathway is shown in Fig. 6–2. Some rather similar cofactors are shown in Figs 6–3 and 6–4.

The details of the reaction in Fig. 6–2 (stage 5) are:

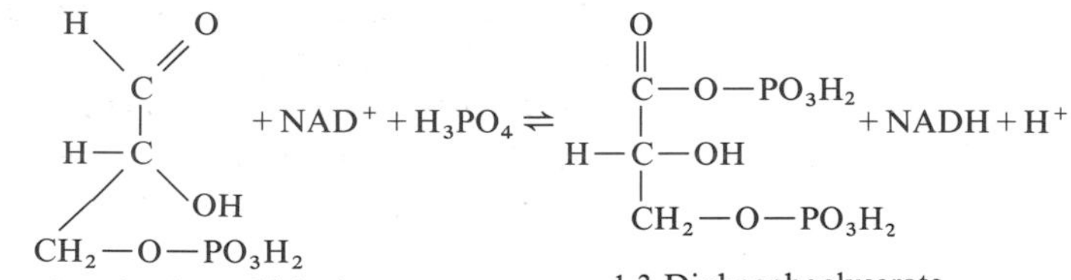

3-Phosphoglyceraldehyde is oxidized to an acid by NAD^+ and an additional phosphate group becomes attached to the molecule. NAD^+ is reduced to NADH. (Another coenzyme which functions in a similar way is $NADP^+$, formed from NAD^+ by the addition of a phosphate group (Fig. 6–3a).)

The steps in the reactions of the pathway are as follows (Fig. 6–2):

1. Glucose is phosphorylated by reaction with a molecule of ATP which is converted into ADP as shown. The reaction is Mg^{2+}-dependent.

2. The glucose 6-phosphate so formed undergoes an *isomeric* transformation to fructose 6-phosphate.

3. This combines with a further molecule of ATP to form fructose 1,6-diphosphate. This reaction is Mg^{2+}-dependent.

4. Fructose 1,6-diphosphate is now split into two 3-carbon molecules: 3-phosphoglyceraldehyde and dihydroxyacetone phosphate. But dihydroxyacetone phosphate is converted by an enzyme into 3-phosphoglyceraldehyde. From this stage there are two 3-carbon molecules taking part in the subsequent reactions.

5. It is in this reaction that the important reduction of the cofactor NAD^+ to NADH takes place with the formation of 1,3-diphosphoglycerate.

6. At the next stage 1,3-diphosphoglycerate is dephosphorylated to form 3-phosphoglycerate with a yield of one molecule of ATP for *each* 3-carbon molecule.

7. 3-Phosphoglycerate is converted into 2-phosphoglycerate which is transformed to phosphoenol-pyruvate.

8. Dephosphorylation of two phosphoenolpyruvate molecules yields two further molecules of ATP.

9. Two molecules of pyruvate are finally produced.

The names of the enzymes are shown on the diagram. The nomenclature will be described in detail in Chapter 7.

The net gain in ATP shows that the degradation of one molecule of glucose yields two molecules of ATP.

In most cells these reactions provide a source of NADH which is utilized later within the mitochondria. In muscle the reactions do not stop at pyruvic acid. Normally, muscle respires aerobically, oxidizing pyruvate via the Krebs cycle (Section 6–2), but during violent exercise oxygen cannot reach the tissues fast enough. In this case muscles obtain extra energy by reduction of pyruvic acid to lactic acid, a molecule of NADH being oxidized.

$$CH_3-CO-COOH+NADH+H^+ \longrightarrow$$
$$CH_3-CH(OH)-COOH+NAD^+$$

The role of NAD^+ as a coenzyme is now clear. In the formation of lactic acid NADH is *oxidized* while pyruvic acid is *reduced*. The NAD^+ now becomes available once more to take part in the pathway (stage 5). Note that oxygen is not required in glycolysis as shown in Fig. 6–2. For this reason the process is called *anaerobic glycolysis*, the net return being two molecules of ATP for one molecule of glucose used. This is very inefficient compared with the 38 produced when the pyruvic acid is used for respiration (oxidation of carbon to CO_2) in the mitochondria. Anaerobic glycolysis, however, is important when a rapid supply of energy is needed—an athlete produces ATP in his muscles in this way during a sprint, but lactic acid is also produced. An excess of lactic acid can reduce the pH of the blood stream to an intolerable level. The muscles have incurred an *oxygen debt*. When violent activity ceases, they continue to use large quantities of oxygen to reconvert lactic acid to pyruvate.

Animal cells use glycogen, a glucose polymer, as the starting point in the pathway.

In yeast, however, which is used for fermentation in the production of beer, the starting point is glucose.

Pyruvic acid is decarboxylated to acetaldehyde as shown:

$$CH_3-CO-COOH \longrightarrow CH_3-CHO$$
$$\searrow CO_2$$

Acetaldehyde is reduced by NADH:

$$CH_3-CHO + NADH + H^+ \longrightarrow CH_3-CH_2-OH + NAD^+$$

The molecule of NADH which is oxidized provides a continuous supply of NAD^+ for the earlier stages in the pathway.

Utilization of fatty acids. Most cells use glucose or its polymer glycogen as a primary source of energy, but some cells such as liver cells in higher animals can use fatty acids. This burning of fats takes place as shown in Fig. 6-5.

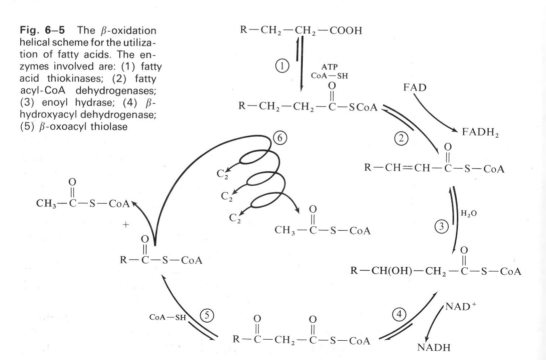

Fig. 6–5 The β-oxidation helical scheme for the utilization of fatty acids. The enzymes involved are: (1) fatty acid thiokinases; (2) fatty acyl-CoA dehydrogenases; (3) enoyl hydrase; (4) β-hydroxyacyl dehydrogenase; (5) β-oxoacyl thiolase

The reactions involve another cofactor, coenzyme A, the structure of which is shown in Fig. 7–16. The SH group of this complicated molecule is the reactive group. The sulphur esters which it forms with pyruvic acid or fatty acids differ from oxygen esters, because sulphur has little affinity for electrons. This characteristic accounts for the biochemical reactivity of acetyl-coenzyme A (Fig. 6–8).

The stages of the reaction are as follows:

1. The fatty acid forms a thioester with coenzyme A. (Thiokinase type of reaction.)

2. Two hydrogen atoms are removed and FAD is reduced to $FADH_2$. (Acyl dehydrogenase type of reaction.)

3. An enzyme brings about the addition of one molecule of water. (Enoyl hydrase type of reaction.)

4. Further loss of hydrogen converts NAD^+ to NADH. (β-Hydroxyacyl dehydrogenase type or reaction.)

5. An enzyme (β-Oxoacyl thiolase) catalyses the reaction, leading to the production of acetyl-coenzyme A plus the thioester of the remaining fatty acid chain.

6. This chain now re-enters the cycle, being combined with coenzyme A but two carbon atoms shorter than before. Each cycle removes a *two-carbon unit* from the fatty acid. This process continues step by step until the molecule of fatty acid has been utilized in these fascinating reactions, commonly called the β-oxidation helix.

Amino acids are also used sometimes as a source of energy; they are first deaminated (losing nitrogen) to form keto acids (see Chapter 7):

$$R-\underset{\underset{O}{\|}}{C}-COOH$$

6–2
Mitochondria

During the latter half of the 19th century scientists working in very different fields studied what is basically the same biological problem: cellular respiration. Cytologists and biochemists had been working on this problem for well over a hundred years before they realized in the 1940s that mitochondria are responsible. The story began in the latter part of the 18th century when Lavoisier demonstrated that living animals require oxygen for respiration.

These early studies on living organisms were of great importance to chemistry. Liebig, the German chemist, studied animal respiration and metabolism, which helped in the development of the law of constant proportions. The mitochondria were first observed by Kölliker in 1850 as granular structures in striated muscle. In 1888, he isolated them from insect muscle (which contains many slab-like mitochondria —see Fig. 6–1) and showed that they swelled in water and possessed a membrane. Flemming, also working in the 1880s, demonstrated the presence of thread-like bodies in the cytoplasm of many other cells, using improved methods of staining. Benda used crystal violet as a stain and was the first to use the term 'mitochondria' (Greek: *mitos*—a thread; *chondros*—a grain).

The really big advance came when, in 1898, Michaelis showed that mitochondria could bring about colour changes in a dye by oxidation–reduction reactions, like those observed in inorganic chemistry when iron salts are changed from yellow to green by reduction. For a number of years biochemists had been studying oxidation reactions in cell extracts without considering their localization in particles. A development which led to much fruitful investigation was made by Claude Bernard, the great French physiologist, between 1857 and 1865, who showed that animal respiration can be inhibited by cyanide and by carbon monoxide.

The first possible link between the early biochemical studies and cytology was made by Kingsbury in 1912, who suggested that mitochondria were the sites of cellular oxidations, but the suggestion was neglected. A closer link was demonstrated by Keilin in 1923, who showed that effects of cyanide and carbon monoxide on respiration could be demonstrated in intact insect muscles. It was the advent of improved methods of centrifugation for the isolation of subcellular particles by Claude and his associates in the 1940s, together with the fine electron microscopy carried out at the Rockefeller Institute in New York by Porter and Palade, which finally bridged the gap between cytology and biochemistry. This has led to great advances in our knowledge of the structure and biological function of these organelles.

Structure. All mitochondria are built on the same basic principle (Fig. 6–6). They possess an outer and an inner membrane. The inner membrane is folded inward at various points to form the cristae which effectively increase the surface for the attachment of enzyme molecules; these can thus be packed more densely in the mitochondria. The cristae are not attached to the inner membrane at all points along their length as might be suggested by the section shown in Fig. 6–6b. Figure 6–1 shows them to consist of sheets of double membrane. They contain windows (fenestrae) which allow diffusion of ATP to the exterior of the mitochondria.

The mitochondrial membranes, like other cell membranes, consist mainly of phospholipid (Chapter 5), but there is apparently insufficient lipid to generate a bimolecular leaflet. Sjöstrand in fact believes that the structure consists of spherical bodies called micelles, having a lipid core and a protein coat. Fleischer has extracted 90 per cent of the lipid from mitochondria with an acetone–water mixture. The basic structure remains intact, indicating that protein molecules provide much of this structure.

The pioneering work of Humberto Fernandez-Moran established that discrete particles exist on the membranes (Figs 6-6a and 6-6b). The particles are about 100 Å in diameter, and those on the outer membrane are attached firmly to the surface while those on the inner membrane are on stalks and project into the interior. The space between the two membranes is filled with fluid. The interior space between the cristae of the mitochondrion contains protein and a small amount of DNA (see Chapter 9).

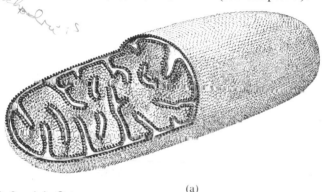

(a)

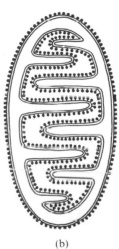

(b)

Fig. 6-6 (a) Cut-away drawing of a typical mitochondrion showing the two membrane layers separated by a fluid-filled space called the intrastructure space. The space within the inner membrane is known as the interstructure space. (From David E. Green in J. Brachet (ed.). Copyright © January 1964 by Scientific American, Inc. All rights reserved.) (b) Diagrammatic representation of the structure of a mitochondrion. The particles on the outer surface of the outer membrane and the inner surface of the inner membrane play an important role in oxidative metabolism. (Reproduced from *Biological Science* by William T. Keeton. Illustrated by Paula DeSanto Bensadoun. Copyright © 1967 by W. W. Norton & Company Inc., New York, N.Y. Used with permission of the publisher.)

Mitochondria were first observed in living cells by Lewis and Lewis in 1914, using time-lapse cinematography. The way in which mitochondria are arranged in cells having a large demand for ATP is demonstrated by the insect muscle mitochondria in Fig. 6-1. Their membrane structure enables ATP produced inside the mitochondria to diffuse rapidly to the functional units. In the insect muscle, the units are the fibrils of contractile protein which convert energy from ATP into mechanical work. Numerous mitochondria also exist in kidney glomerular cells, which are actively engaged in the transport of blood plasma (minus proteins) into the kidney tubules. In most cells mitochondria are generally distributed throughout the cytoplasm. The numbers vary with the cell type; in liver cells there are about 1,000 mitochondria per cell.

Krebs or citric acid cycle. The round particles on the outside of the outer mitochondrial membrane contain a group of enzymes which are capable of utilizing pyruvate in the Krebs cycle. This is a cycle of biochemical reactions in the course of which pyruvate is degraded to CO_2, NADPH being produced.

The NADPH is used for further reactions within the mitochondria which eventually lead to the production of ATP.

The discovery of this cycle was a landmark in the history of biochemistry. Szent-Györgi and Martinus worked on the problem, but Sir Hans Krebs finally elucidated the nature of the cycle in 1937 and later received the Nobel Prize for his discovery. It is called a cycle because citrate or citric acid * in Fig. 6–7), the nominal starting point of the reaction, is produced again at the end of the pathway from oxaloacetate, utilizing acetyl-coenzyme A. The reaction then restarts.

The various stages of the Krebs cycle are listed below:

in cytoplasm

Are all the enzymes for whole reaction in only on each particle?

1. Pyruvate produced in the Embden–Meyerhof pathway is first converted to acetyl-coenzyme A by combination with coenzyme A. (Catalysed by pyruvate oxidase.)

2. Oxaloacetate combines with acetyl-coenzyme A to form citrate, the nominal starting point of the cycle. (Catalysed by a condensing enzyme.)

3. Citrate loses one molecule of water to form aconitate. (Catalysed by aconitase.)

4. On addition of water this is converted to isocitrate. (Catalysed by isocitrate dehydrogenase.)

5. The oxidation of isocitrate to oxalosuccinate reduces $NADP^+$ to NADPH. (Catalysed by an isocitric enzyme.) The molecule of NADPH takes part in other mitochondrial reactions leading to the production of ATP.

6. Oxalosuccinate loses carbon dioxide and forms α-oxoglutarate. (Catalysed by the same enzyme as 5.)

7. A reaction again involving coenzyme A leads to the formation of NADH from NAD^+ with liberation of CO_2. (Catalysed by α-oxoglutarate oxidase.) Succinyl-coenzyme A is formed.

8. The conversion of succinyl-coenzyme A to succinate leads to the formation of one molecule of guanosine triphosphate (GTP) from guanosine diphosphate (GDP) and release of coenzyme A. (Catalysed by succinate thiokinase.)

9. Part of the succinate so produced is also utilized in later stages for the formation of ATP.

The remainder of the succinate is converted to fumarate. (Catalysed by succinate dehydrogenase.)

10. And then to malate, by addition of water. (Catalysed by fumarase.)

11. Malate is oxidized to oxaloacetate with the formation of one more molecule of NADH from NAD^+. (Catalysed by malate dehydrogenase.)

And so the cycle is repeated, the oxaloacetate again combining with acetyl-coenzyme A to produce citrate.

The reactions in a number of cases involve more steps than are shown in Fig. 6–7; there are, for instance, several mitochondrial enzymes involved in the formation of acetyl-coenzyme A from pyruvate. (See the bibliography for references to the detailed biochemistry.)

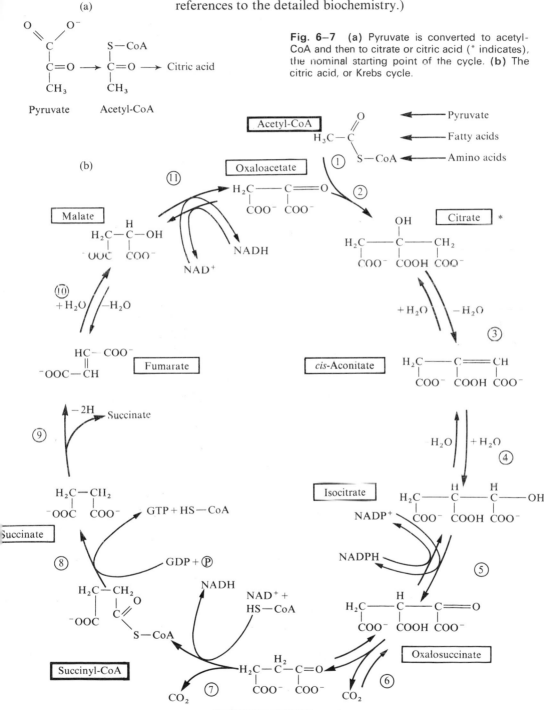

Fig. 6–7 (a) Pyruvate is converted to acetyl-CoA and then to citrate or citric acid (* indicates), the nominal starting point of the cycle. (b) The citric acid, or Krebs cycle.

$$RCH-\overset{O}{\underset{\delta-}{C}}-S-CoA \leftrightarrow RCH_2-\overset{O}{C}-S-CoA \leftrightarrow RCH_2-\overset{O^{\delta-}}{\underset{\delta+}{C}}-S-CoA$$

nucleophilic   electrophilic
character   character

Fig. 6-8 Thioesters, such as acyl-coenzyme A molecules, show dual reactivity. They tend to have a fractional positive charge on the carbonyl carbon atom (i.e., it is an electrophilic centre), and a fractional negative charge on the adjacent carbon atom (a nucleophilic centre)

The structure of mitochondria is most important in relation to their function. Glycolysis takes place in the cell cytoplasm because the product of one reaction merely needs to be passed on to the next enzyme. It can equally well occur in the test tube. But an examination of Fig. 6-7 shows that the Krebs cycle leads to the gross production of NADH and succinate $^-OOC-CH_2-CH_2-COO^-$. These are small molecules in a reduced state and must be protected from the oxidative conditions in the cytoplasm. They pass through the space between the two membranes to reach the inner particles which carry out respiration. The compartment between the two membranes is therefore in a reduced state and a large potential difference (redox potential) exists between this space and the cell cytoplasm. This can only be maintained by the protective action of the lipid membranes. Because the series of reactions is a cyclic process there are clear advantages in having an organized group of enzymes in which the products can be passed from one enzyme to another. This is achieved in the outer-membrane particles of the mitochondria.

During the various stages of the Krebs cycle we see that for one molecule of pyruvate, two molecules of NADH, one of NADPH, one of GTP, and one of succinate are produced. All

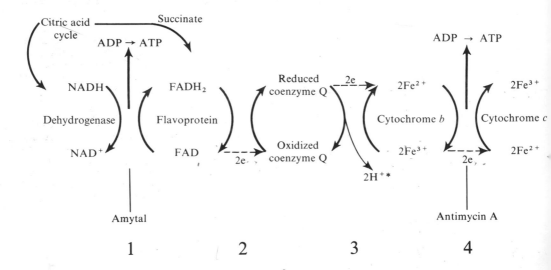

of these are utilized as energy carriers. Ultimately the energy is locked in ATP. For example GTP converts ADP to ATP by transfer of phosphate.

Respiratory chain. The respiratory chain takes in succinate and NADH from the Krebs cycle enzymes; together with oxygen, the chain produces many molecules of ATP and finally carbon dioxide and water. As the electrons carried by NADH and succinate travel down the chain they give up their energy, which is harnessed for the conversion of ADP to ATP.

The enzymes which carry out these oxidation reactions are localized in the stalked particles of the mitochondria. They are probably arranged in a definite sequence like a mass production factory, as shown in Fig. 6–9. The start of the chain is localized at the base of the stalks, where it can collect NADH molecules diffusing across the outer membrane space from the Krebs cycle. Reduced molecules enter the chain while those leaving it are fully oxidized to CO_2 and H_2O. Each step along the chain involves a partial oxidation, and a change of redox potential. The percentage of reduced molecules is indicated at various stages in a living cell in Fig. 6 10. To maintain these electrical redox potentials at the steps in the chain a highly organized structure is needed in the stalk particles, involving lipid molecules as permeability barriers (Chapter 8), as in the Krebs cycle.

Work by electron microscopists, particularly Fernandez-Moran, and by the biochemists Keilin, Hartree, Lehninger, and King, has greatly advanced our understanding of the respiratory chain. The stalked particles have been partially broken down into their constituents and in some cases reconstituted from the partially separated enzymes. When broken off the mitochondria, the stalked particles become spherical and have a diameter of 150 Å. This corresponds to a molecular weight of 1 4 million, which would be expected for

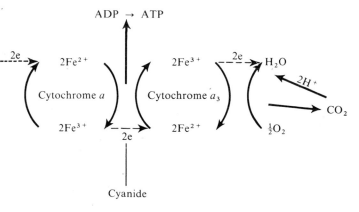

Fig. 6–9 Diagrammatic representation of the successive oxidation–reduction reactions in the respiratory chain. The exact sites of formation of ATP are not yet clearly known. *Indicates hydrogen ions liberated early in the chain; these are used at the final stage in the formation of water. (Amytal, antimycin A, and cyanide are poisons, the use of which is described in text.)

6   7   8

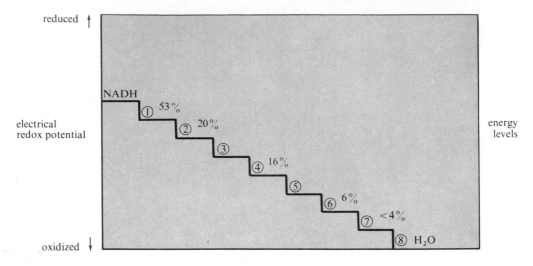

Fig. 6–10 The percentage of reduced molecules in the various stages of the respiratory chain (not to scale), ①–⑧ correspond to the numbered stages of the chain in Fig. 6–9

an assembly of the enzymes which catalyse the reactions. The isolated particles can carry out respiration and electron transport. If the lipid is removed from the particles by gentle treatment at low temperature they cease to function; on replacement of the lipid they are again able to carry out electron transport. This shows clearly the importance of a structure containing compartments, so that the steps in the chain can be separated both in chemical and electrical potentials.

Note, however, that the steps shown in Fig. 6–9 are not equivalent to an electric current. In this respect the use of the term 'electron transport' is misleading. Each step does not differ in basic chemical terms from the oxidation–reduction reactions encountered in inorganic chemistry. What is special about these biochemical reactions is the way the enzyme molecules are arranged within the stalk particles together with lipid carriers. This enables oxidation to proceed in a series of controlled steps, each step being used to transfer energy to molecules of ATP, so that it may later be employed by the cell.

The stages in the respiratory chain reactions are:

1. NADH and succinate ($^-OOC-CH_2-CH_2-COO^-$) molecules pass through the space between the inner and outer membranes of the mitochondria—they act as links between the Krebs cycle enzymes which are on the outside and the respiratory chain enzymes which are on the inside. NADH is oxidized to NAD^+ and returns to the Krebs cycle. The reduction of another coenzyme, flavin–adenine dinucleotide (FAD), thus takes place. The structure of the oxidized and reduced form of this coenzyme is shown in Fig. 6–3e and 6–3f. It is permanently bound to the enzyme catalyst with which it is associated in its specific reaction. This enzyme is therefore called a flavoprotein, because FAD contains flavin (Fig. 6–3d).

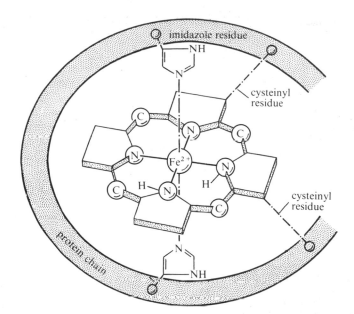

Fig. 6–11 Diagrammatic representation of a flat porphyrin ring structure inside the tertiary structure of a protein. Imidazole residues from histidines in the protein presumably coordinate with the iron of the porphyrin

2. The next stage involves coenzyme Q (also known as ubiquinone) but the exact nature of this stage is still unknown. The structure is shown in Fig. 6–4.

3. The third stage involves the cytochromes—the iron-containing enzymes. The iron atom lies at the centre of a porphyrin ring as shown in Fig. 6–11. Proteins of this type are involved in oxidation reduction reactions and depend on the change $Fe^{3+} + e \rightleftharpoons Fe^{2+}$. The porphyrin ring has the structure:

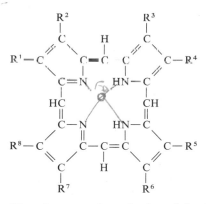

The rings can be substituted in the following way. In cytochrome c, R^1 is CH_3, R^2 is $CH{=}CH_2$, R^3 is CH_3, R^4 is $CH{=}CH_2$, R^5 is CH_3, R^6 is $CH_2{-}CH_2{-}COOH$, R^7 is $CH_2{-}CH_2{-}COOH$. The porphyrin ring is flat and the R groups associate with the side-chains of the protein, as also occurs in myoglobin (Chapter 4). This association gives the particular character to the molecule. The rings are of a resonating type with conjugated double bonds. Various arrangements of the bonds correspond to equal energy and the molecule is said to resonate between these conformations. For this reason we can consider the iron atom to be linked simultaneously to all the nitrogen atoms, which are equivalent. The presence of the protein together with the resonance of the rings gives the iron atom unusual properties so that it can be reduced and re-oxidized under specific biological conditions.

There are in fact at least five types of cytochrome in animals, known as cytochromes b, c_1, c_2, a, and a_3 (see Fig. 6–9). Each differs slightly in its redox potential because of the arrangement and structure of the porphyrin rings and protein. Cytochrome a_3 carries out the final stage of transferring the electrons to oxygen and combining it with the hydrogen ions liberated earlier in the chain, to form water.

This is the only stage in aerobic respiration at which oxygen is needed. The isolation and characterization of the complex molecules involved has been carried out by Green, Chance, Keilin, and others.

Oxidative phosphorylation.   Figure 6–9 shows that the energy liberated at various stages along the respiratory chain is used to produce ATP from ADP. This is an endergonic reaction and energy is stored in ATP. The exact mechanism of these reactions is still not fully understood. One method of investigation has been to 'poison' certain stages with Amytal, antimycin A, or cyanide. These poisons appear to act by combining with specific groups in the chain responsible for electron transport. The process of ATP formation is called *oxidative phosphorylation*, because phosphate is added to ADP using energy from oxidation.

$$ADP + P_i + energy \longrightarrow ATP \qquad (i = inorganic)$$

These reactions involve several intermediate stages, which have not been identified.

Wadkins and Lehninger have postulated the following mechanism:

$$Ⓟ—X + E \rightleftharpoons Ⓟ—E + X$$
$$Ⓟ—E + ADP \rightleftharpoons ATP + E$$

where E is the terminal enzyme of a series of reactions; X is the unidentified intermediate.

We see from Fig. 6–10 that there is a stepwise decrease in the number of reduced molecules of the various coenzymes and cytochromes as we pass along the respiratory chain. The change in the redox potential at each step is an indication of the net energy change occurring in the synthesis of ATP at various stages.

Figure 6–12 summarizes the various stages in the utilization of glucose by the cell to produce ATP. The net yield of 38 mol of ATP per 1 mol of glucose is shown.

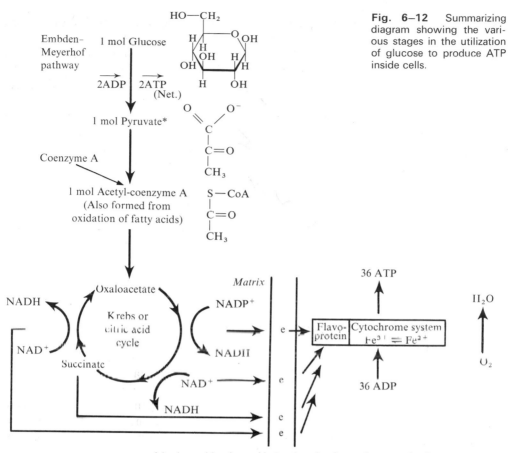

Fig. 6–12 Summarizing diagram showing the various stages in the utilization of glucose to produce ATP inside cells.

*Amino acids after oxidative deamination and transamination may also be metabolized by this pathway.

6–3
Chloroplasts and
photosynthesis

Structure of chloroplasts. Nearly 200 years ago, Joseph Priestley discovered that green plants 'instead of affecting the air in some manner with animal respiration, reverse the effort of breathing'. In 1779 the Dutch physicist Jan Ingenhousz suggested that a green plant splits apart the carbon dioxide that it has absorbed from the air. It throws out 'oxygen alone and keeps the carbon to itself as nourishment'.

Photosynthesis requires light, but Winogradsky discovered in 1880 that certain bacteria, which obtain their energy from chemical sources in the absence of light, can utilize carbon dioxide in the dark. Further, Engelmann in Utrecht found that purple bacteria, which carry out a kind of photosynthesis, do *not* liberate oxygen. Consequently photosynthesis remained mysterious until Van Niel (1949) put forward a new hypothesis which has proved most fruitful. He proposed that the action of light was to split water molecules and not carbon dioxide. The work of Arnon and Calvin, Benson and Bassham (since 1948) in the United States has led to a greatly increased understanding of photosynthesis.

The important stages of photosynthesis as they occur

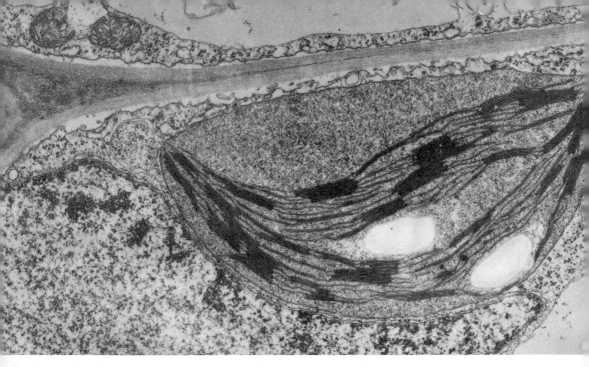

Fig. 6–13 A low-power electron micrograph of a chloroplast. The clear regions are starch-containing plastids. Magnification ×22,000 (With permission from A. Greenwood.)

The way aren't plants to real since most lights.. "
Since most ?
↳ green.

in photosynthetic bacteria and green plants will be outlined below. The main steps are:

Light reaction

1. Absorption of a red light quantum by the green pigment chlorophyll.

2. Transfer of this energy by the transfer of an electron which reduces a cofactor.

3. Utilization of the energy in an electron-transfer chain similar to that in mitochondria to form ATP from ADP.

In bacteria the reactions are cyclic (cyclic phosphorylation). In green plants non-cyclic phosphorylation occurs because there is a flow of electrons (derived from OH radicals in water).

Dark reaction

This involves various biochemical pathways leading to the formation of glucose utilizing the chemical energy from NADPH (green plants), FADH (bacteria), and ATP, produced during the light reaction.

There are striking similarities between the functions of chloroplasts and mitochondria, as will be shown later, but this has only become clear in comparatively recent years.

From the earliest days of plant cytology it was realized that cells which are capable of photosynthesis contain chloroplasts (organelles which contain a green pigment). Meyer in 1883 and Schimper in 1885 suggested that chloroplasts were finely granulated—that is, they consisted of a colourless background material (*stroma*) in which minute granules (*grana*) were imbedded; the grana were on the boundary of microscopic visibility. Later studies of the

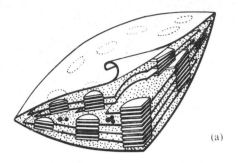

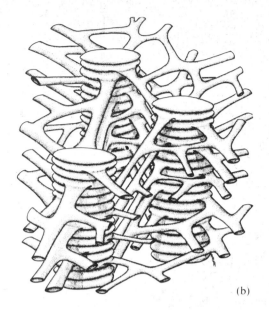

Fig. 6–14 **(a)** Diagram of a chloroplast showing the inner structure with the grana in stacks perpendicular to the surface. (With permission from D. von Wettstein.) **(b)** Diagram of the ultrastructure of three grana showing the anastomosing tubules which join some of the membrane components. (With permission from T. Elliot Weier.)

(a)

(b)

colloidal properties of chloroplasts did not support this view. But Heitz in 1935 photographed chlorophyll grana in the actual living cells of the leaves of higher plants. The presence of chlorophyll granules was made even more conspicuous by photography in red light, which is absorbed by chlorophyll. Heitz was able to show that the grana were platelets; in side view they are seen as sheets in the light microscope. But, as with the other cytoplasmic organelles, the use of the electron microscope—originally by Kausche and Ruskey in 1940— led to the final clarification of the structure.

Figure 6–13 is a low-power electron micrograph showing that sheet-like structures lie along the length of the grana. Strugger in 1951 carried out swelling experiments similar to those on giant chromosomes described in Chapter 3. He showed that the whole chloroplast swells sideways and is composed of lamellae rather like a pile of pennies. The number and shape of the lamellae vary greatly in different cell types. In *Euglena gracilis* there are 20 parallel lamellae in each granum and the lamellae are 180–320 Å thick with 300–500 Å spaces between them. In *Amphidinium elegans* there are long bundles of lamellae which traverse the whole section of the chloroplast. Usually in the algae, grana are absent and the lamellae are long and parallel, about 30 Å thick.

Notice that the lamellae are joined together by expansion of certain regions of their double membranes to form an anastomosing network (generally called *fret* membranes), as in Fig. 6–14b. The space between is filled with a *stroma* (shown stippled in Fig. 6–14a). All the chloroplast lamellae are composed of *double* membranes, unlike the plasma membrane of the cell which is a single membrane. The membranes are composed mainly of phospholipid.

Fig. 6–15 Freeze-etched picture of quantasomes on the chloroplast membrane of *Nicotiana tobaccum.* Magnification ×120,000. (With permission from K. H. Moor.)

Composition of chloroplasts. Chloroplasts contain protein, lipid, and chlorophyll. An examination of plant cells in the ultraviolet microscope shows fluorescence caused by chlorophyll only in the grana. Modern electron microscopy using very high magnification up to 300,000, by Moor and Weier, and by Benson, has revealed the presence of fine granules in the grana membranes. These have been called quantasomes because they are thought to be the basic units of photosynthesis (Fig. 6–15). Moor used the method of freeze-etching in which successive layers are broken off the chloroplasts rather like peeling an onion.

There are at present two different ideas about the basic structure of the grana membranes. In the first, the granular model, shown in Fig. 6–16a (Frey-Wyssling), the membranes consist of a string of particles. These are dense on the outside, probably with a lipid coat, and protein inside. The other idea is that the particles are attached to and penetrate a unit membrane as shown in Fig. 6–16b (Mühlethaler; Park and Pon). In this model the particles are widely spaced on the membrane and those on opposite membranes overlap each

Fig. 6–16 (a) Grana membranes consisting of a string of particles; suggested structure by A. Frey-Wyssling. (With permission from A. Frey-Wyssling.) **(b)** The particles b are attached to a unit membrane a; suggested structure by R. G. Park and N. G. Pon. (With permission from R. B. Park.)

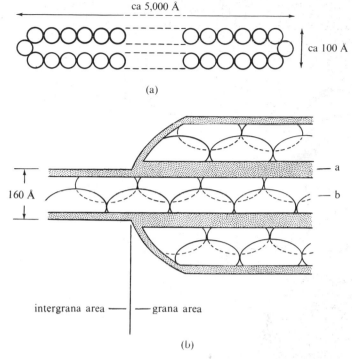

ca 5,000 Å

ca 100 Å

(a)

160 Å

intergrana area ——|—— grana area

a

b

(b)

other. Interpenetration is more fully shown in Fig. 6 17b. Grana are thought to be composed of layers of orientated chlorophyll molecules sandwiched between electron donors and acceptors. Absorption of light by chlorophyll results in a steady flow of electrons between the two.

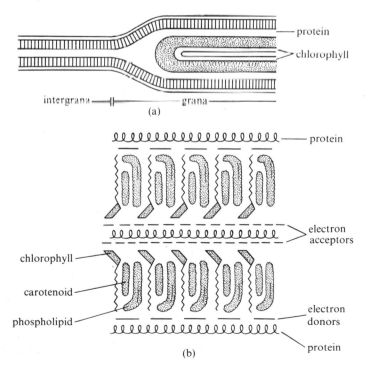

protein

chlorophyll

intergrana ——|—— grana

(a)

protein

electron acceptors

electron donors

protein

chlorophyll

carotenoid

phospholipid

(b)

Fig. 6–17 (a) The location of chlorophyll within the grana region. **(b)** Penetration of phospholipid by chlorophyll molecules, showing one possible way in which electron donors and acceptors could be arranged. (Both after J. Paul, with permission.)

Fig. 6–18　The structure of chlorophyll a and of its phytol side-chain

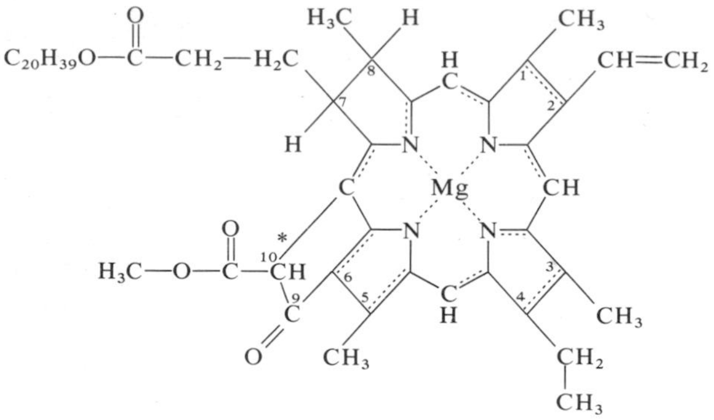

Chlorophyll a

$$C_{20}H_{39}OH = CH_3-\overset{\overset{\displaystyle CH_3}{|}}{CH}-CH_2-[CH_2-CH_2-\overset{\overset{\displaystyle CH_3}{|}}{CH}-CH_2]_2-CH_2-CH_2-\overset{\overset{\displaystyle CH_3}{|}}{C}=CH-CH_2-OH$$

Phytol

Chemical structure and function of chlorophyll. There are two types of chlorophyll, chlorophyll a and chlorophyll b. On examining the structure of chlorophyll a we immediately notice its relationship to the cytochromes just described in mitochondria. There is a porphyrin ring with alternating arrangements of the double bond leading to resonance (Fig. 6–18). A magnesium atom replaces the iron atom of the cytochromes. One of the side-chains of the ring is extremely long, giving the chlorophyll molecule its tadpole-like appearance. The complexity of the ring system is further complicated by condensation at the point marked * with acetic acid and formaldehyde. Chlorophyll b differs from chlorophyll a only in the substitution of methyl group on the porphyrin ring.

　　Chlorophyll absorbs light in the red region of the spectrum (Fig. 6–19) which causes the green colour of green plants (white minus red gives green). The other pigment molecule present in green plants is carotene. This consists of a carbon chain with alternate single and double bonds, the terminal groups of which are ionone rings instead of porphyrin. Their main function appears to be to modify the absorption spectrum of chlorophyll:

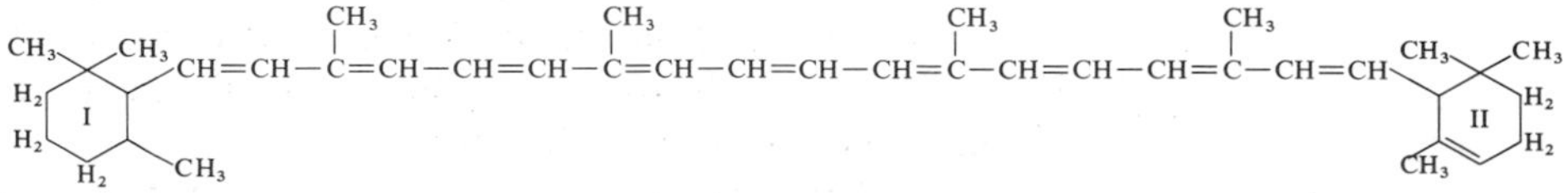

I　β-ionone ring.
II　α-ionone ring.

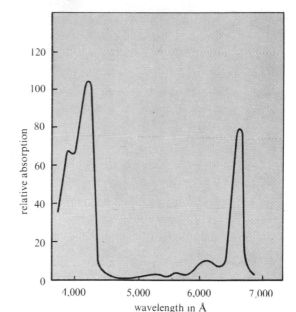

Fig 6–19 The absorption spectrum of chlorophyll

Stages of photosynthesis. The work of Van Niel, which demonstrated that the primary chemical event in photosynthesis leads to the splitting of water molecules in green plants, began a great advance in our knowledge of the light reaction. Basically, the electron-transfer chain described for mitochondria is similar in chloroplasts, and cytochromes are also involved, but for chloroplasts the electrons which enter the chain are emitted from chlorophyll molecules by light quanta, instead of being derived from the Krebs cycle. That all these reactions occur within the chloroplasts and not throughout the cell was demonstrated by the British botanist R. Hill. He showed that isolated chloroplasts can evolve oxygen when exposed to light (Hill reaction).

The quantasomes are thought to be the units which absorb light quanta and contain chlorophyll in association with lipid. The general scheme for photosynthesis is probably as shown in Fig. 6–20a and b. The various stages are summarized below.

Light reaction
(a) Cyclic phosphorylation
In photosynthetic bacteria, photosynthesis occurs in a cyclic process. Frenkel showed that photosynthesis could occur in the isolated chromatophores (regions containing chlorophyll) from the bacterium *Rhodospirillum rubrum*. These stages are shown in Fig. 6–20a. The reactions resemble in certain respects those seen in mitochondria. ATP is generated, but the energy is provided initially from absorbed light quanta instead of glucose. The stages are as follows:

1. Chlorophyll absorbs a light quantum and an electron is raised to a higher energy level.

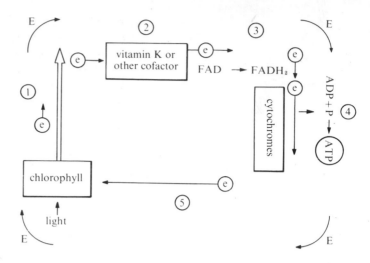

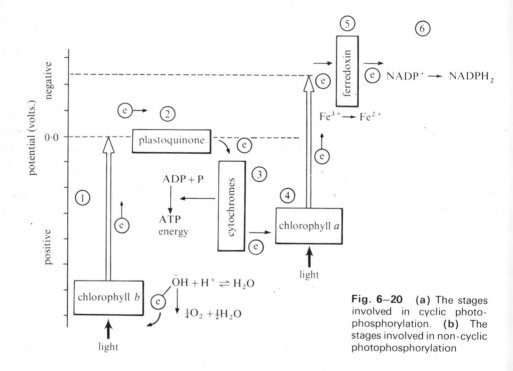

Fig. 6–20 (a) The stages involved in cyclic photophosphorylation. (b) The stages involved in non-cyclic photophosphorylation

2. This electron is transferred to vitamin K or another cofactor of the type shown in Fig. 6–4.

3. The electron reduces FAD to $FADH_2$.

4. The electron is transferred to a cytochrome chain as in mitochondria, whence cyclic phosphorylation occurs. As the energy of the electron is reduced, this energy is available to produce ATP from ADP.

5. The electron is returned to the vacated low energy level of chlorophyll. (Actually, a net flow of electrons occurs: one should not speak of an individual electron.)

(handwritten margin notes): Mechanism? H: similar process to energy accumulation in mito. Where are the enzymes?

Oxydation = loss of electrons ; oxidizing agent accepts electron

Reduction = gain of electrons — Reducing ag. gives up electron

Does losing a /H =

losing a electron >>

(b) Non-cyclic phosphorylation

This type of reaction occurs in green plants and involves the absorption of light quanta both by chlorophyll *a* and by chlorophyll *b*. The most important difference is that a water molecule provides electrons to replace those which are being continually used to produce NADPH from $NADP^+$. The stages are as follows:

1.   A molecule of chlorophyll *b* absorbs a light quantum. The electron which has been raised to a high energy level leaves the chlorophyll molecule and the system probably behaves like a semiconductor. The electron is replaced in the low energy level by electrons derived from the OH^- ions of water. In other words, $chlorophyll^+$ (minus an electron) acts as an electron receptor. OH^- ions are always present in water at pH 7 because of the dissociation equilibrium $(H_2O \rightleftharpoons H^+ + OH^-)$. The OH radicals left after transfer of electrons to $chlorophyll^+$ form oxygen and water $(4OH \rightarrow 2H_2O + O_2)$.

2.   The high-energy electron just mentioned reduces plastoquinone.

3.   The electron is passed on to a cytochrome chain where its energy is used to produce ATP (2 mol).

4.   The low-energy electron so formed enters the vacant ground energy level of an excited chlorophyll *a* molecule. The $chlorophyll^+$ *a* has previously absorbed a light quantum and the excited electron has entered a semiconductor region.

5.   The electron of chlorophyll *a* in the semiconductor region has reached an extremely high energy level (0·423 V expressed as an oxidation–reduction potential). This electron reduces ferredoxin, the most electro negative carrier found in nature.

6.   Ferredoxin reduces $NADP^+$ to NADPH.

There is therefore a net flow of electrons from ^-OH groups of water to NADPH, which becomes available for the dark reaction.

Note that removal of lipid from chloroplasts at low temperature also prevents electron transfer, as in mitochondria. Electron transfer can be restored by replacing the lipid. The chain of reactions involving redox potentials and electron transfer again requires a highly organized and compartmental structure for its biochemical function.

Dark reaction

The NADPH and ATP produced in the light reaction now pass to a cycle of CO_2 fixation. Here again we note a series of reactions somewhat similar to the Krebs cycle except that now the reactions are being driven *backwards* by the energy from sunlight, harnessed in the light reaction. The energy balance is as follows:

$$6CO_2 + 6H_2O \longrightarrow C_6H_{12}O_6 + 6O_2 \qquad (\Delta F' = +685,000 \text{ cal})$$

Ruben and Kamen in 1940 began the elucidation of the dark reaction and the fixation of carbon. They first used the isotope ^{14}C to label the CO_2 fed to the chloroplasts. Then Calvin, Benson, and Bassham used the single-celled algae *Chlorella pyrenoidosa* and *Scenedesmus obliquus*, dropping the cells into alcohol for fixation so that reactions were suddenly arrested at various stages of synthesis. They found that, 30 seconds after introducing labelled CO_2, activity was concentrated in the carbonyl group of 3-phosphoglyceric acid (Fig. 6–21). We now know that 3-phosphoglyceric acid (PGA) is not formed from a C_2 compound but from a C_5 compound, D-ribulose 1,5-diphosphate, which is split in two (Fig. 6–21) in reactions involving combination with CO_2 and water.

The ATP and NADPH produced by the *light* reaction are now used to produce a sugar molecule, *triose phosphate*, from 3-phosphoglyceric acid. 3-Phosphoglyceric acid in fact takes part in a carbon cycle as shown in Fig. 6–22. Note that in this cycle both the 2ATP molecules and the NADPH synthesized in the light reaction have been used. Sugar molecules are produced which are stored in plants as starch molecules.

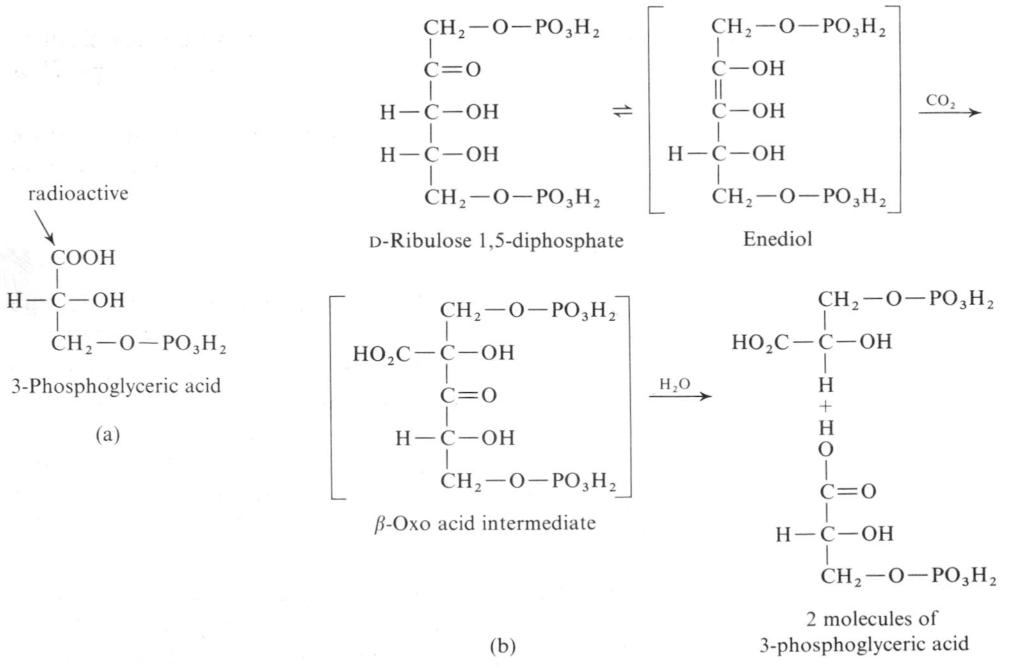

Fig. 6–21 (a) The structure of 3-phosphoglyceric acid.
(b) Its formation from D-ribulose 1,5-diphosphate

Synthesis of fats, amino acids, and carboxylic acids. Figure 6–22 shows in more detail the various other dark reactions within the plant cell. They occur as a cycle: part of the triose phosphate formed during the cycle is converted into fructose phosphate and later into starch, part enters a pathway leading to the formation of *glycerol phosphate*, while the remainder gives rise to ribulose 5-phosphate which re-enters the cycle. That part of the 3-phosphoglyceric acid which does not continue along the cycle by further phosphorylation to form 1,3-diphosphoglyceric acid is diverted to other pathways leading to *fatty acid*, *amino acid*, and *carboxylic acid synthesis*.

6–4
Energy turnover
in cells

Certain bacteria obtain their energy by the oxidation of inorganic salts in solution but it is unlikely that these salts would be available without the photosynthetic activity of the green plants which produces free oxygen molecules and carbon dioxide in the atmosphere.

The turnover of energy in photosynthesis in bacteria, Fig. 6–20a, occurs in the following stages:

1. The light energy enters the chlorophyll as photons of energy hv, where h is Planck's constant and v is the frequency. For light of wavelength 6,500 Å:

$$hv = 3 \cdot 05 \times 10^{-12} \text{ ergs.}$$

One electron volt, the work required to move one electron across a potential of one volt, is $1 \cdot 601 \times 10^{-12}$ ergs. Thus $hv = 1 \cdot 9$ eV.

2. Two molecules of water are split to produce molecular oxygen, hydrogen ions, and electrons as in the light reaction.

$$2H_2O \longrightarrow O_2 + 4H^+ + 4e$$

This generates a redox potential of 0·82 eV per electron when measured at standard temperature and pressure.

3. For the reaction $FAD^+ \rightarrow FADH_2$ the potential change is $-0 \cdot 28$ eV, that is the transfer of the electron to $FADH_2$ requires $+0 \cdot 82 - (-0 \cdot 28) = 1 \cdot 1$ eV.

The energy in the light quantum (1·9 eV) is adequate to bring about this change. The exact quantum efficiency is difficult to determine because the free energy of a photon is not known.

4. A further transfer of energy now takes place in the cytochrome system of enzymes leading to the synthesis of ATP from ADP. The free energy change is $\Delta F' = +8,900$ cal.

5. In the reactions shown in Fig. 6–22 it is found that 18 ATP and 12 NADPH molecules are required for the

synthesis of one molecule of glucose. $\Delta F'$ for the formation of NADPH from $NADP^+$ is $+52,600$ cal, so the total free energy decrease in the utilization of these molecules of ATP and NADPH is

$$-160,200 - 631,200 = -791,400 \text{ cal}$$

As mentioned above, to form glucose directly from CO_2 and water would require a free energy change $\Delta F'$ $+685,000$. Clearly, therefore, the utilization of the 18 ATP and 12 NADPH molecules is more than sufficient for the synthesis of one molecule of glucose.

6. Glucose is available in various forms as a source of energy both in plant and animal cells. The most efficient utilization takes place during glycolysis, the Krebs cycle, and oxidative phosphorylation as follows:

(a) *Glycolysis*

$$C_6H_{12}O_6 + 2NAD^+ \longrightarrow 2CH_3-CO-COOH + 2NADH + 2H^+$$
  glucose   pyruvic acid
$$(\Delta F' = -37,000 \text{ cal})$$

In these reactions there is a net yield of 2 mol of ATP $(\Delta F' = +17,800 \text{ cal})$ and the efficiency at this stage is about 48 per cent.

(b) *Krebs cycle and oxidative phosphorylation*
For every pair of hydrogen atoms oxidized to water, three molecules of ATP (Fig. 6–9) are produced:

$$NADH + H^+ \longrightarrow NAD^+ + 2H \qquad\qquad (\Delta F' = -52,000 \text{ cal})$$
$$2H \xrightarrow{+\frac{1}{2}O_2} H_2O$$

The formation of three molecules of ATP requires $3 \times 8,900 = 26,700$ cal. Here also the efficiency is about 50 per cent.

In photosynthesis, glycolysis, and respiration the importance of the phosphate esters cannot be over-emphasized. As in Chapter 2, we can write ATP in the form

Adenosine————(P)————(P)————(P)
                    ($\Delta F' =$ ($\Delta F' =$ ($\Delta F' =$
                    2,200)   6,500)   8,900)

If ATP is used for the formation of an ester, say of a sugar such as ribulose diphosphate (Fig. 6–22), the energy released from ATP becomes stored in the sugar molecule. A further series of steps in photosynthesis can then occur, using this stored energy gained from ATP.

Similarly, with the Embden–Meyerhof pathway shown in Fig. 6–2, the esterification of glucose to form glucose 1-phosphate from ATP provides energy which makes the

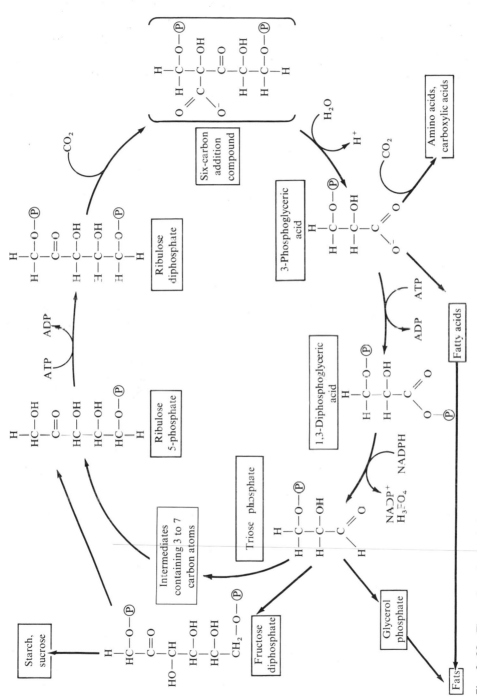

Fig. 6–22 The end products of photosynthesis include carbohydrates and in addition, fatty acids, fats, carboxylic acids, and amino acids. The diagram shows a highly simplified carbon cycle

glucose ring less stable; it is transformed into glucose 6-phosphate and then into fructose 6-phosphate without further supply of ATP. When glycogen is the source of energy, the energy stored due to the polymerization becomes available for phosphorylation, and ATP is not required at this stage. Further stages in the breakdown of glucose later begin to pay dividends in the form of a large production of ATP, until glucose is finally oxidized to carbon dioxide and water.

Bibliography   Buvat, R., Electron microscopy of plant protoplasm. *Int. Rev. Cytol.,* **14**, 41, 1963.

Conn, E. E. and P. K. Stumpf, *Outlines of Biochemistry.* John Wiley, New York, 1964.

Green, D. E. and H. Baum, *Energy and the Mitochondrion.* Academic Press, New York, 1970.

Lardy, H. A. and S. M. Ferguson, Oxidative phosphorylation in mitochondria. *Ann. Rev. Biochem.,* **38**, 991, 1969.

Lehninger, A. L., *The Mitochondrion.* W. A. Benjamin, New York, 1964.

Park, R. E., Subunits of chloroplast structure and quantum conversion in photosynthesis. *Int. Rev. Cytol.,* **20**, 67, 1966.

7 Biochemical synthesis of building units

It is unfortunate, although perhaps inevitable, that so much biochemistry in the past has been studied analytically by chemists and even by molecular biologists without considering the structure and function of living cells as integrated units. Biochemistry should now be regarded for our purpose as a branch, albeit an important branch, of cell biology; this is our aim in this chapter. Particular aspects are discussed in other chapters (nucleic acid synthesis in Chapter 3, protein synthesis in Chapter 4, and sources of energy for the cell in Chapter 6)

Enzyme-catalysed biochemical reactions enable the cell to produce macromolecules rapidly, by lowering the amount of energy needed in their formation. The macromolecules—nucleic acids, proteins, lipids, and polysaccharides—are used in growing and dividing cells to provide new material, and in those differentiated adult cells which do not divide to maintain structure and function and to store energy. Purines and pyrimidines are required to make nucleic acids, amino acids to make proteins, fatty acids for the synthesis of neutral lipids and phospholipids, and sugars to make polysaccharides.

For maximum efficiency, and increased capacity to survive, cells make double use of certain intermediates; for example, purines and pyrimidines are used not only for nucleic acid synthesis but also in the formation of other important nucleotides, such as adenosine triphosphate (ATP) and nicotinamide–adenine dinucleotide (NAD^+). The multiple steps involved in the pathways of synthesis, together with the utilization of these intermediates, make cellular biochemistry especially complex. Nevertheless, each basic step is a simple chemical reaction, catalysed by a specific enzyme molecule. Only a few of the more important reactions are described here.

7–1 Enzymes

Since the time of the Buchner brothers great advances have been made in the isolation and purification of enzyme proteins: Willstätter and his colleagues carried out extensive purification in the 1920s; Dixon and Kodama isolated xanthine oxidase in 1926; and Northrop and his coworkers obtained some proteolytic enzymes in pure crystalline form in the late

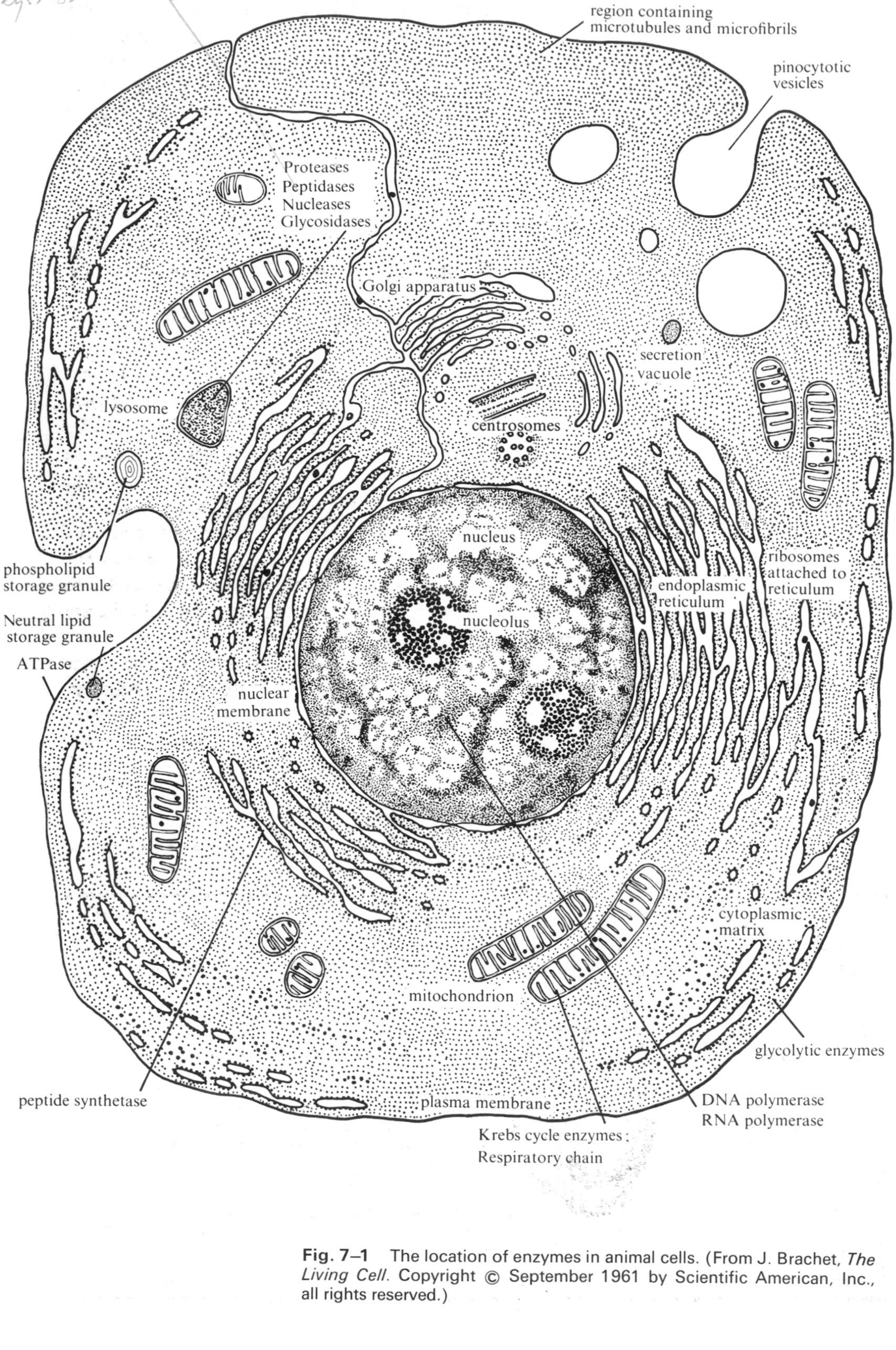

Fig. 7–1   The location of enzymes in animal cells. (From J. Brachet, *The Living Cell*. Copyright © September 1961 by Scientific American, Inc., all rights reserved.)

1920s. The crystallization of globular proteins allowed X-ray diffraction methods to be used. The biochemical determination of the amino acid sequence of the protein chains of various enzymes, and the application of crystal structure analysis, have led to the determination of the complete structure of several enzymes. Figure 7-5, for example, shows the structure of the enzyme lysozyme which we consider below.

Many of the biochemical reactions in living cells occur extremely rapidly. For instance, within a few minutes, a bacterial cell can synthesize a protein made necessary by a change in its environment. This would not be possible at a practical rate without enzyme catalysts.

The speed of enzyme reactions depends on the concentration of the substance, known as the *substrate*, on which the enzyme acts. The high rate of synthesis depends on the efficient localization of the reaction products of one enzyme-catalysed reaction for immediate transfer to an adjacent enzyme, the next step in the particular biochemical pathway. Such an assembly line system has already been described for the respiratory chain (Chapter 6). A high concentration of substrate molecules can also be produced by the localization of enzymes at particular cellular sites. Figure 7-1 shows the localization of some important enzymes in animal cells. In green plants certain enzymes, required for photosynthesis, are localized in chloroplasts as described in Chapter 6.

Another way in which enzyme reactions can be made more efficient is by the involvement of *cofactors*. These are unchanged at the end of the reaction, and so may be regarded as part of the catalytic process. They fall into three main groups:

Fig. 7-2   A list of some important coenzymes and prosthetic groups showing the essential nutritional factors concerned.

Coenzyme or prosthetic group	*Enzymic and other functions*	*Essential nutritional factor or vitamin*
Nicotinamide–adenine dinucleotide (NAD$^+$)	As hydrogen acceptor of dehydrogenases	Nicotinic acid
Nicotinamide–adenine dinucleotide phosphate (NADP$^+$)	As hydrogen acceptor of dehydrogenases	Nicotinic acid
Adenosine triphosphate (ATP)	Transphosphorylation	None
Pyridoxal phosphate	Transaminases, amino acid decarboxylases, racemases, etc.	Pyridoxine
Thiamin pyrophosphate	Oxidative decarboxylation	Thiamin or vitamin B$_1$
Flavin mononucleotide (FMN)	As hydrogen acceptor of dehydrogenases	Riboflavin
Flavin–adenine dinucleotide (FAD)	As hydrogen acceptor of dehydrogenases	Riboflavin
Coenzyme A (CoA)	Acetyl or other acyl group transfer; fatty acid synthesis and oxidation	Pantothenic acid
Iron-protoporphyrin	In catalase, peroxidase, cytochromes, haemoglobin	None
6,8-Dithio-*n*-octanoic acid (lipoic acid)	Oxidative decarboxylation; as hydrogen and acyl acceptor	Required by some micro-organisms
Tetrahydrofolic acid	One-carbon transfer	Folic acid
Biotin	CO$_2$ transfer	Biotin
Cobamide	Group transfer	Cobalamine

coenzymes—small organic molecules which may be readily dissociated from enzymes, and which often act as carriers of some particular group; *prosthetic groups* on the enzyme itself; and *metal activators*. A list of some important coenzymes and prosthetic groups is given in Fig. 7–2, together with a brief account of their function. Coenzymes generally contain a vitamin sub-unit, which probably accounts for the essential role of vitamins in animal diets.

An example of a coenzyme is NAD^+, nicotinamide-adenine dinucleotide (Fig. 7–14). In man, nicotinic acid or its amide (one of the B group of vitamins) has to be supplied in the diet. NAD^+ acts as a hydrogen carrier in enzyme-catalysed oxidation–reduction reactions (see Chapter 6).

Prosthetic groups are usually considered to be carriers which are firmly bound to the enzyme itself. An example is FAD, flavin–adenine dinucleotide, which combines with protein to give the flavoprotein enzymes which catalyse certain oxidation–reduction reactions (Fig. 6–3).

Metal activators are uni- or bi-valent metallic cations such as Na^+, K^+, Mg^{2+}, Zn^{2+}, Ca^{2+}, Fe^{2+}, and Co^{2+}. The nutritional need for trace metals appears to arise from the many enzymes which need metallic cofactors. The size of the cation appears to be important; most of the metal activators fall into a narrow size range, having ionic crystal radii lying within the limits of 0.78 Å to 1.0 Å. These metal activators are not interchangeable in enzyme reactions, and similar ions may act as antagonists; for example Na^+ often acts as an inhibitor of K^+ activation, and similarly Ca^{2+} inhibits a number of Mg^{2+} activated reactions. Metal activation occurs in a variety of ways and in some cases the exact mechanism has been established. The metal cation may form an essential part of the active centre of the enzyme molecule (see below) and may act as a carrier for oxidation–reduction reactions. It may activate by forming a binding link between the enzyme and substrate, or it may shift the equilibrium of the reaction, possibly by removal of one of the products, or it may remove possible inhibitors. In some cases the action may be an alteration of the surface charge of the enzyme protein.

Some enzymes are synthesized in an inactive form known as a *zymogen* (or proenzyme). For instance, as already mentioned in Chapter 5, the important digestive enzymes trypsin and chymotrypsin, which catalyse the hydrolysis of certain peptide bonds of proteins, are synthesized in the pancreas in the form of their zymogens trypsinogen and chymotrypsinogen. They are converted into the active forms in the intestinal tract. This generally involves the removal of a short peptide group, followed by a rearrangement of the protein chains of the molecule, the process being catalysed by an enzyme. Pepsin, another digestive enzyme, is produced in the gastric juice as its zymogen pepsinogen. It is converted, in the presence of hydrochloric acid, into its active form which catalyses the hydrolysis of internal bonds of polypeptide chains.

Effect on energy requirements and reaction rates.   The function of an enzyme is to lower the energy barrier to a chemical reaction. The enzyme does not affect the course of the reaction, or the products of the reaction, or the total energy change involved. The energy barrier may be caused, for example, by repulsive forces between the interacting molecules. The comparison between an enzyme-catalysed reaction and one in which the enzyme is absent is shown in Fig. 7–3. The initial energy level of the reactants is at A. The activation energy, which must be achieved for the reaction to proceed, is at D in the absence of enzyme, but in the presence of enzyme is lowered to C. The final energy level of the reaction products, B, is the same in both cases. Since this is lower than the initial energy level at A, the net energy change in this case is negative—the reaction is exergonic. If the level at B were higher than the initial level at A, the reaction would be endergonic.

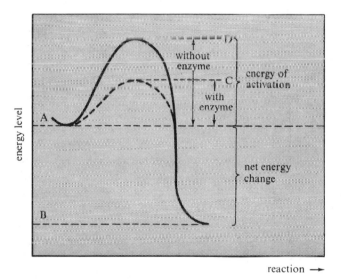

Fig. 7–3 The energy of activation of a chemical reaction in the presence and absence of a specific enzyme

reaction ➝

The catalytic function of enzymes is therefore to lower the activation energy of reactions and hence to speed them up. Enzymes are more efficient than inorganic catalysts because they lower energies of activation to a greater extent than catalysts, so the reactions concerned go more quickly. They are also more specific in that the complex structure of an enzyme is such that it will accept only a few types of molecule, or even a single type, as its substrate. Enzymes vary widely in their substrate specificity; many will act as catalysts for a group of similar compounds taking part in a particular type of reaction. There is one enzyme, for example, which phosphorylates a series of aldohexose sugars (see Section 7–4).

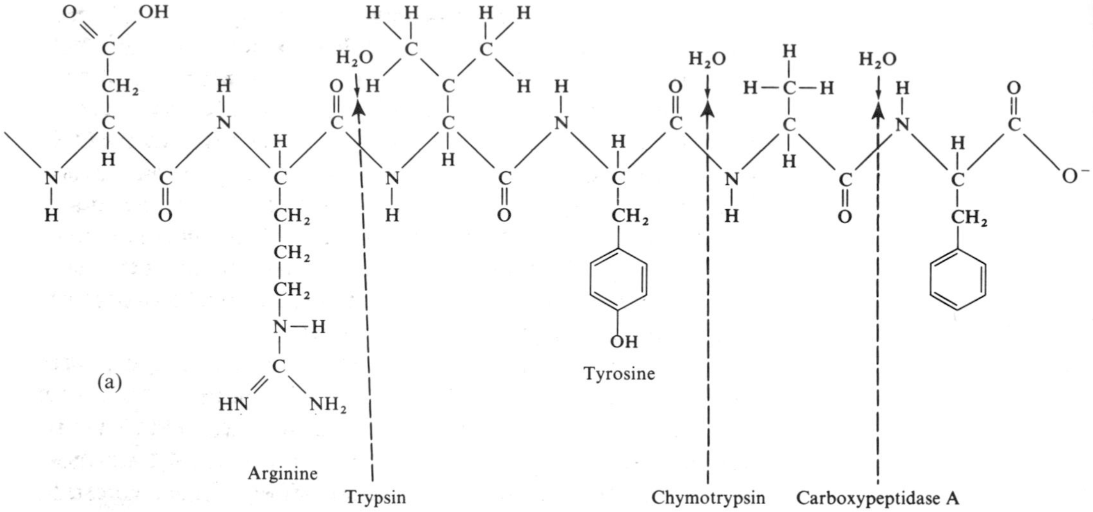

(a)

Arginine   Tyrosine

Trypsin   Chymotrypsin   Carboxypeptidase A

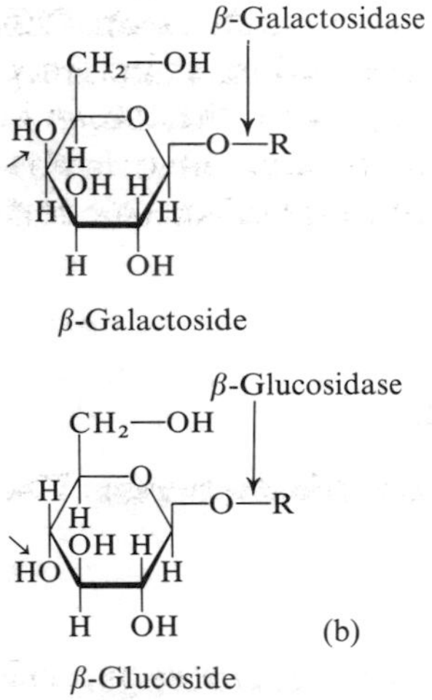

β-Galactosidase

β-Galactoside

β-Glucosidase

β-Glucoside

(b)

Fig. 7–4 **(a)** The specificity of peptide bond cleavage by various enzymes. **(b)** β-Galactosidase and β-glucosidase are highly specific enzymes in spite of a difference in positioning of only one OH group in their substrates

Structure and active sites. All known enzymes are proteins. The amino acid sequence of proteins is determined by the coding in the DNA molecules, and is characteristic for each individual protein. Folding of certain regions of the polypeptide chains into an α-helix (secondary structure) takes place. Interactions between side-chains, again determined by the amino acid sequence, leads to further folding of the molecule, until a very complicated three-dimensional (tertiary) structure is obtained. This unique tertiary structure is a characteristic feature of each kind of protein molecule, and the irregular surface of a protein confers a high degree of chemical affinity for specific substrate molecules.

Much work has been devoted to the study of enzyme action. The first stage, the combination of the enzyme with the substrate molecule, can be studied by varying the relative concentrations of the enzyme and substrate and seeing how the speed of the reaction is affected. The region on the surface of the enzyme molecule which combines with, or adsorbs onto, the substrate is known as the *active site*. There may be several active sites on the same enzyme molecule. Information is gradually being obtained about the nature of these sites, and the way in which they combine with the substrate. One approach is to break down the enzyme molecule carefully, for instance by controlled digestion with other enzymes, and to remove gradually the non-essential parts, leaving the active sites intact. This method is necessarily laborious.

Figures 7–4a and b illustrate the great specificity shown by enzymes in their association with substrate molecules. The enzymes which break down proteins by hydrolysing peptide bond linkages are specific because they break linkages between particular amino acid residues (Fig. 7–4a). For instance, trypsin acts on peptide linkages involving the carboxyl group of arginine and lysine: chymotrypsin acts on bonds involving phenylalanine, tyrosine, and tryptophan, while the carboxypeptidases attack the peptide bond next to a terminal carboxyl group. A different example is shown in Fig. 7–4b; the β-galactoside and β-glucoside linkages differ only in the way that galactose and glucose do—in the position of the OH group on the C-4 atom. Yet the enzyme β-galactosidase catalyses only the hydrolysis of the β-galactoside linkage, and a different enzyme, β-glucosidase, is required to split the β-glucoside linkage. Also, in the case of certain optically active isomeric compounds, one enzyme is required for the D-isomer, and a different enzyme for the L-isomer.

It is clear from this type of evidence that enzymes are extremely sensitive to small changes in the stereochemistry of the substrate, and this suggests a close steric relationship between the two. This was the basis of the original 'lock-and-key' theory framed by Emil Fischer (1894), which suggested that enzyme and substrate are spatially complementary to one another and fit together sterically as a key fits into a lock. This type of complementarity at active sites would result in a very high degree of specificity. Further support for the theory is that compounds similar to the substrate can often inhibit enzyme activity. For instance, the conversion of succinic acid

$$HOOC \quad CH_2 \quad CH_2 \quad COOH$$

into fumaric acid

$$HOOC-CH=CH-COOH$$

is catalysed by the enzyme succinate dehydrogenase. The presence of malonic acid.

$$HOOC-CH_2-COOH$$

will inhibit enzyme activity, but increasing the concentration of succinic acid will restore it. This implies that malonic acid is sufficiently similar both sterically and chemically to compete with succinic acid and to occupy, and thus inactivate, the active site on the enzyme molecule.

Work on the behaviour of lysozyme, an enzyme found in tears and in other secretions and tissues of the human body, and also in egg-white, has confirmed a close steric complementarity between lysozyme and its substrate. Lysozyme is so called because of its capacity to lyse various bacterial cells, which it does by catalysing the hydrolysis of certain structural polysaccharide molecules in the cell walls. Phillips and others (1965), using the X-ray crystallographic methods described earlier, determined the three-dimensional structure of the lysozyme molecule (Fig. 7–5). They found that the non-polar hydrophobic side-chains are situated inside the protein molecule, and the polar ionizing side-chains are on the surface, as in the myoglobin molecule. The spatial arrangement of the atoms is such that the folded protein chains on one side of the

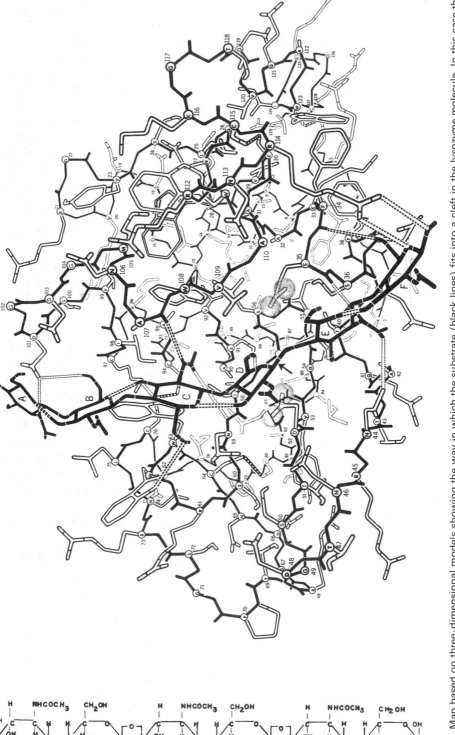

Fig. 7–5 Map based on three-dimensional models showing the way in which the substrate (black lines) fits into a cleft in the lysozyme molecule. In this case the substrate is a synthetic hexamer of *N*-acetylglucosamine (inset). The substrate is attached by hydrogen-bonding to various positions on the enzyme chain. A, B, C, D, E and F are the sugar rings of the substrate. Cleavage of the substrate takes place between rings D and E (arrowed in the diagram). The stippled spheres represent van der Waals surfaces of the oxygen atoms of glutamic acid (E) at position 35, and of aspartic acid (D) at position 52 which are believed to be involved in the activated enzyme-substrate complex. (With permission from Margaret O. Dayhoff and Richard V. Eck (eds), *The Atlas of Protein Sequence and Structure, 1967–68,* National Biomedical Research Foundation, Silver Spring, Maryland.)

molecule form wings with a deep cleft running between them.

By a process of model-building, combined with X-ray methods, and the gradual accumulation of chemical evidence, the cleft was found to be the active site of the molecule. A portion of the substrate, consisting of six amino sugar rings (labelled A to F in Fig. 7–5), fits into this cleft and is held to the enzyme by hydrogen bonding. Cleavage of the substrate occurs between rings D and E, at the point marked in the diagram. It was found from models that the substrate fits closely into the cleft of lysozyme only with considerable distortion of ring D, which is next to the cleavage point. One of the theories put forward by enzyme chemists to explain the actual breakage of a substrate bond is that the enzyme molecule activates the substrate by distortion, thus making bond cleavage easier. The evidence obtained in the case of lysozyme certainly appears to confirm this view.

Work on the structure of the plant enzyme papain has shown that the active site again lies at the surface of a cleft dividing the molecule into two distinct parts, and formed by folding of the single polypeptide chain.

Reaction rates. The rate or velocity of an enzyme-catalysed reaction depends directly on the concentration of the enzyme. Figure 7–6a shows the effect on reaction rate of increasing the enzyme concentration in the presence of an excess of substrate. The curve obtained with a fixed enzyme concentration and an increasing substrate concentration is shown in Fig. 7–6b. At first there is a rapid rise in reaction rate as substrate concentration increases, but with further increments the rate gets slower, and eventually levels off at high substrate concentration. At this stage the maximum velocity possible under these conditions is reached.

The substrate concentration required for the reaction to

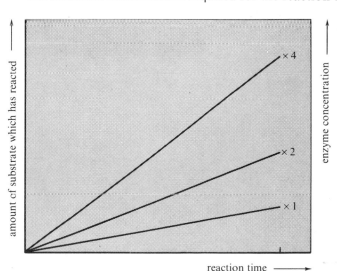

Fig. 7–6   (a) The effect of increasing enzyme concentration in the presence of an excess of substrate

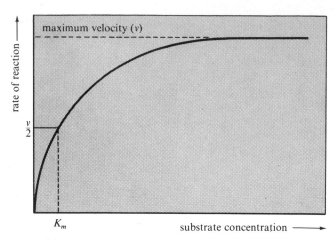

Fig. 7–6 **(b)** The effect of increasing the substrate concentration at a fixed concentration of enzyme

proceed at half the maximum velocity is known as the *Michaelis constant*, K_m. It is of great importance, since its reciprocal, $1/K_m$, gives a measure of the affinity of an enzyme for its substrate. For instance, if a high concentration of substrate is needed to give half the maximum velocity, then a low affinity of enzyme for substrate is indicated, and vice versa. In biochemical terms this means that enzyme reactions which require a relatively high concentration of substrate will proceed at a low rate in the cell, unless there is a localized high concentration of substrate at the site at which the enzyme operates.

Temperature. Enzyme-catalysed reactions, like all chemical reactions, are affected by changes in temperature—increasing temperature will increase the rate of reaction. But since enzymes are proteins, they are subject to denaturation by heat; that is, at temperatures above about 45°C their complex three-dimensional structure, which depends on interactions between side-chains, tends to be broken down. The effects of denaturation will therefore oppose the tendency for the rate to increase as the temperature is raised, and will predominate more and more until about 55°C, at which point most enzymes are inactivated so rapidly that they are quite ineffective as catalysts.

pH. Enzymes contain many ionic side-groups which may be basic or acidic in character; clearly, pH changes will considerably alter the ionic nature of these groups. Extremes of pH in either direction will tend to denature the protein structure and so destroy enzyme activity, just as does thermal denaturation. In practice, there is generally found to be a narrow pH range, known as the optimal pH, inside which an enzyme functions most efficiently. As yet little is known about variations in pH at the ultrastructural level.

Inhibition of enzyme reactions. Enzymes can be prevented from acting on their substrates in two ways: by competitive inhibition and irreversible inhibition. In competitive inhibition, a substance which is similar chemically and sterically to the substrate competes with the substrate for the active site on the enzyme molecule. An example quoted above was the succinate dehydrogenase enzyme system, where malonic acid is a molecule sufficiently similar to the true substrate, succinic acid, to compete with it for the enzyme active site, and thus to inhibit the reaction. The addition of further succinic acid, however, will tend to reverse this effect and increase the reaction rate. It is true in general of competitive inhibitors of enzyme reactions that the association between enzyme and inhibitor is easily broken and the inhibition reversed by increasing the concentration of substrate.

By contrast, in irreversible inhibition the inhibitor forms a strong bond with the enzyme, which is not broken by increasing the substrate concentration. If enough inhibitor is present to react with all of the enzyme, the inhibition will proceed until it is complete. An example of this type of irreversible inhibition is the action of the so-called nerve gases on the cholinesterase enzymes which play a specific role in nerve function. The salts of heavy metals such as mercury and lead inactivate most enzymes in high concentrations, and a few enzymes are extremely sensitive to even low concentrations of the metals.

Inhibition may also be brought about by combination of a substance with the substrate, with a coenzyme, or with a metal activator. For instance, cyanide and hydrogen sulphide can form stable complexes with heavy-metal catalysts, such as iron in the cytochrome oxidase system. Substances such as cyanide are inhibitors of the respiratory processes of many tissues.

The study of various enzyme inhibitors has increased rapidly and has applications not only to the determination of the intermediates involved in various metabolic pathways, but also to fundamental problems of pharmacology. Many drugs function as inhibitors of specific enzyme reactions, and a study of the inhibitory effects of toxic substances is valuable for the development of specific antidotes.

Nomenclature of enzymes.   Since their function is to catalyse specific reactions, it is convenient to name enzymes in terms of these particular reactions. The general method is to add the suffix *-ase* to the name of the substrate; thus lipid is attacked by lipase, urea by urease, and so forth. But names such as trypsin and pepsin, which were given to certain enzymes before a more systematic notation was developed, are still in common use.

A Commission on Enzymes (International Union of Biochemistry, 1961) has devised a complete but rather complex system for classification and nomenclature. The system classes enzymes in six main groups, according to the type of reaction which they catalyse. The groups are briefly as follows:

1. *Oxidoreductases.* These enzymes, which catalyse oxidation–reduction reactions, include oxidases, peroxidases, and dehydrogenases (e.g. malate dehydrogenase).

2. *Transferases.* The enzymes in this category catalyse the transfer of a chemical group from one molecule to another—for example, transaminases.

3. *Hydrolases.* This is a large group of enzymes which catalyse the hydrolysis of various substrates—for example, peptidases.

4. *Lyases.* In this group are enzymes which catalyse the removal of groups from substrates—for example, decarboxylases.

5. *Isomerases.* This group includes those enzymes which catalyse different types of isomerization reaction, such as racemases, epimerases, and mutases.

6. *Ligases.* The function of these enzymes is to link together two molecules, the reaction being coupled with the breaking of a pyrophosphate bond. Examples are DNA ligase, which links sections of DNA polynucleotide chain, and acetate thiokinase, the enzyme which catalyses the formation of acetyl-coenzyme A from acetate and coenzyme A, a reaction requiring ATP, which is itself converted into AMP.

7–2
Amino acids

The synthesis of proteins by peptide bond formation between amino acid building units was described in Chapter 4. The structures of the twenty common amino acids found in proteins are given in Fig. 4–1a, and the importance of proteins both structurally and functionally in living material has been frequently emphasized. In animal tissue proteins are quantitatively the main material present, accounting for about three quarters of the dry weight.

The cellular amino acid pool is chiefly used in protein synthesis. But many other metabolically important compounds are formed from amino acids; examples are purines and pyrimidines (Section 7–3), used in nucleotide synthesis, and the porphyrins (this section). Amino acid structures also occur in substances such as the vitamin pantothenic acid (Section 7–3), which contains β-alanine, H_2N—CH_2—CH_2—COOH, and thyroxine, or 3,5,3',5'-tetraiodothyronine:

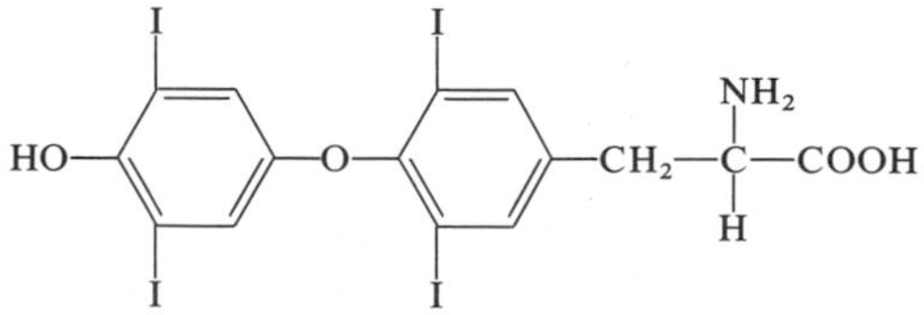

which is the important iodine-containing compound of the thyroid gland.

Amino acids act in general as suppliers of carbon dioxide in cellular metabolism, with the formation of the corresponding amine:

$$R-\underset{\underset{H}{|}}{\overset{\overset{NH_2}{|}}{C}}-COOH \longrightarrow R-\underset{\underset{H}{|}}{\overset{\overset{NH_2}{|}}{C}}-H + CO_2$$

Methionine is the source of a methyl group in various biosyntheses, a process known as transmethylation, and glutamic acid acts as the supplier of an amine group in transamination reactions (see below).

All cells can use amino acids present in their external environment. In higher animals the main source of nitrogen is amino acids from dietary proteins; the acids enter cells by one of the transport mechanisms described in Chapter 8. Plant cells and micro-organisms can synthesize amino acids using ammonia (as the ammonium ion) as a nitrogen source when suitable carbon compounds are available. Ammonia is obtained in plants by bacterial reduction of the nitrogen of the atmosphere, or of nitrate from the soil. It is used in plant cells and micro-organisms for three main enzyme-catalysed reactions, which lead to the formation of glutamic acid, glutamine, and carbamoyl phosphate (Fig. 7–7).

For plants and many micro-organisms (except Gram-positive bacteria which cannot use ammonia as their sole source of nitrogen) all of the amino acids in proteins may be synthesized, since these organisms are able to make the basic carbon structures of each of the amino acids. Animal cells, however, lack or have lost the ability to make about half of the necessary carbon structures involved and hence the ability to make the corresponding amino acids. This means that these

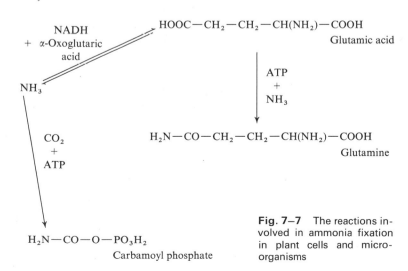

Fig. 7–7 The reactions involved in ammonia fixation in plant cells and micro-organisms

particular amino acids, known as *essential amino acids,* must be present in the animal diet, either free or as constituents of proteins which are hydrolysed in the gastrointestinal tract with the release of their amino acids. (The term 'essential' refers specifically to this dietary requirement, and not to any particular role of these amino acids in cellular metabolism, where they are neither more nor less necessary than the amino acids which can be synthesized in the animal's own cells.) Nutritional studies on adults have shown that in man the following amino acids are essential dietary components in the sense that human cells cannot synthesize them: histidine, leucine, isoleucine, lysine, methionine, phenylalanine, threonine, tryptophan, and valine.

Synthesis of amino acids in higher animals. Animals can synthesize rather more than one half of the amino acids they need. These *non-essential* amino acids are made from the α-oxo acids, such as pyruvic acid and oxaloacetic acid, which are products of carbohydrate metabolism; the necessary amine group is transferred from another amino acid to the carbon chain. The process is known as *transamination,* the general reaction being:

$$R^1-\underset{\underset{NH_2}{|}}{CH}-COOH + R^2-\underset{\underset{O}{||}}{C}-COOH \rightleftharpoons R^1-\underset{\underset{O}{||}}{C}-COOH + R^2-\underset{\underset{NH_2}{|}}{CH}-COOH$$

It was shown in 1937 by Braunstein and Kritzmann that there are enzymes—widely distributed, but chiefly present in the liver—known as the transaminases, which specifically catalyse these reactions. Once ammonia is fixed into glutamic acid, transamination provides a mechanism for the synthesis of the non-essential amino acids. In animal tissues, glutamic acid and α-oxoglutarate are almost invariably one of the pairs of acids involved. For example, the formation of alanine by interaction between glutamic acid and pyruvate from the Embden–Meyerhof pathway is as follows:

$$\begin{array}{cccc}
\text{COOH} & & \text{COOH} & \\
| & & | & \\
H_2N-C-H & \text{COOH} & C=O & \text{COOH} \\
| & | & | & | \\
CH_2 & +C=O & \rightleftharpoons CH_2 & +H_2N-C-H \\
| & | & | & | \\
CH_2 & CH_3 & CH_2 & CH_3 \\
| & & | & \\
\text{COOH} & & \text{COOH} &
\end{array}$$

| L-Glutamic acid | Pyruvic acid | α-Oxoglutaric acid | L-Alanine |

These reactions are readily reversible, and so provide a means for the redistribution of nitrogen. For instance, an animal may ingest a mixture of amino acids very different from the optimal mixture required for its metabolism. Transamination then provides a means of converting one amino acid, present in large quantities, into a second amino acid, initially present in small quantities, by the following reactions:

Amino acid 1 + α-oxoglutaric acid ⇌ keto acid 1 + glutamic acid
Glutamic acid + keto acid 2 ⇌ α-oxoglutaric acid + amino acid 2

Another method by which the nitrogen of one amino acid may be made available for the synthesis of other amino acids is by *deamination* reactions in which ammonia is produced and subsequently fixed, as mentioned above. These deaminations may be either oxidative or non-oxidative. In the oxidative category there are two subdivisions: deaminations catalysed by an enzyme which requires NAD^+ as a coenzyme; and deaminations catalysed by flavoproteins. The most important example of the first group is the reaction catalysed by the enzyme glutamate dehydrogenase:

$$
\begin{array}{ccc}
\text{COOH} & & \text{COOH} \\
| & & | \\
\text{H}_2\text{N--CH} & & \text{C=O} \\
| & & | \\
\text{HCH} & +\,\text{NAD}^+ + \text{H}_2\text{O} \rightleftharpoons & \text{HCH} \quad +\,\text{NADH} + \text{H}^+ + \text{NH}_3 \\
| & & | \\
\text{HCH} & & \text{HCH} \\
| & & | \\
\text{COOH} & & \text{COOH} \\
\end{array}
$$

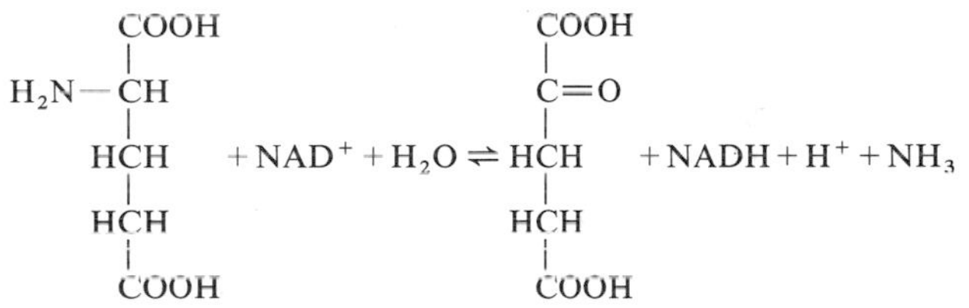

Glutamic acid   α-Oxoglutaric acid

This reaction is readily reversible, and in its reverse form shows how ammonia is fixed in the presence of α-oxoglutarate glutamate dehydrogenase, and NADH to give glutamic acid (see Fig. 7–7). The conversion into NH_3 of the α-amino groups of most amino acids probably occurs first by transamination with α-oxoglutarate to give the corresponding α-oxo acid and glutamic acid, as shown above, followed by oxidation of the glutamic acid so formed.

The second method of oxidative deamination is catalysed by the enzyme L-amino acid oxidase (a flavoprotein with **FAD** as the prosthetic group). This enzyme has, however, a low order of activity in liver and kidney, so this method of amino acid oxidation is probably not of primary importance under normal conditions. The general reaction is:

$$\text{R--CH(NH}_2)\text{--COOH} + \text{H}_2\text{O} + \text{O}_2 \rightleftharpoons \text{R--CO--COOH} + \text{NH}_3 + \text{H}_2\text{O}_2$$

The H_2O_2 is decomposed by catalase, an enzyme present in all cells, to give H_2O and O_2.

A non-oxidation type of deamination is also significant in ammonia production. For serine, cysteine, threonine, and possibly other amino acids, the group of enzymes known as dehydratases act as specific catalysts, and require the presence of pyridoxal phosphate as coenzyme. In each case the keto acid is formed, sometimes by rearrangement of the molecule, by way of various intermediates, as in the case of serine:

$$CH_2(OH)-CH(NH_2)-COOH \longrightarrow CH_3-CO-COOH+NH_3$$

L-Serine   Pyruvic acid

This, then, gives an indication of the general processes involved in amino acid biosynthesis in animal cells. There may in addition be specific mechanisms which operate for certain individual amino acids. We go on to consider the synthesis of some of these acids in more detail.

Synthesis of non-essential amino acids in higher animals. We have already described the synthesis of the key compound, *glutamic acid*, and the subsequent formation of *alanine* from glutamic acid by transamination. *Aspartic acid* is also formed from glutamic acid by transamination with oxaloacetic acid, obtained from the Krebs cycle:

$$HOOC-CH_2-CH_2-CH(NH_2)-COOH+HOOC-CH_2-\underset{\underset{O}{\|}}{C}-COOH \rightleftharpoons$$

Glutamic acid

Oxaloacetic acid

$$HOOC-CH_2-CH_2-\underset{\underset{O}{\|}}{C}-COOH+HOOC-CH_2-CH(NH_2)-COOH$$

Aspartic acid

α-Oxoglutaric acid

These three amino acids are generally the most abundant in cells if one considers amino acid content apart from that present in protein.

Glutamic acid is also the direct precursor for the formation of *glutamine* (Fig. 7–7) and for the synthesis of *proline*, *hydroxyproline*, and *ornithine*. The first step in the formation of these three amino acids is the conversion of one of the carboxyl groups of glutamic acid into an aldehyde group (Fig. 7–8). Direct transamination then gives ornithine, while ring closure by means of a condensation reaction, followed by reduction with NADH, leads to proline; from this hydroxyproline is obtained by an oxidative reaction.

Arginine,

$$H_2N-\underset{\underset{NH}{\|}}{C}-\overset{\overset{H}{|}}{N}-CH_2-CH_2-CH_2-\underset{\underset{NH_2}{|}}{C}H-COOH,$$

is in an intermediate position as regards the classification into essential and non-essential amino acids in animals. Animal cells have some capacity for arginine biosynthesis, but cannot generally produce it in sufficient quantities to meet the demands

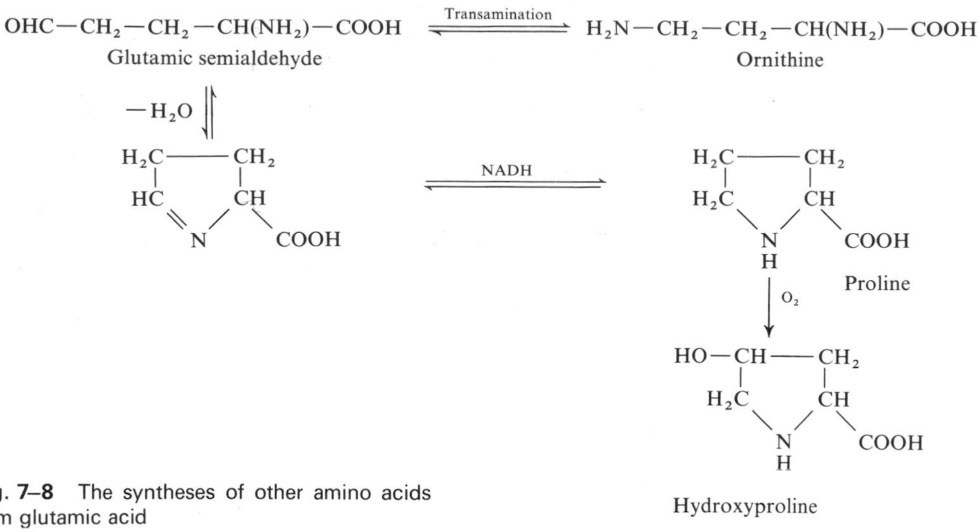

$$OHC-CH_2-CH_2-CH(NH_2)-COOH \quad \underset{\text{Transamination}}{\rightleftharpoons} \quad H_2N-CH_2-CH_2-CH(NH_2)-COOH$$

Glutamic semialdehyde   Ornithine

Fig. 7–8   The syntheses of other amino acids
from glutamic acid

Hydroxyproline

of normal growth. The starting points for arginine synthesis
are ornithine (from glutamic acid, as shown above) and
carbamoyl phosphate. Aspartic acid (again from glutamic
acid, by transamination) is required in the reaction process at
a later stage. So the production of arginine is again dependent
upon supplies of glutamic acid.

The carbon chain of *serine*,

$$CH_2(OH)-CH(NH_2)-COOH,$$

comes from 3-phosphoglyceric acid formed during glycolysis.
Serine in turn is used to form the carbon chain and the amino
group of the sulphur-containing amino acid, *cysteine*,
$HS-CH_2-CH(NH_2)-COOH$. The sulphur itself comes
from the essential amino acid methionine. Serine is also used
in the biosynthesis of *glycine*, by the loss of the hydroxymethyl
group in the β-position:

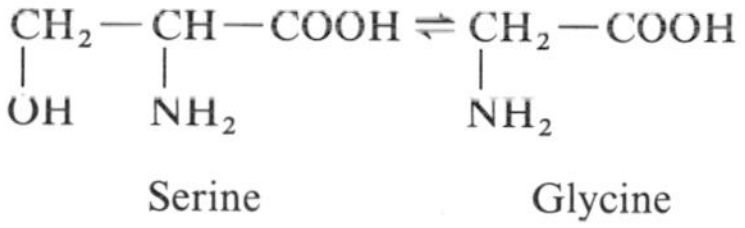

$$\underset{\underset{\text{OH}}{|}\quad\underset{\text{NH}_2}{|}}{CH_2-CH-COOH} \rightleftharpoons \underset{\underset{\text{NH}_2}{|}}{CH_2-COOH}$$

Serine   Glycine

Tyrosine, which contains a benzene ring, is formed by
hydroxylation of the essential amino acid, phenylalanine:

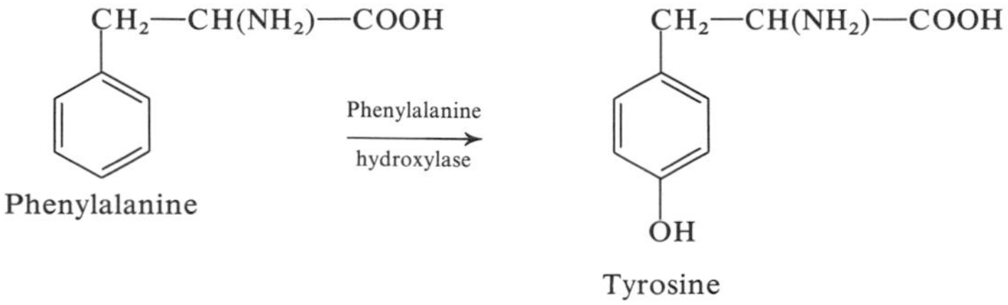

Phenylalanine
hydroxylase

Phenylalanine

Tyrosine

Apart from protein synthesis, this is the only way in
which phenylalanine is used in normal metabolism. If the

specific enzyme phenylalanine hydroxylase is missing, an hereditary defect which occurs in certain individuals as the result of the presence of a recessive gene, alternative pathways of phenylalanine metabolism take over. Transamination occurs to give phenylpyruvic acid, the formation of which in large quantities (phenylketonuria) in early childhood is accompanied by serious mental retardation. This may be largely prevented by restricting the dietary intake of phenylalanine.

Synthesis of amino acids in plants and micro-organisms. The biosyntheses of the various non-essential amino acids in animals may follow different pathways in plants and micro-organisms. Just as glutamic acid plays a key role in mammalian amino acid synthesis, aspartic acid or alanine (or both) may play a similar central role in various micro-organisms.

The method by which *asparagine*

$$H_2N-CO-CH_2-CH(NH_2)-COOH$$

is synthesized in animal cells is not yet known, but in plants and micro-organisms it appears to be similar to the way glutamine is synthesized; that is, from aspartic acid, ammonia, and ATP:

$$HOOC-CH_2-CH(NH_2)-COOH \xrightarrow[ATP]{NH_3}$$
Aspartic acid

$$H_2N-CO-CH_2-CH(NH_2)-COOH$$
Asparagine

Asparagine acts in plants as a reservoir of ammonia and of aspartic acid.

Threonine is also formed from aspartic acid as starting point. First, homoserine is formed by reduction with NADH of the second carboxyl group of aspartic acid (via the phosphate intermediate) to the aldehyde and then to the alcohol:

$$
\begin{array}{ccccc}
COOH & & CH_2-OH & & CH_3 \\
| & & | & & | \\
CH_2 & \xrightarrow{NADH} & CH_2 & \longrightarrow & CH-OH \\
| & & | & & | \\
CH-NH_2 & & CH-NH_2 & & CH-NH_2 \\
| & & | & & | \\
COOH & & COOH & & COOH \\
\text{Aspartic} & & \text{Homoserine} & & \text{Threonine} \\
\text{acid} & & &
\end{array}
$$

Migration of the hydroxyl to the β-position then occurs, again by way of a phosphate intermediate, in the presence of a specific enzyme and coenzyme.

Lysine is synthesized by two different pathways. In higher plants, yeast, and *Neurospora*, the starting point is α-oxoglutarate. In the series of reactions shown in Fig. 7–9, α-oxoglutarate is first converted into α-oxoadipate (the mechanism involved is not clearly understood), which is transaminated to give α-aminoadipic acid. The semialdehyde of this

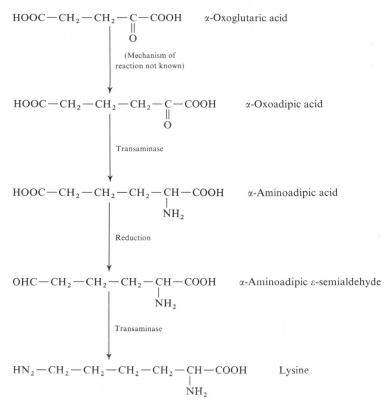

$$HOOC-CH_2-CH_2-\underset{\underset{O}{\|}}{C}-COOH \qquad \alpha\text{-Oxoglutaric acid}$$

(Mechanism of
reaction not known)

$$HOOC-CH_2-CH_2-CH_2-\underset{\underset{O}{\|}}{C}-COOH \qquad \alpha\text{-Oxoadipic acid}$$

Transaminase

$$HOOC-CH_2-CH_2-CH_2-\underset{\underset{NH_2}{|}}{CH}-COOH \qquad \alpha\text{-Aminoadipic acid}$$

Reduction

$$OHC-CH_2-CH_2-CH_2-\underset{\underset{NH_2}{|}}{CH}-COOH \qquad \alpha\text{-Aminoadipic }\varepsilon\text{-semialdehyde}$$

Transaminase

Fig. 7–9 Synthesis of lysine in higher plants, yeast, and *Neurospora*

$$HN_2-CH_2-CH_2-CH_2-CH_2-\underset{\underset{NH_2}{|}}{CH}-COOH \qquad \text{Lysine}$$

is again transaminated, giving lysine. In *E. coli*, however, a very complex pathway, with aspartic acid as starting point, has been found.

The relationship between *cysteine* and *methionine* in animals is reversed in plants and in many micro-organisms. In the latter, inorganic sulphur is first fixed to an organic carbon chain in cysteine, and later transferred to a different carbon chain for the synthesis of methionine. Sulphur occurs naturally chiefly as inorganic sulphate; reduction of this occurs, before fixation, in plants and many micro-organisms, but not in most animals. The actual mechanism of reduction is not clear, but the first step appears to be a reaction between sulphate and ATP with the formation of adenosine phosphate sulphatophosphate.

Many organisms can reduce sulphate via sulphite to sulphur, and even to H_2S. It has been suggested that sulphur fixation may occur by the type of reaction shown below, a specific enzyme and coenzyme being involved:

$$\underset{\text{Pyruvic acid}}{CH_3-CO-COOH} + NH_3 + H_2S \rightleftharpoons \underset{\text{Cysteine}}{HS-CH_2-\underset{\underset{NH_2}{|}}{CH}-COOH}$$

Methionine is formed first by the transfer of the sulphur of cysteine to homoserine, formed from aspartic acid (Fig. 7–10). The thioether so formed is then split by a specific enzyme

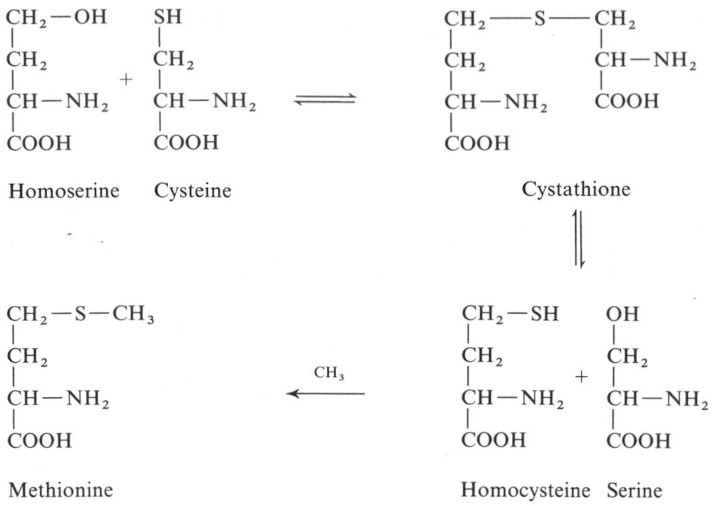

$$
\begin{array}{ccc}
\begin{array}{c}
CH_2-OH \\
| \\
CH_2 \\
| \\
CH-NH_2 \\
| \\
COOH
\end{array}
&
\begin{array}{c}
SH \\
| \\
CH_2 \\
| \\
CH-NH_2 \\
| \\
COOH
\end{array}
&
\begin{array}{c}
CH_2-S-CH_2 \\
| \qquad\qquad | \\
CH_2 \qquad\; CH-NH_2 \\
| \qquad\qquad | \\
CH-NH_2 \quad COOH \\
| \\
COOH
\end{array}
\end{array}
$$

Homoserine   Cysteine   Cystathione

$$
\begin{array}{ccc}
\begin{array}{c}
CH_2-S-CH_3 \\
| \\
CH_2 \\
| \\
CH-NH_2 \\
| \\
COOH
\end{array}
&
\xleftarrow{\quad CH_3 \quad}
&
\begin{array}{c}
CH_2-SH \qquad OH \\
| \qquad\qquad\quad | \\
CH_2 \qquad\qquad CH_2 \\
| \qquad\qquad\quad | \\
CH-NH_2 \quad + \; CH-NH_2 \\
| \qquad\qquad\quad | \\
COOH \qquad\quad COOH
\end{array}
\end{array}
$$

Fig. 7–10 Synthesis of methionine, an essential amino acid in animals

Methionine   Homocysteine Serine

to give homocysteine and serine. Finally a methyl group is transferred to homocysteine.

The way in which the synthesis of the cyclic structure of *phenylalanine* occurs in micro-organisms, and probably also in higher plants, has been determined with the help of ^{14}C-labelled precursors. The carbon atoms are all derived from erythrose 4-phosphate, an intermediate in photosynthesis and also in the oxidative metabolism of glucose, and phosphoenol-pyruvic acid, a product of glycolysis (Fig. 7–11). The phosphate derivative of shikimic acid (Fig. 7–11) is a cyclic intermediate in the series of reactions, as it is also in the synthesis of *trypto-phan* in E. coli. The mechanism of synthesis of tryptophan in higher plants is not yet determined.

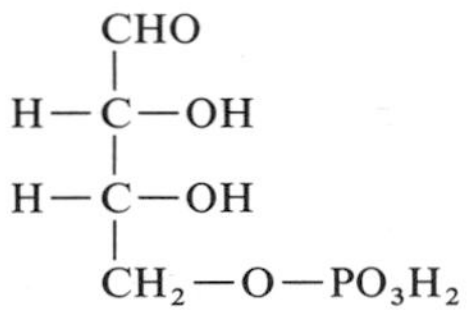

$$
\begin{array}{c}
CHO \\
| \\
H-C-OH \\
| \\
H-C-OH \\
| \\
CH_2-O-PO_3H_2
\end{array}
$$

Erythrose 4-phosphate

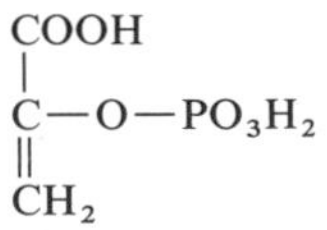

$$
\begin{array}{c}
COOH \\
| \\
C-O-PO_3H_2 \\
\| \\
CH_2
\end{array}
$$

Phosphoenolpyruvic acid

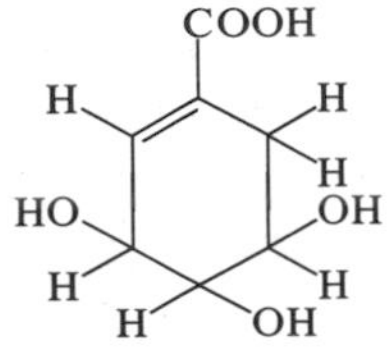

Shikimic acid

Role of glycine in porphyrin biosynthesis. The biologically important compounds chlorophyll, haemoglobin, and the cytochromes all contain a common cyclic structure, called a *porphyrin*, which consists of four pyrrole rings linked by methine bridges ($-CH=$) (Figs 6–18 and 7–12a). The bio-synthesis of this basic structure is therefore of considerable interest. Succinyl-coenzyme A from the Krebs cycle and the amino acid glycine are the starting points. They react to form α-amino-β-oxoadipic acid which readily loses CO_2 to give δ-aminolaevulinic acid (Fig. 7–12b). This reaction occurs in mitochondria, where succinyl-coenzyme A is available, and is catalysed by a specific synthetase which seems to be bound to mitochondria; pyridoxal phosphate must also be present. Then two molecules of δ-aminolaevulinic acid condense, again in the presence of a specific enzyme, to give the porphobilinogen unit. Four of these units finally undergo a rather complicated series of reactions to give the porphyrin structure. Thus all of the nitrogens in porphyrin come from the amino group of glycine. Isotopic studies have also shown that eight methylene

Fig. 7–11 Compounds involved in the biosynthesis of phenylalanine in micro-organisms and higher plants

carbon atoms derived from glycine are used in the synthesis of each porphyrin nucleus.

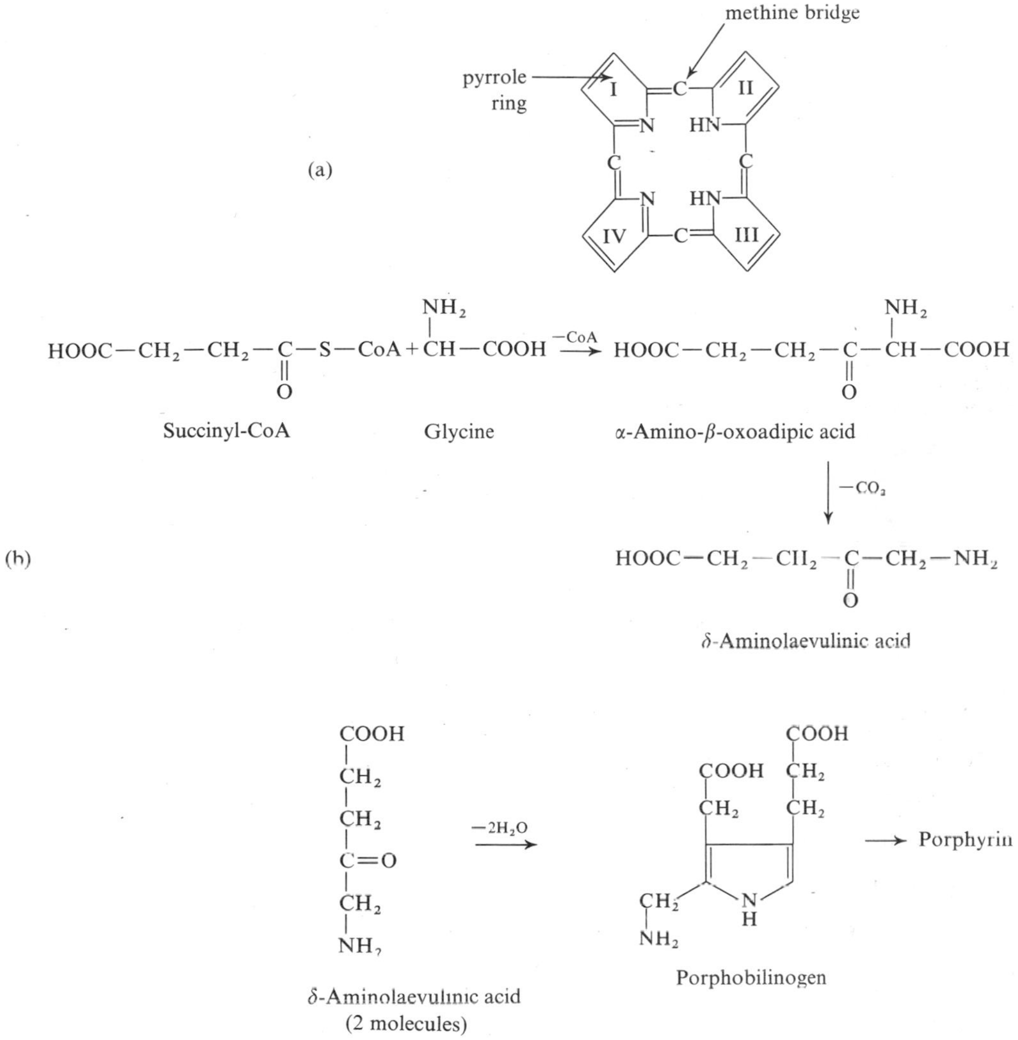

(a)

(b)

$$HOOC-CH_2-CH_2-\underset{\underset{O}{\|}}{C}-S-CoA+\overset{\overset{NH_2}{|}}{CH}-COOH \xrightarrow{-CoA} HOOC-CH_2-CH_2-\underset{\underset{O}{\|}}{C}-\overset{\overset{NH_2}{|}}{CH}-COOH$$

Succinyl-CoA   Glycine α-Amino-β-oxoadipic acid

$$\downarrow -CO_2$$

$$HOOC-CH_2-CH_2-\underset{\underset{O}{\|}}{C}-CH_2-NH_2$$

δ-Aminolaevulinic acid

$$\begin{array}{c} COOH \\ | \\ CH_2 \\ | \\ CH_2 \\ | \\ C=O \\ | \\ CH_2 \\ | \\ NH_2 \end{array} \xrightarrow{-2H_2O}$$

δ-Aminolaevulinic acid
(2 molecules)

Porphobilinogen

$\longrightarrow$ Porphyrin

Fig. 7–12   (a) The basic structure of porphyrin. (b) The biosynthesis of porphobilinogen units

7–3
Nucleotides

The name nucleotide was originally given to a class of compounds obtained by partial hydrolysis of nucleic acids. We have described the way in which DNA and RNA are synthesized in the cell from nucleotide building units, consisting of a purine or pyrimidine base, a pentose sugar, and a phosphate group (Chapter 3). The nucleic acids so formed occur in every living cell, and not only are they the means by which genetic information is transferred, but they also direct the synthesis of the many proteins made in cells.

However, there are other important nucleotides which do not occur in nucleic acids, and which may contain bases other than purines or pyrimidines. More generally, then, we can define a nucleotide as a compound containing a phosphorylated sugar linked glycosidically to a base. A nucleoside consists of a base and sugar component only and is obtained by the alkaline hydrolysis of a nucleotide, together with inorganic phosphate:

$$\text{base-sugar-phosphate} \xrightarrow[\text{hydrolysis}]{\text{alkaline}} \text{base-sugar} + H_3PO_4$$

Nucleotide   Nucleoside

Figure 7–13 shows the naturally occurring nucleotide formed from adenine (6-aminopurine), ribose, and a phosphate group attached in the 5′ position. The nucleoside of adenine and ribose is called adenosine, and the nucleotide itself is known as adenosine 5′-monophosphate (AMP), or 5′-adenylic acid. This is one of the most important naturally occurring nucleotides, in the form of its diphosphate (ADP) and triphosphate (ATP) derivatives. ADP and ATP play an essential role in the conservation and utilization of energy in many biochemical reactions in cells, their unique importance depending upon their ability to accept or donate a phosphate group.

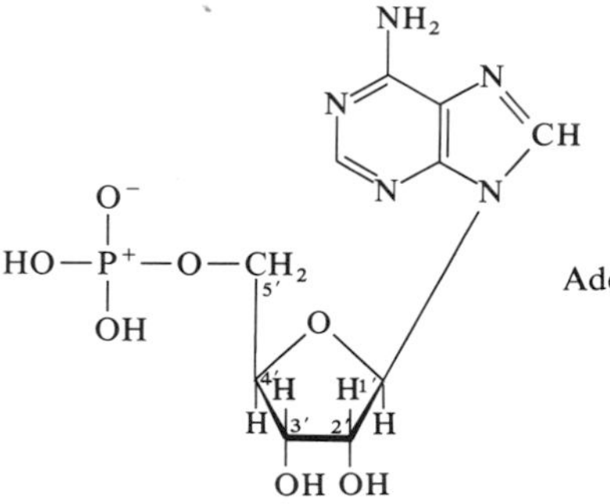

Adenosine 5′-monophosphate (AMP)
(adenylic acid)

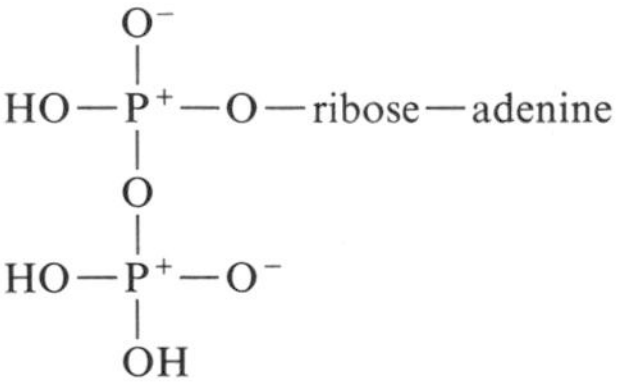

Adenosine diphosphate (ADP)

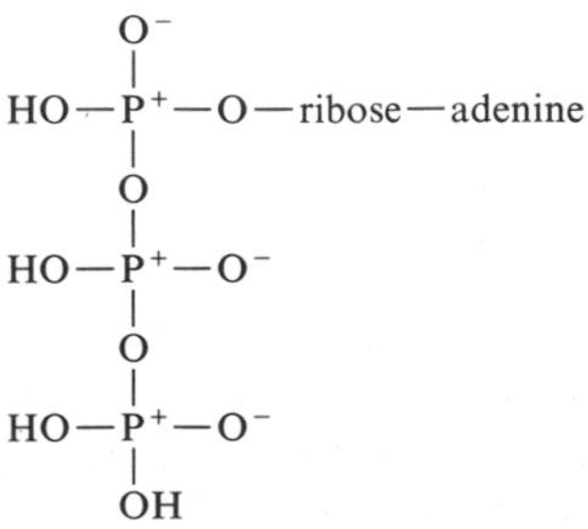

Fig. 7–13 The mono-, di-, and triphosphates of adenosine. Phosphate is attached to the 5′ position of the sugar ring

Adenosine triphosphate (ATP)

Another important nucleotide is nicotinamide–adenine dinucleotide (NAD$^+$) (Fig. 7–14). Here, an adenine mononucleotide is linked to a nicotinamide mononucleotide through an anhydride bond between the two phosphate groups. Nicotinamide, or niacin, is the vitamin component of the molecule, and is widely found in plant and animal tissues. The phosphate derivative of nicotinamide–adenine dinucleotide (NADP$^+$) is also shown in Fig. 7–14. Both of these coenzymes are widely

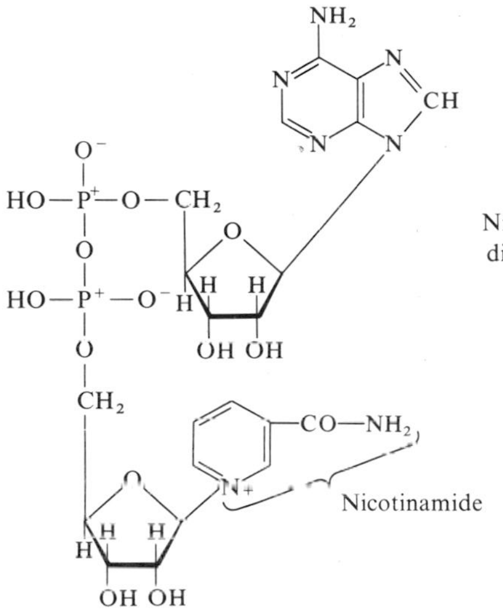

Nicotinamide–adenine dinucleotide (NAD$^+$)

Nicotinamide

(a)

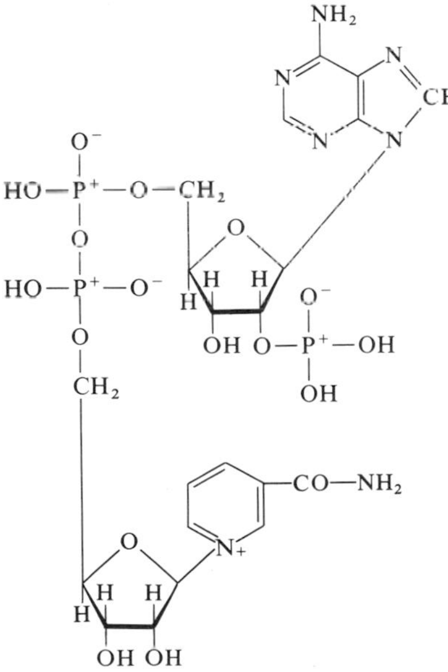

Nicotinamide–adenine dinucleotide phosphate (NADP$^+$)

Fig. 7–14 (a) Nicotinamide–adenine dinucleotide (NAD$^+$). (b) Nicotinamide–adenine dinucleotide phosphate (NADP$^+$)

(b)

distributed in cells because of their role in biological oxidation reactions. Their reduced forms (NADH or NADPH) result from reduction of the pyridine ring of the nicotinamide part of the molecule (Fig. 7–15).

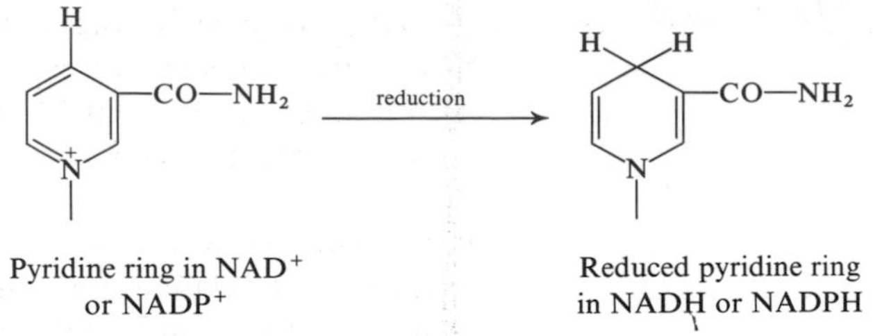

Fig. 7–15 NAD$^+$ and NADP$^+$ catalyse oxidation reactions and are themselves reduced

Pyridine ring in NAD$^+$ or NADP$^+$ reduction   Reduced pyridine ring in NADH or NADPH

Another important compound which contains a nucleotide sub-unit is coenzyme A (Fig. 7-16). Here, an adenine nucleotide is linked, again by a bond between two phosphate groups, to the non-cyclic molecule pantotheine; the vitamin component of the molecule is in this case pantothenic acid. Coenzyme A reacts with carboxylic acids to form thioesters, which are important in many biochemical reactions, in particular in the synthesis and the β-oxidation of fatty acid chains.

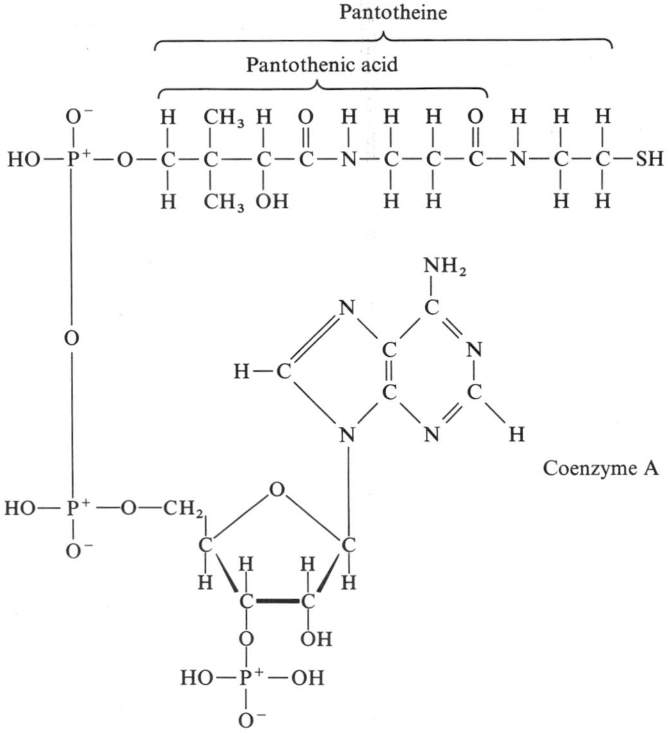

Fig. 7–16 Coenzyme A. The biochemically important reactions take place at the sulphur atom of the terminal SH group

The phosphate component of a nucleotide comes from ATP; the sugar component, whether it is a ribose sugar or a deoxyribose sugar in the completed nucleotide, is always derived from ribose, as will be shown below. The origin of ribose in cells is discussed in Section 7–4; here we are concerned with the origin of the bases in nucleotides, and first of all with the purine and pyrimidine bases. These are the only ones used in building the nucleotides of nucleic acids, and they also occur in various other important nucleotide compounds in cells. Purines and pyrimidines are not required in the diet of animals, and so are evidently synthesized *in vivo*. On the other hand, many micro-organisms require certain purines or pyrimidines, or even nucleosides such as thymidine (the deoxyriboside of thymine).

Pyrimidine synthesis. Work with isotope precursors in the 1940s showed that biosynthesis of the pyrimidine ring involves relatively simple substances (Fig. 7–17a)—ammonia, carbon dioxide, and aspartic acid. However, the way in which they are assembled is not at all simple; an indication of the steps involved is given in Fig. 7–17b. Orotic acid is formed as an intermediate; this then couples with ribose 5'-phosphate (formed from glucose by oxidative or anaerobic pathways) to

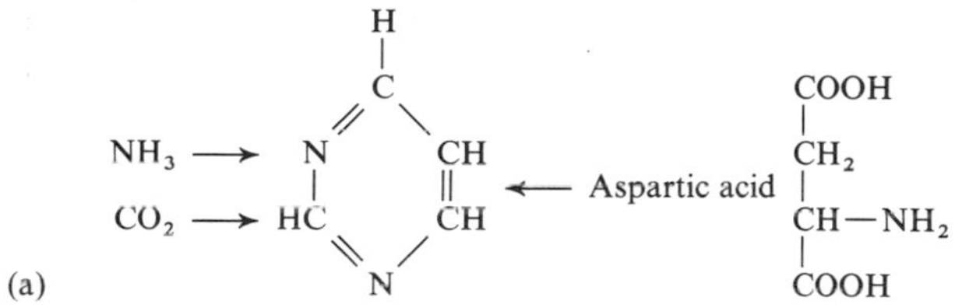

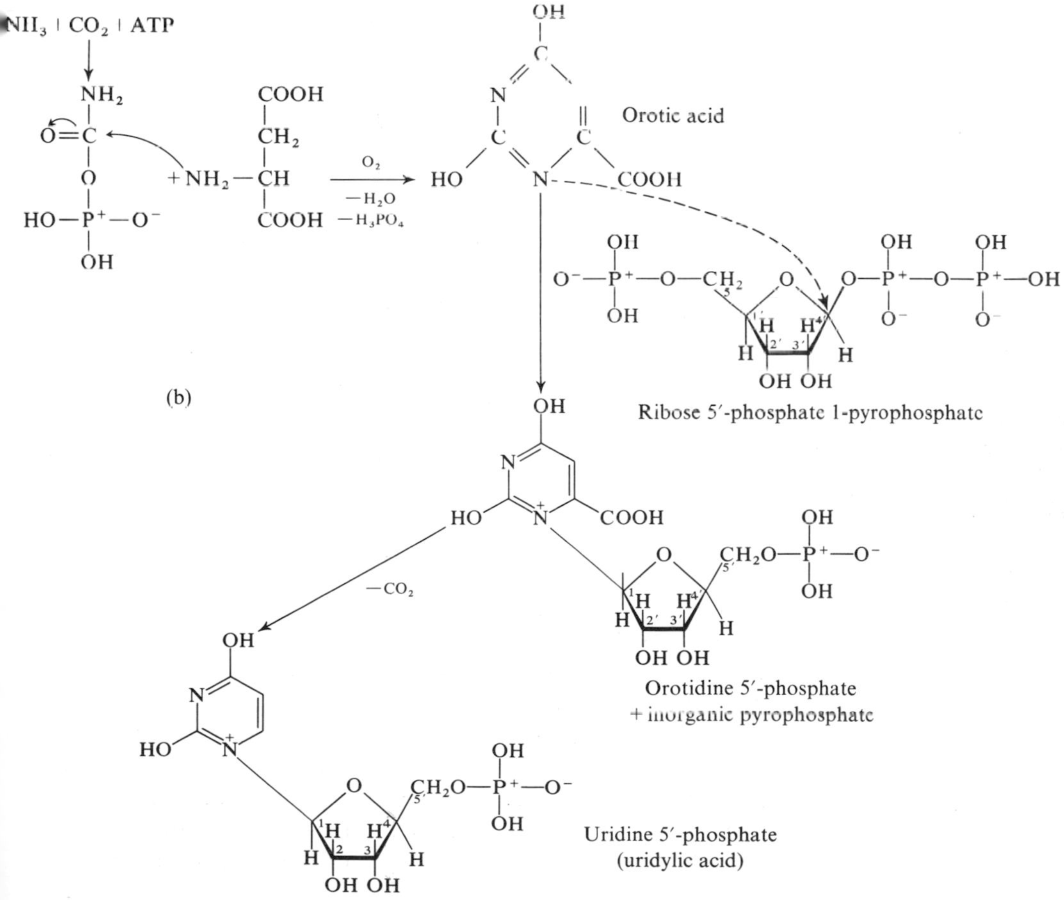

Fig. 7–17 **(a)** The biosynthesis of the pyrimidine ring from simple precursors. **(b)** An outline of the steps involved in the biosynthesis of the pyrimidine nucleotide, uridine 5'-phosphate

which a pyrophosphate group has been transferred from ATP. The pyrophosphate group is eliminated, and a glycosidic linkage is formed between the nitrogen of the base and the 1' position of the sugar, forming the nucleotide orotidine 5'-phosphate. This is then decarboxylated to form uridine 5'-phosphate or uridylic acid, one of the pyrimidine nucleotide building units for RNA, and the key nucleotide from which all other pyrimidine nucleotides for nucleic acid synthesis are made.

The only known pathway for the formation of a cytidine nucleotide is the amination of uridine triphosphate, formed by the addition of phosphate groups from ATP to uridine 5'-phosphate, in the presence of magnesium ions:

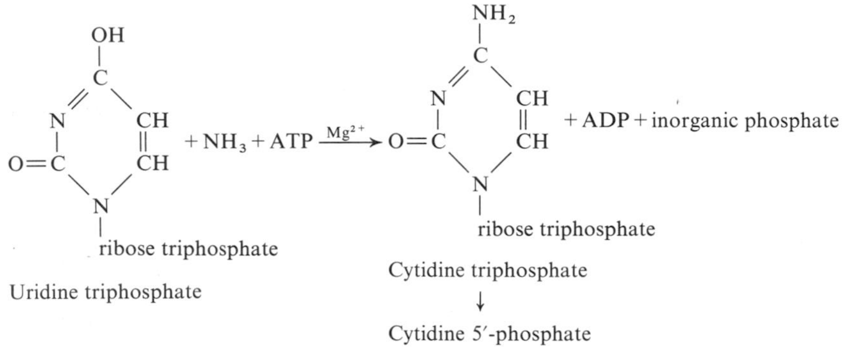

Uridine triphosphate

ribose triphosphate

$+ NH_3 + ATP \xrightarrow{Mg^{2+}}$

Cytidine triphosphate

ribose triphosphate

$+ ADP +$ inorganic phosphate

$\downarrow$

Cytidine 5'-phosphate

This then supplies both pyrimidine nucleotide units for RNA synthesis.

It seems improbable that DNA pyrimidine nucleotides are formed in the same way from deoxyribose. The evidence indicates that deoxygenation of the sugar occurs at the nucleotide level, the presence of vitamin B_{12} being necessary (Fig. 7–18). For instance,

Cytidine 5'-phosphate $\xrightarrow{\text{vitamin } B_{12}}$ Deoxycytidine 5'-phosphate

The other deoxypyrimidine nucleotide, thymidine 5'-phosphate, is apparently formed by the deoxygenation of the sugar in uridine 5'-phosphate, again in the presence of vitamin B_{12}, as in Fig. 7–18, followed by methylation of the pyrimidine ring:

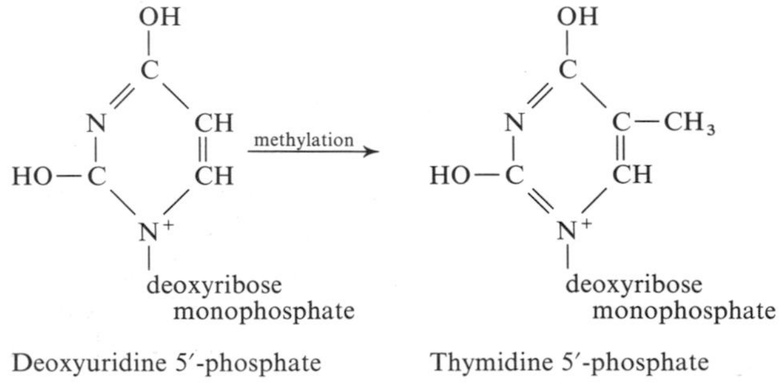

Deoxyuridine 5'-phosphate

deoxyribose monophosphate

$\xrightarrow{\text{methylation}}$

Thymidine 5'-phosphate

deoxyribose monophosphate

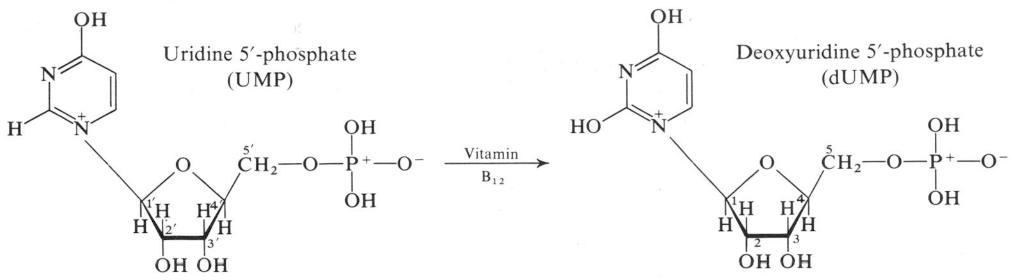

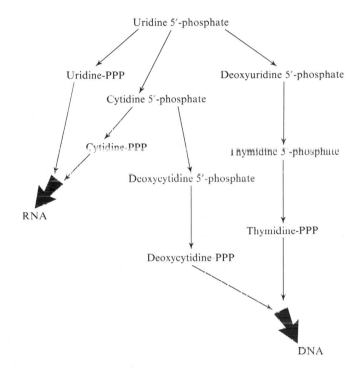

Fig. 7–18 The deoxygenation of ribose for DNA synthesis

The formation from the key nucleotide of uridylic acid of all the pyrimidine nucleotides used in the synthesis of DNA and RNA is summarized in Fig. 7–19.

Fig. 7–19 Synthesis of the pyrimidine nucleotide precursors of RNA and DNA from the key nucleotide uridine 5'-phosphate (uridylic acid). PPP represents triphosphate

Purine synthesis. The origin of the atoms in the purine ring was also established by the use of labelled precursors (Buchanan, 1959, and others). The building units in this case are carbon dioxide, formic acid, glycine, aspartic acid, and glutamine (Fig. 7–20a). Again, the series of reactions is somewhat complicated; an outline of the steps involved, which seem to be essentially the same for all organisms, is given in Fig. 7–20b. The purine ring system is built up in a stepwise manner on the C-1 position of ribose 5-phosphate, leading directly to the formation of a purine ribonucleotide; that is, free purine does not appear as an intermediate. Once more a key nucleotide is formed, in this case inosinic acid, which is then converted into the other purine nucleotides.

Inosine 5'-phosphate is converted into adenosine 5'-

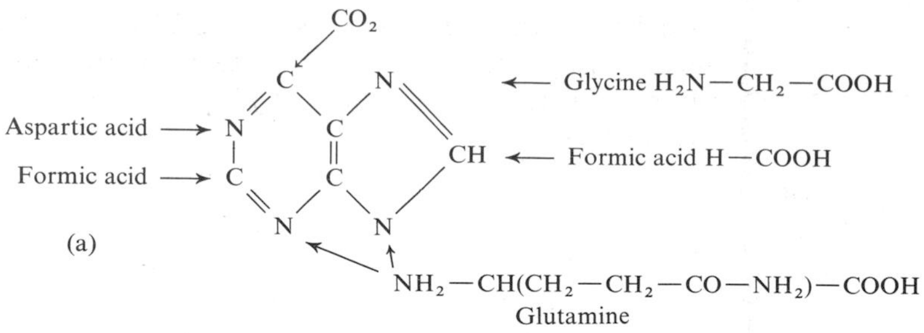

Fig. 7-20   **(a)** The biosynthesis of the purine ring system from simple precursors. **(b)** An outline of the steps involved in the biosynthesis of the purine nucleotide, inosine 5'-phosphate

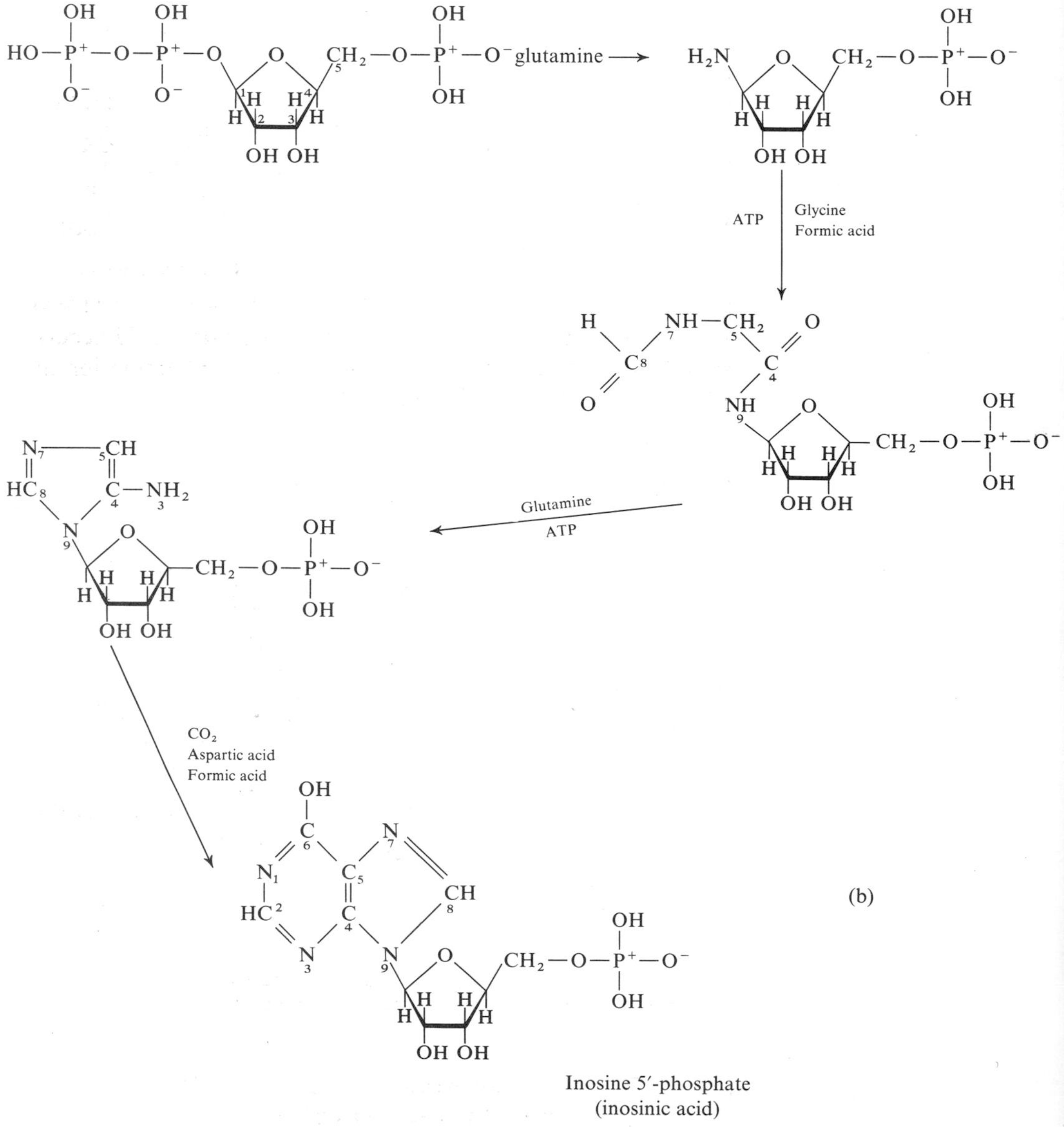

Inosine 5'-phosphate
(inosinic acid)

phosphate in the presence of aspartic acid and GTP (guanosine triphosphate, another nucleoside triphosphate which releases considerable energy on hydrolysis), via the succinic intermediate:

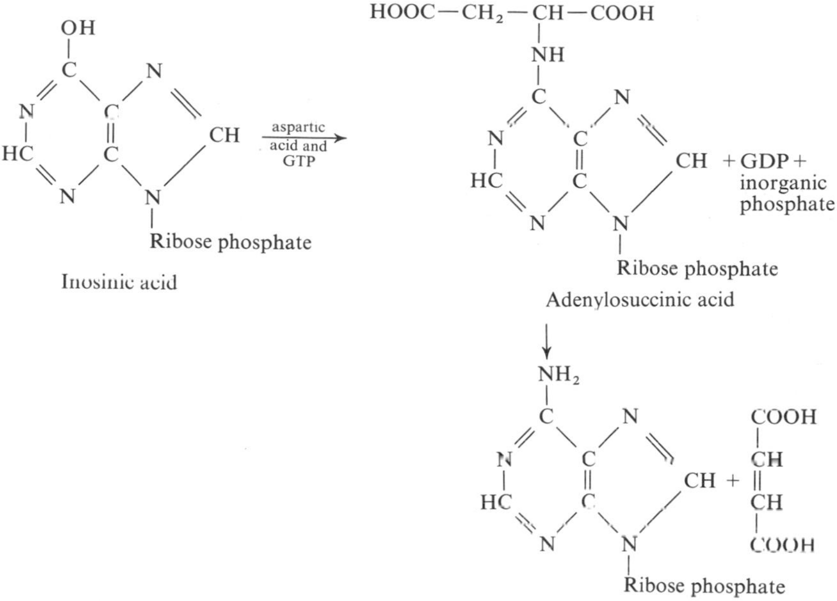

$$HOOC-CH_2-CH-COOH$$

Inosinic acid $\xrightarrow{\text{aspartic acid and GTP}}$ Adenylosuccinic acid $+\ GDP +$ inorganic phosphate

Inosinic acid *(structure: OH–C / N / HC / N / N / CH, Ribose phosphate)*

Adenylosuccinic acid *(structure with HOOC—CH₂—CH—COOH / NH, Ribose phosphate)*

↓

Adenylic acid (adenosine 5′-phosphate) *(structure: NH₂, + COOH / CH ‖ CH / COOH, Ribose phosphate)*

The synthesis of guanylic acid from inosinic acid occurs first by oxidation to xanthylic acid, followed by amination in the presence of glutamine and ATP:

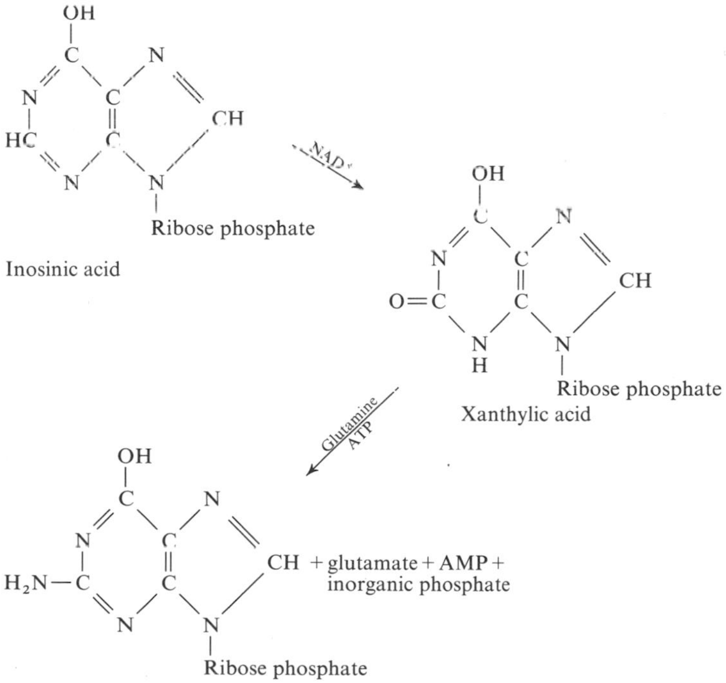

Inosinic acid *(structure: OH–C / N / HC / N / N / CH, Ribose phosphate)*

$\xrightarrow{\text{NAD}}$

Xanthylic acid *(structure: OH–C / N / O=C / N–H / N / CH, Ribose phosphate)*

$\xrightarrow{\text{Glutamine ATP}}$

Guanylic acid (guanosine 5′-phosphate) *(structure: OH–C / N / H₂N—C / N / N / CH, Ribose phosphate)* $+\ \text{glutamate} + \text{AMP} +$ inorganic phosphate

Again, the deoxyribonucleotide derivatives are obtained by reduction of the ribose sugar, involving vitamin B_{12}, at the nucleotide stage. The various stages involved in the synthesis of all the purine nucleotides are summarized in Fig. 7–21.

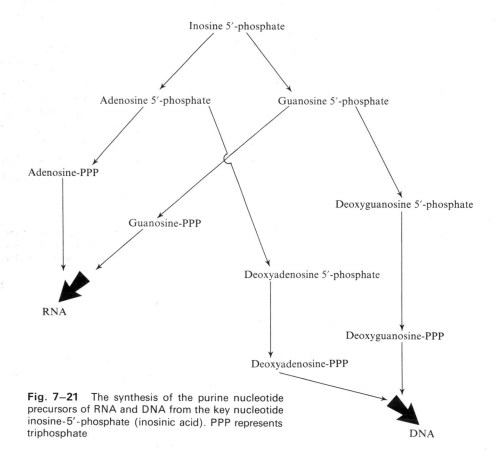

Fig. 7–21   The synthesis of the purine nucleotide precursors of RNA and DNA from the key nucleotide inosine-5′-phosphate (inosinic acid). PPP represents triphosphate

Synthesis of coenzyme nucleotides.   All of the ribonucleotides used in the synthesis of RNA are important in other metabolic processes, sometimes in combination with components not found in nucleic acids; for example, the coenzymes nicotin-amide–adenine dinucleotide (NAD$^+$) and coenzyme A.

Using NAD$^+$ (Fig. 7–14), the nicotinamide mono-nucleotide derivative is obtained by a reaction between nicotinamide, an important vitamin constituent of mammalian diet, and the pyrophosphate derivative of ribose 5′-phosphate. This then reacts with the other half of the molecule, adenosine 5′-phosphate, in the form of the triphosphate (ATP), a specific enzyme being required as catalyst:

Nicotinamide mononucleotide + adenosine triphosphate

$\downarrow$ enzyme

Nicotinamide–adenine dinucleotide + inorganic phosphate
(NAD$^+$)

Although nicotinic acid may be synthesized *in vivo*, the evidence indicates that it is not possible to prepare nicotinamide directly from this. The synthesis of nicotinic acid in fact leads to the formation of nicotinic mononucleotide, which then reacts with ATP to give the dinucleotide; finally the amide group of nicotinamide is added by means of a reaction involving glutamine and ATP.

With coenzyme A, the pantothenic acid part of the molecule is essential in the mammalian diet, although it may be synthesized in micro-organisms. Pantotheine (Fig. 7–16) is an intermediate in coenzyme A formation, as shown in the outline of the reaction processes (Fig. 7–22). First, pantothenic acid is phosphorylated on the terminal methyl group. This phosphate derivative then reacts at the carboxyl end of the molecule with cysteine in the presence of ATP to give phosphopantotheine, which finally reacts with ATP in the presence of magnesium ions to give coenzyme A (Fig. 7–22).

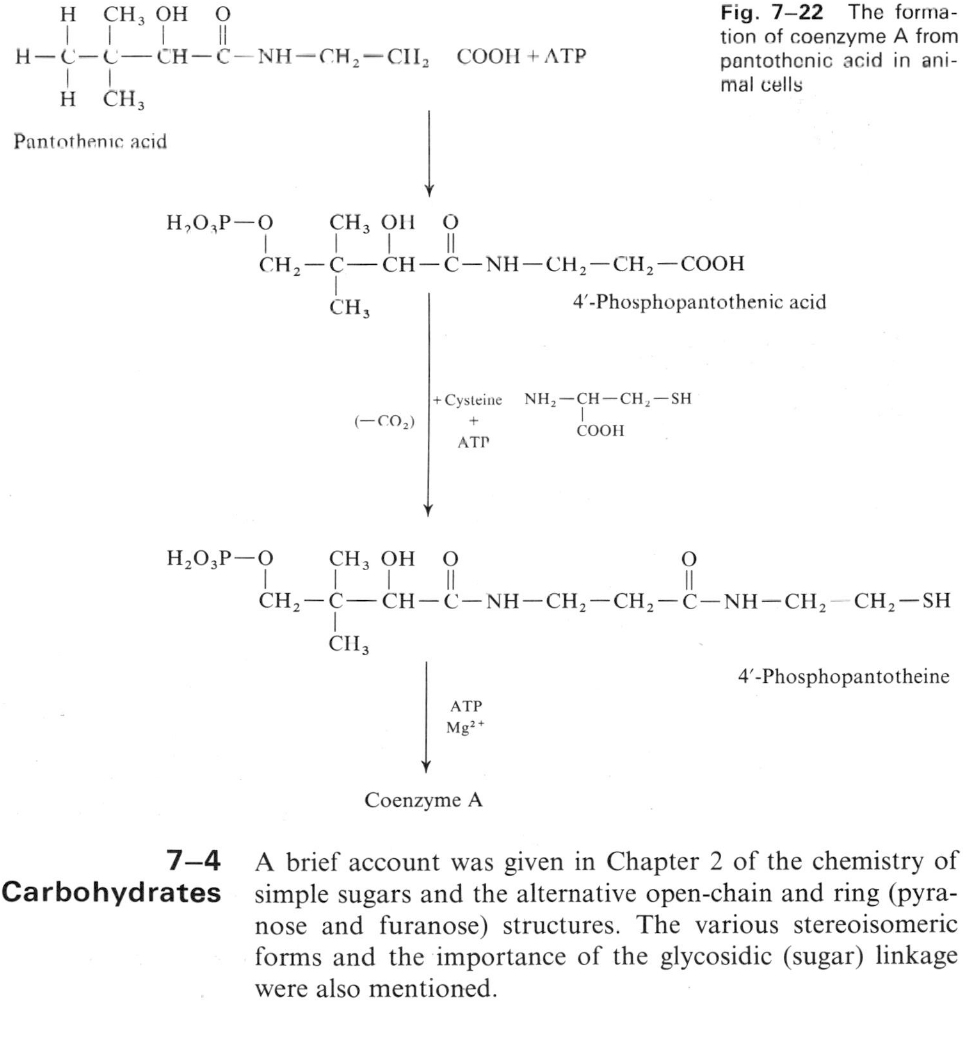

Fig. 7–22 The formation of coenzyme A from pantothenic acid in animal cells

7–4 Carbohydrates

A brief account was given in Chapter 2 of the chemistry of simple sugars and the alternative open-chain and ring (pyranose and furanose) structures. The various stereoisomeric forms and the importance of the glycosidic (sugar) linkage were also mentioned.

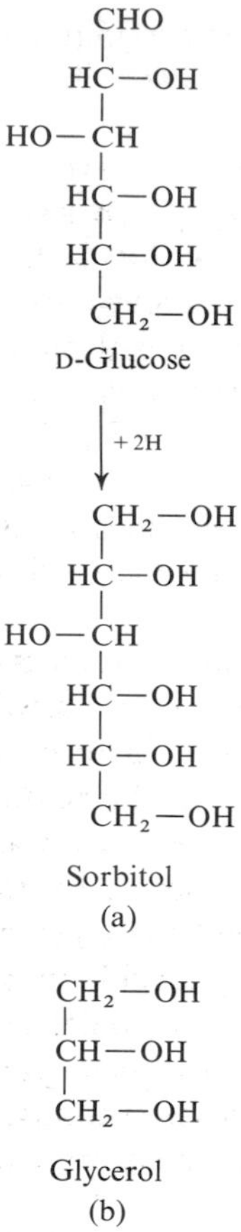

CHO
|
HC—OH
|
HO—CH
|
HC—OH
|
HC—OH
|
CH$_2$—OH

D-Glucose

| +2H
↓

CH$_2$—OH
|
HC—OH
|
HO—CH
|
HC—OH
|
HC—OH
|
CH$_2$—OH

Sorbitol
(a)

CH$_2$—OH
|
CH—OH
|
CH$_2$—OH

Glycerol
(b)

Fig. 7–23 The structure of sugar alcohols. **(a)** The reduction of glucose to sorbitol. **(b)** Glycerol

The production of glucose and fructose in photosynthesis, where carbon dioxide and water provide the necessary carbon, oxygen, and hydrogen atoms, was described in Chapter 6. Glucose is a source of energy in the cells of higher plants and animals, and it is also converted into other sugars which are important structurally and functionally. For instance, the pentose sugar ribose is of particular importance in cellular biochemistry because of its role in the synthesis of nucleotides.

Sugar alcohols are formed by the reduction of the aldehyde or ketone group of a sugar to the corresponding alcohol group. For instance, sorbitol is formed by the reduction of glucose (Fig. 7–23). Although sorbitol occurs widely in plants, it and the other C_6 compounds of the series are not of particular biochemical interest. But the C_3 compound, glycerol, is of very great importance in lipid biosynthesis (Section 7–5).

Sugar acids are also known, the most important group being the uronic acids (Fig. 7–24), some of which are components of structural polysaccharides (see below). These sugar acids tend to form lactones, or internal esters, usually with five-membered (γ-lactone) or six-membered (δ-lactone) rings. A sugar acid of great biological importance is vitamin C (ascorbic acid) which is widely distributed in nature in both animals and plants. As can be seen from Fig. 7–24, it is the γ-lactone of a C_6 acid with an unsaturated linkage between C-2 and C-3.

Amino sugars are formed by the replacement of a hydroxyl group by an amino group. Examples of this are D-glucosamine and D-galactosamine, which are substituted aldohexoses (Fig. 7–25), and which are important components of structural polysaccharides.

Most of the naturally occurring oligosaccharides occur in plants rather than in animals. The best known is sucrose—ordinary domestic sugar—which is the chief constituent of beet and cane sugar. It is a heterodisaccharide, giving rise on hydrolysis to one molecule of D-glucose and one molecule of

Fig. 7–24 The structure of uronic acids: **(a)** general formula, **(b)** γ-lactone, **(c)** δ-lactone, **(d)** D-glucuronic acid, **(e)** D-galacturonic acid, **(f)** ascorbic acid

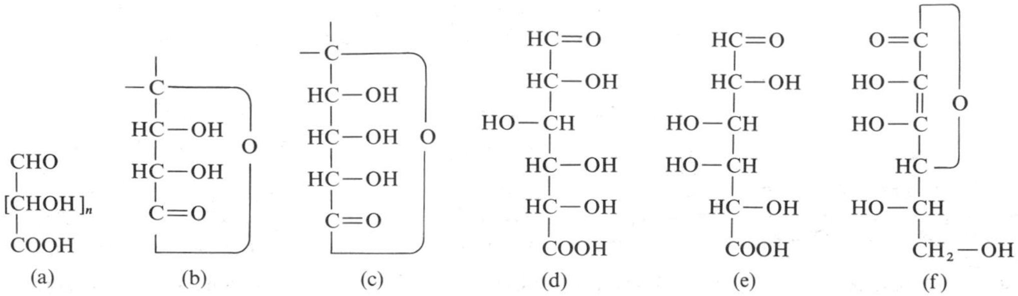

(a)
CHO
|
[CHOH]$_n$
|
COOH

(b)
—C—
|
HC—OH
|
HC—OH
|
HC—OH
|
C=O
  O

(c)
—C—
|
HC—OH
|
HC—OH O
|
HC—OH
|
C=O

(d)
HC=O
|
HC—OH
|
HO—CH
|
HC—OH
|
HC—OH
|
COOH

(e)
HC=O
|
HC—OH
|
HO—CH
|
HO—CH
|
HC—OH
|
COOH

(f)
O=C
|
HO—C
| O
HO—C
|
HC
|
HO—CH
|
CH$_2$—OH

CH₂—OH

HO

H   NH₂

D-Glucosamine

(a)

CH₂—OH

HO

H   NH₂

D-Galactosamine

(b)

CH₂—OH

HO

H   NH—CO—CH₃

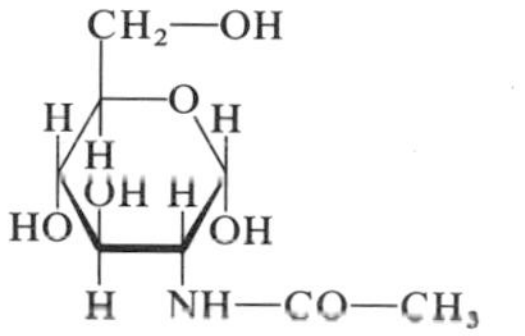

N Acetyl D glucosamine

(c)

Fig. 7–25 The structure of amino sugars: **(a)** D-glucosamine, **(b)** D-galactosamine, **(c)** N-acetyl-D-glucosamine

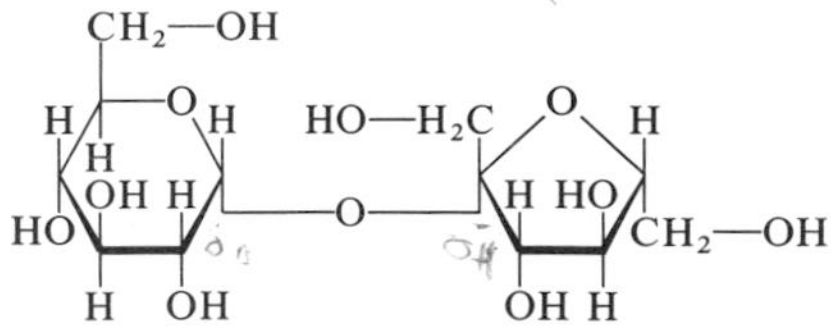

D-fructose. The two halves of the molecule are joined by a glycosidic linkage between the C-1 of glucose and the C-2 of fructose (Fig. 7–26).

Maltose is not a common naturally occurring disaccharide but it is the major product of the enzymic hydrolysis of starch (Fig. 7–27a). Maltose consists of two molecules of glucose joined between C-1 and C-4 by an α-linkage (Section 2–2). Cellobiose, which is the disaccharide formed from the hydrolysis of cellulose, is identical to maltose except that the two glucose molecules are joined by a β-linkage (Fig. 7–27b).

Another oligosaccharide of interest is lactose, a disaccharide in milk. On hydrolysis it gives one molecule of D-glucose and one of D-galactose. The two monosaccharides are joined by a β-linkage.

Other oligosaccharides contain varying numbers of simple carbohydrates. In addition there are compounds which are glycosides, since the two parts of the molecule are joined by a glycosidic linkage, but in which only one of the parts is a sugar molecule. The second, non-sugar, part is known as an aglycon. Most of these glycosides occur in plants. Examples are indican (a compound related to the dye indigo), a glucoside derivative of indoxyl, and amygdalin, the glycoside in oil of bitter almonds, which on hydrolysis gives two molecules of glucose, one of benzaldehyde, and one of hydrogen cyanide. The drugs of the digitalis group are glycosides composed of di- and tri-saccharides linked to aglycons which are closely related to steroids.

Most naturally occurring carbohydrates exist in the form of high molecular weight polysaccharides, which serve either a nutrient or structural function. The main nutrient polysaccharides are starches in plant cells and glycogen in animal cells. The fundamental building unit in both cases is D-glucose, which is the only monosaccharide obtained on complete hydrolysis.

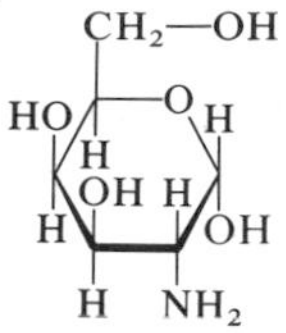

The starches consist of two types of compound, present

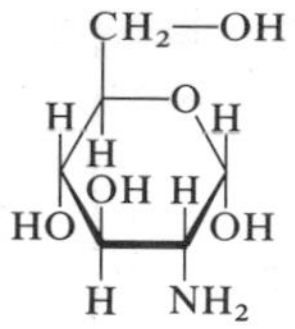

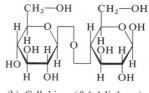

Fig. 7–26 The structure of sucrose, a disaccharide

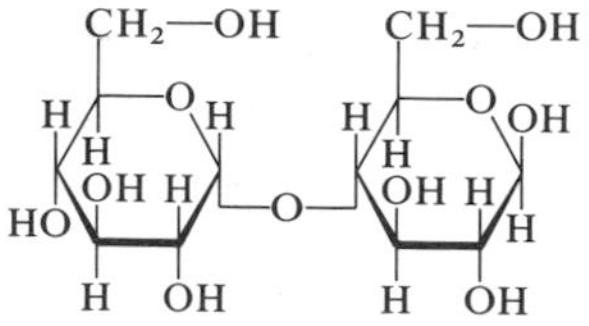

CH₂—OH

HO—H₂C

HO

H   OH

OH H

Glycosidic linkage, α-configuration

Fig. 7–27 **(a)** Maltose, the repeating unit of starch (α-1,4-linkage). **(b)** Cellobiose, the repeating unit of cellulose (β-1,4-linkage)

CH₂—OH   CH₂—OH

HO

H   OH   H   OH

(a) Maltose (α-1,4-linkage)

CH₂—OH   CH₂—OH

HO

H   OH   H   OH

(b) Cellobiose (β-1,4-linkage)

in varying amounts. The first, amylose, is a long straight-chain molecule composed of repeating maltose units—that is, the linkage between the glucose molecules is always of the α-1,4 type (Fig. 7–27a). The molecular weight of potato amylose varies from 4,000 to 150,000. The other component, amylopectin, is a branched-chain polysaccharide. In amylopectin, chains of 24 to 30 glucose units, with α-1,4-linkages, are joined to each other by α-1,6-linkages; at the point of branching, the glucose residue is substituted at the C-6 as well as at the C-4 position (Fig. 7–28a). One of the product of hydrolysis, as would be expected, is the α-1,6-disaccharide, isomaltose (Fig. 7–28b). The molecular weight of amylopectin may be 500,000 or more. The polymer chains are not straight but form hollow helices (Fig. 7–28c and d).

Reservoirs of starch in cells, in the form of starch granules, are made available for various metabolic processes by the action of certain enzymes, the amylases. These enzymes,

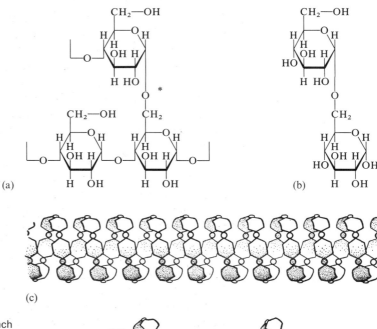

(a)

(b)

(c)

Fig. 7–28 **(a)** A branch point in amylopectin (or glycogen) at an α-1,6-linkage.* **(b)** The isomaltose unit obtained on the incomplete hydrolysis of amylopectin or glycogen. **(c)** Molecular structure of polymer chains in starch—unbranched form (amylose) and **(d)** the branched form (amylopectin). (With permission from *Principles of Plant Physiology* by James Bonner and Arthur W. Galston. W. H. Freeman and Company. Copyright © 1952.)

(d)

which catalyse the hydrolysis of starches, may be divided into two classes. The first group splits the α-1,4-linkages in an apparently random way, leading to a mixture of glucose and maltose; the second group successively removes maltose units from polysaccharide chains. The α-1,6-linkage is not attacked, so that for amylopectin the hydrolysis is incomplete, and when carried to the limit a polysaccharide fragment known as a limit dextrin still remains.

The counterpart of starch is animal cells is the nutrient polysaccharide glycogen which occurs chiefly in liver and muscle. Its structure is very similar to that of amylopectin; it is a branched molecule, with α-1,4-linkages between the glucose molecules of the straight chains, and α-1,6-linkages at the branches (Fig. 7–28a). The chain length, however, is shorter than that of amylopectin, and varies from 8 to 20 glucose molecules, so that the structure is even more branched. The molecular size is also much larger than that of amylopectin, and molecular weights of up to 100,000,000 have been reported. Glycogen is present in animal cells in particles which are much smaller than starch granules in plant cells. Enzymic attack of glycogen in animal cells by phosphorylases does not lead to the formation of glucose, but directly to the phosphorylated form of glucose 1-phosphate. This is then easily converted by enzymes into its isomer, glucose 6-phosphate, which enters directly into the Embden–Meyerhof pathway without further energy being required. The whole process is thus extremely efficient.

There are other less important nutrient polysaccharides, such as inulin, a polysaccharide found in certain plants, which is formed of fructose units.

Of the many structural polysaccharides, the commonest is cellulose which, in fact, accounts for about one half of the carbon present in all vegetation. It is a major component of higher plants and is found in its purest form in cotton, which is approximately 90 per cent pure cellulose. Unlike starch, cellulose is a straight-chain polysaccharide formed from glucose units joined by β-1,4-linkages (Fig. 7–27b). Its molecular weight can vary from 200,000 to 2,000,000. In large aggregates the chains may be arranged parallel to each other. Partial hydrolysis gives the cellobiose units mentioned earlier, while complete hydrolysis yields D-glucose.

Another structural polysaccharide of plants is xylan, which is associated with cellulose in wood. It is a polymer of units of D-xylose, a pentose sugar,

$$OHC-[CH(OH)]_3-CH_2(OH),$$

and probably has a branched chain structure.

Many plant tissues, and especially fruits, contain inter-cellular materials known as pectins, which are polymers of galactose (a hexose sugar), arabinose (a pentose), and galac-

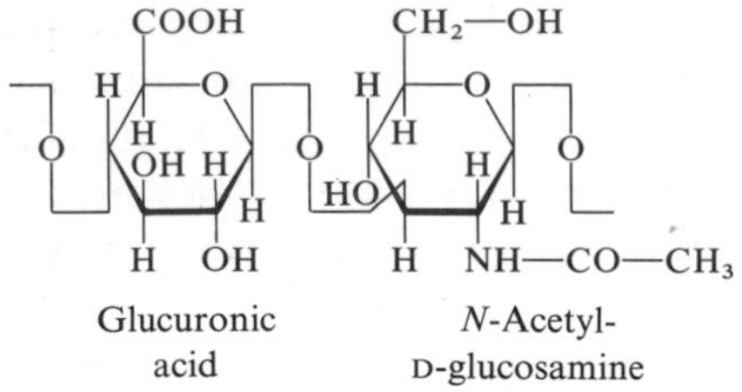

Glucuronic acid   N-Acetyl-D-glucosamine

Fig. 7–29 The building unit of hyaluronic acid

turonic acid. The pectic acid component of pectin consists of chains of the acidic sugar D-galacturonic acid (Fig. 7–24).

The important structural polysaccharide of invertebrates is chitin, which is found in large amounts in the shells of crustaceans and insects, and is probably the most abundant polysaccharide after cellulose. Chitin is a polymer of N-acetyl-D-glucosamine (Fig. 7–25), in which the amino group on the C-2 of glucosamine is substituted with an acetyl group.

Widely distributed throughout animal connective tissue there are large quantities of a group of polysaccharides, each of which contains a repeating disaccharide unit of an amino sugar and a uronic acid. The most abundant of these is hyaluronic acid (Fig. 7–29), which appears to act as a cementing substance in the subcutaneous tissue of animals. Solutions of hyaluronic acid are extremely viscous; the synovial fluid of joints has a lubricating action which is largely due to its hyaluronic acid content. The term mucopolysaccharide is often applied to this and related compounds. The building units for hyaluronic acid are D-glucuronic acid and N-acetyl-D-glucosamine, joined by alternate β-1,3- and β-1,4-linkages (Fig. 7–29).

Chondroitin is another animal polysaccharide, differing from hyaluronic acid only by the presence of a substituted galactosamine instead of a glucosamine. It occurs chiefly as the two sulphate derivatives, found in cartilage and tendons. These two derivatives, chondroitin sulphate A and chondroitin sulphate C, differ in the position in which the sulphate group is esterified to the galactosamine molecule (Fig. 7–30).

Another important member of the group of sulphated mucopolysaccharides is heparin, a powerful anticoagulant, present chiefly in the liver, the lungs, and the walls of large arteries. In heparin both the glucuronic acid and the glucosamine sub-units are sulphated.

Fig. 7–30 (a) The building unit of chondroitin sulphate A. **(b)** The building unit of chondroitin sulphate C

(a) Glucuronic acid   Chondroitin sulphate A

(b) Glucuronic acid   Chondroitin sulphate C

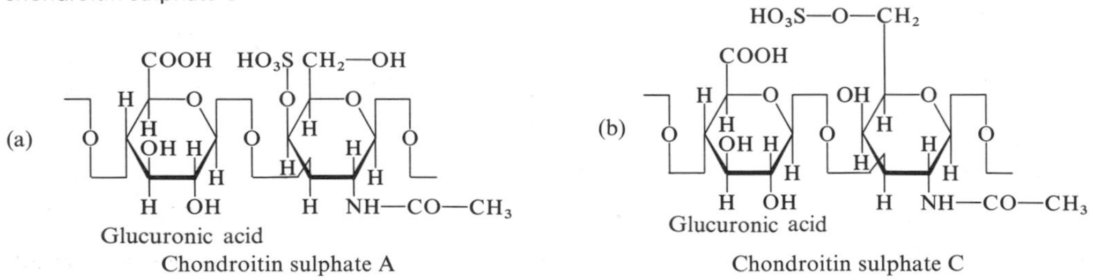

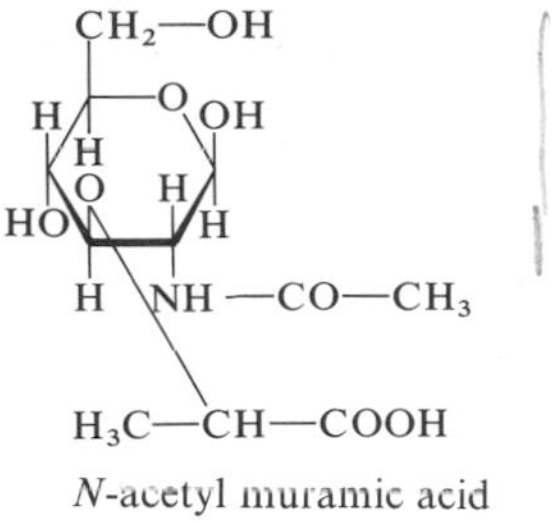

CH₂—OH structure (N-acetyl muramic acid)

N-acetyl muramic acid

Fig. 7–31 The structure of muramic acid

In tissues, acid mucopolysaccharides appear to be associated with proteins, forming mucoproteins. The cell surface complex of mammalian cells contains mucoproteins in which the acid sugar occurring most frequently is sialic acid (*N*-acetylneuraminic acid). Glycolipids, or gangliosides—compounds of oligosaccharides linked to lipid—occur chiefly in nerve tissue and in the spleen. Polysaccharides are also involved in the specificity of the blood group substances, and the antigenic properties of animal tissue cells and bacterial cells. Several of these compounds are discussed further in Chapter 8.

Finally, the rigidity and strength of bacterial cell walls is due to the presence of an unusual polysaccharide, formed of units of *N*-acetylglucosamine and a complex sugar derivative known as muramic acid (Fig. 7–31). Muramic acid is also an *N*-acetylamino sugar which is joined to the C3 compound lactic acid. In bacterial walls the two amino sugars are joined by β-1,4-linkages. This is the linkage which is broken by the enzyme lysozyme, as already described.

Certain bacteria and some yeasts also elaborate polysaccharides known as dextrans. These consist of glucose units, joined mostly by α 1,6 linkages, but with some branching.

There is another group of polymers, found both in bacterial cell walls and also intracellularly, known as teichoic acids. These are not polysaccharides, but polyols, in which the sugar alcohol phosphates are linked by phosphodiester bonds. These linear chains carry alanine molecules and carbohydrates linked to the hydroxyl groups. Bacterial walls will be discussed further in Chapter 14.

Biosynthesis of carbohydrates. *Hexoses.* The various hexose sugars are all derived from glucose by a few general interconversion reactions. The initial step is always the phosphorylation of glucose by ATP to give glucose 6-phosphate. The interconversion of glucose 6-phosphate and fructose 6-phosphate in glycolysis is a reversible reaction catalysed by the enzyme phosphoglucose isomerase. Mannose, a hexose sugar in which the positions of the —OH and —H on C-2 are reversed

Fig. 7–32 Methods of interconversion of various hexose sugars

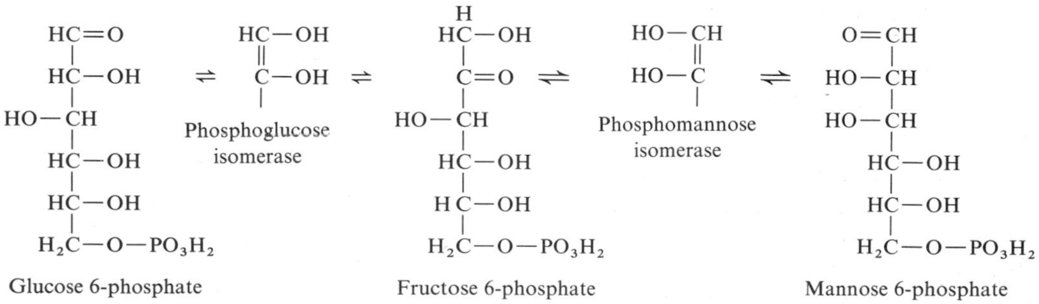

Glucose 6-phosphate   Phosphoglucose isomerase   Fructose 6-phosphate   Phosphomannose isomerase   Mannose 6-phosphate

Fig. 7–33 The formation of the important intermediate, uridine diphosphate glucose (UDP-glucose)

Glucose 1-phosphate

Uridine triphosphate

Uridine diphosphate glucose (UDP-glucose)

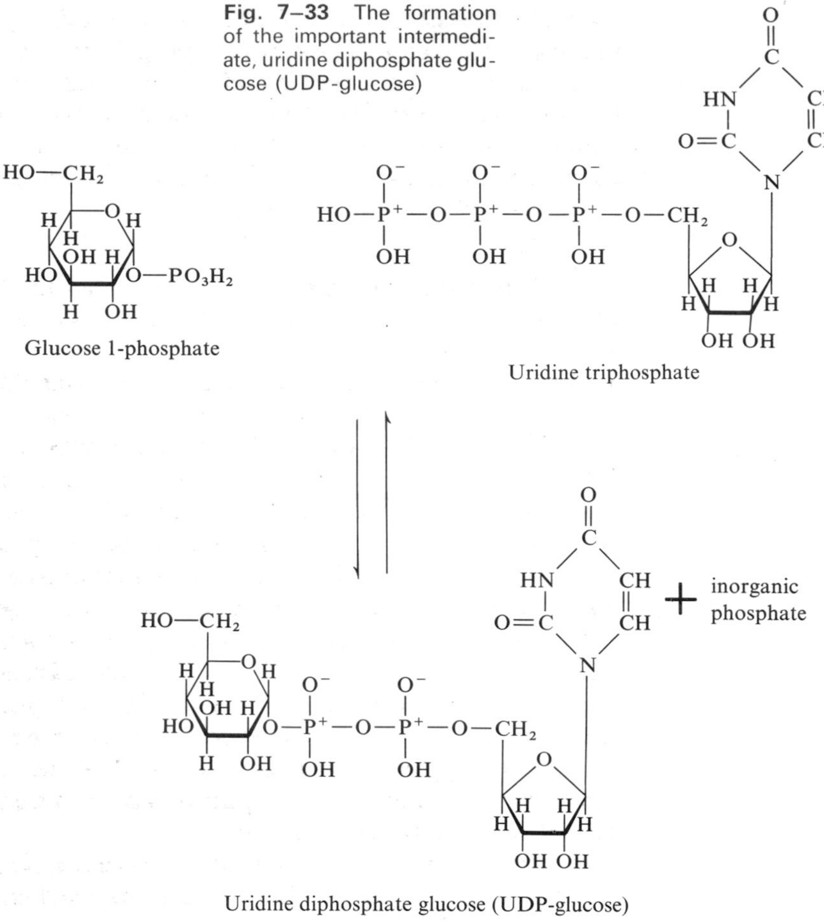

with respect to glucose, is formed by a similar conversion from fructose 6-phosphate (Fig. 7–32). The specific enzyme in this case is phosphomannose isomerase.

Many of the important reactions in carbohydrate metabolism involve the nucleoside diphosphate esters of sugar molecules. The formation of uridine diphosphate glucose (or

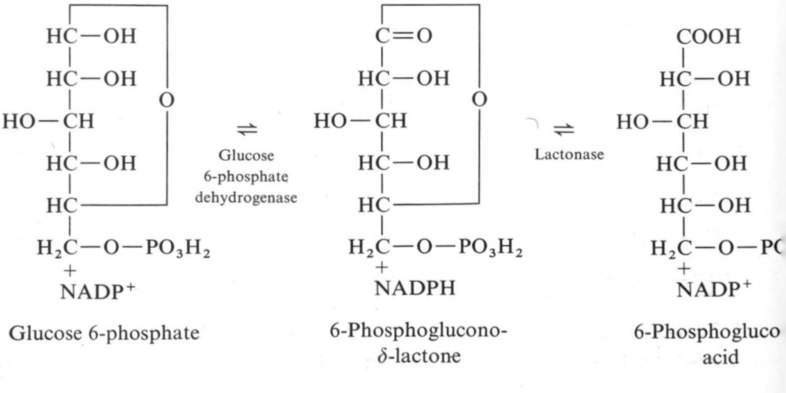

Fig. 7–34 The oxidative pathway of glucose metabolism

Glucose 6-phosphate

6-Phosphoglucono-δ-lactone

6-Phosphogluco acid

UDP-glucose) is shown in Fig. 7–33. A specific pyrophos-phorylase is required. The hexose sugar galactose, an isomer of glucose in which the positions of the —OH and —H on carbon atom 4 are reversed with respect to glucose, is formed from UDP-glucose. An inversion (known as a Walden inversion) of the groups about C-4 occurs through a specific enzyme called an epimerase:

UDP-glucose ⇌ UDP-galactose

There are other interconversions between the hexoses, which are catalysed by enzymes known as mutases and trans-ferases.

Ribose. The pentose sugar ribose is formed from glucose by an oxidative pathway known as the hexose monophosphate shunt. This is an alternative to the glycolytic Embden–Meyer-hof pathway of glucose metabolism. The steps involved in the oxidative pathway have been elucidated by Warburg and other workers. The first step is the formation of glucose 6-phosphate, which then undergoes the series of reactions shown in Fig. 7–34, each of which requires the presence of a specific enzyme. The keto-phosphogluconic acid is enclosed in a bracket as it has not been isolated, and is probably a transient intermediate which is rapidly decarboxylated to give ribulose 5-phosphate. This then undergoes isomerization to ribose 5-phosphate, which, when pyrophosphorylated by ATP, is the form in which ribose combines with purines and pyrimidines in nucleotide synthesis (Section 7 3).

Ribose may also be produced by an anaerobic process from fructose 6-phosphate. The various steps involved require the presence of the usual glycolytic enzymes, transketolase, transaldolase, and the enzymes necessary for pentose inter-conversions. The net reaction is:

Fructose 6-phosphate (2 molecules) +
glyceraldehyde 3-phosphate (1 molecule) ⟶
ribose 5-phosphate (3 molecules)

Uronic acids. Synthesis of uronic acids affords another example of reactions in which the nucleoside diphosphate sugars are the active intermediates. A specific enzyme is

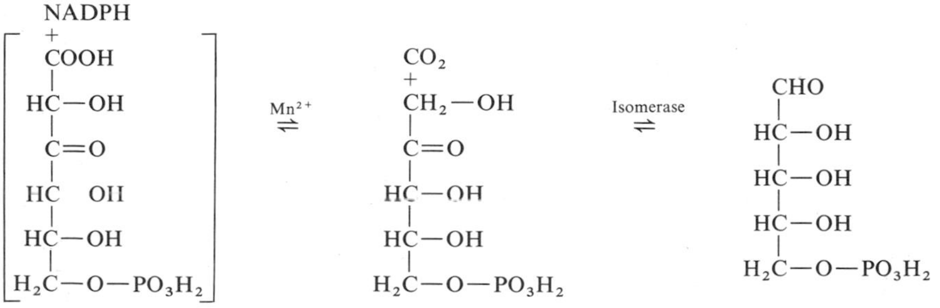

NADPH
+
$$\begin{bmatrix} COOH \\ | \\ HC-OH \\ | \\ C=O \\ | \\ HC\ \ OII \\ | \\ HC-OH \\ | \\ H_2C-O-PO_3H_2 \end{bmatrix} \underset{Mn^{2+}}{\rightleftharpoons} \begin{array}{c} CO_2 \\ + \\ CH_2-OH \\ | \\ C=O \\ | \\ HC-OH \\ | \\ HC-OH \\ | \\ H_2C-O-PO_3H_2 \end{array} \underset{Isomerase}{\rightleftharpoons} \begin{array}{c} CHO \\ | \\ HC-OH \\ | \\ HC-OH \\ | \\ HC-OH \\ | \\ H_2C-O-PO_3H_2 \end{array}$$

3-Keto-6-phospho-gluconic acid | D-Ribulose 5-phosphate | D-Ribose 5-phosphate

required to promote the oxidation by NAD^+ of the alcohol group of the sugar to the carboxyl group of the uronic acid. For instance:

$$UDP\text{-glucose} + NAD^+ \longrightarrow UDP\text{-glucuronic acid} + NADH$$

Amino sugars. These are formed by the transfer of the $-NH_2$ of the amide group of glutamine to the 6-phosphate esters of the keto sugars. For instance, in glucosamine synthesis:

$$Fructose\ 6\text{-phosphate} + glutamine \longrightarrow$$
$$glucosamine\ 6\text{-phosphate} + glutamic\ acid$$

The *N*-acetylamino sugars which occur frequently in polysaccharides are formed by acetylation of the 6-phosphate esters of the corresponding amino sugars by acetyl-coenzyme A.

Disaccharides. The most common method of disaccharide formation appears to be via the nucleoside diphosphate derivative of one of the simple sugars. For instance, sucrose synthesis has been observed, with various plant extracts, to take place as follows:

$$UDP\text{-glucose} + fructose\ 6\text{-phosphate} \rightleftharpoons sucrose\ 6'\text{-phosphate} + UDP$$

There is another type of reaction, catalysed by enzymes known as phosphorylases, in which a sugar phosphate reacts with another sugar to form a disaccharide and inorganic pyrophosphate:

$$\alpha\text{-D-Glucose}\ 1\text{-phosphate} + \text{D-fructose} \rightleftharpoons sucrose + inorganic\ phosphate$$

This is a reversible reaction, as shown, and in practice the free energy is such that the equilibrium is well to the left; thus the phosphorolysis of sucrose is more significant than sucrose formation.

For glycosides consisting of a mono- or disaccharide combined with an aglycon moiety, the evidence indicates that formation takes place via an enzyme-catalysed reaction between the sugar nucleoside diphosphate and the aglycon.

Polysaccharides. Many bacteria can synthesize linear polysaccharides from existing disaccharides. For example, *E. coli* can synthesize amylose from maltose by the following reaction, catalysed by the enzyme amylomaltase:

$$Maltose\ (n\ molecules) \rightleftharpoons amylose + glucose\ (n\ molecules)$$

Only the glucose unit which gives its carbon atom 1 to the glycosidic bond of the disaccharide can be used in this way for polysaccharide synthesis; the other glucose unit is released.

In plants and animals, most polysaccharides are formed directly from the appropriate hexose and pentose units rather than from disaccharides. Synthesis involves the nucleoside diphosphate (or occasionally monophosphate) sugar. For instance, synthesis of the amylose chain of glycogen in liver proceeds by the following reaction, catalysed by the enzyme glycogen synthetase:

$$\text{UDP-glucose } (n \text{ molecules}) \rightleftharpoons (\text{glucose})_n + \text{UDP } (n \text{ molecules})$$

Amylose

A trace of 'primer' is necessary for the reaction, preferably in the form of polysaccharide, of dextrin derived from polysaccharide, or of glucose oligosaccharide containing not less than four glucose units. This is generally true of polysaccharide biosynthesis; formation cannot take place from monosaccharides alone, but requires the presence of a larger carbohydrate unit as a primer.

The above reaction results in the synthesis of the straight chains of glycogen, with α-1,4-linkages only between the glucose groups. The branched structure, resulting from the presence of α-1,6-bonds, is due to the activity of a transglycosylase, or 'branching enzyme', which occurs in liver, muscle, and brain tissue. This enzyme cleaves specific fragments of straight-chain polysaccharide at the α-1,4-linkages and transfers them to the same or another glycogen molecule in the α-1,6-linkage.

The synthesis of starch in plants appears to be analogous to that of glycogen in animals. The building unit is adenosine diphosphate glucose, and the action of a synthetase enzyme results in the formation of straight-chain amylose. The branched chain of amylopectin is again formed by means of a transglycosylase.

Little is known about the reactions involved in the synthesis of heteropolysaccharides. Here also the nucleoside diphosphate derivatives of the monosaccharide sugars are involved, and the final assembly of the disaccharide repeating units by means of a synthetase enzyme probably also occurs via the nucleoside diphosphate disaccharide derivative.

7–5
Lipids

The important part played by lipid molecules in the structure and function of cellular membranes has already been described. Lipids are, too, the most concentrated source of fuel in animals; weight for weight they yield twice as many calories as do carbohydrates or proteins. Most of the lipid in animal diets is triglyceride of animal or vegetable origin. The way in which the carbon chains of the fatty acids of lipids are degraded in the cell to yield carbon dioxide, water, and a supply of energy was described in Chapter 6. In mammals, lipid is mostly in the form of triglyceride; it is distributed in various organs and particularly in adipose tissue, in which a large proportion of the cytoplasm of highly specialized connective tissue cells appears to be replaced by lipid droplets, these regions forming reservoirs of stored energy. Animal adipose tissue may contain 80% or more of liquid. But in plants, it is only in seeds and fruits that lipid is found in large amounts. Here it is primarily a food store for the developing embryo.

We have already noted that lipid is important as an electrical insulator in membrane structure. It is also a thermal insulator: with mammals, much of the body lipid is located under the skin, preventing excessive heat loss to the environment. The most impressive example of this is the whale, with its thick layer of adipose tissue, known as blubber. Another function of subcutaneous lipid, insulation against mechanical trauma, is exemplified in the reservoir of highly specialized lipid known as spermaceti (cetyl palmitate), in the head of the sperm whale, which enables it to make destructive blows with its head. The lipid in plant seeds and fruits presumably provides thermal insulation against the environment, and also prevents moisture loss.

Apart from certain unsaturated fatty acids, lipid is apparently not essential in the diet; it is well known that animals, including man, can become obese on a high carbohydrate diet which is low in lipid. This means that the body must be able to synthesize lipids from the degradation products of other nutrients. The manufacture of lipids inside cells, which occurs mainly in the liver, depends upon two processes: the build-up of fatty acid chains, and the formation of an ester linkage between these chains and an alcohol, to form neutral fats (triglycerides) and phospholipids.

Fatty acid synthesis.   Fatty acids of natural origin are mostly straight-chain, unbranched, monocarboxylic acids containing an even number of carbon atoms; odd-numbered chains are found only rarely. In typical animal lipids the most common saturated fatty acids are palmitic acid with 16 carbon atoms and stearic acid with 18 carbon atoms. Shorter-chain acids containing 12 and 14 carbon atoms occur in smaller quantities, and so also do longer chains of up to 24 carbon atoms. Because these naturally occurring fatty acids almost always have an even number of carbon atoms, it seemed probable earlier that their biosynthesis involved the stepwise addition of a unit containing two carbon atoms.

In 1926, Smedley–Maclean demonstrated that, for yeast, fat could be synthesized from the C_2 molecule of acetic acid. Later, Gurin and coworkers showed that long-chain fatty acids could be synthesized from acetic acid by soluble extracts of pigeon liver, and Popják and coworkers made the same observation in mammary gland extracts. This explained how the consumption of carbohydrates and of proteins, as well as fats, could give rise to lipid deposits, since all of these foodstuffs may be broken down in the body to give the acetic acid building unit.

Knoop suggested in 1904 that during the β-oxidation of fatty acids, degradation occurred two carbon atoms at a time, with the removal of an acetic acid derivative on each occasion. Similarly it was suggested that fatty acid synthesis might be merely a reversal of the β-oxidation breakdown process. This idea received support in the early 1930s from studies on the bacterium *Clostridium kluyverii* which can build up fatty acids from acetic acid, or break them down again to acetic acid, depending on whether the conditions are anaerobic or aerobic. But it was pointed out that for bacteria only short-chain fatty acids are involved, whereas animal systems synthesize much longer chains. It had been observed, too, that in an isolated animal system such as liver cells the synthesis and degradation of fatty acids could occur simultaneously, which would be most improbable if each system were merely the reverse of the other.

Then, in 1953, Lynen in Germany and Green in the United States worked out the details of the mechanism by which fatty acids are degraded. They found that the free fatty acid is first converted to a thioester by combination with coenzyme A (CoA), the fatty acid R—COOH and coenzyme A forming the corresponding acyl-coenzyme A. Although CoA is commonly used as an abbreviation for coenzyme A, it is necessary when writing chemical reactions to show its terminal —SH group (Fig. 7-16). In this case it is indicated as CoA—S—H.

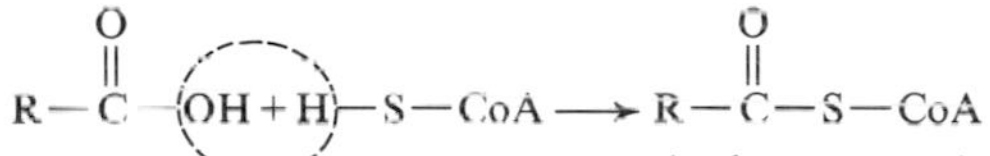

$$R - \overset{\overset{\text{O}}{\|}}{C} - (OH + H) - S - CoA \longrightarrow R - \overset{\overset{\text{O}}{\|}}{C} - S - CoA$$

Acyl-coenzyme A

The thioesters, R—CO—S—CoA, have a relatively large negative free energy of hydrolysis and are thus energy-rich compounds. After formation they undergo the series of reactions shown in cyclic form in Fig. 6-5. Each completion of a cycle removes a two-carbon unit and releases a molecule of acetyl-coenzyme A. Green and Lynen succeeded in isolating the five enzymes necessary for this process. But attempts in 1954 to synthesize long-chain fatty acids from acetyl-coenzyme A and these five enzymes were unsuccessful and it became clear that the method by which acetic acid was used to build up long-chain fatty acids was not a simple reversal of β-oxidation.

In 1959 Wakil and other workers showed that the actual unit that is condensed is in fact not acetyl-coenzyme A but malonyl-coenzyme A, formed by the addition of carbon dioxide to acetyl-coenzyme A (the acetyl-coenzyme A is produced from glucose via pyruvate in the Embden–Meyerhof pathway—see Figs 6-2 and 6-7). A specific enzyme complex with

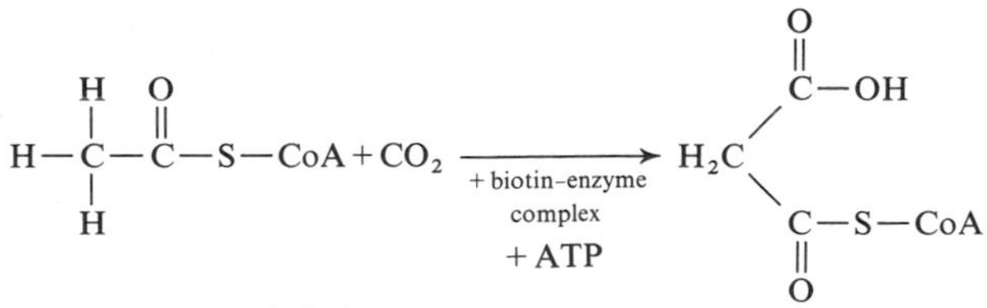

$$H - \overset{\overset{\text{H}}{|}}{\underset{\underset{\text{H}}{|}}{C}} - \overset{\overset{\text{O}}{\|}}{C} - S - CoA + CO_2 \xrightarrow[\substack{\text{+ biotin–enzyme} \\ \text{complex} \\ \text{+ ATP}}]{} H_2C \overset{\diagup \overset{\overset{\text{O}}{\|}}{C} - OH}{\diagdown \underset{\underset{\text{O}}{\|}}{C} - S - CoA}$$

Fig. 7–35 The formation of malonyl-coenzyme A

Acetyl-CoA

Malonyl-CoA

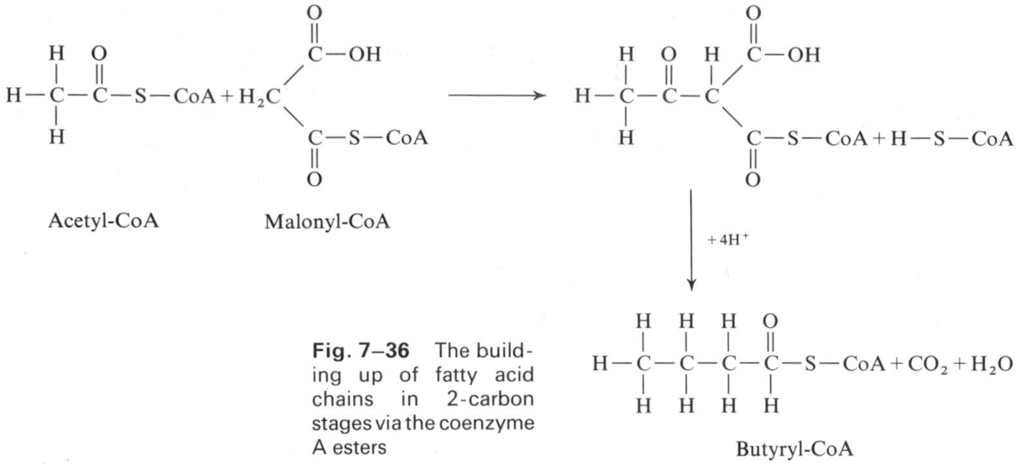

Acetyl-CoA   Malonyl-CoA

+4H$^+$

Fig. 7-36 The building up of fatty acid chains in 2-carbon stages via the coenzyme A esters

Butyryl-CoA

biotin (one of the B group of vitamins) is required to catalyse the reaction (Fig. 7-35), and a supply of ATP is also necessary. The malonyl-coenzyme A so formed then condenses with, for example, a molecule of acetyl-coenzyme A to form a β-oxoacyl-enzyme complex and coenzyme A (Fig. 7-36). Reduction, dehydration, and loss of carbon dioxide occur to give the coenzyme A ester of a C_4 fatty acid, in this case butyryl-coenzyme A. This can subsequently condense with a molecule of malonyl-coenzyme A to give the ester of the C_6 fatty acid, and so forth, with synthesis occurring in steps of two carbon atoms at a time. The molecule malonyl-coenzyme A is thus the source of the C_2 unit, and combines only with even-numbered fatty acid coenzyme A esters. The complete synthetic process is thought to involve six enzymes. Four cofactors are also necessary; these are ATP, NADPH, manganese ions, and carbon dioxide. The reason why the synthetic process stops sharply in most cells at the C_{16} acid stage, and with only traces of C_{12} and C_{14} acids remaining, is not yet clear.

In contrast to the fatty acid degradation process at the mitochondria, the synthesis of fatty acids appears to be associated with the cytoplasmic membranes which are obtained in the microsomal fraction.

Formation of glycerol, and triglyceride (or phospholipid) synthesis. The usual source of glycerol for lipid biosynthesis is dihydroxyacetone phosphate obtained from the anaerobic degradation of glucose (the Embden–Meyerhof pathway, Fig. 6-2). This may then be reduced by means of the enzyme glycerol phosphate dehydrogenase in the presence of NADH to give L-α-glycerophosphoric acid.

Dihydroxyacetone phosphate   L-α-Glycerophosphoric acid

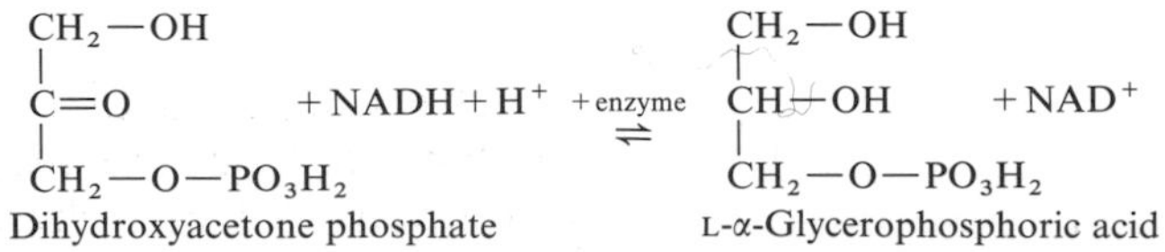

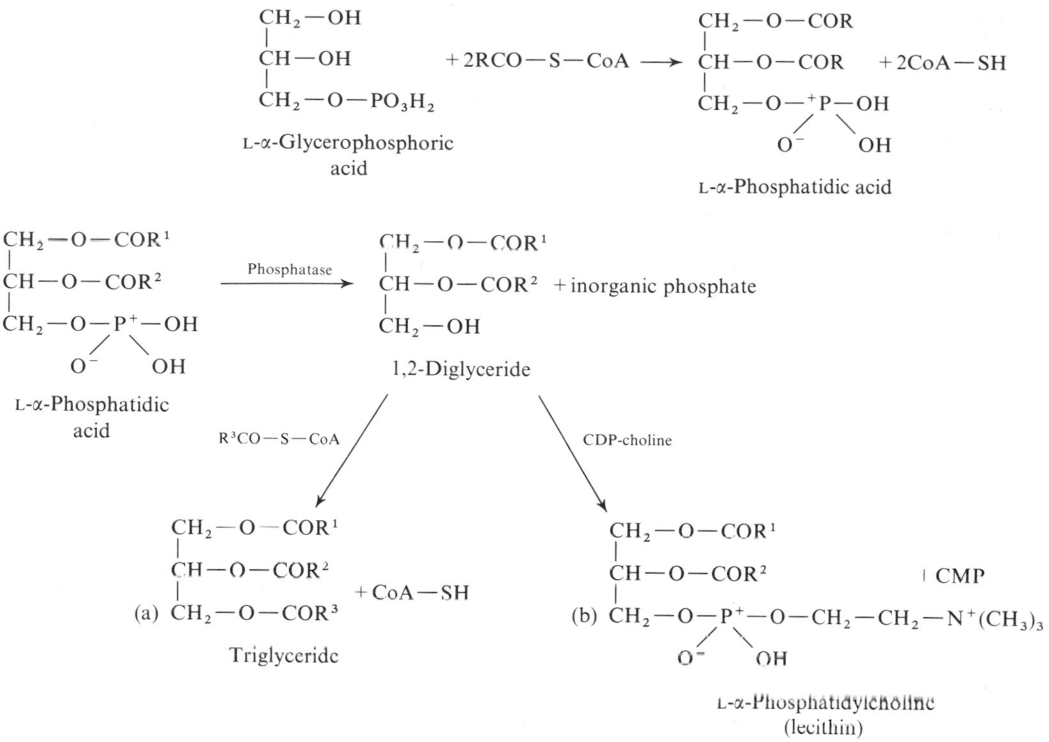

$$
\begin{array}{l}
\text{CH}_2\text{—OH} \\
\text{CH—OH} \\
\text{CH}_2\text{—O—PO}_3\text{H}_2
\end{array}
\quad + 2\text{RCO—S—CoA} \longrightarrow
\begin{array}{l}
\text{CH}_2\text{—O—COR} \\
\text{CH—O—COR} \\
\text{CH}_2\text{—O—}^+\text{P—OH}
\end{array}
\quad + 2\text{CoA—SH}
$$

L-α-Glycerophosphoric
acid

L-α-Phosphatidic acid

L-α-Phosphatidic acid + Phosphatase → 1,2-Diglyceride + inorganic phosphate

(a) Triglyceride + CoA—SH (from R³CO—S—CoA)

(b) L-α-Phosphatidylcholine (lecithin) + CMP (from CDP-choline)

Fig. 7–37 The formation of lipids from glycerol and the coenzyme A esters of fatty acids: **(a)** neutral lipid; **(b)** phospholipid

The coenzyme A esters of the various fatty acids then react with L-α-glycerophosphoric acid to give L-α-phosphatidic acid, which is the central intermediate for both triglyceride and phospholipid biosynthesis (Fig. 7–37). The alkyl groups R, may of course be different in the 1 and 2 positions, R^1 and R^2. The hydrolysis of L-α-phosphatidic acid by a phosphatase gives a 1,2-diglyceride; this may then react either with another molecule of the coenzyme A ester of a fatty acid to form a neutral triglyceride, or with a phosphorylated derivative to form a phospholipid, as for instance in the formation of lecithin (phosphatidylcholine). Here the intermediate involved is the nucleotide derivative cytidine diphosphate choline (CDP-choline in Fig. 7–37):

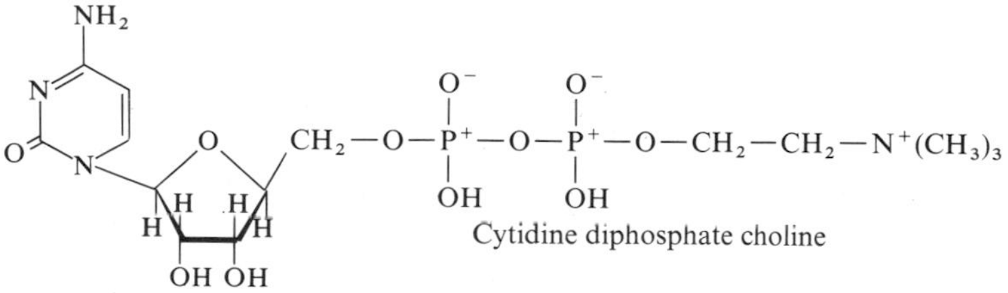

Cytidine diphosphate choline

This compound forms an ester linkage between the phosphoric acid group already attached to choline and the free hydroxyl of the 1,2-diglyceride, with the elimination of

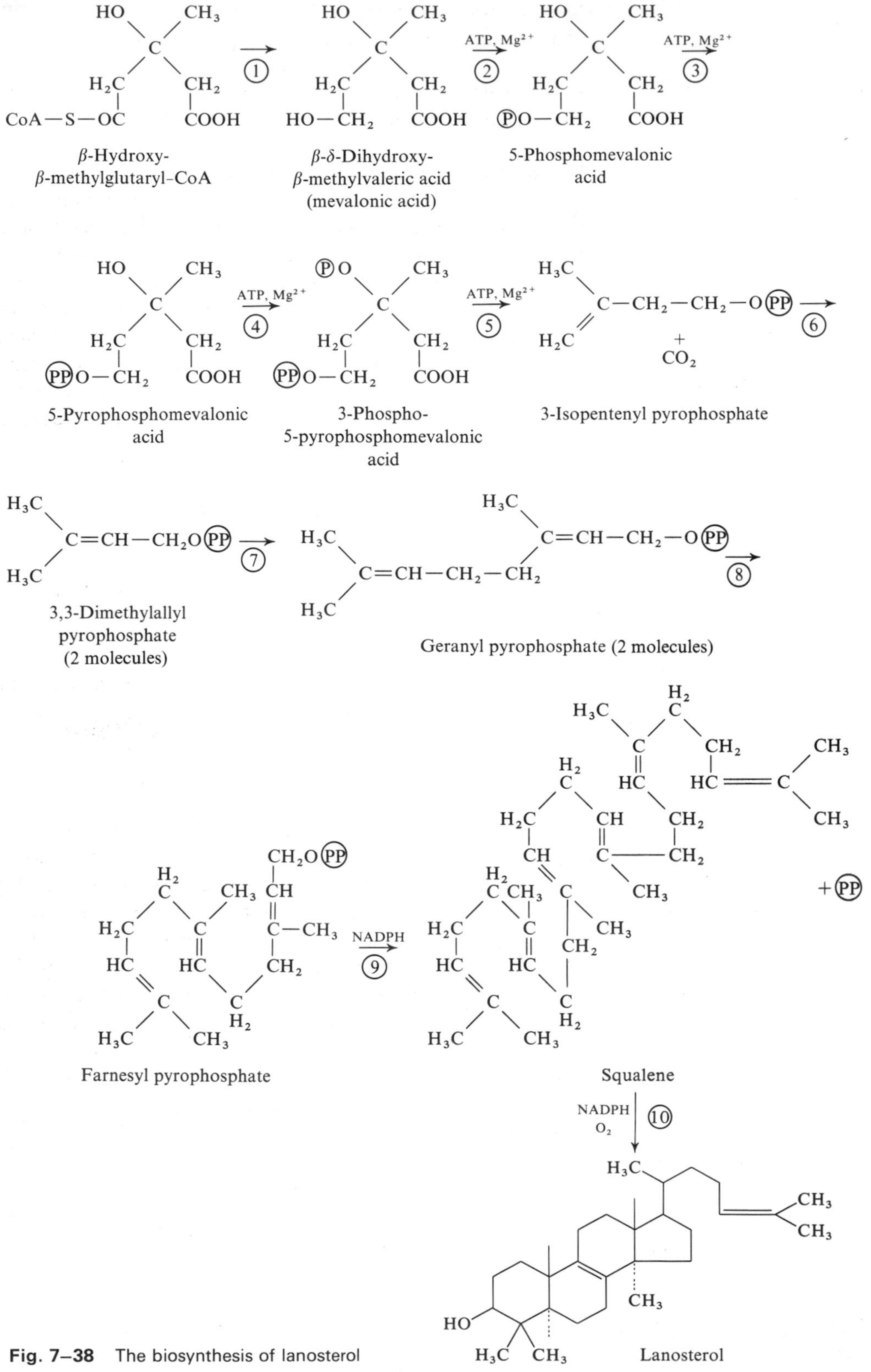

Fig. 7–38   The biosynthesis of lanosterol

cytidine monophosphate (CMP in Fig. 7–37), thereby completing the synthesis of lecithin.

Other types of lipid are assembled in the same way. It seems probable that mitochondria play a role in the final assembly of the component parts of lipids.

Biosynthesis of cholesterol.   As we have seen, certain animal lipids are formed by esterification of fatty acids with cholesterol, not glycerol, and cholesterol is contained in all animal tissues. It is synthesized in various tissues, but chiefly in the liver.

Liver tissue *in vitro* can make cholesterol from acetic acid, the building unit of fatty acids. The series of reactions involved was determined in the 1950s by Gurin, Lynen, Popják, and others. Here also, the first step is the formation of acetyl-coenzyme A. Three molecules of acetyl-coenzyme A then combine to form β-hydroxy-β-methylglutaryl-coenzyme A:

$$2CH_3 - CO - S - CoA \rightleftharpoons$$
$$CH_3 - CO - CH_2 - CO - S - CoA + CoA - SH$$

Acetoacetyl-coenzyme A

$$CH_3 - CO - CH_2 - CO - S - CoA + CH_3 - CO - S - CoA \longrightarrow$$

$$HOOC - CH_2 - \overset{\overset{\displaystyle OH}{|}}{\underset{\underset{\displaystyle CH_3}{|}}{C}} - CH_2 - CO - S - CoA + CoA - SH$$

β-Hydroxy-β-methylglutaryl-coenzyme A

Writing the formula of this compound as in Fig. 7–38, to emphasize its 5-carbon unit structure, we see that six of these 5-carbon units are eventually able to form the ring structure of lanosterol, by a complicated series of reactions, outlined in Fig. 7–38, involving many enzymes and cofactors.

Cholesterol, the formula for which is shown in Fig. 7–39, is formed from lanosterol by further highly specific reactions, which involve the removal of three methyl groups, the saturation of two double bonds, and the introduction of a double bond in a different position. The enzymes involved in the biosynthesis of cholesterol are associated with the microsomal fraction of cells.

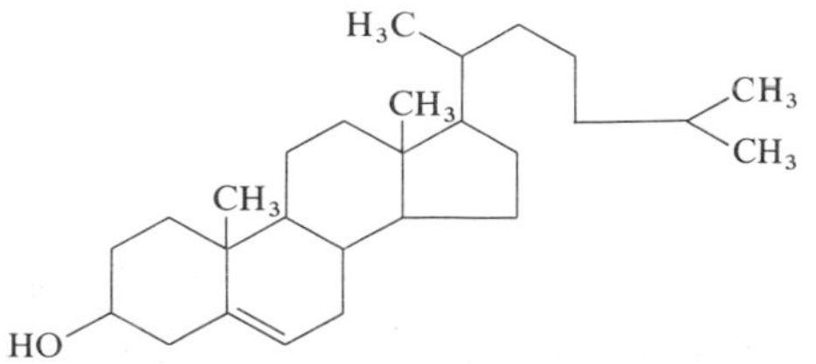

Fig. 7–39   The structure of cholesterol

We have set out to describe cellular biochemistry in terms of the integrated biological function of the cell. Biochemical reactions are mainly concerned with the synthesis of cellular and extracellular macromolecules of which the lipids, amino acids, nucleotides, and sugars form the building units. Cellular biochemistry is complicated by the interactions between the pathways leading to the synthesis of these building units. The pathways can interact in the following ways:

A product of one pathway may be used in two or more other pathways. Pyruvate, for example, which is formed in the Embden–Meyerhof pathway, is also used in the Krebs cycle for amino acid synthesis and for fatty acid synthesis.

Some molecules of an intermediate may be shunted into a new synthetic pathway. This happens, for example,

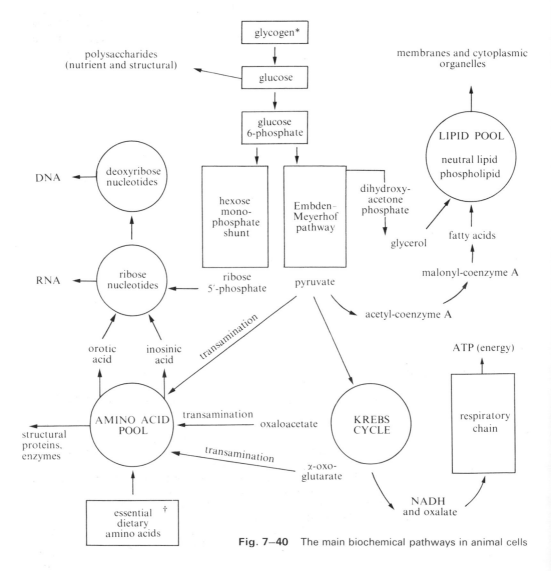

Fig. 7–40 The main biochemical pathways in animal cells

in the case of dihydroxyacetone phosphate, an intermediate of the Embden–Meyerhof pathway, part of which is shunted into a new pathway for the synthesis of glycerol.

Figure 7-40 indicates in broad terms the relationship between various metabolic pathways in mammalian cells. The detailed picture is extremely complicated and can be found in the textbooks of biochemistry mentioned below.

In plant cells the source of stored energy (* in Fig. 7-40) is starch, while in bacteria the sugar molecules are absorbed directly from the medium. The dietary amino acids († in Fig. 7-40) are not required by plant cells and many microorganisms, while the lactobacilli, organisms found in milk, are almost incapable of synthesizing their own amino acids and depend upon the milk proteins in their environment to provide amino acids for metabolism and growth.

Cells of higher plants and animals carry out the synthesis of hormones, respiratory proteins such as haemoglobin, and so forth. For simplicity these are not shown in the diagram.

Bibliography   Conn, E. C. and P. K. Stumpf, *Outlines of Biochemistry*. John Wiley, New York, 1964.

Davidson, J. N., *Biochemistry of the Nucleic Acids*, 6th edn. Methuen, London, 1969.

Dixon, M. and E. C. Webb, *Enzymes*. Longmans Green, London, 1964.

Green, D. E., The synthesis of fat. *Scientific American*, February 1960.

Mahler, H. R. and E. H. Cordes, *Biological Chemistry*. Harper & Row, New York and John Weathcrill Inc, Tokyo, 1966.

Phillips, D. C., The three-dimensional structure of an enzyme molecule. *Scientific American*, November 1966.

Ramsay, J. A., *The Experimental Basis of Modern Biology*. Cambridge University Press, London, 1965.

White, A., P. Handler, and E. L. Smith, *Principles of Biochemistry*. McGraw-Hill, New York, 1964.

8 Properties of the cell surface complex

The outer cell membrane, or plasma membrane (sometimes known as the cell membrane, or plasmalemma), has a unique role, since the cell interacts with its environment through it. As early as the 1850s it was recognized that this membrane is essentially a permeability barrier, permitting some molecules to enter the cell, and excluding others. Later studies showed that it also controls the osmotic and ionic balance between the exterior and the cell cytoplasm. Control of the passage of ions causes an electrical potential across the membrane, which is of great physiological importance.

Early investigators of plasma membranes concentrated on permeability characteristics. Overton (1895) studied the rate of permeation of solutes into plant cells, and Grijns (1896) and Hedin (1897) experimented on erythrocytes. Their results indicated that the permeability of molecules is related to their lipid solubility. Later, Collander and his coworkers (1933) showed that small molecules pass through the plasma membrane more easily than one would expect on the basis of their lipid solubility. This led to the suggestion that the membrane functions as a molecular 'sieve', having pores of molecular size in the lipid structure. Detailed studies of the permeability of membranes to gases, water, non-electrolytes, and ions were interpreted physico-chemically by Davson and Danielli (1943).

Work was also carried out on the surface tension of plasma membranes (Harvey, 1931) and the analysis of membrane extracts (Gorter and Grendel, 1925). Surface films of lipids were prepared at an air–water interface (Adam, 1938) and later at an oil–water interface. Studies of these by physical chemists were of great help in explaining why phospholipids can form membranes.

All of this work gradually led to the bimolecular lipid layer model of the plasma membrane (Fig. 5–2b). The lipid layer, 30–50 Å thick, is modified by the presence of outer layers of protein, the charged ends of the lipid molecules being orientated towards the protein. This concept, of a predominantly continuous lipid layer with protein and other non-lipid components, is still the most acceptable, but it has been considerably modified as the result of studies carried out since the early 1940s.

Of the various more recent techniques used in these studies, one of the most exciting has been that of electron microscopy (Sjøstrand, 1956 and onwards; Robertson, 1958 and onwards). This has provided evidence for a triple-layered membrane structure which agrees remarkably well with the Danielli model (see Fig. 5-7). Emmelot and Benedetti and other workers have obtained indications of a micellar arrangement for at least parts of the plasma membrane of certain cells (Fig. 5-9). In combination with various tissue culture methods electron microscopy has also been valuable in studying replicas of the actual surfaces of cells of various types, revealing new detail about their topography.

Much work has been done on the antigenic nature of the cell surface and the chemistry of antigens. With Metazoan cells most of this has been on blood group substances (Morgan, 1949 onwards), which are mucopolysaccharides of various sizes. Antigens concerned with transplantation immunity are also being studied (Möller, 1961).

Disappointingly little progress has been made towards isolating and determining the composition of membranes of nucleated mammalian cells, as distinct from work on erythrocyte ghosts (the membrane structure remaining after the haemolysis of red blood cells). The main problem is the great difficulty in obtaining a clean membrane preparation, free from contamination with cell contents.

Measurements of cell surface charge by cell electrophoresis have also contributed considerably to knowledge of cell surface structure (Bangham *et al.* 1958 onwards; and others).

The methods of approach outlined above have helped us to formulate our present ideas on the nature of the plasma membrane. In particular, the membrane is now regarded as being in a highly dynamic state; the basic structure of the plasma membrane can probably change easily from the lamellar bimolecular lipid structure to a micellar form, although the latter may only exist transiently.

Time-lapse cinemicrography of cultured cells *in vitro*— of great value in investigating cell movements and interactions —has revealed that the free membranes of cells in culture are in a highly active state. The cells undergo considerable changes of shape, to which the membrane must adapt, indicating that it is very flexible. The presence in the cytoplasm of sub-membrane fibrils known as *microtubules* has been demonstrated by electron microscopy. The role which these play in cell movements will be discussed in Chapter 11.

A major problem in studying the cell surface is to define its outer limits. We cannot always be certain if various components are situated inside the membrane lipid structure or on the membrane surface. Some substances are strongly attached to the surface, while others are rather loosely bound

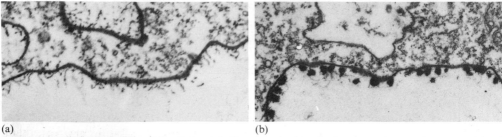

(a)   (b)

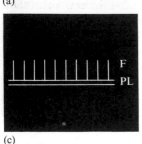

(c)

Fig. 8–1 **(a)** Electron micrograph of an amoeba membrane, unstained, showing the normal polysaccharide–protein complex. (With permission from E. H. Mercer and R. J. Goldacre, in *Proceedings of the Sixth Inter-International Congress for Electron Microscopy*, Kyoto, 1966. Maruzen Ltd, Nihonbashi, Tokyo.) Magnification × 100,000 **(b)** Membrane stained with Alcian blue showing coagulated mucous fringe. (With permission from E. H. Mercer and R. J. Goldacre, *in Proceedings of the Sixth National Congress for Electron Microscopy*, Kyoto, 1966. Maruzen Co. Ltd, Nihonbashi, Tokyo.) **(c)** Diagram of the cell membrane complex of *Amoeba proteus:* PL phospholipid–protein complex; F surface fringe of polysaccharide–protein complex, up to ca 2,000 Å in depth

—for instance, proteins, which are readily adsorbed on to cell surfaces. In Section 8–10 we consider whether in living animal tissues a cell 'coat' of material lies outside the plasma membrane. Certain large amoebae have a surface 'coat' in the form of a fringe, probably consisting of a polysaccharide-protein complex (Figs. 8–1 a, b, and c).

In plant and bacterial cells there is a rigid structure, the cell wall, outside the plasma membrane. This provides extra strength and osmotic protection for the cell, and may impose a non-spherical shape on the cell. The special properties and functions of bacterial walls are discussed in Chapter 14.

**8–1
Permeability and
transport**

After the plasma membrane was recognized to constitute a permeability barrier, one of the largest and most fruitful fields of study was that of membrane permeability to various solutes. Many of the early studies were carried out on red blood cells and plant cells. Direct chemical determination of concentrations inside and outside the cell was extremely difficult and tedious, so that for the erythrocyte the method of haemolysis (Fig. 8–2a) was used. This depends on the fact that if the cells are placed in a solution of a penetrating non-electrolyte they will swell until they burst, or haemolyse, due to penetration first of the electrolyte and then of water to equalize the osmotic pressure. As haemolysis proceeds, the opacity of the suspension (due to the presence of whole cells) decreases, and light is more easily transmitted through it. By this method Jacobs (1930) determined how long it took to reach a given level of haemolysis in different solutions, and so obtained a measure of the relative amounts of penetration of various solutes.

For plant cells, the method of plasmolysis (Fig. 8–2b) was used. If a plant cell is placed in a hypertonic saline solution,

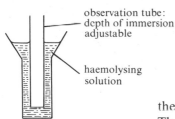

observation tube:
depth of immersion
adjustable

haemolysing
solution

(a)

lamp

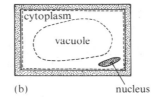

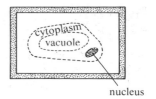

(b) nucleus

nucleus

Fig. 8–2 (a) The apparatus used by Jacobs for the quantitative determination of haemolysis. **(b)** Plasmolysis of a plant cell: *left*, a normal cell in isotonic saline; *right*, a plasmolysed cell in hypertonic saline

the difference in osmotic pressure drives water out of the cell. The volume of the cell sap diminishes, and the cell membrane detaches itself from the rigid cell wall; this is known as plasmolysis. If the external medium is made hypertonic by a substance which penetrates the cell slowly, then this substance will tend to enter the cell to equalize the osmotic pressure. So the sequence of events is plasmolysis, followed by recovery of sap volume (de-plasmolysis). The rate at which these processes occur gives a measure of the rate of penetration of the added substance into the cell.

Volume changes have also been studied in the unfertilized egg of the sea urchin, *Arbacia punctulata*. These eggs are almost perfect spheres, and volume changes may be followed easily by various physical methods. More recently the technique of labelling with radioactive isotopes has often been used to measure rates of penetration of various compounds into cells.

The plasma membranes of all cells studied have revealed very similar properties. It now seems clear that substances entering the cell may be divided into three groups: lipid soluble substances; certain very small molecules and ions; and substances for which special mechanisms appear to be necessary—certain cations and anions, sugar molecules, and amino acids.

Studies on lipid soluble compounds showed a direct correlation between their ease of passage into cells and their lipid solvent–water partition coefficient (Fig. 8–3) (see Section 2–13). The actual size of the molecule, provided it is

Fig. 8–3 The ease of permeation through cell plasma membranes related to lipid solubility

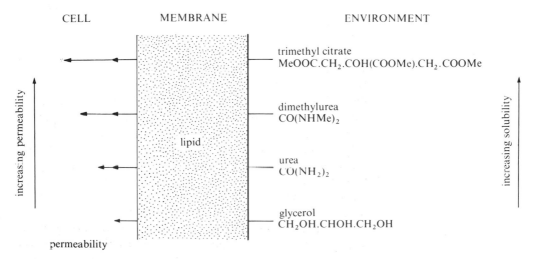

CELL MEMBRANE ENVIRONMENT

increasing permeability

increasing solubility

trimethyl citrate
$MeOOC.CH_2.COH(COOMe).CH_2.COOMe$

dimethylurea
$CO(NHMe)_2$

lipid

urea
$CO(NH_2)_2$

glycerol
$CH_2OH.CHOH.CH_2OH$

permeability

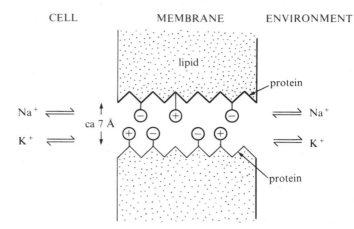

CELL   MEMBRANE   ENVIRONMENT

Fig. 8–4 A possible arrangement of plasma membrane pores associated with ion permeability

fairly large, does not appear to affect its permeability. This is what would be expected if the plasma membrane is a bimolecular lipid layer.

The observation that small molecules [water, methyl alcohol, and formamide ($HCONH_2$)] and small solvated ions (Na^+, K^+, and Cl^-, all generally of less than 7 Å in diameter) can easily pass inside the membrane led to the suggestion that there must be small pores of molecular dimensions in the otherwise continuous lipid layer (Fig. 8–4). These pores are probably lined with protein with the polar groups orientated towards the aqueous phase. Mueller, and Bangham independently, prepared artificial membranes *in vitro*. Bangham, using micelles of phospholipid, found in 1968 that if he incorporated a cyclic polypeptide such as gramicidin with the phospholipid then the rate of passage of K^+ ions through the model membranes was much greater than that through phospholipid membranes alone. This is presumably because gramicidin has a proton acceptor region in the centre of the molecule. If only one molecule of gramicidin is incorporated in 10^6 molecules of phospholipid the flux of K^+ is increased a hundredfold. An arrangement of protein of this type in the lipid layer could account for the pore properties of plasma membranes. The fact that there appears to be a fairly sharp cut-off in the size of substances which can enter the cell in this way indicates the operation of some steric restriction; that is, there is a geometric limit to the size of the pores.

Factors which affect pore size are: solvation of charged groups on protein side-chains; interactions between oppositely charged groups; and linkage of two negatively charged groups by bivalent cations such as Ca^{2+}. It is thought now that the pores are not necessarily permanent gaps in the plasma membrane, but may be essentially short-lived routes of entry.

We have already discussed the theory that the lamellar lipid form of the membrane may be able to change (or partly change) transiently to a micellar form. The membrane protein molecules must be able to unfold in some way to line any

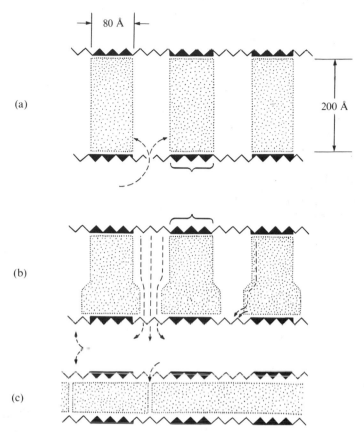

Fig. 8–5 Kavanau's model of cell surface structure. **(a)** Lipid micelles (stippled), which are about 80 Å in diameter and 200 Å in height, are separated by a hydrophilic matrix which has flowed into the surface complex from the cytoplasm. The protein layers are indicated by zig-zag lines. **(b)** The hydrophilic matrix is starting to flow back into the cytoplasm and the micelles are changing shape. **(c)** The matrix has completely flowed back, and the membrane has become a bimolecular lipid layer with protein layers on either side. (With permission from J. L. Kavanau, *Nature*, **198**, 1968.)

newly formed pores and so stabilize the micellar arrangement. Kavanau has proposed a model (Fig. 8–5) in which he demonstrates how a continuous lipid layer with protein on its outer surfaces could form a micellar structure of blocks of lipid with accompanying extension of the protein chains.

The third group of substances which enter the cell includes the cations Na^+ and K^+. Although the plasma membrane appears to be freely permeable to these cations, the necessity to postulate a special transport mechanism is clear when the composition of the intracellular fluid in animals is compared with that of the extracellular body fluid. The extracellular fluid diffuses through the walls of the circulatory system, fills all of the minute spaces between tissue cells, carries with it the nutrients and metabolites required by cells, and removes waste products such as urea and carbon dioxide. It has a relative salt composition very similar to that of sea water, which means that it has high concentrations of Na^+ and Cl^- and much smaller amounts of K^+, Mg^{2+}, Ca^{2+}, and PO_4^{3-}. The interior of the cell, on the other hand, contains large amounts of potassium but comparatively little sodium. With a few exceptions this is true of all cells, both plant and animal. These observations could not be explained in terms of simple diffusion and osmosis, because cells obviously have the

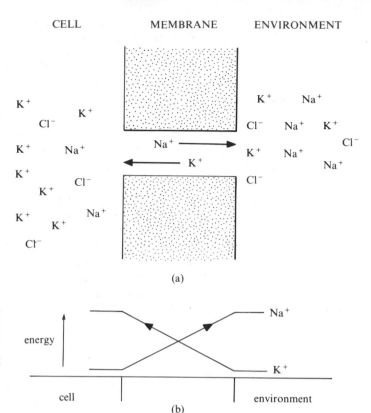

Fig. 8–6 (a) K+ ions moving into a cell against an existing concentration gradient. **(b)** Energy is required to pump ions against a concentration gradient

capacity to go on accumulating K^+ ions against the existing concentration gradient (Fig. 8-6).

Other examples of this are known. Cells can accumulate substances such as amino acids, for instance, so that their concentration inside the cell becomes higher than the external concentration by a factor of 10 or more. Also, naturally occurring amino acids are allowed to enter the cell but artificial ones, although chemically very similar, are excluded. The classic example of accumulation against a concentration gradient is that of certain marine algae which can maintain an internal concentration of iodine 10^6 times greater than that of the surrounding seawater.

The plasma membrane, then, far from being a passive filtering device, must play an active part in pushing these substances in the opposite direction to that in which they would go if left to diffuse freely. Also, a supply of energy is necessary to carry out the work being done in transporting substances across the plasma membrane. That energy is required for Na^+ and K^+ transport has been demonstrated *in vitro* by taking slices of brain or kidney and blocking their energy supply by various means—keeping them cold, removing supplies of oxygen or glucose, or adding metabolic poisons such as cyanide or dinitrophenol. Under these conditions the high K^+ concentration inside the cell falls, and the Na^+

Question: Charge difference between inside membrane & outside membrane — same as between inside & outside of cell membrane? Analogue of Sodium pump?

small particles containing group of enzymes which utilize pyruvate in Kreb's cycle + pass NADPH through membrane wall

NADPH

reduction

Substrate being oxidized along these enzymes

Mitochondria

Stalk particle made up of enzymes (v 4 cytochromes — iron centered enzymes separated by fatty particles)

concentration rises, until the concentrations inside and outside the cell are equalized, the membrane behaving as if it is freely permeable to both types of ion. When an energy supply is restored by reversing the blocking processes (that is by warming the cells up to 37°C, or by adding oxygen or glucose, or by washing away the metabolic inhibitor), the membrane resumes its activities of concentrating potassium inside the cell, and transporting sodium out of it. The process is called *active transport*, and since the plasma membrane pumps ions against an existing gradient it is said to behave as an *ionic pump*.

It is now known that the energy source for the ionic pump is the universal cell fuel, ATP. Hodgkin, Caldwell, and their collaborators (1960) showed that, by injecting ATP into the poisoned giant nerve of a squid which was unable to synthesize its own ATP, the nerve membrane began at once to pump sodium and potassium ions, and this continued until all the ATP was used up. The application of ATP to the outside of the nerve axon did not, however, restore the pumping activity. Other workers using red cell ghosts demonstrated that the membrane would recommence pumping in the presence of internal sodium and external potassium only if ATP was present inside the cell. The amount of energy which the cell has to use for the ion pumping mechanism is considerable; McIlwain has calculated that up to 18 per cent of the total ATP required by the brain may be expended in pumping activities.

From this evidence, the plasma membrane must contain special sites which can distinguish between Na^+ and K^+ and use ATP. Further, if these sites become inactive due to lack of an energy supply, they can only be reactivated from inside the membrane. The number of these special sites may be small; Glynn (1957) calculated that for the red blood cell no more than 1,000 sites per cell are necessary. Danielli suggested that at these sites the ion to be transported combines with a carrier substance on the plasma membrane, and that it is then carried across the membrane and released on the other side.

The nature of the carrier substance has given rise to much speculation. The method of approach has been to look for a component of the cell membrane which is phosphorylated by ATP while the pump is functioning, and the phosphorylation of which is blocked by substances which also block the pump. Heald (1956) obtained evidence, and Rose (1964) later confirmed, that phosphorylation by ATP increases, with increasing pump activity, in the phosphoprotein of the plasma membrane of slices of brain tissue. (Phosphoproteins contain phosphate radicals bound to the amino acid serine, which is part of the protein chain found in the plasma membrane—see below.) Later it was shown that brain tissue contains two enzymes—kinase which can transfer phosphate from ATP to phosphoprotein, and phosphatase which can

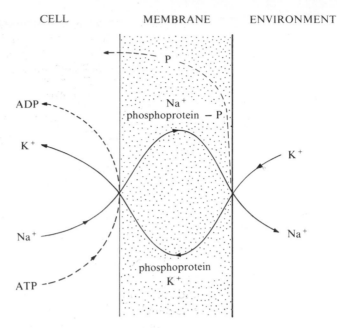

CELL   MEMBRANE   ENVIRONMENT

Fig. 8–7 The suggested model for the Na^+–K^+ ionic pump

selectively remove this transferred phosphate. Thus there exists a coupled system which is apparently functionally related to the ionic pump.

A significant advance was made by Skou (1957) who demonstrated the presence in crab nerve membrane of an enzyme system capable of breaking down ATP into ADP and phosphate very rapidly. This system, known as ATP-ase, was found to work very much faster in the presence of sodium and potassium. Every tissue with an ion pump that has been examined has been found to have an ATP-ase system in its plasma membrane, the activity of which is proportional to the pump activity. Later work has indicated that ion transport in higher plant cells may also be correlated with a membrane-bound ATP-ase system. For the enzyme system to be active, the presence of some organized structure, such as the phospho-protein mentioned above, within the membrane seems to be necessary, probably as a transitory intermediate. The trans-ferred phosphate group finally goes into the inorganic phosphate fraction of the membrane.

The complete system suggested for the sodium–potassium ionic pump in brain cells is shown diagrammatically in Fig. 8–7. A specific phosphoprotein site is phosphorylated by ATP (a process known to be stimulated by sodium) at the inner surface of the plasma membrane, where it binds to sodium ions. The phosphoprotein–sodium complex is then transferred in some way across the membrane to the external surface, where the sodium is released. Then potassium is picked up instead, dephosphorylation (stimulated by potassium) occurs, and the complex returns to the inner surface where potassium is released. The whole process then begins again. Two or more ions are probably involved in each transfer. The model may have to be modified in the light of further experience and it may not apply exactly to all cells with ionic pumps. For instance, work carried out by Hokin and Hokin (1959) indicates that for the salt gland of the albatross (a tissue having a very active ion pump), and for certain mammalian tissues, phosphatidic acid, a component of phospholipid, is the phosphorylated intermediate which transports the ions across the plasma membrane.

How does the carrier transport the ions across the membrane? Danielli has suggested various possibilities—the carrier may itself diffuse across the membrane; it may rotate about an axis through the centre of the membrane; or it may hand its ion along a chain of similar carriers, so transferring the ion across.

Various other ionic pumps, anionic as well as cationic, are known to exist, for example, the Ca^{2+}, Cl^-, and I^- pumps, and their mechanisms of operation are gradually being elucidated.

Glucose and other sugars are also substances for which special transport mechanisms across the plasma membrane appear to be necessary. The actual mechanism by which glucose enters cells is not known, nor is it known if a supply of energy is needed for the process. In mammalian cells energy is probably not needed, unless work is being carried out against an uphill gradient—in other words, the mechanism can be one of *passive transport*.

In the transport of glucose and other monosaccharides across human red blood cell and rat fat cell plasma membranes, the work of Langdon and Sloan (1967) indicates that the carrier site on the membrane is a protein molecule. The sugar molecules form imine linkages ($=NH$) with the partially dissociated protein complexes.

Mitchell (1963) has suggested that glucose is transported across bacterial cell membranes by what he has called a translocation process. The glucose molecule first becomes attached to a site of special configuration on the plasma membrane, as in the ionic pumping mechanism, but in this case the specific site is located on the outside of the membrane. It is then transported on this carrier site through the membrane by an unknown mechanism and released on the inside. The process is facilitated by certain enzymes, the system being known as the permease system. This is an active transport process and energy is necessary for its completion. (The passage of substances through bacterial cell walls is discussed in Chapter 14.)

In certain mammalian cells the active transport of sugar against a concentration gradient is linked in some way to the pumping of sodium ions. In intestinal cells Crane (1961) found that the accumulation of glucose inside the cell is dependent

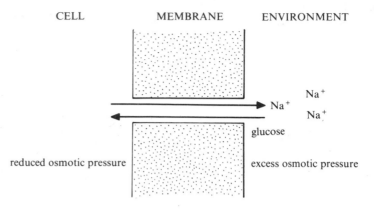

Fig. 8–8   The accumulation of intracellular glucose due to an osmotic gradient

on the transport of sodium out of the cell. He suggested a mechanism involving a common carrier for the sodium ion and the sugar molecule, and using osmotic rather than chemical energy to drive the sugar uphill into the cell, against the gradient. The pumping of sodium out of the cell steepens the osmotic gradient for the transport of sugar into the cell, and the final result is an accumulation of intracellular glucose (Fig. 8–8).

Other examples of the interrelationship of two transport mechanisms have been recorded. For slices of thyroid tissue *in vitro*, the accumulation of I^- in the cells is linked to sodium transport, and in the presence of substances which inhibit sodium transport the accumulation of iodide is also blocked. Again, it has been shown that the transport of amino acids into tumour cells (Christensen, 1960) appears to be coupled to loss of potassium previously accumulated in the cells. The flow of potassium with the concentration gradient provides the energy which drives the transport of the amino acids.

Probably these are examples of a more general phenomenon. In view of the very powerful control mechanisms maintaining and protecting the various cellular functions it would seem reasonable to suppose that many of the transport mechanisms for various substances across the plasma membrane are interdependent.

8–2 Membrane potential

The Donnan equation states that when a membrane permeable to small ions such as Na^+ and Cl^- is placed between a solution of diffusible ions on one side and a solution of diffusible and non-diffusible ions (such as the sodium salts of proteins) on the other side, then at equilibrium the products of the concentrations of the diffusible ions will be the same on both sides. The consequences of this for osmotic imbalance have been considered in Chapter 2. Here, we are concerned with the fact that the laws of simple diffusion no longer apply when non-diffusible ions are present, and that ions may be maintained at different concentrations on either side of semipermeable membranes under conditions of thermodynamic equilibrium.

CELL   MEMBRANE   ENVIRONMENT

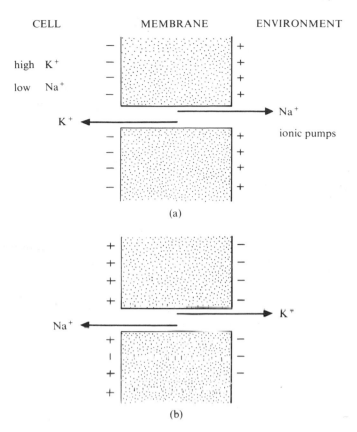

Fig. 8–9 (a) Membrane potential due to differences in concentration of Na+ and K+ ions across the membrane. **(b)** The temporary free passage of ions across the membrane may be accompanied by charge reversal

The imbalance of concentration across a membrane means that an electrical potential is generated across the membrane, as a result of the difference in the rates of diffusion of the various ions. Because ions tend to move down a concentration gradient and also down a potential gradient, equilibrium is possible when a stable concentration difference across the membrane is exactly balanced by a potential difference, known as the equilibrium potential.

The electrical properties of the cell are also partly determined by the contribution to the membrane potential of the concentration differences across the plasma membrane resulting from the ionic pump mechanisms. In certain large cells, such as the giant axons of squid, it is possible to measure the potential across the plasma membrane. The inside of the cell is normally about 80 millivolts negative compared to the external surface of the membrane. When the ionic concentrations on both sides of the membrane can be determined, it is possible to compare the calculated equilibrium potential with the measured potential. When the differences recorded between the two diverge from thermodynamic equilibrium this must be due to some form of active ion transport across the plasma membrane (Fig. 8–9a).

The potential across the plasma membrane is of great physiological importance, particularly in animal tissue such as nerve and muscle. It enables a nerve cell to conduct an impulse, or a muscle cell to contract, if a suitable stimulus is applied. The stimulated part of the plasma membrane is momentarily depolarized or even undergoes charge reversal, due to a temporary free passage of ions across it (Fig. 8–9b). This movement of ions causes a current to flow along the membrane to an unstimulated region, which then becomes depolarized and in turn stimulates the next part of the membrane, and so forth. The result is the movement of an electrical impulse across the cell surface; the impulse is known as the *action potential*. The depolarization is only transient, because the ionic pumps restore the balance to its original value in a few milliseconds. Clearly the maintenance of an electrical potential across the membrane is essential to the physiological activity of nerve, brain, and muscle tissues. It is not so immediately obvious what function it serves in erythrocytes, for example.

8–3
Isolation and
analysis of
plasma
membranes

Much experimental work has been done on the preparation of plasma membranes in a pure state, but the inherent difficulties are such that the results on the whole have been disappointing. For direct chemical analysis to be valid, the membrane must be isolated in a pure state, with none of its components missing, and uncontaminated by other cellular material. Criteria must also be devised for demonstrating unequivocally that the material isolated does in fact correspond to the cell surface complex. As its appearance in the electron microscope is similar to that of other intracellular membranes, identification is extremely difficult.

Least difficulty is experienced in isolating the membrane of the mammalian erythrocyte. The red cell 'ghost' obtained after the cell contents have escaped during haemolysis still retains its characteristic biconcave shape; but even here the final product depends on the method of lysis and the pH (Ponder, 1952; Waugh and Schmitt, 1940). Myelin has also been used as a source of plasma membrane material (Robertson, 1958), but all analyses have been carried out on bulk tissue, and not on myelin first stripped from the nerve fibres, and so must be modified because of the presence of some nuclear and cytoplasmic materials.

Various attempts have been made to isolate the plasma membrane from other types of cell. Ehrlich mouse ascites tumour cells (that is, tumour cells which grow in an ascitic or suspension form in the peritoneal cavity) have been used by Rajam and Jackson (1958), who disrupted the cells by ultrasonic methods. They found that antigens (presumably located on or near the cell surface) were present in their isolated material. Wallach and coworkers (1962 onwards) disintegrated Ehrlich ascites cells by suddenly releasing the pressure on cells supersaturated with nitrogen. They identified the membrane

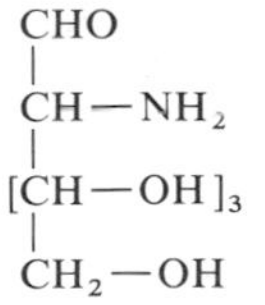

CHO
|
CH—NH₂
|
[CH—OH]₃
|
CH₂—OH

Hexosamine

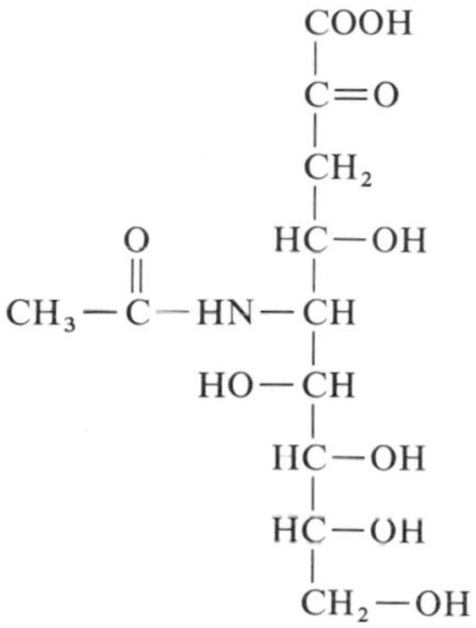

COOH
|
C=O
|
CH₂
|
O   HC—OH
‖ |
CH₃—C—HN—CH
|
HO—CH
|
HC—OH
|
HC—OH
|
CH₂—OH

Sialic acid
(N-Acetylneuraminic acid)

Fig. 8–10 The structures, shown in linear form, of hexosamine and sialic acid or N-acetylneuraminic acid

fraction by the presence of ATP-ase activity and again by certain (presumably surface) antigens. They also reasoned that the presence of sialic acid (N-acetylneuraminic acid, Fig. 8 10) in their material indicated that it was obtained from the plasma membrane. Mouse and rat liver cells have also been used as starting material by various workers (Neville, 1960; Emmelot and Bos, 1962 onwards; and others). In this case, as well as antigenic markers, the remains of bile canaliculi were used for identification, and also formations were obtained in the membrane preparation which possessed the distinctive morphology of two or more liver cells stuck together.

O'Neill and Wolpert (1961) prepared surface fractions from *Amoeba proteus* by a glycerol treatment which causes the cell contents to shrink away from the plasma membrane, which retains its characteristic shape. The fractions were identified as cell surface by the appearance in electron micrographs of the long hairlike processes of the cell surface coat (Fig. 8-1b).

The outer wall and capsule in many bacteria can be digested away by enzyme treatment leaving the *protoplast*, which is surrounded by a membrane still retaining the main permeability characteristics of the original bacterium. The membrane material of protoplasts may be obtained, as for red cell ghosts, by bursting the cells in hypotonic solution; chemical analyses have been done on these (Mitchell, 1959).

A fluorescent marker substance (Maddy, 1964) has been attached to the membrane of the intact erythrocyte before haemolysis, and fluorescent ghosts have been obtained; however it has proved difficult so far to apply this marking technique to other plasma membranes. Warren and coworkers (1966) treated intact L-cells (a strain of mouse cells grown in tissue culture) with various reagents which react with the membrane and so stabilize it during isolation procedures. Negative criteria, such as the absence of nuclear or cytoplasmic components, have been used to judge the purity of membrane preparations. The recognition of contaminants is probably one of the most difficult aspects of the problem; for instance, in erythrocytes the question of whether haemoglobin is a true component of the membrane is still unresolved. Extraction of intact cells at pH 6·0 can give a membrane preparation containing up to 50 per cent of haemoglobin, while at pH 7·4 this figure is reduced to about 1 per cent. It is not yet clear whether raising the pH facilitates the removal of haemoglobin which is non-specifically adsorbed on the membrane, or whether it breaks a bond between haemoglobin and other membrane components which is an integral part of the intact membrane structure.

Analyses of membrane preparations from animal and bacterial cells indicate that the main constituents are lipid, with a value of 20–40 per cent of the total dry weight in erythrocytes (depending upon the method of preparation and the species of origin), and protein, with a value of 70–60 per cent. The carbohydrate component is of the order of 5 per cent in erythrocytes, distributed between the protein and lipid fractions, and has been found to be somewhat lower in nucleated animal cells—for example, about 1 per cent in mouse liver cells.

8–4
Lipid fraction of
plasma
membranes

The main lipid constituents of plasma membranes are phospholipids (in particular phosphatidylcholine or lecithin), and cholesterol, a steroid (Fig. 7–39).

The phospholipids, as their name implies, contain phosphate groups and usually also glycerol, fatty acids, and a nitrogenous base. The chemical formulae of some of the most important of these lipids are shown in Fig. 5–4. The lengths of the hydrocarbon side-chains may vary. It can be seen from Fig. 5–4 that the phospholipids, with their long hydrocarbon chains and polar amino and phosphate groups at one end, could equally well fit into a bimolecular layer model of the plasma membrane, or form a micellar arrangement. The available analyses (Fig. 5–3) suggest that there is more cholesterol in plasma membranes than in cytoplasmic membranes; for instance the values in Fig. 5–3 for both bovine erythrocytes and myelin sheath membranes are at least ten times greater than the figure quoted for mitochondrial membranes. The higher proportion of cholesterol should strengthen and stabilize the plasma membrane. (The way phospholipid and cholesterol molecules may be closely packed together in membranes, thus increasing their stability, was considered in Section 5–2.)

With the improvement of techniques of paper chromatography and the development of gas chromatography, more detailed analysis of membrane lipids has been possible. In addition to the large fraction of phospholipids, and the somewhat smaller fraction of neutral lipids (mostly consisting of cholesterol in the free state, with a small amount esterified), there are relatively small amounts of various lipids, the structures of which are now being investigated. In addition to triglyceride (Fig. 5–4) which is present in small quantities, a glycolipid fraction has also been identified and investigated in erythrocyte ghost membranes by Japanese and German workers. Glycolipids are compounds containing sugar and lipid components. The fatty acid component consists mainly of chains which are 24 carbon atoms long, but there is wider variation in the chain length than in that of the fatty acids of the sphingomyelins from the same range of species. There is also considerable species variation in the sugar component; for instance, in human erythrocyte membranes the glycolipid fraction contains hexosamine but no sialic acid (Fig. 8–10), while the reverse is true of horse glycolipid. A sub-fraction of glycolipid has been found to carry A and B blood group activity (see below).

The plasma membrane, then, contains a complex mixture of lipids, and the function of all of the components is not yet fully understood. The relationship of membrane permeability to the nature of the binding of lipids to the plasma membrane is of considerable interest. If we assume that the degree of binding is inversely proportional to the ease of extraction with various organic solvents, we may group membrane lipids into three categories:

Loosely bound, extractable with dry ether.
Weakly bound, extractable with wet ether.
Strongly bound, extractable with alcohol–ether.

The loosely bound fraction contains most of the neutral lipids of the membrane and a mixture of phospholipids, and the amount present tends to vary from one species to another. Species which have a higher membrane permeability to compounds such as ethylene glycol contain a greater percentage of loosely bound material than those with low permeability. It has been suggested that the loosely bound material interacts only through van der Waals forces and the more tightly bound material is held in addition by polar forces (Fig. 8–11), so that the presence of a high proportion of loosely bound material may indicate a more open configuration of the membrane, and so a higher permeability.

8–5
Protein fraction
of plasma
membranes

The observations of Danielli and Harvey on the unexpectedly low surface tension of cell plasma membranes led to the Danielli model, in which the layer of protein on the outside surface of the bimolecular leaflet modifies the properties of a continuous lipid layer. Danielli and his collaborators proposed that this surface protein is present in the form of extended chains, and that other protein components may exist in the membrane structure in a globular, unextended configuration.

Attempts over many years to characterize the protein component of plasma membranes have met with relatively little success, largely because the protein is associated with the various lipid components in the form of a lipoprotein complex. What has been definitely established by means of histochemistry and electron microscopy is the presence of enzymes on the surfaces of cells. In particular ATP-ase, which plays an important part in ionic pump mechanisms, is present on the surfaces of various cells. There have also been reports of the presence of other enzymes on the surfaces of some mammalian cells and also of yeast cells (Rothstein and coworkers, 1954).

In bacteria, it is clear from the work of Weibull and Mitchell that the respiratory chain enzymes present in the mitochondria of mammalian cells are located in the protoplast membrane (Fig. 8–17).

Presumably in these cases, since enzymic activity is still present, the protein must still largely retain the three-dimensional structure of its active site. This could be accounted for if the protein is adsorbed, without any change in its structure, onto the polar regions of the lipids, or alternatively if it is more closely incorporated into the membrane lipids with only partial loss of its three-dimensional structure.

Attempts have been made to characterize the protein of erythrocyte ghost membranes. Removal of lipid from ghost membranes by organic solvents leaves first a lipid–carbohydrate–protein complex, known as elenin, and finally a residue of denatured protein, stromatin. Since red cell ghost membranes contain a relatively large amount of protein (see Fig. 5–3) it was thought that not all of this protein could be present on the membrane surface, and that some of it must be incorporated within the lipid structure. It has been suggested that the protein is completely unfolded, and that the peptide chains lie randomly interwoven among the lipid molecules in a state of continuous change. This arrangement could contribute greatly to the stability of the membrane structure, but the proteins concerned, having lost their secondary and tertiary structure, would have no enzymic activity.

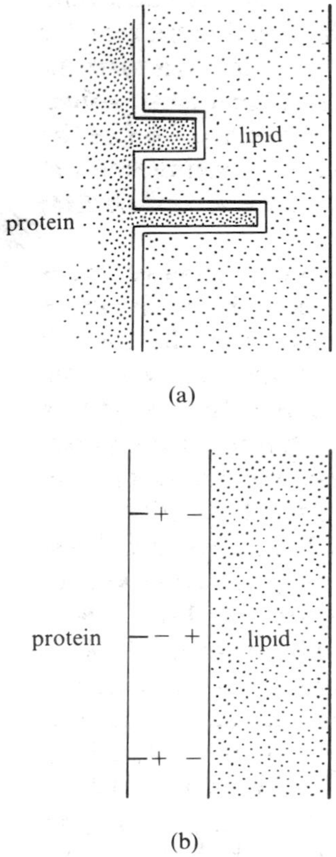

Green, after working on the structural protein of mitochondria, extended his work to other membrane systems, including erythrocyte ghost and chloroplast membranes (Green and coworkers, 1963). In each case he isolated a protein fraction with the properties of the mitochondrial structural protein. All plasma membranes may contain variable amounts of structural protein.

X-ray diffraction studies on nerve myelin suggest that there is more than one type of lipid–protein binding (Finean et al., 1956). Since the lipid can be completely separated from the protein components of plasma membranes by organic solvents, it is unlikely that covalent linkages exist between them. Forces likely to contribute to the stability of the lipoprotein membrane complex are, first, electrostatic interactions between the polar groups of the lipids and proteins and, second, van der Waals forces of attraction between the hydrocarbon chains of the lipid and the non-polar side-chains of the proteins. Which of these is the more important would seem to depend on whether the protein chains lie on the surface of the phospholipid layer or lie interwoven through the bimolecular lipid layer. In the former case electrostatic interactions could take place, but the protein side-chains would be too short to extend beyond the polar phospholipid groups to the hydrocarbon lipid chains. In the latter case non-electrostatic forces could apply (Fig. 8–11).

Peptide chains also exist at or near the surface of mammalian cells at various antigenic sites; they are present, for example, in the blood group substances, which are

Fig. 8–11 Suggested forms of lipid–protein binding in cell membranes: (a) due to van der Waals forces; (b) due to polar forces

mucopolysaccharide peptides. (Mucopolysaccharides are defined as polysaccharides which contain hexosamines.)

8–6
Blood group substances and transplantation antigens

Chemical studies on the nature of the antigenic ABO blood group substances are particularly relevant to theories of cell surface structure, because these substances have been shown by various immunological techniques to be located on the surface of the plasma membrane.

Early work on the ABO surface antigens of human erythrocytes (Morgan, 1949) proved unrewarding, and when it was found that substances with high ABO activity, assumed to be of essentially the same composition, could be obtained in a water-soluble state from urine, saliva, gastric juice, and ovarian cyst fluid, these sources were used to supply the material for analysis (Morgan, 1960). The antigenic activity of the substances resides in the polysaccharide portion of the molecules. The polysaccharide components of all members of the group are constituted from the same four sugars: D-galactose, L-fructose, D-glucosamine, and D-galactosamine, the specificity being due to different amounts and structural arrangements of these molecules. The peptide part of the molecule is formed from the following amino acids: aspartic and glutamic acids, arginine, lysine, serine, threonine, proline, glycine, alanine, valine, leucine, and isoleucine. All of the substances also contain sialic acid, but the quantities of this have proved difficult to determine owing to the labile nature of the bonding.

As well as the A and B blood group substances, the H substance and the Lewis (Le) substance have also been studied. The H substance may be a precursor of the A and B substances; Le substance is found in those individuals who do not secrete water soluble A, B, or H substance—about 25 per cent of the population. Two further blood group antigens, M and N, have been identified and shown to differ in their carbohydrate contents (Cook and Eylar, 1965).

Another well-studied group of antigens are those concerned with transplantation immunity. In particular, the histocompatibility (H-2) antigens of mice have been investigated by Möller and Möller from 1961 onwards. Little is known yet about the chemical structure, but some at least of the antigenic substances are located on the cell surface. Antigenic activity has also been detected in other cell fractions.

From preliminary work on the location of other cellular antigens, it seems that not only are there surface antigenic differences between cells of different species, but also between cells from genetically different individuals of the same species, and even between cells from different organs. Consequently there are differences in cell surface composition which arise from the differentiation of cells.

8–7
Cell
electrophoresis

The negatively charged groups on the cell surface attract, in saline solution, counter ions of opposite charge, so that an electrical double layer is formed (Fig. 8–12a). The concentration of oppositely charged ions decreases with increasing distance from the cell, and so also does the potential of the electrostatic field in the medium surrounding the cell. When an electric field is applied to a suspension of cells, the cells move in one direction and the oppositely charged ions associated with the double layer move in the opposite direction (Fig. 8–12b). We assume that the cell carries a thin layer of medium with it, and that there is a surface of shear between the outside of this layer and the rest of the medium. The potential at this surface of shear is called the zeta potential.

The equation derived by Debye and Hückel for a small charged particle in an electric field is

$$\mu = C\frac{\zeta D}{\eta}$$

where μ is the rate of migration (mobility) of the particle, ζ is the zeta potential, D is the dielectric constant, C is a constant independent of ζ but related to the particle shape, and η is the viscosity.

This equation involves various assumptions and approximations for a biologist dealing with cells, but from the observed mobility of a cell we can calculate the zeta potential. Using the further derived equation:

$$\zeta = \frac{4\pi\sigma}{D\kappa}$$

where κ is the Debye–Hückel factor, we can find σ, the uniform surface charge density of a cell.

Cell electrophoresis thus gives a measure of the surface charge of various cells. The conditions under which the measurements are carried out must be carefully controlled; in particular the pH and the ionic strength of the medium can affect the type and the density of charged groups on the cell surface, the adsorption or desorption of macromolecules, the

Fig. 8–12 (a) Diagram illustrating the charge on the cell surface, with associated oppositely charged ions. (b) In a charge field, the cell and its associated ions move in opposite directions

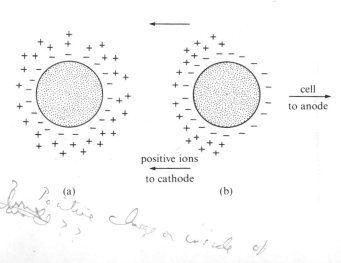

thickness of the double layer, and so on, all of which will affect the electrophoretic mobility. In fact, controlled variations of the conditions of measurement, and treatment of cells with enzymes or other specific reagents which react with and inactivate charged groups, are methods which have been used to determine the chemical nature of these charged groups on the cell surface.

Much of the cell electrophoretic work has been carried out on mammalian erythrocytes (Bangham, Pethica, and Seaman, 1958; Heard and Seaman, 1960). These, like all other cells so far investigated, have a negative surface charge at physiological pH values.

Studies on the effects of enzymes on cell surface charge indicate that much of the surface charge on erythrocytes, and perhaps on all mammalian cells, is due mainly to sialic acids (in particular N-acetylneuraminic acid) and not, as was thought earlier, to the phosphate groups of phospholipids. Sialic acids are substituted sugar molecules present as the polysaccharide part of a mucoprotein—that is, a protein molecule combined with multiple hexosamine-containing oligosaccharides, the carbohydrate portion comprising more than 4 per cent of the molecule. The negative charge of sialic acid is due to the carboxyl group in the molecule (Fig. 8–10).

Hirst (1948) suggested that the sites of virus adsorption on the erythrocyte surface are mucoprotein in character, and Hanig showed in the same year that the mobility of red cells is markedly reduced by virus adsorption and subsequent elution. This fall in mobility was attributed by Klenk (1958) to the liberation of sialic acid from the cell surface. Work by Eylar and coworkers (1962) showed that pre-treatment of red cells with neuraminidase, an enzyme which attacks and hydrolyses sialic acid, greatly reduces the amount of surface charge. Since neuraminidase does not enter the cell and since it removes most or all of the cell's sialic acid, we conclude that sialic acid is present on the cell surface.

Workers such as Cook (1962) and Ruhenstroth (1962) have studied the mobility of other mammalian cells, and have also found a drop in surface charge after neuraminidase treatment of mouse Ehrlich ascites tumour cells and HeLa cells (a line of cells derived from a human tumour). The results indicate that much of the surface charge of mammalian cells in general may be due to the presence of sialic acid. In Ehrlich ascites cells there is another acidic group, probably the carboxyl group of an amino acid or of membrane lipid, and also a basic amine group. The surface of human erythrocytes lacks any basic amino groups, although free amine groups occur in the red cell surface of other species, and work on bacteria (Hill and coworkers, 1963), in which the tough bacterial cell wall permits of a wider range of treatments than the more fragile animal cells, has revealed that amino groups are present on the surface of the wall of some bacteria at least.

Work on tumour cells indicates that malignancy may in general be accompanied by an increase in cell surface negative charge. Forrester and coworkers (1962) found that cells from a line of hamster kidney fibroblasts, transformed *in vitro* by polyoma virus to malignant cells, produce clones (colonies of cells grown from a single cell), about half of which consist of cells with a higher negative charge than the original fibroblasts. Ruhenstroth (1963) showed that, in human leukaemias, myeloid cells of characteristic mobility in normal persons are replaced by cells of higher mobility. This may be a general characteristic of tumour cells, along with increased sialic acid on the cell surface, lack of cellular adhesiveness, and inability to form normal contacts with other cells.

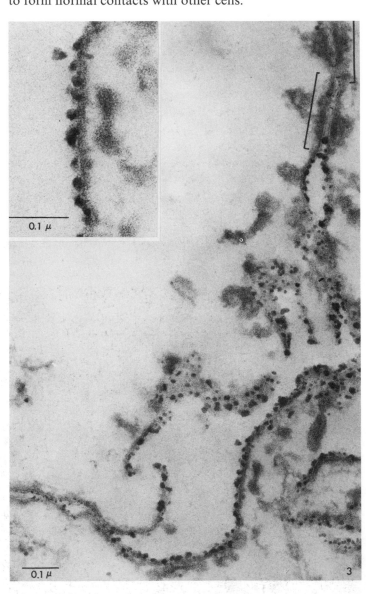

0.1 μ

0.1 μ

3

Fig. 8–13 Isolated rat liver plasma membranes stained with Hale stain. The electron-dense granules are on the outside of the membrane only (see inset). Where junctional complexes are present (see brackets) there is no staining. Magnification ×100,000 (With permission from E. L. Benedetti and P. Emmelot.)

Electrophoretic studies reveal that sialic acid lies on the outside, or very close to the outside, of the plasma membrane. As it cannot be removed by washing cells but only by enzyme treatment it is presumably a true membrane component. However, its bonding and steric arrangement in the membrane complex are not yet known. Gasic and Gasic and their colleagues (1962 onwards), detecting mucopolysaccharides on the surface of HeLa and other cells by means of Hale stain (which contains iron particles), confirmed the siting of sialic acid. Using electron microscopy, they found that the iron particles of the stain were visible as large scattered bodies attached to one side of the plasma membrane; and also the staining of cells was largely abolished by treatment with neuraminidase. This work was extended by Benedetti and Emmelot in 1967 (Fig. 8–13), who demonstrated that cells which do not show Hale staining can absorb mucopolysaccharides from the medium in which they are suspended and then take up stain. The observed staining of cells may sometimes be due to adsorption of components of body fluids on the surface.

**8–8
Surface
topography
of membranes**

Electron microscopy has been used to examine replicas of the cell surface in attempts to elucidate what Maddy has called the surface topography. Workers in the field have been mainly concerned with investigating possible differences between normal and tumour cells (Coman and Anderson, 1955; Nowell and Berwick, 1958; and Easty and Mercer, 1960). Replicas of hamster kidney epithelial cells and the corresponding hormone-induced tumour cells were prepared by Easty and Mercer. Figure 8–14 shows a surface replica of HeLa tumour cells grown in culture. The culture was washed

Fig. 8–14 A replica of the surface of a HeLa tumour cell grown in vitro. Magnification × 27,000. (With permission from G. C. Easty.)

and fixed, then shadowed with carbon and gold–palladium. Numerous minute spikes or microvilli can be seen on the free surfaces of the cells. These occur on both normal and tumour cells, but are often more frequent on the latter. They are approximately 2,000 Å in diameter, and may be up to 30,000 Å in length. Their presence has also been observed in sections of solid tumours grown *in vivo*. It is thought that these microvilli may in some way be connected with nutrition, as they are more frequent on the free surfaces of cells, and their formation obviously leads to an increase in cell surface area.

8–9
Membrane turnover

Palade (1968) studied the biosynthesis of lipids in the cytoplasm of liver cells and found that there is a continuous turnover of membrane lipid components, production within the cell being followed by degradation into the lipid building units again. He demonstrated that the lipid and protein components of membranes are not produced synchronously. The lipid structures have a shorter half-life, about 2 days, while that of the membrane proteins is about $3\frac{1}{2}$ days. This would tend not to support the theory of Green (1967) that the formation of membranes involves the assembly of repeating units of lipid and protein. Perhaps the observed asynchrony of lipid and protein production is a stabilizing mechanism— phospholipid may be injected into the plasma membrane, for instance, without interrupting the protein continuity.

Sometimes components are exchanged between the plasma membrane and the cell environment. For instance, the erythrocyte membrane is in a dynamic relationship with certain components of blood serum; cholesterol labelled with radioisotopes is taken up from serum and incorporated in the red cell membrane, and the phospholipid fraction of the membrane also exchanges with components of the serum.

8–10
Is there a cell surface coat?

A layer of material may or may not loosely cover the surface of the cell and lie outside the plasma membrane. Such material would easily be removed without causing damage to the plasma membrane or a marked change in its properties.

Moscona (1962 and 1963) found that when cells are disaggregated with enzymes or chelating agents (substances which bind bimolecular cations such as Ca^{2+} and Mg^{2+}), macromolecules of a mucoid nature appear in the medium. These, he suggested, comprise the surface coat on cells, were stripped off during treatment, and are essential for cell-to-cell adhesion. In other words, they form the 'extracellular cement'. However, these substances might in fact be intracellular material leaking from damaged cells. Similarly, the exudate laid down by trypsinized cells on glass surfaces, described by Rosenberg (1960), may be cell coat material, but could equally well be material lost from cells because of damage to the membrane.

Electron microscopy reveals that there is generally a gap of about 100 Å between the outermost stained regions of the plasma membranes of two apposed tissue cells. Some workers believe that the gap only appears when the tissue is prepared for study under the microscope (Sjøstrand, 1962); others believe that the gap exists but disagree about the nature of the material contained in it. For example, Robertson (1958 and 1963) has suggested that there really is a gap between living cells, and that it is filled with some colloidal material which does not take up electron-dense stains. Epstein and his coworkers (1964), and others, have sometimes observed faint staining outside the cell plasma membrane, and Epstein found that virus particles adhere to this faintly staining coat, and not directly to the plasma membrane. But Morgan and coworkers (1961) found no evidence of this, and consider that viruses attach directly to the plasma membrane. Possibly the staining zone is an artefact, and represents products leached out of the cell during fixation. Mercer (1957) obtained electron-dense staining in the intercellular gap with a lead hydroxide reagent. However, the action of this stain is not clear, and it may indicate not the chemistry of a tissue but its physical shape, with heavily staining regions representing hollows.

What can we make of all this conflicting evidence? It does seem fairly certain that components of body fluids, particularly colloids such as collagen, occur in intercellular spaces. Also, there may be traces of waste material from cells outside the plasma membrane. But we lack clear cut evidence for a coating of rather loosely attached material outside the plasma membrane, although we know that proteins are readily adsorbed from the suspending medium onto the surface of isolated mammalian cells *in vitro*.

8–11 Plant cell plasma membrane and wall

The plasma membrane of plant cells also seems to be a bimolecular leaflet structure of lipid and protein. Its structure probably varies with the metabolic state of the cell, and in particular with the synthesis and transport of material for cell wall formation (Northcote, 1968). The outer wall (Fig. 5–20) is the outstanding characteristic of the plant cell, and determines the manner of cell division and of growth. The composition and morphology of the wall are related to the age and function of the cell. In young cells, during what is called the primary stage of wall growth, it is elastic, about 1–3 μ thick, and increases greatly in area as the cell grows. During secondary growth, when expansion has ceased, the wall becomes thicker (about 5–10 μ) and more rigid and gives great tensile strength to the cell.

The wall is a matrix of polysaccharide material in which is embedded a highly organized interwoven system of microfibrils. These microfibrils are cellulose, chains of glucose

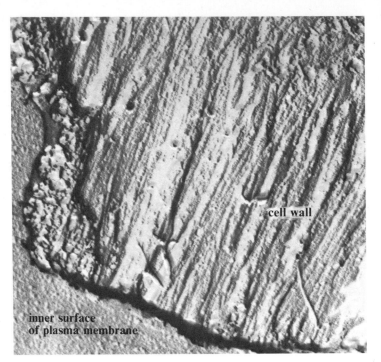

Fig. 8–15 A freeze-etch preparation of a plant cell plasma membrane and wall. Magnification × 68,000. (With permission from D. H. Northcote.)

molecules joined by β-1, 4- linkages (Fig. 7–27b). The continuous matrix contains hemicellulose (in which polysaccharides of pentose, as well as hexose, sugars occur) and pectic substances, which are long chains of repeating units of uronic acid derivatives of hexoses (Fig. 7–24). Lignin also occurs in secondary walls, particularly in woody plants. This is again formed of repeating units, this time of a cyclic carbon compound, probably with unsaturated linkages, the structure of which is not completely understood. The linkage of these repeating units confers great tensile strength, and lignified cell walls are strong and rigid.

Northcote (1968 and 1969) has obtained some beautiful electron micrographs of plant cells using the freeze-etch technique, in which cells are frozen at −180°C and then fractured while frozen. The fracture runs through the cell at different levels and exposes various internal surfaces. The technique of etching and shadowing renders the structures of these surfaces more clearly visible in the electron microscope. Northcote has found by this technique that there are particles of 90–130 Å in diameter on both the inner and outer surfaces of the plasma membrane. Figure 8–15 shows the microfibrils of the cell wall, among which are particles of the same size as those on the plasma membrane. These particles may be synthetic units which can migrate from the plasma membrane surface of the microfibrillar structure of the wall.

Similar studies indicate that the hemicellulose of the wall matrix is synthesized in the cytoplasm and transported out of the cell to the wall by vesicles from the Golgi apparatus

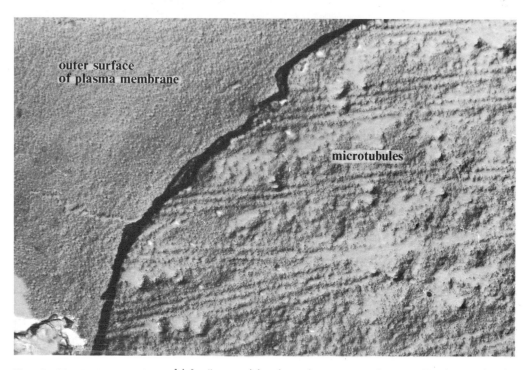

Fig. 8–16 A freeze etch preparation of a plant cell plasma membrane and the underlying cytoplasmic structures. Magnification × 39,000. (With permission from D. H. Northcote.)

which fuse with the plasma membrane. During mitosis, material for cell plate formation is also carried by vesicles from the Golgi apparatus. Microtubules exist in the cytoplasm underlying the plasma membrane (Fig. 8–16): they are 260–300 Å in diameter and have a distinct substructure along their length. One of their functions may be the organized movement of the vesicles carrying the matrix material of the wall to the plasma membrane.

8–12 Summary

In Fig. 8–17 we summarize the structure and properties of the plasma membranes of mammalian, bacterial, and plant cells. The main points are:

The plasma membrane, because of its permeability properties, must be a continuous or near-continuous lipid layer; the properties of this layer are modified by the presence of protein.

In mammalian cells this lipid layer has pores (possibly transient), probably lined with protein polar groups, through which small ions may easily pass.

Protein occurs in cell plasma membranes as various enzymes, notably ATP-ase. In bacterial cells the respiratory chain enzymes are also associated with the plasma membrane.

Mucoprotein is present on the mammalian cell surface —sugar molecules form the specific blood group substances, various surface antigens, and also account for most of the cell surface negative charge.

The plasma membrane, far from being an inert boundary, is highly dynamic. At physiological temperatures the mixture of various lipids in membranes probably forms liquid lipid layers. This means that the hydrocarbon chains are constantly moving and altering their configurations, and only the polar groups of the lipids are restricted by the need to be orientated at the lipid–aqueous interface. The plasma membrane can easily and rapidly change from a lamellar bimolecular lipid structure to a micellar structure, although the latter form may only exist transiently.

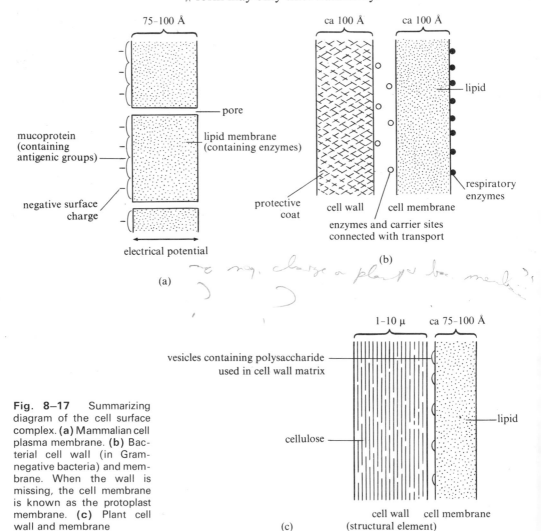

Fig. 8–17 Summarizing diagram of the cell surface complex. (a) Mammalian cell plasma membrane. (b) Bacterial cell wall (in Gram-negative bacteria) and membrane. When the wall is missing, the cell membrane is known as the protoplast membrane. (c) Plant cell wall and membrane

Bibliography   Curtis, A. S. G., *The Cell Surface: its Molecular Role in Morpho-Genesis*. Logos Press, 1967.

Davson, H. and J. F. Danielli, *The Permeability of Natural Membranes*, 2nd edn. Cambridge University Press, London, 1952.

Finean, J. B., The molecular organization of cell membranes. *Prog. Biophys. Mol. Biol.*, **16**, 145, 1966.

Heard, D., G. V. F. Seaman, and I. Simon-Reuss, Electrophoretic mobility of cultured mesodermal tissue cells. *Nature*, **190**, 1009, 1961.

Korn, E. D., Structure of biological membranes. *Science*, **153**, 1491, 1966.

Maddy, A. H., The chemical organization of the plasma membrane. *Int. Rev. Cytol.*, **20**, 1, 1966.

Northcote, D. H., Fine structure of cytoplasm in relation to synthesis and secretion in plant cells. *Proc. Roy. Soc. B.*, **173**, 21, 1969.

Rose, S., Ion pumps in the living cell. *Science J.*, **1**, 54, 1965.

Wallach, D. F. H. and E. M. Eylar, Sialic acid in the cellular membranes of Ehrlich ascites carcinoma cells. *Biochem. Biophys. Acta*, **52**, 594, 1961.

PART 3 The life of cells

Now we discuss the integrated function of the various intracellular components; that is, the entire cell is considered as a highly coordinated functional unit. Here, too, the function of the nucleus in relation to the rest of the cell is still of dominant interest. Chapter 9 deals with division in somatic cells, with the interactions between the cytoplasm and the nucleus during cell division, and with the transfer of genetic material from parent to daughter cells in growing tissues during mitosis. The synthesis of proteins and nucleic acids during the cell cycle is examined in relation to various control mechanisms, including hormonal control by the environment; the factors governing cell proliferation are also described.

Meiotic division in the germ line cells, which leads to the transfer of genetic material from one generation to the next, is described in Chapter 10, together with the way in which the genetic complement, or genome, of a developing organism determines the properties of the adult individual. The molecular biology of genetics in bacteria is also described in this chapter.

Mitosis and control of the cell cycle

Chromatid is chromosome completely duplicated except (for centromere which duplicates in metaphase)

Cells grow and divide, the period between the end of one division and the end of the next being known as the cell cycle. This chapter discusses the various stages of the cell cycle (Fig. 9–1), including the interphase, or period between divisions and mitosis, the delicate operation in which the chromosomal material is doubled and exactly divided between the two incipient daughter cells. In growing cells, nuclear function changes with time. This is illustrated in Fig. 9–1; actual observations of mitosis in a plant cell are shown in Fig. 9–2b. The growing cell increases in size and then divides into two daughter cells. The complete process could perhaps be more appropriately called the replication cycle. The growth phase is divided into three preiods: G_1, which starts immediately after the last cell division, S; and G_2, as shown in Fig. 9–1. These stages of the interphase nucleus will be considered in detail. At the end of G_2 the various changes which lead to cell division begin. The function of mitosis is to ensure that the DNA contained in the chromosomes is divided equally between the two daughter cells after replication. Provided that the coded sequences in the DNA are identical in the two daughter cells, these cells will be genetically identical with the parent cell, a

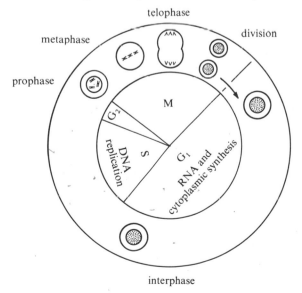

Fig. 9–1 The life cycle of a cell

condition which is essential for the preservation of the character of the organism. Mitosis depends on a coordinated interaction between nucleus and cytoplasm as the cell approaches division.

9–1 Historical introduction

Cell division was first observed in the salamander *Triturus maculosa* by Flemming in 1882; van Beneden studied cells of the worm *Parascaris* in 1883. Both observed threadlike bodies in dividing cells. The term chromosome (derived from the Greek, 'coloured body') was introduced by Waldmeyer in 1888, because the threads could be stained with dye solutions. Much of the early work on cell division was done by botanists, partly because the chromosomes are extremely clear and conspicuous in fixed plant material. Darlington, la Cour, Revell, and others have given detailed accounts of the way in which chromosomes behave during the stages of prophase, metaphase, anaphase, and telophase. But cell division involves the whole cell, not only the chromosomes; for an integrated study of mitosis it is essential to work with living cells. Much of this work has involved the use of time-lapse filming, the earliest film being made by Comandon and Jolly in 1913. Lewis and Lewis filmed cells a little later, and cultured cells were also studied by Canti. The development of the phase-contrast microscope has led to pioneering advances by Hughes. Bajer and Bajer (1950) produced fine films of mitosis in endosperm, the tissue which surrounds the embryo on the seeds of plants. Abercrombie, Ambrose, and Easty (1954) have since developed time-lapse filming with the interference microscope, a method which allows quantitative measurements of mass charges during cell growth and mitosis.

Detailed analyses of the actual mechanics of cell division were first made by Bĕlăr in the 1920s. A particularly suitable material proved to be the hairs on the stamen of the plant *Tradescantia*. Studies of the spindle with the polarizing microscope and with the electron microscope have since given information about the transient orientation of molecules which is responsible for drawing the chromosomes apart towards the two daughter cells (Swann and Mitchison; Mazia; Inoué; and others). But despite intensive study of nuclear and cytoplasmic changes during mitosis, the nature of the process is still mysterious and theories of mitosis are almost as numerous as the investigators in the field.

9–2 Stages of mitosis

The various stages of the cell cycle (Fig. 9–1) are as follows:

Interphase, which is divided into phases G_1, S, and G_2. G_1 follows the previous mitosis and precedes the onset of DNA synthesis. DNA synthesis occurs during the S phase and is followed by the G_2 phase. At the end of G_2 the cell begins to enter mitosis.

The stages of mitosis are as follows:

Prophase
Metaphase
Anaphase
Telophase

Chromosomes in different species vary in length, size, number, and other characteristics. Nevertheless, there is a clear similarity about mitosis in all plant and animal cells. Only in the later stages, when cell division involves for example the formation of a new cell wall or cell plate in plant cells, are there significant differences between plant and animal cells.

We now give a generalized account of the various stages of mitosis, and follow this by a more specific description of how mitosis proceeds in various types of cell.

Interphase. During the interphase period the nucleus generally exhibits remarkably little structure either before or after fixation. The two most distinct structures are the nucleolus and the part of the chromosome structure called heterochromatin. The nucleolus contains RNA and protein, which means that it can be stained with the combination of methyl green and pyronin, which gives a red colour with RNA. The heterochromatin contains DNA and stains with Feulgen stain.

The exact function of heterochromatin is still not fully understood. It represents a region of the chromosome which remains condensed during interphase and is not swollen by hydration to the same extent as the remainder of the chromosome, which is known as euchromatin. The nucleolus acts as the centre for the synthesis of ribosomal material as described in Chapter 3. The appearance of the interphase nucleus does not change greatly throughout G_1, S, and most of G_2, but undergoes striking changes immediately before prophase (Fig. 9–2, 2) Although threads corresponding to the chromosomes are just visible in some interphase nuclei under the interference or phase-contrast microscope, they become strikingly clear in later interphase. This is partly because the chromosomes have been replicated during the S phase; each chromosome is now double, although the strands are so closely associated that the separate strands cannot be distinguished. During this condensation stage, the heterochromatin and nucleolus are still visible, the heterochromatin being mainly localized near the nuclear membrane.

Early prophase. In prophase, most dramatic changes take place both in the nucleus and cytoplasm (Fig. 9–2, 2). The chromosomes condense and this continues throughout the successive stages of mitosis. The condensation seems to be brought about by two factors: first the loss of water from a highly hydrated gel into which the chromosomes have

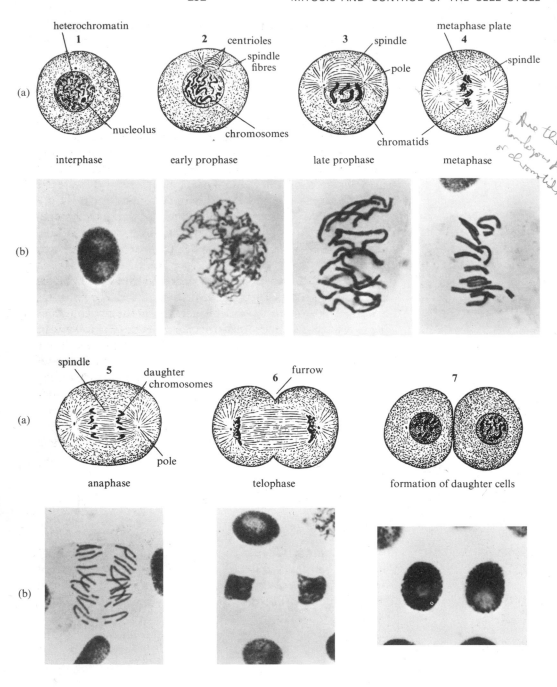

Fig. 9–2 **(a)** Diagrams showing the stages of mitosis from interphase to formation of two daughter cells (used by permission of Dodd, Mead & Company, Inc., from *Biology in Action* by N. J. Berrill. Copyright © 1966 by Dodd, Mead & Company, Inc.). **(b)** Photographs of mitosis in *Vicia faba* cells as seen in the microscope (with permission from S. H. Revell). The process is one of constant, integrated change, with each stage flowing smoothly into the next: **1** interphase nucleus; **2** prophase; **3** late prophase; **4** metaphase; **5** mid-anaphase; **6** telophase; **7** two daughter cells

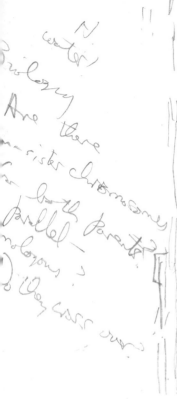

expanded during interphase, and second, the progressive spiralization of the threads. In a hydrated gel many molecules of water are associated with each nucleoprotein molecule. This condensation is also accompanied by the deposition of material, possibly RNA, on the chromosome strand. The spiralization of chromosomes, as demonstrated in plant material by Darlington, la Cour, and others, occurs in two stages—primary and secondary coiling (Figs 9–3a and 9–3b). In primary coiling, the long strands of the chromosome form a helical structure. Once this has been completed the spiralized chromosome can coil up in large secondary coils. Because the chromosomes are doubled at this stage, the coiling must not prevent the separation of the daughter chromosomes at metaphase. Three possible arrangements are shown in Fig. 9–3.

The chromosomes are first seen to be double stranded in prophase because of the replication that occurred during interphase. The two strands spiralize independently and lie beside one another. The separate strands of the chromosome are called *chromatids*. In some plant material the chromosomes in the prophase nucleus lie in particular positions in relation to each other, positions which are identical to those seen in the new nucleus formed after the previous division. This is further evidence for the *continuity* of the chromosome threads during interphase. In some plants, although the two chromatids spiralize separately and two strands are seen, the chromatids are coiled round each other.

As the chromosomes continue to condense the distinction between heterochromatin and the remainder of the

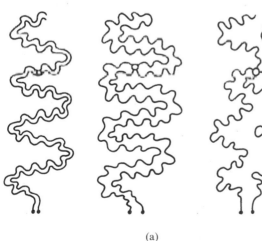

(a)

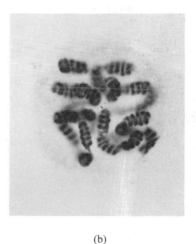

(b)

Fig. 9–3 (a) Three possible interpretations of a first metaphase chromosome in relation to its major and minor spirals (a three-dimensional helical structure represented in two dimensions). (After C. D. Darlington, 1935.) (b) Spiralization of plant cell chromosomes. Magnification × 1,200. (With permission from C. D. Darlington and L. F. LaCour, *The Handling of Chromosomes*, 5th edn. George Allen and Unwin, London, 1969.)

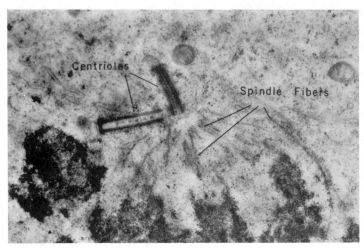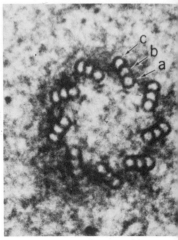

Left: **Fig. 9–4** An electron micrograph of centrioles. Magnification × 14,800. (With permission from D. W. Fawcett (ed.), *The Cell, An Atlas of Fine Structure.* W. B. Saunders Co., Philadelphia, 1966.) *Right:* **Fig. 9–5** An electron micrograph of a transverse section of a centriole from embryonic chick pancreas, showing the bundles consisting of sets of three small rods (a, b, and c). Magnification × 130,000. (With permission from D. W. Fawcett (ed.), *The Cell, An Atlas of Fine Structure.* W. B. Saunders Co., Philadelphia, 1966.)

Fig. 9–6 Diagrammatic representation of centriole structure as observed in electron micrographs. There are nine paired small rods and encircling belts of macromolecules. The length of a centriole is about 0·2 μ. (With permission from E. H. Mercer, *Cells and Cell Structure.* Hutchinson, London, 1961.)

chromosome (euchromatin) becomes less striking and finally disappears. Darlington has shown that when plants are maintained at low temperature the regions of heterochromatin are relatively deficient in DNA; this is shown by lack of Feulgen staining at prophase. This nucleic acid 'starvation' does not affect the subsequent viability of the plant, which suggests that, for genetic continuity, an exact 2:1 replication of heterochromatin may not be so important as exact replication of euchromatin. During prophase the nucleolus shrinks and finally disappears. Possibly part of the RNA so released becomes attached to other regions of the chromosomes. The nucleolus is a functional unit of the cell during interphase; it does not appear to be a highly organized structure.

While these changes are taking place in the nucleus, changes have also begun in the cell cytoplasm (these differ slightly between plant and animal cells). Adjacent to the nuclear membrane, but lying just within the cytoplasm, are bodies known as *centrioles.* These *corpuscules polaires* were first described by van Beneden in 1880 in cells of certain parasites of cephalopods, and the electron microscope allows us to observe them in great detail. They consist of hollow cylinders, or pairs of cylinders, somewhat like open beer cans (Fig. 9–4). Viewed in cross-section, they are clearly bundles of small rods grouped in sets of three, there being nine sets of three altogether (Fig. 9–5). Interestingly, there is a corresponding arrangement of nine outer bundles in cilia and flagella (see Fig. 11–15) and also in sperm tails. Centrioles differ from cilia in having three fibres to a bundle instead of two; also they do not possess a central bundle.

Each dividing cell has two centrioles. In ciliated epithelial cells, which are differentiated and no longer dividing, the centrioles replicate and give rise to *basal* bodies from which cilia develop. The spindle (Fig. 9–2, 2), which plays a major role in mitosis, is first formed round the centrioles; the

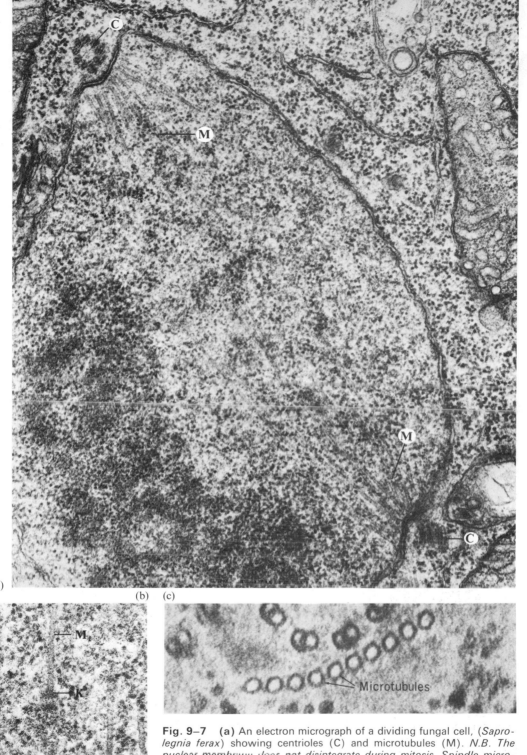

(a)

(b)   (c)

Fig. 9–7 (a) An electron micrograph of a dividing fungal cell, (*Saprolegnia ferax*) showing centrioles (C) and microtubules (M). *N.B. The nuclear membrane does not disintegrate during mitosis. Spindle microtubules do not penetrate nuclear membrane.* Magnification ×41,000. (With permission from J. B. Heath and A. D. Greenwood.) (b) The same showing a microtubule (M₁) attached to a centromere (K) (?) (Traction or chromosomal fibre?) and two continuous microtubules M₂. Magnification ×80,000. (c) Cross-section through microtubules. (With permission from D. W. Fawcett (ed.), *The Cell, An Atlas of Fine Structure.* W. B. Saunders Co., Philadelphia, 1966.)

295

structures from which it develops are rounded bodies lying around the centriole (Fig. 9–6). Once spindle fibres have begun to form, they radiate out from the centriole, the assembly of radiating spindle fibres being called an *aster* (from the Greek word meaning 'star'). Plant cells differ from animal cells in not containing centrioles. In the early stages of its formation the spindle is less sharply defined than in animal cells and does not become organized until the nuclear membrane breaks.

Late prophase. In late prophase (Fig. 9–2, 3) the spindle develops and the asters move apart. Additional spindle fibres appear between the asters, and the nuclear membrane disintegrates. In endosperm the nuclear membrane appears to break up into small liquid droplets. The nuclear sap and cytoplasm then mix together, a process which appears to play a role in the later development of spindle material. The structure of the mitotic spindle has been extensively studied with the polarizing and the electron microscope. For many years, fibres which could be seen with the light microscope after fixation were thought to form part of the spindle, but we now recognize they are merely an artefact of fixation produced by precipitation. Nevertheless, there is good evidence that orientated structures are present in the mitotic spindle in living cells.

Electron microscopy has revealed that the spindle fibres are formed by the aggregation of much smaller fibres called microtubles. Microtubule structures are preserved with glutaraldehyde, a recently developed fixative, but were poorly preserved by the older fixatives such as osmium tetroxide. They are shown very clearly in the beautiful electron micrographs of Fig. 9–7. The microtubles are straight and lie in parallel bundles; they have an outer diameter of 200–270 Å and the tubule wall is 50–70 Å thick. The technique of negative staining shows that they consist of 13 longitudinal filaments spaced 55–60 Å apart. The radiating microtubules develop from the centriole and associate with other microtubules in late prophase. There is some evidence that still finer fibres, about 40 Å in diameter, are associated with the microtubules. The spindle of living cells is clearly visible under a sensitive polarizing microscope (Fig. 9–8a and Fig. 9–8b). The sea urchin egg is particularly suitable for this study and its complete spindle can in fact be separated from disrupted cells (Mazia, Swann, and others, 1955).

Metaphase. At metaphase the spindle and the chromosomes interact to form the metaphase plate (Fig. 9–2, 4). The behaviour of individual chromosomes is shown diagrammatically in Fig. 9–9. Each chromosome carries a distinct region know as a *centromere* or *kinetochore* which plays a fundamental role in chromosome movements during mitosis. In fixed preparations it appears as a region of constriction in the spiralized

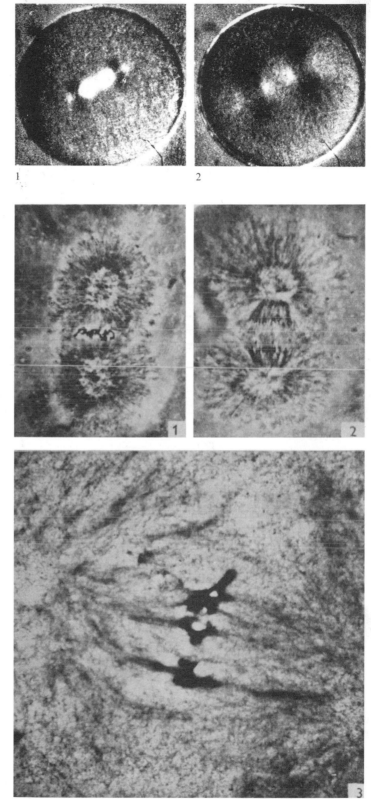

Right: **Fig. 9–8 (a)** The appearance of the spindle in the egg of the sea urchin, *Psammechinus miliaris,* as seen in a polarizing microscope: (1) 56 minutes after fertilization (metaphase), (2) 62 minutes after fertilization (anaphase). Magnification ×470. (With permission from A. Hughes, *The Mitotic Cycle.* Butterworth, London, 1952.)

Right: **Fig. 9–8 (b)** (1) and (2), an isolated and purified mitotic apparatus in a fertilized sea urchin egg, showing spindle fibres and astral rays at stages of metaphase and anaphase. Phase contrast microscopy, magnification × 1,080. (3) Electron micrograph of the same. Magnification × 6,500. (With permission from D. Mazia.)

chromosomes. A chromosome which does not carry a centromere is not carried to the daughter nucleus; this sometimes happens after radiation damage. The centromere is the region of the chromosome which apparently becomes attached to the spindle fibres and is therefore able to carry the remainder of the chromosome as it moves along the spindle towards the poles. When the centromere lies near the middle, the chromosome is called metacentric (Fig. 9–9 (1a)); when it is terminal the chromosome is called telocentric (Fig. 9–9 (1b)). In Fig. 9–7b microtubule M_1 is probably attached to a centromere; serial sectioning indicates that other microtubules such as M_2 stretch from pole to pole. Sliding of M_1 within sets of M_2 may occur in anaphase.

At metaphase, the chromosomes first have the appearance shown in (1a) of Fig. 9–9. To begin with the centromere cannot be seen to be double (2a), but later it splits into two separate centromeres which separate as shown in (3a) of Fig. 9–9. Some repulsive force seems to operate between the divided centromeres. In normal mitosis spindle fibres begin to be associated with the centromeres on the metaphase plate as shown in (3a) of Fig. 9–9. The chromosomes at metaphase (Fig. 9–2, 4) are in a plane in the centre of the spindle and at

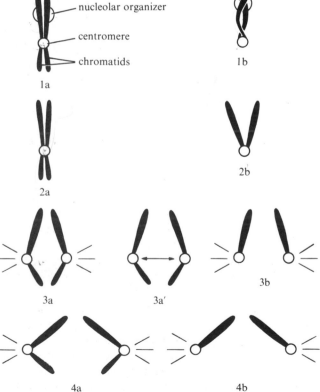

Fig. 9–9 The behaviour of individual chromosomes at metaphase shown diagrammatically: 1a–4a stages in the separation of the daughter chromosomes in a metacentric chromosome (centromere near middle); 3a′ separation of centromeres by mutal repulsion in the absence of spindle fibres; 1b–4b similar to 1a–4a for case with telocentric chromosome (centromere at end)

nucleolar organizer

centromere

chromatids

1a

1b

2a

2b

3a   3a′   3b

4a   4b

right angles to the spindle axis. Although for convenience the chromosomes are shown in the diagram to lie in this plane, it is in fact the centromeres or kinetochores which become orientated on the plane. In living cells they can be seen to move and oscillate about this position.

Colchicine, an alkaloid of great importance in studying chromosomes, specifically inhibits the formation of spindle fibres. Dividing cells treated with colchicine are arrested at metaphase. The centromeres separate as shown in 3b of Fig. 9–9. The repulsive force still operates between the centromeres, but spindle fibres do not become attached and the cell cannot move into anaphase.

Anaphase. In anaphase the pairs of centromeres move apart along the spindle, and carry one daughter chromosome of each pair to opposite poles (Fig. 9-2, 5). At the same time the spindle grows longer. The birefringence is as shown in Fig. 9–8b with the maximum refractive index parallel to the length of the spindle; the magnitude, measured along the axis of the spindle, is shown in Fig. 9–10, in anaphase.

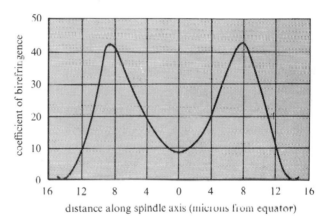

Fig. 9–10 The coefficient of birefringence measured along the axis of the mitotic spindle in a living egg of *Psammechinus millaris.* (With permission from M. Swann.

Telophase. By telophase the chromosomes have been drawn towards the poles in a highly contracted state; shortening and thickening has been occurring progressively throughout mitosis. At the same time, in animal cells at least, the division furrow appears at the cell surface. Initially this takes the form of a condensation of material at the cell membrane, followed by a flow of membrane material towards the centre of the cytoplasm. Cell division now occurs, and the two daughter nuclei re-form. The nuclear membranes appear, the chromosomes become dispersed within the nucleus, and the nucleoli are re-established.

The stages of mitosis are long in the telling but time-lapse films show them as a continuous process, perhaps the most beautiful natural phenomenon seen in cell biology.

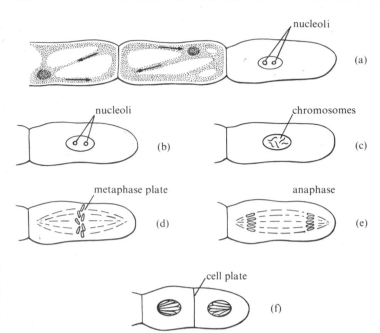

Fig. 9–11 Diagram showing the various stages of mitosis in a staminal hair of *Tradescantia*. Division of terminal cell. Remaining cells show cytoplasmic streaming (indicated by arrows).

9–3
Cell division in plant cells

Plants mainly grow at the tips of shoots and roots, usually by division of a terminal cell into two daughter cells, produced by the formation of a cell plate across the cell diameter. A new daughter cell at the tip again divides, while its neighbour develops into a mature cell, in which streaming movements of the cytoplasm occur (Fig. 9–11). The cellulose wall or cell plate forms between the two daughter cells (Fig. 9–11, and the electron micrograph of Fig. 9–12). The diagrams are drawn from photographs of mitosis in living staminal hairs of *Tradescantia*, the material used by Bělař in his classic work on mitosis. The cell plate forms during late anaphase. Vesicles containing polysaccharide material are secreted, as shown in Fig. 9–12, from the Golgi apparatus, which appears to be a general mechanism for secretion of cell wall matrix (Chapter 8). These form the so-called *phragmaplast* which later develops into the two plasma membranes that separate the two daughter cells, with a cellulose wall lying in between ((f) in Fig. 9–11).

9–4
Theories of mitosis

It is easy to propose a theory that accounts for mitosis in a particular kind of cell, but hard to find a general explanation that is true of mitosis in all cells. The way in which chromosomes seem to move immediately suggests that they are being pulled towards the poles by the spindle fibres, as was proposed by van Beneden (1883). Schmidt (1937, 1939) noted that the birefringence of the spindle decreases during anaphase, which also happens in the myofibrils of muscle during contraction. Swann (1952) noted that birefringence was first reduced near the centromere, which he attributed to secretion from the centromere causing the spindle fibres to shorten progressively. A most

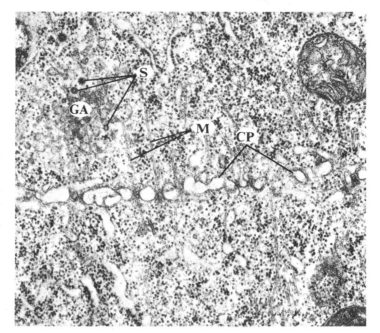

Fig. 9–12 Electron micrograph showing the relation of the Golgi apparatus to the formation of the cell wall in mitosis in spinach cells: GA Golgi apparatus (sectioned parallel to sheets of membrane, Fig. 5–15 shows section perpendicular to membranes). Magnification ×26,000; CP cell plate or wall; S Golgi secretions which become incorporated into the cell plate; M microtubules. (With permission from A. D. Greenwood.)

serious objection to the simple contraction model is that sometimes the chromosomes are carried right beyond the centriole, to which the chromosome fibres are attached.

Note too that during cell division in mammalian cells the whole cell elongates; in single layer cultures of cells the entire cell is stretched and the two daughter cells appear to pull each other apart (Fig. 9–13a). This is compatible with a cytoplasmic contraction theory because a mechanical force needs a point of attachment from which to be exerted. The side which wins a tug of war is not the side which simply pulls the hardest, but the side which is able to keep its feet firmly on the ground. The point of attachment shown in Fig. 9–13a could be the place where the two daughter cells are anchored to the glass. But cells can divide equally well in suspension, where no such attachments exist. Possibly the spindle fibres themselves provide a sufficiently rigid framework to give an attachment point for the contractile mechanism within the cytoplasm of the dividing cell. The work of Dan on the dividing egg of the sea urchin *Pseudocentrosus depressus* has shown that, when the egg is compressed in a direction along the spindle axis, the polar length still increases against this resistance during the first half of the division process; during the latter half it yields to the applied force and becomes shorter.

An alternative model for chromosome movements is the expansion or pushing model originally due to Watase (1891). According to this model, fibres growing out from the poles exert pressure on the nucleus, and the chromosomes become flattened on the metaphase plate. Specific fibres become attached to chromosomes which are then pushed towards the

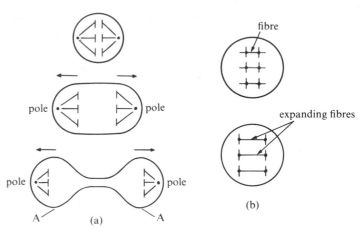

Fig. 9–13 Suggested models for chromosome movements in mitosis. **(a)** Traction model (A points of cell attachment to glass in monolayer cultures). **(b)** Bělăr expanding fibre model

opposite poles. The most serious problem with this model lies in explaining how the individual chromosomes on the metaphase plate become separated into daughter chromosomes which move to opposite poles. Bělăr produced a model for mitosis (Fig. 9–13, b) which has formed the basis for much research and in a modified form is still acceptable. The main points may be summarized briefly as follows:

1. The basic framework of the spindle is constituted of continuous fibres which stretch from one pole to the other.

2. Each chromosome becomes attached to a continuous fibre by a fluid secreted from the centromere. This fluid flows or spreads out on its continuous fibre which moves to the nearest pole. This coated fibre constitutes Bělăr's *traction fibre*.

3. The initial separation of the daughter chromosomes is an autonomous process (see Fig. 9-9, separation of centromeres).

4. The part of the continuous fibre located between the separating daughter chromosomes, as well as the non-fibrous substance of that region, begins to expand (Fig. 9–13b). The region acts as a *pushing body* or *Stemmkörper*.

5. Each traction fibre glides on a network of continuous fibre (which serves as a guide or track) towards the pole. The traction fibres may also shorten.

Bělăr's basic ideas fit in well with some of the most recent developments in this field. Much work has been carried out on the endosperm of plants, the tissue which surrounds the developing germ cells of plants and contains free nuclei lying within a mass of cytoplasm. The prophase nuclei can be maintained in culture and studied by time-lapse filming (Môlé Bajer, and Bajer). Studies of this and other material by Bajer, Inoué, and others, using electron microscopy, Normarski interference microscopy, and sensitive polarizing microscopy, have revealed that the spindle structure contains two main groups of microtubule assemblies. These are the chromosome fibres which are attached to and radiate from the centromere and continuous fibres which appear as sheets in thin sections in the electron microscope. These continuous fibres associate in the form of bundles, which can just be resolved in the light microscope using Normarski interference contrast (Allen and Bajer). Additionally, a few neocentric fibres are formed occasionally which penetrate regions of the chromosome other than the centromere.

During mitosis these microtubule assemblies change. At prometaphase, the centromere fibres are about 10 μ long, and their length decreases during anaphase to about 3–5 μ. But during this shortening of the centromere fibres the whole spindle *elongates* and the distance between the pole increases (Fig. 9–13b). The continuous fibres appear to grow by assembly of new material from a pool of microtubule material during growth of the spindle. The centromere fibres probably decrease in length by breakdown of their polar ends as the chromosomes are pulled towards the poles (4a and 4b in Fig. 9–9).

Inoué and Sato have suggested that the movement of the chromosomes follows from the growth of continuous fibres and the breakdown of the ends of the centromere fibres. However, Bajer, Östtergren and others think that the movement of chromosomes is an active process in which the centromere fibres *glide* as 'sail boats' between the continuous fibres. Ambrose has proposed that the motive force for these movements could be electro-osmosis and electrophoresis. Both these phenomena occur when the electric field exists along a direction of flow. With fixed particles it causes water to flow along a charged surface, whereas with free electrically charged particles it causes the particles to move. Such forces of interaction between the assemblies of microtubules due to microscopic regions of differing ionic balance generated by breakdown of ATP could be of the magnitude required to generate spindle movements.

Recent electron microscopy and other work by McIntosh, Helper, and van Wie, and also by Krishan and Buck, suggests that actual mechanical bridges can exist between microtubules. These bridges have in fact been observed in the electron microscope and they may be somewhat similar to the ratchet model for muscle contraction shown in Fig. 11–27. According to this model it is the making and breaking of these ratchet connections by an *active* process associated with the breakdown of ATP which causes the centromere fibres to glide between the continuous fibres and draw the chromosomes apart. In combination with assembly and disassembly of the fibres from the microtubule pool, most of the characteristics of mitotic movements can be explained on this model. The difficult problem of explaining how the ratchets can open and close synchronously to produce a continuous movement is discussed in Section 11–5.

Electrical phenomena very probably play some part in the control of mitosis. This was recognized many years ago by Lillie, who proposed that changes in membrane potential due to local changes in permeability occurred near the poles of the cell and round the nuclear membrane. Electrophoretic studies, by applying electric fields to dividing cells attached to a glass surface, have shown that prophase chromosomes carry a net negative charge, but this varies during mitosis. According to Lillie's model and those of Koller (1934) and Darlington (1936) changes of electrical potential in various regions of the cell produce electrical fields in which the charged particles of the chromosomes are able to migrate. The presence of microtubules and evidence for microtubule bridges were not known at that time. Nevertheless it is difficult to explain the co-ordinated *control* of spindle development and mitosis without invoking an electrical phenomenon; the generation of minute regions of *local* potential difference at the surface of microtubules and their associated ratchets which could interact along the length of the tubule could overcome some of the difficulties in reconciling the two basic concepts of a structural and an electrical model for chromosome movements during mitosis. Teorell (1937) demonstrated that such local regions of potential difference within a solution can be generated simply by *ionic* diffusion.

9–5
Synthesis during
the cell cycle

Once the two daughter cells have been formed, with their interphase nuclei as shown in Fig. 9–2, 7, a new cell cycle may begin for one or both cells. In some cases a cell may go on to form a differentiated cell and will not divide further. Figure 9–1 indicates that the G_1 phase immediately follows mitosis in growing cells. In G_1, all the complex phenomena of synthesis of messenger RNA, transfer RNA, and ribosomes takes place, as described in Chapter 3. The most simple cytological indicator of this process is the rapid growth of the nucleolus as new ribosomal material is synthesized.

Direct investigations of the rate of RNA synthesis in the nucleus and in the cytoplasm have been made with the help of radioactive labels, usually uridine labelled with tritium. Labelling of the nucleolus can then be demonstrated by auto-radiography. General labelling of the cell nucleus can also be seen at this stage. The labelling is due to synthesis of messenger RNA on the chromosomes, in other words to transcription. But not all the RNA is used for protein synthesis; some of it does not enter the cytoplasm but is broken down again within the nucleus. This probably provides a very adaptable system for changes in protein synthesis. Very soon after nuclear labelling of RNA in G_1, labelled material begins to appear in the cytoplasm where translation and protein synthesis take place. Protein synthesis has presumably already begun on pre-existing ribosomes transferred from the parent cells. The various enzymes required for cell growth are synthesized at this stage.

Protein synthesis can be followed by seeing how radio-actively labelled amino acids are incorporated into cells. Most of the cytoplasmic proteins are slightly acid in composition; phenylalanine is incorporated into these proteins quite actively and labelled phenylalanine is commonly used to observe synthesis. The basic proteins of the nucleus (histones) contain a higher proportion of basic amino acids. Labelled arginine is commonly used as a marker in this case. But the difference between basic and acidic proteins is only quantitative; acidic proteins incorporate a smaller proportion of arginine while basic proteins incorporate a smaller proportion of phenylalanine.

Most cytoplasmic proteins are being made throughout interphase, but the rate of synthesis falls considerably during G_2. As with mRNA, there is a large turnover of protein material; cell growth occurs when synthesis exceeds break-down. The histones are synthesized during the S phase when DNA synthesis occurs. Their rate of breakdown is only one tenth that of the cytoplasmic proteins.

The onset of DNA synthesis at the beginning of the S phase was first demonstrated by Caspersson, Walker, and others on fixed cultures. Walker measured the nuclear content of DNA after Feulgen staining and by ultraviolet absorption spectrometry, a microscope being attached to a spectrometer (microspectrophotometer) for these measurements. Modern culture methods to produce synchronized cultures (Puck) have greatly simplified the analytical method. In synchronized cultures, all the cells divide at approximately the same time. One of the simplest ways of producing synchronized cultures is to exploit the fact that tissue cells become less adhesive during mitosis. By shaking a growing culture it is possible to remove the cells in mitosis while leaving the interphase cells firmly attached to the glass. The dislodged cells may be harvested and placed in a new culture vessel. They then remain in reasonable synchrony for one or two subsequent mitoses. In rapidly grow-ing cultures of bacteria the S phase begins immediately after cell division, and DNA synthesis can be observed throughout the cell cycle. The length of G_1 varies widely with different plant and animal cells.

To study the timing of the S phase, tritium-labelled thymidine, which becomes incorporated in DNA, is ad-ministered to synchronized cells in tissue culture. Cultures are fixed at various stages of the cell cycle after labelling and auto-radiographs are prepared. Cells are labelled neither in the G_1 nor in the G_2 phase. Early work in whole tissues in this field was carried out by Howard and Pelc on the roots of growing broad

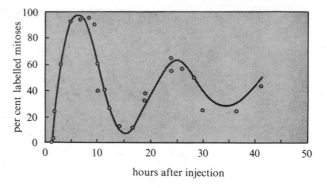

Fig. 9–14 Peaks of mitosis observed after labelling cells with radioactive thymidine give a measure of the mean doubling time. (With permission from D. D. Steel and L. F. Lamerton.

beans. *In vivo* studies with mammalian tissues have also been made, particularly by Lamerton and Steel. If a pulse label is used in which the labelled thymidine is introduced to the tissues for only a short period ($\frac{1}{2}$ hour or more) then only those cells which are in the S phase of DNA synthesis will be labelled. The labelled cells will continue through G_2 and enter mitosis. Daughter cells will be produced which also pass through the G_1, S, and G_2 phases and mitosis.

A group or cohort of cells can thus be followed through successive mitoses by fixing the tissues at various time intervals after the first pulse label. Peaks are observed in the number of labelled cells (Fig. 9–14) and these give a measure of the *mean doubling time* of the tissue. Examples of tissues which can be studied in this way in mammals are the epithelial cells of the skin, the wall of the intestine, and tumours. In the normal skin, division occurs in the basal layer of epithelium; these cells then move up towards the surface, become hardened by the formation of the insoluble protein keratin, and are shed continuously by abrasion as dead cells from the surface of the body. A persistent cell turnover also exists in the wall of the intestine and in the bone marrow, from which the circulating cells of the blood are continuously derived. Such normal tissues divide rapidly, in some cases with a doubling time of 8–10 hours. This is faster than almost any growing tumours; the idea widely held in the past, that tumours constantly divide, whereas normal tissues in adults do not, was misleading. The failure of control mechanisms in cancer cells is of more significance than rapid growth.

During the S phase both DNA and the other material of the chromosomes are replicated. The replication of DNA in terms of the complementary strands of the Watson and Crick double helix was described in Chapter 3. In the semi-conservative replication of the genetic material of the bacterium *E. coli*, the structure consists of one double helix, so that the replication is basically the same as that of the individual helices. Besides containing basic histones and other acidic proteins, the chromosomes of green plants and animals are in general much larger than the bacterial genetic material. In exceptional

where? on the chromosomes?

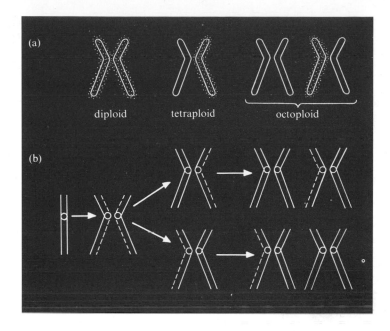

Fig. 9–15 Diagram of the experiment carried out by Taylor on the chromosomes of *Vicia faba* in root tips grown in tritiated thymidine for eight hours. (a) The drawings represent chromosomes at metaphase. After colchicine treatment the two chromatids of each chromosome are held together only at the centromere. The dots represent the position of tritium incorporated into DNA as observed in autoradiographs. (b) Diagrammatic interpretation of the results. The lines represent the two longitudinal sub-units assumed to exist in each chromatid. Unbroken lines are without tritium, and broken lines contain tritium. In each group of four lines the two outer ones represent old sub-units and the two inner ones new sub-units. After J. H. Taylor.

cases they can become giant chromosomes as in dipteran larvae where the cells continue to grow without cytoplasmic and mitotic division.

A tissue widely used for studying the replication of histones and acidic nuclear proteins is regenerating rat liver; this tissue regenerates rapidly after part of it has been removed and there is a partial synchrony of cell division. Once regeneration is complete, the growth is switched off by an accurately controlled regulator mechanism. Evans and others showed that a maximum incorporation of basic amino acids into histones occurred at 20 hours after part of the liver had been cut away, and another peak occurred at 28–43 hours. Histone synthesis reached its peak shortly before the peak of DNA synthesis. The acidic nuclear proteins show their peak of synthesis slightly earlier than do the histones.

The stages of actual chromosome replication have been mainly studied in growing plants, particularly in the root tips of the broad bean, *Vicia faba*. Taylor has introduced the drug colchicine into the water surrounding the roots to prevent the formation of the mitotic spindle. Colchicine has this effect without producing major changes in chromosome replication and spiralization. The chromosomes reach metaphase and become arrested at this stage; after a time a new resting nucleus may form which has all the replicated chromosomes. The somatic cell nucleus is *diploid*; it contains two of each chromosome—one derived from each parent. The cells after colchicine treatment can become tetraploid and even octoploid.

Such an experiment by Taylor on *Vicia faba* is illustrated in Fig. 9–15. After the introduction of tritium-labelled thymidine at the S phase both the *chromatids* which would become daughter chromosomes in normal division are labelled. At metaphase, the two chromatids remain joined at the centromere as shown in Fig. 9–15a. In the tetraploid nucleus only one out of two chromatids is labelled, while in the octoploid only one of four is labelled. An interpretation of these results is shown in Fig. 9–15b. If we imagine that each chromosome is assumed to replicate *semi*-conservatively like the single DNA helix, the results can be easily explained. The broken lines in Fig. 9–15b indicate tritium-labelled chromatids.

A more detailed study of these events can be made. Each chromatid consists of at least one Watson and Crick double helix. At replication, one of the strands is labelled and one unlabelled; Taylor has demonstrated that in the root cells of *Bellevalia*, which has four large chromosomes, exchanges occur between parts of the sister chromatids. Because the two chains of the Watson and Crick model run in opposite senses, after replication exchange must occur between pairs which run in the same sense. This will mean that at the succeeding metaphase a segment of one chromatid may be labelled while another part is unlabelled. This is basically similar to the phenomenon of crossing-over which is seen as chiasmata in meiosis (Chapter 10).

In seeking a model for the chromosome to explain replication during the S phase we note the following possibilities.

(a) First, the somatic chromosomes may each consist of one strand of double helix. There is much evidence in favour of this view, including work on the structure of lampbrush chromosomes (which appear to consist of one continuous DNA double helix), X-ray studies, labelling experiments, and so forth. The problem of chemical mutagenesis, which need only involve a reaction with one group on a DNA molecule, could be understood on the basis of this simple model. Objections to this model are that with some chromosomes the DNA would be about 1 metre long! Clearly, it must be tightly coiled to fit into the nucleus, which is only about 10μ in diameter. The replication of a strand of this length would also require a considerable time, far longer than the S phase, in fact far longer than the life of the cell.

(b) Second, electron microscopists at one time believed that the somatic chromosomes might be multistranded because this would explain their appearance; the giant chromosomes of the salivary glands are also clearly multiple-stranded in structure, but on a bigger scale. The main objection to this model for the somatic chromosome is that it is difficult to explain why a chemical reaction with only one DNA chain affects the whole assembly of multiple strands at a given point (or genetic *locus*).

(c) Third, as suggested by Ambrose and Gopal-Ayengar, the chromosomes may be composed of discrete chromomeres joined together by secondary forces of cohesion such as hydrogen bonds. Chromosomes can be dispersed by agents which break hydrogen bonds, as well as by agents which remove calcium ions (Kauffmann). But it is difficult to reconcile this model with the fact that lampbrush chromosomes can be dispersed with deoxyribonuclease, suggesting the presence of a continuous strand of DNA.

9–6
The replicon

Does replication begin at one point only along the chromosome, or does it proceed simultaneously in several regions? In *E. coli*, replication begins at one point and continues round the circular strands as shown in the beautiful electron micrograph of Fig. 3–8. A unit which replicates in this way is called a replicon. Experimental evidence shows that chromosomes of cultured cells contain a number of replicons (Fig. 9–16). If the cells are labelled for short periods during the S phase, using tritium-labelled thymidine, some parts are labelled early in the S phase while others are labelled late in the S phase.

It is hard to explain the replication of a whole chromosome in terms of a single replicon; as the DNA molecule would have to be several centimetres long. Keyl and Pelling have suggested that the unit of replication in the chromosome is in

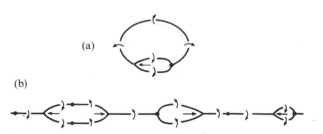

(a)

(b)

Fig. 9–16 (a) Diagram showing replication of the *E. coli* genetic material, which takes place in one direction around the circular molecule. (b) Suggested units of replication, or replicons, in a plant or animal chromosome. Dots indicate replicon junctions where chains can rotate independently (swivel points). Curved arrows show the direction of rotation and straight arrows the direction in which chain separation and synthesis are occurring. (With permission from H. K. L. Whitehouse, *Towards an Understanding of the Mechanism of Heredity*, 2nd edn. Edward Arnold, London, 1969.)

fact the chromomere. Since inversions of parts of chromosomes are known to occur (Chapter 10) it is likely that the replicons do not all replicate in the same direction (Fig. 9–16b). The replication of the bacterial genome necessitates a region of swivel, where the strands can rotate to compensate for the unwinding of the double helix during replication. These regions of swivel are also shown in Fig. 9–16 as dense circles.

Taylor has proposed several models for the replication of DNA in the chromosomes of plants and animals. One of these is shown in Fig. 9–17. It was pointed out in Chapter 3 that each chain of the complementary strands of DNA is replicated in the 3′ to 5′ direction. This involves a discontinuous synthesis of one strand which is synthesized 'backwards' (Fig.

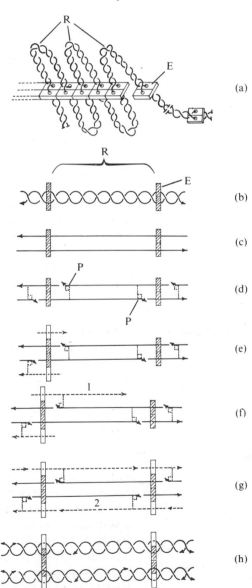

Fig. 9–17 (a) *Right*, chromosome consisting of single DNA helix with replication guides (E) attached at intervals between replicons (R); *left*, the same during replication with guides associated in a linear sequence. (b) Single replicon. (c) The same shown uncoiled for simplicity. (d) Priming of both chains at P before replication from 3′ ends. (e) Initiation of replication of alternate strands of successive replicons. (f) Replication of the 1st set of strands. (g) Replication of 2nd set. (h) Replicated chromatids. (With permission from J. H. Taylor.)

3–7). In the model shown in Fig. 9–17 the chromosome is composed of one continuous double helix of DNA. This double helix is subdivided into replicons somewhat similar to those of Figs 9–16 a and b. The replicons are subdivided by replication guides (E) from neighbouring replicons. These guides control the priming of the strands for replication. During replication they become associated throwing the DNA chain into loops as shown in Fig. 9–17a. At this time the chromosome has a form similar to lampbrush chromosomes.

The separate replicons are now primed by enzymatic breaking of the chains as shown (P). Replication then proceeds from 3′ ends. Set 1 is first synthezised. Set 2 is synthezised later. At the same time the replication guides are replicated. The two strands are replicated in opposite directions as in Fig. 3–7, but in this case both the old DNA strands are broken enzymatically before replication. Both old and new strands must subsequently be joined by a ligase.

Note that any model for chromosome structure must allow for a considerable degree of parallel orientation of double helices in some cells, particularly sperm cells, because the sperm heads of certain species shows considerable double refraction and X-ray diffraction patterns which indicate that the DNA chains lie parallel to one another. This could occur from opening of replication guides.

After the S phase, the cell enters the G_2 phase. The materials required for spindle formation, such as the precursors of microtubules, are probably formed at this time. The cell is known to contain more reduced sulphur, namely -SH groups, which may play some role in the energetics of mitosis. In division in sea urchins, this reaches a maximum concentration at metaphase.

**9–7
Cytoplasmic
division**

As can be seen in Fig. 9–2a and Fig. 9–2b cell division involves division of the cytoplasm as well as the nucleus between the two daughter cells. Most cytoplasmic organelles are present in large numbers—there may be many hundreds of mitochondria, ribosomes, lipid storage granules, and so on. A random distribution between the two daughter cells might lead to slight variations, such as 120 mitochondria in one daughter cell and 80 in the other. But these are sufficient to supply energy immediately after telophase to the daughter cells. Since the synthesis of all these compounds can be brought about by synthesis controlled by nuclear DNA, the cell can rapidly make good any deficiencies. In recent years evidence has been obtained to show that some cytoplasmic organelles are self-replicating, at least in part. The first demonstration of this was made by Sonneborn (1948) who studied the so-called kappa particles of the cytoplasm of *Paramecium*, which were shown to be self-replicating. Since that time it has been shown that several cytoplasmic particles are self-replicating and are at least in part independent of nuclear DNA as a source of template for protein synthesis.

As a result of the work of Chévremont on mammalian cells in tissue culture, of Chayen and others on plant cells, and Saiger on Protozoa, we now know that DNA is localized in mitochondria, in chloroplasts, and in the basal regions of cilia. This DNA is capable of controlling protein synthesis within the cytoplasmic organelle in which it is located and does so independently of the nucleus; nevertheless this DNA is probably to some extent regulated in its function by nuclear control.

Chévremont treated mammalian cells in culture with the enzyme DNAase, which breaks down DNA, and with an isomer of adrenochrome, trihydroxy-N-methylindole. Both these substances appear to act as inhibitors at the mitotic stage, so preventing the cell from entering mitosis. After this treatment DNA can be demonstrated in mitochondria by the Feulgen reaction. The DNA in mitochondria carries genetic information which codes for the synthesis of certain but not all mitochondrial enzymes. The most complete studies of cytoplasmic DNA have been carried out on mitochondrial DNA. Thomas and Wilkie have studied erythromycin-resistant mutants in yeast and have compared the Mendelian inheritance of resistance with the non-Mendelian (cytoplasmic). The cytoplasmic factor is called the *petite mutation* (P factor, assumed to be the mitochondrial DNA). Loss of this factor resulted in a permanent loss of resistance but a *petite mutation* in *gene* (nuclear) resistant mutants had no effect on the inheritance of resistance. The mitochondrial DNA controls, in the main, synthesis of enzymes located and bound to the inner membrane in *Saccharomyces cerevisiae*. These are cytochromes *a* and *b* and certain dehydrogenases.

9–8 Regulation of growth

Cellular control mechanisms have already been discussed in Chapter 4, where we pointed out that the capacity of bacteria to synthesize enzymes is affected by the composition of the growth medium. Addition of a particular substrate as a source of carbon may induce the cell to synthesize the particular enzyme which is capable of reacting with the new substrate. The Jacob and Monod hypothesis concerning gene suppressors and regulator genes was described in Section 4–12. Whatever the precise mechanism in a given instance, whether it operates at the level of the nucleus or the cytoplasm, the net effect is to cause the cell to synthesize new types of protein or to change the rate of synthesis of proteins already being produced, in response to a change in the environment. In cells of higher plants and animals other factors also control cellular activity, many of which involve the phenomenon of homeostasis.

The simplest example of homeostasis is the speed governor which is used on certain types of steam and diesel

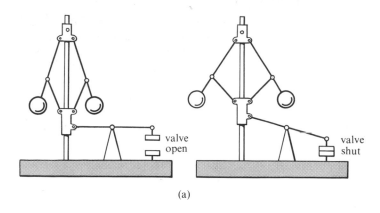

(a)

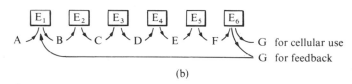

(b)

Fig. 9–18 (a) The speed governor on certain types of engine as a simple example of homeostasis—*left* at low speed, *right* at high speed. (b) An example of intracellular homeostasis. A self-regulating biochemical pathway is controlled by end-product inhibition

engines required to run at constant speed (Fig. 9–18a). As the engine goes faster, centrifugal force causes steel balls to move outwards against the force of the spring controlling the supply of steam or fuel for the engine. This movement closes the throttle and the speed is reduced. This control of the opening of the throttle by the engine speed is known as feedback. A similar type of feedback is involved in controlling the temperature of thermostats. A rise in temperature produced by the electric heater causes expansion of a metal strip or mercury column. This operates a relay which switches off the heater. In homeostatic mechanisms there is a time delay and the system fluctuates about a mean value, the magnitude of this fluctuation depending on the sensitivity of the regulator and the time lag before it comes into operation.

A simple example of homeostasis in living cells is shown in Fig. 9–18b. This represents a biochemical pathway of the type described in detail in Chapter 6. Metabolite A reacts at the surface of enzyme E_1 to give product B; product B is passed to enzyme E_2 to give reaction product C; and so on, along the enzyme chain. Finally the end product G is reached; this might be a secretory product of the cell such as a serum protein or procollagen. Part of the end product is secreted by the cell but the remainder combines with enzyme E_1 or with molecules of A. This reduces the rate of production of B and hence finally of G. Such a self-regulating enzyme chain will be able to adapt to the final rate at which the end product G is secreted or utilized by the cell. In this case the feedback process involves the interaction of G with E_1. It is called end product inhibition.

A feedback process is essential in a homeostatic control mechanism. One of the simplest control mechanisms which has been found in recent years involves physical contact between the cells. (The problem of the control of other cellular processes by intercellular contacts (contact inhibition) will be discussed in Chapter 13). Not only is the movement of cells controlled by contact with neighbouring cells, but the growth rate of cultures is affected by close contact.

An important factor in control of growth in higher plants and animals is the quantity of certain hormones in the circulating fluids. In higher plants the hormone indolyl-3-acetic acid (IAA).

$$\text{indole ring}-CH_2.COOH$$

and other hormones, known as auxins, promote growth. They occur in the most rapidly growing tissues, such as the coleoptile tip, apical bud, young leaf, and root tip. In mammals, growth hormone is secreted by the anterior pituitary gland. It is a protein which can be crystallized, and has a molecular weight of about 44,000. Besides affecting the growth of muscle and bone, mammalian growth hormone affects cellular retention of protein and metabolism of carbohydrate and fat. Other hormones stimulate particular organs, called target organs. For example, after fertilization the follicular tissue surrounding the ovum is converted into a corpus luteum, which secretes the steroid hormone progesterone. This hormone stimulates the activity of the mammary glands. A similar secretion of the steroid hormone testosterone by the testis maintains the functioning of the male reproductive organs.

Equally important are the homeostatic control mechanisms concerning growth in higher organisms. In an adult, the size of the various organs is precisely controlled. If the tissue is damaged the wound is healed and growth ceases when the organ has regained its original size.

A well-known hormone is thyroid-stimulating hormone (TSH) which can be isolated from the pituitary gland. TSH is a mucoprotein of molecular weight 26,000–28,000, and besides amino acids it contains the sugars galactosamine and glucosamine which include nitrogen. It is active in minute quantities. To study its action one can take a slice of calf thyroid tissue containing intact cells and place it in a salt solution with some serum. Thyroid cells need iodine for the synthesis of thyroid hormones, and the rate at which they take up iodine added to the medium can be followed with the isotope ^{131}I. On adding TSH to a thyroid slice or to a whole animal, uptake of iodine by the thyroid cells can be stimulated as much as 500 times. This is an example of the complicated interactions which occur within mammals, a hormone from one gland being able to stimulate another gland to produce hormone. Little is known about the targets on which such hormones act. There is

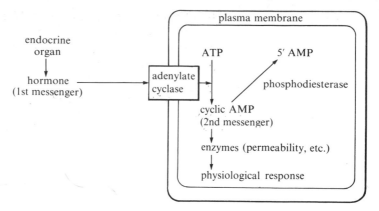

Fig. 9–19 Diagram to illustrate the second messenger hypothesis of the action of a hormone (the first messenger) on a cell. (With permission from T. Underwood. *New Scientist*, 27 March, 1969.)

evidence that they can operate on the transcription stage of protein synthesis, and receptor sites have also been found in the nucleus.

In adult tissues of mammals, the cells are differentiated to carry out particular functions and do not usually proliferate. However, under suitable stimulation, such as in wound healing and in responce to growth hormones, cell growth can recommence. The stage when the cells are differentiated and not in a stage of growth has been called the G_0 phase, which may last for hours, days or years depending on the type of cell. Protein turnover may still take place, and the cells can enter the G_1 phase under suitable stimulation. During G_0 many cells and tissues are particularly responsive to hormones. Sutherland has shown that the hormone adrenaline acts on liver slices to convert an inactive form of the enzyme phosphorylase into the active form. This leads to the conversion of glycogen into glucose and thus raises the blood sugar level. More detailed analysis has shown that the hormone acts according to the stages shown in Fig. 9–19.

Located at or near the plasma membrane are enzymes called adenylate cyclases. These enzymes are characteristic for particular tissues. For example, the cyclase from liver will

Fig. 9–20 The structure of cyclic AMP (adenosine 3′,5′ monophosphate)

respond to adrenaline, the cyclase from the adrenal cortex will respond to adrenocorticotrohic hormone (ACTH) but not to adrenaline, and vice versa, although it is not certain whether the hormone combines directly with the adenylate cyclases.

Adenylate cyclase converts ATP into a cyclic form of adenosine monophosphate, known as cyclic AMP (Fig. 9–20). This has been called the 'second messenger', the hormone molecules being the first messenger. Cyclic AMP affects cellular processes such as enzyme activity and permeability. First messenger hormones probably act at or near the cell surface, giving rise to second messenger changes and activating already synthesized enzyme molecules. Such a system might perhaps be called 'a quick action' system—certainly adrenaline has a quick action. Growth hormones probably act at a more profound level of cellular activity, controlling gene activity and protein synthesis.

The rate of mitosis in somatic tissues of higher animals depends on the need for new cells. Where the differentiated tissues are being continuously lost, as in keratinized skin, a mitosis of the epithelium occurs continuously to provide a steady supply of cells, that is, the rates of mitosis and of cell differentiation are subject to mutual control, as will be discussed further in Chapters 12 and 13.

Bibliography

Fautrez-Firlefyn, N. and J. Fautrez, The dynamism of cell division during early stages of the egg. *Int. Rev. Cytol.,* **22,** 171, 1967.

Harris, R. J. C. (ed.), *Cell Growth and Cell Division,* Symposium of the International Society for Cell Biology (vol 2). Academic Press, New York, 1963.

Hughes, Arthur, *The Mitotic Cycle,* Butterworth, London, 1952.

Locke, M. (ed.), *Reproduction: Molecular, Subcellular, and Cellular,* 24th Symposium of the Society for the Study of Developmental Biology. Academic Press, New York, 1965.

Mazia, D., Mitosis and the physiology of cell division, in J. Brachet and A. E. Mirsky (eds.), *The Cell,* vol 3. Academic Press, New York, 1961.

Schnader, Franz, *Mitosis.* Columbia University Press, New York, 1953.

10 Cellular aspects of heredity

The extent to which the characteristics of parents are transmitted to their offspring has interested mankind since the earliest times. Attempts to breed particularly useful varieties of domestic animals or plants gave an added stimulus to investigations of the inheritance of certain characters.

The theory of heredity which was still widely held until the mid-19th century was that the factors inherited by an offspring are a blend of those of its parents—that is, that the parental factors combine with and modify each other, producing a factor in the offspring which is intermediate between those of its parents. Certain observations, however, were never satisfactorily explained by the blending theory. As early as 1669, Becher had observed that when a black cock pigeon is mated with a white hen pigeon, the offspring are usually a mixture of entirely black and entirely white birds. None of the offspring are grey, or black and white spotted, as would be expected on the basis of the blending theory. It is only in the second generation, obtained by mating the all-black with the all-white birds, that black and white spotted offspring are produced.

Darwin, in the mid-19th century, accepted the blending theory as correct in some cases, but pointed out that there were serious objections to it, such as the existence of throw-backs. A throw-back is said to occur when a member of one generation shows an obvious resemblance to a predecessor in his grandparents' or an earlier generation, although the resemblance did not appear in his parents. This situation implies not a blending or combination of inherited factors, but rather the existence of particles of heredity which can survive unchanged during transmission from one generation to the next.

The first scientific studies of inheritance were begun in the early 19th century by means of plant breeding experiments. In the 1820s, Goss, a horticulturist, crossed garden pea plants giving green peas with others giving yellowish-white peas. In the first generation he obtained only yellowish-white peas, but by breeding again from these he obtained in the second generation three kinds of plant—one with green peas only, one with yellowish-white peas only, and the third with both green and yellowish-white peas in the same pod. Other workers, including Darwin, carried out similar experiments and obtained similar results. Darwin, who put forward his theory of evolution in collaboration with Wallace (who arrived at similar conclusions independently) in 1858, realized that evolutionary development must depend upon the inheritance of certain characteristics which would favour the survival of certain individuals in a particular environment. He carried out breeding experiments whith Antirrhinums (commonly known as snapdragons) as well as with garden peas, but did not succeed in devising a theory to account for his observations.

The classic work, published in 1866, which established the principles of particulate inheritance and laid the foundations of the modern science of genetics was carried out by an Austrian monk, Mendel. However, his observations aroused no interest amongst the biologists of his day, and were ignored until 1900, when they were rediscovered simultaneously by no fewer than three independent workers.

10–1
Work of Mendel

Mendel worked with the garden pea, as others had done before, but he was the first (in his beautifully designed experiments) to use quantitative methods for the study of inheritance. He chose his experimental material with care, selecting strains of peas which differed in well-defined characteristics, such as seed shape and colour, stem length, pod shape, and colour of the cotyledons. Since the pea is a self-fertilizing plant, crosses between different strains can usually be obtained only by artificial means. Mendel first ensured that the strains he had chosen bred true to type, by growing them for a period of 2 years. He then produced crosses by transferring pollen, containing male *gametes* or germ cells, from the anthers of one strain to the stigma of a second strain, so that the female gametes of the second strain were fertilized by the male gametes of the first strain. He studied the appearance of the offspring that developed from the *zygotes* produced. Such offspring are known in modern terminology as the first filial generation, or F_1 hybrids. Mendel then grew the F_1 seeds into plants and allowed them to self-pollinate in the usual way. He classified the progeny so obtained, known as the F_2 generation, in various categories. It was his quantitative approach in determining the relative numbers of each category of offspring which laid the basis for his success.

In his first experiments Mendel made crosses between plants which differed in a single characteristic only, such as seed shape *or* seed colour, so as to simplify his observations. For instance, he applied the pollen from a plant producing round seeds to the stigmata of a plant producing wrinkled seeds, and vice versa. He found that all of the progeny were *round* seeded, irrespective of whether the cross was made between male round and female wrinkled, or between male wrinkled and female round. Mendel termed the round character as dominant, and the wrinkled character, which disappeared in the F_1 generation, as recessive. He then planted out the hybrid seeds and reared plants from them. These he allowed to self-fertilize and found that of the 7,324 seeds of the F_2 generation so produced, 5,474 were round and 1,850 were wrinkled. That is, the ratio of round to wrinkled was 2·96 : 1, which, as Mendel noted, is a close approximation to 3 : 1. Thus the second filial generation was not uniform, but contained dominant and recessive forms in the ratio of approximately 3 : 1.

The experiment was carried further and plants were grown from the seeds of the second generation. These plants were self-fertilized and produced seeds in their turn (the third generation). Mendel found that all plants grown from the wrinkled F_2 seeds, which formed a quarter of all the seeds of the second generation, produced only wrinkled seeds; in other words, they bred true like the original parental wrinkled strain, or, in Mendel's terminology, they were *constant*. Of the 565 plants grown from the round seeds, 193, that is, one third of the round seeds (one quarter of the total F_2 number), were constant in giving round seeds. The remainder, constituting one half of the individuals of the second generation, gave both round and wrinkled seeds, again in the approximate ratio of 3 : 1 as in the second generation. Mendel continued the experiment as far as the sixth generation, and obtained essentially the same result.

Mendel interpreted his results to mean that the characteristics he was investigating were determined by a pair of factors, one inherited from each parent. The factors were not blended in the F_1 hybrids, but retained their individuality, and were *segregated*—that is, separately distributed to the germ cells of the hybrids. The sequence of events is summarized

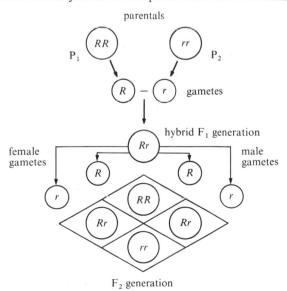

Fig. 10–1 How Mendel's first law (of segregation) explains the 3:1 ratio of dominant to recessive factors among the F_2 progeny. *R* represents the dominant factor and *r* the recessive factor. P_1 and P_2 are the parents

in Fig. 10–1, where R represents the round factor (dominant), and r the wrinkled factor (recessive). One parental strain produced a gamete containing the R factor, and the other a gamete containing the r factor. When these gametes combined the resultant zygote contained one R and one r factor. Hence, since R is dominant, the entire F_1 hybrid generation took the form of rounded seeds.

If we now assume that the R and r factors remain unchanged during the development of the adults and that this generation produces 'round' and 'wrinkled' germ cells, both male and female, in equal numbers; and if we also assume that if is a matter of chance which of the two sorts of pollen will combine with each of the two sorts of egg-cells, there are three possibilities in the second generation:

> (i) The offspring may inherit the R dominant factor from both parents, giving rise to the true-breeding strain of round peas (RR), as in one of the original parents (P_1).
> (ii) Similarly, the progeny could derive the r factor from both parents, giving rise to the constant wrinkled strain (rr) as in the second parent (P_2).
> (iii) The R factor could be inherited from one parent and the r factor from the other, it being immaterial which factor was derived from the male and which from the female parent; the offspring (Rr) would then appear as round peas in the F_1 hybrid generation.

On the basis of mathematical probability, the expected ratio of round to wrinkled peas in the second generation would be $3:1$. (In practice, however, this would only be approximate, as it is a statistical probability; it will approach the limiting value of $3:1$ as the number of experimental cases studied increases.) A shortened way of indicating the composition of this generation is $RR + 2Rr + rr$.

The 50 per cent of peas with Rr factors would behave like the F_1 hybrid generation, and their progeny would consist of round to wrinkled peas in the proportion of $3:1$.

In modern terminology, individuals in groups (i) and (ii), the true-breeding strains with RR or rr factors, are called *homozygous*; those in the third group with both factors Rr are called *heterozygous*.

Mendel's work was thus convincing evidence that inherited factors, whether dominant or recessive, are not altered by the development of the parents, but are segregated and transmitted independently to the progeny. Although Mendel did not state his conclusions in the form of laws, other biologists did so subsequently, and the principle of segregation is often referred to as *Mendel's first law*.

In Mendel's experiments one of each pair of characteristics was dominant and one recessive. As Mendel himself was aware, dominance is not universal: the principles of his discoveries are, however, still applicable in the absence of dominance. An example is the crossing of a pure strain of red-flowered antirrhinum with a pure strain of white antirrhinum. Since neither red nor white is dominant, the F_1 hybrid generation is an intermediate pink colour. Of the second generation a quarter are pure strain red flowers, a quarter pure strain white flowers, and a half intermediate pink flowers. If R represents the red characteristic and W the white characteristic the F_2 generation may be described as

$$RR \quad + \quad 2RW \quad + \quad WW$$
red   pink   white

Here it is easy to determine the $1:2:1$ ratio of homozygous (red) : heterozygous : homozygous (white) by means of the intermediate pink colour of the heterozygotes.

Mendel extended his crossing experiments in peas to more than one pair of differentiating characters. He took seeds from a strain having round peas and yellow cotyledons and crossed them with a strain having wrinkled peas and

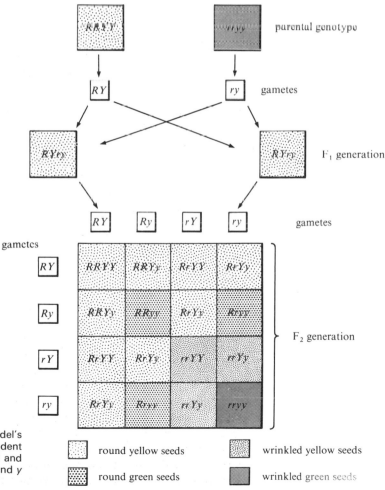

Fig. 10–2 How Mendel's second law (independent assortment) operates. R and Y are dominant, and r and y are recessive

□ round yellow seeds   ▨ wrinkled yellow seeds

▨ round green seeds   ▩ wrinkled green seeds

green cotyledons. The yellow cotyledon colour, like the round pea shape, was dominant. The peas of the F_1 generation were all round with yellow cotyledons. When these were raised as plants and allowed to self-fertilize, among the seeds obtained two new categories appeared—round peas with green cotyledons and wrinkled peas with yellow cotyledons.

In Fig. 10–2, R and Y represent the dominant, and r and y the recessive factors. The peas of the F_1 hybrid generation all contained the factors $RrYy$, which meant—since R and Y were dominant—that they were all round, with yellow cotyledons. Assuming that the R and Y, and r and y factors are transmitted independently of one another, the four possible combinations of pairs of factors in the reproductive cells of the F_1 generation would be as shown in Fig. 10–2.

Mendel showed that these are formed in equal quantities by crossing the F_1 generation with the doubly recessive strain and obtaining approximately equal numbers of the four different kinds of pea (see Fig. 10–3). Then, since their combination in the F_2 generation is presumably determined solely by chance, the expected outcome is that shown in the table of Fig. 10–2 (known as a contingency table).

		Gametes of F_1 generation		
	RY	Ry	rY	ry
Gametes of wrinkled green strain — ry →	$RrYy$	$Rryy$	$rrYy$	$rryy$
	Round, yellow	Round, green	Wrinkled, yellow	Wrinkled, green
Numbers of offspring observed by Mendel	31	26	27	26

Fig. 10–3 Effect of crossing gametes of the F_1 generation with the doubly recessive strain in Mendel's experiments with two differentiating characters

Those classes which differ from the original parental classes are known in modern terminology as *recombinants*. There are 16 possible combinations of the pairs of factors in the male and female gametes of the F_1 generation. It is immaterial which factors are derived from which parent, so some of these recombinants give the same final result (see table). For example, there are four classes of offspring with the $RrYy$ combination. There are in fact nine different categories, as Mendel showed by further breeding experiments. These are

$RRYY$	$RRYy$	$RRyy$
$RrYY$	$RrYy$	$Rryy$
$rrYY$	$rrYy$	$rryy$

Of every 16 seeds, the nine which contained a dominant R and a dominant Y factor will be round with yellow cotyledons, the three which have at least one R factor with y factors only will be round with green cotyledons, the three seeds with

r factors and at least one Y factor will be wrinkled with yellow cotyledons, and one with the r and y factors will be wrinkled and green. In fact Mendel found that of 556 seeds of the F_2 generation the numbers of the four different types were:

round, yellow	315
round, green	108
wrinkled, yellow	101
wrinkled, green	32

which approximates to this $9:3:3:1$ ratio.

Mendel's experiments thus clearly showed that factors inherited from one parent do not remain in association when transmitted to their offspring, but are randomly rearranged. This was later formulated as *Mendel's second law* or the *law of independent assortment*.

Mendel's method of studying inherited factors by means of carefully controlled crossing experiments is essentially the same as that used today in much more detailed analytical studies. His conclusions have stood the test of time, with one modification; that is, later experiments have shown that the law of independent assortment is not universal. In some cases inherited factors tend to remain in their existing combinations, a phenomenon known as *linkage* (see below).

Mendel's work took eight years to complete. He presented the results and conclusions to the Brünn (now Brno) Society for the Study of Natural Sciences in 1865, and his lectures were published the following year in the proceedings of the society. His work, however, was unnoticed during his lifetime. He died in 1884, and it was not until 16 years later that three scientists independently carried out similar experiments and reformulated his conclusions. In 1900, de Vries in Holland, Correns in Germany, and Tschermak in Austria, by means of plant breeding experiments, developed similar ideas about dominant and recessive traits and the separation in the reproductive cells of the factors involved. Each realized later that Mendel had anticipated him in his conclusions.

When Mendel's work was repeated and rediscovered in this way, it stimulated a great deal of interest. Bateson, in England, who invented the name *genetics* for the science of heredity, became a great supporter of Mendelian ideas. He felt that genetic factors were possibly material particles, but did not consider that they could be equated with chromosomes. Johannsen in Denmark had similar ideas, although he considered that there existed units, which he named *genes*, in the reproductive cells which determined inherited characteristics.

In modern nomenclature, hereditary factors are still called *genes*, although the concept of what constitutes a gene has altered considerably (see Section 10–9). Also, we now know that it is necessary to distinguish between the full complement of inherited factors in an individual, known as the *genotype*, and the extent, depending upon the environment, to which these factors function in practice to produce the various characteristics which make up the *phenotype* of the same individual.

From the point of view of cell biology and genetics, considerable progress had been made by cytologists by 1900. By then they knew that chromosomes appear from nuclear material at cell division and they also knew the essential facts of mitosis and meiosis. Sutton in 1902 was the first to point out the similarity of the behaviour of Mendel's segregating characters to that of chromosomes in dividing cells. Others, including de Vries, also drew attention to this point. From the bringing together of the discoveries of cytologists and geneticists the new subject of cytogenetics developed.

10–2
Meiosis

The process of mitosis, in which single cells divide in animals or plants to give daughter cells which are exact replicas of the parent cells, has been described in Chapter 9. The way reproduction occurs in whole organisms varies considerably from one type of organism to another. In *asexual* or *vegetative* reproduction the parent divides into two or more parts, each of which grows, by a process of mitotic division, into a new individual. In sexual reproduction the male and female reproductive cells unite to form a zygote, the single cell from which a new individual develops. In 1883, van Beneden observed in his studies of the horse thread-worm, *Parascaris equorum*, that there were twice as many chromosomes visible at the mitoses in the fertilized egg as there had been in the egg and sperm nuclei. This implied that the male and female gametes had each contributed half the chromosome number to the zygote. Weismann suggested in 1887 that in each generation there must at some stage be a *reduction division*, involving a halving of the chromosome number, otherwise the body cells of each generation would contain twice the chromosome complement of those of the parents.

It was observed by Flemming (1887), Strasburger (1888), and others that the nuclear divisions just before the formation of mature sperm and eggs in animals (and of pollen grains and embryo sacs in flowering plants) were different in that two nuclear divisions occurred in quick succession. It is in fact during gamete formation that reduction division occurs. The whole process was given the name *meiosis* (or reduction) in 1905. The chromosome complement in the zygote is known as the *diploid* number, whereas the halved number in the gametes is called the *haploid* number. The diploid number is the normal complement in the individual cells of higher plants and animals, but in lower plants, such as the green algae, in fungi, and in bacteria the normal state is often haploid. In the green algae and fungi, fusion of the sex cells is usually followed

In normal cells are different chromosomes from father & father functioning?

1st MEIOTIC DIVISION

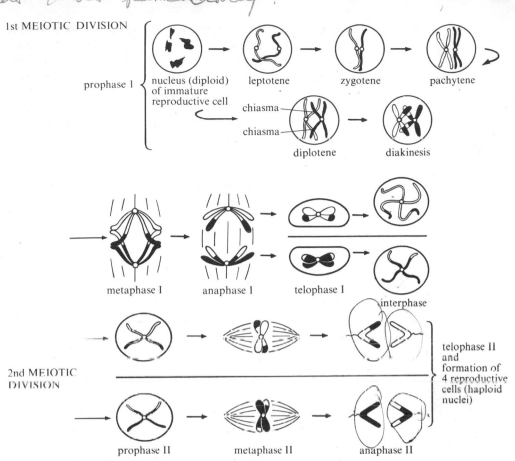

Fig. 10–4 Stages of meiosis

almost immediately by a reduction division to the haploid state. Reproduction in bacteria is discussed in Section 10–6.

The process of meiosis is essentially the same in all of the different types of cell in which it occurs. It begins in immature reproductive cells and involves, as mentioned above, two successive nuclear divisions. In the first division, prophase is considerably longer than the prophase of mitosis and can be divided into recognizable stages, shown diagrammatically in Fig. 10–4.

Prophase 1. (i) *Leptotene.* The cell nuclei at this stage contain chromosomes in the form of very fine single threads, hence the name leptotene, meaning 'slender ribbon'. Later the chromosomes were found characteristically to possess small densely staining granules of unequal size spaced at irregular intervals along their length. These granules are known as *chromomeres.*

(ii) *Zygotene.* During this stage a process of chromosome pairing begins. Matching pairs, known as *homologous* pairs, consist of one chromosome from the haploid set derived from one parent, and a similar chromosome from the other parental

Highland →

Duplicating DNA
Becoming chromatids
?

haploid set. These two resemble each other in length, in the position of the centromere, and even in the position of the chromomeres arranged along them (Fig. 10–6). The result of this process, which is known as *synapsis* or *zygotene pairing*, is that there is now a haploid number of chromosome pairs, which are called *bivalents*. The mechanism by which this specific pairing takes place, and the nature of the attractive forces involved, are not yet understood.

(iii) *Pachytene*. At this point the two chromosomes of each bivalent shorten and thicken, hence the stage was given the name pachytene, meaning 'thick ribbon'.

(iv) *Diplotene*. The attractive forces operating during synapsis, bringing and holding the homologous chromosomes together, seem now to weaken, and the chromosomes in each pair begin to come apart. It can now be seen that each chromosome has itself divided into two daughter chromatids, and so each dissociating unit now becomes visibly quadrivalent. Separation of the homologous chromosomes, each consisting of two daughter chromatids, is, however, incomplete; at one or more points along their length two of the chromatids still remain in contact. This results in the characteristic appearance of a cross or *chiasma* (from the Greek letter chi, χ) in the case of one point of contact, or a ring in the case of more than one point of contact. Janssens proposed in 1909 that in chiasma formation an exchange of lengths of chromatid takes place by breakage at the point of contact and reunion with the other chromatid involved (Fig. 10–4).

This crossing-over cannot be observed directly, but genetic evidence in support of Janssen's theory was obtained by Morgan (1911) in his experiments with *Drosophila*, and later by other workers. In 1937 fixed preparations revealed that chromosomal abnormalities in *Drosophila* had interchanged, giving visual evidence of crossing-over. Belling in 1928 modified Janssen's theory by suggesting that crossing-over is at the earlier pachytene stage in which the homologous chromosomes are closely associated, and not at diplotene as Janssens had proposed. He argued that the consequences of these exchanges would then be observed later at the diplotene stage, since crossing-over is between non-sister chromatids (derived from different parents), while chromatid pairing at the diplotene stage is between sister chromatids.

The work of Bridges (1916) and Anderson (1925) on breeding experiments in *Drosophila* indicated that chromosome duplication takes place before crossing-over, which only occurs between two chromatids at the four-strand stage. This has been confirmed by later experiments (see below).

(v) *Diakinesis*. The final stage of the first prophase of meiosis is sometimes called diakinesis (or 'moving apart'). Here the chromosomes have become shorter and thicker as the two bivalents move further away from each other.

Metaphase, anaphase, and telophase of first meiosis. At the end of the first prophase the nucleolus and nuclear membrane disappear, spindle formation occurs as in mitosis, and metaphase I begins. However, the metaphase stage differs from that of mitosis in that the centromeres of the two component chromosomes (each consisting of two chromatids) of a homologous pair lie, in each case, on either side of the equator, instead of in the equatorial plane (Fig. 10–4). Again, in contrast to mitosis, the centromeres do not divide at anaphase I, but the two centromeres of a homologous pair move towards opposite poles of the spindle, dragging their chromosomes with them. At this stage each chromosome may easily be seen to consist of two chromatids. A nuclear membrane forms at telophase I round the haploid set of chromosomes at each pole, and the cell divides into two daughter cells.

After a brief interphase when the chromosomes are difficult to see, each of the daughter cells enters a second meiotic division. This resembles mitosis, in that the units involved are chromosomes, each consisting of two chromatids joined at a single centromere, and that the centromeres take up a position on the equatorial plate at metaphase II. It differs from mitosis, however, in that only half the normal number of chromosomes is present, there being one member of each homologous pair, and that no further duplication of chromosomes has taken place since that of prophase I. At anaphase II the centromeres divide and the daughter chromosomes move towards the opposite poles with the centromeres leading the way as usual. At telophase II nuclear membranes form around the four haploid nuclei which have resulted from the original diploid parent cell.

Figure 10–5 shows some of the stages of meiosis in the locust, *Schistocerca gregaria*. These photographs were taken using interference contrast, which is particularly good for the study of nuclear behaviour, on flattened living cells.

To summarize, the actual events of meiosis are that the chromosomes undergo one duplication, one segregation, and one division. The cellular processes underlying these events have yet to be explained. For instance, the nature of the forces responsible for the pairing of homologous chromosomes in the early stages of prophase I, and for their separation at the diplotene stage, is not yet understood. Pritchard (1960) has suggested, on the basis of data from crossing-over experiments, that the beginnings of chromosome pairing are at an earlier stage, in pre-meiotic interphase. He proposed that at this stage chromosomes move around at random in the nucleus, but that, when matching regions of homologous chromosomes come into contact by chance, they tend to adhere (Fig. 10–6), which would solve the problem of how partner finding takes place. Pritchard further suggested that crossing-over occurs at these points as soon as contact is made, the crossing-over

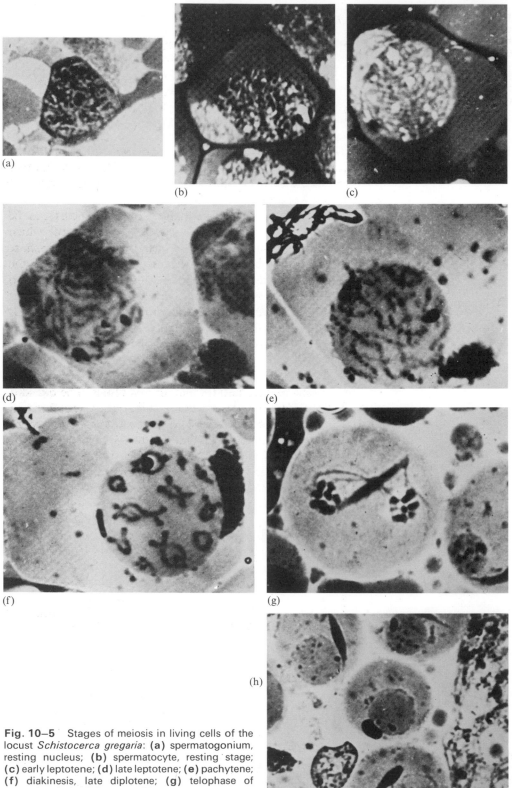

Fig. 10–5 Stages of meiosis in living cells of the locust *Schistocerca gregaria*: **(a)** spermatogonium, resting nucleus; **(b)** spermatocyte, resting stage; **(c)** early leptotene; **(d)** late leptotene; **(e)** pachytene; **(f)** diakinesis, late diplotene; **(g)** telophase of second meiosis; **(h)** spermatids

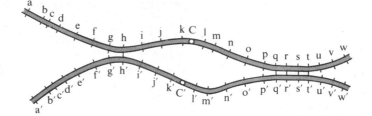

Fig. 10–6 The beginning of pairing between two homologous chromosomes during the first meiotic division. C and C′ mark the centromeres. a and a′, b and b′, etc., are matching regions on the homologous chromosomes

providing a firm foundation for the later lining up of the chromosomes in apposition along their entire length. In other words, crossing-over may be a cause rather than an effect of chromosome pairing.

Darlington (1931) has suggested that meiosis can be regarded as a precocious mitosis, and that the force which keeps the daughter chromatids together in mitotic prophase also keeps the homologous chromosomes paired in the zygotene and pachytene stages. This suggestion now seems to be improbable, as the work of Swift (1950) and Howard and Pelc (1951) on DNA synthesis in animal and plant cells has indicated that in both mitosis and meiosis the synthesis of DNA is completed in the interphase before nuclear division occurs.

There has been much speculation concerning the mechanisms of chromatid breakage and repair. The old idea was that the paired chromosomes twist tightly together and that torsional forces are responsible for breakage, but modern concepts involve enzymatic mechanisms. Equivalent regions of chromatids are exchanged without gain or loss of material, with rare exceptions, in either resultant chromatid.

The formation of the sex cells in animals is shown diagrammatically in Fig. 10–7. In the male, the immature

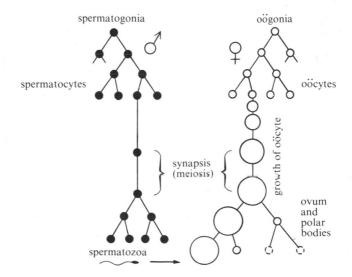

Fig. 10–7 The different processes involved in gametogenesis in the production of spermatozoa and ova

reproductive cells, or spermatogonia, give rise by mitotic division to spermatocytes. These undergo meiotic division, each spermatocyte finally producing four fully differentiated spermatozoa. In the female the oögonia, corresponding to spermatogonia, give rise to oöcytes which, however, unlike spermatocytes, virtually complete their differentiation as ova before they undergo reduction division. Unlike sperm formation, each oöcyte gives rise to one ovum only. The three very small cells, known as polar bodies, which result from unequal divisions of the oöcytes cannot be fertilized and are discarded. The importance of these differences, and the processes involved in fertilization and development, are discussed in Chapter 12.

The process of meiosis keeps the chromosome number constant from one generation to the next. Equally important is the reshuffling of hereditary characteristics which arises from the reassortment of the parental chromosomes and from the exchange of pieces of chromosomal material in crossing-over. It was shown by Carothers (1913) and others that in the first metaphase of meiosis it is usually a matter of chance whether the centromere of the paternal or the maternal bivalent lies towards a particular pole. Since this is true for each homologous pair, the complete set of chromosomes obtained at each pole is a random mixture of paternal and maternal origin. In heterozygotes there are consequently many possible combinations. In fact, since each chromosome can go to either pole, the total possible number of arrangements will be 2^n, where n is the haploid chromosome number of the cells concerned. In addition, the parental genetic material is re-shuffled by crossing-over, resulting in reconstituted chromosomes in the gamete which differ from those of the parent. The mechanisms of reassortment and crossing-over between the chromosomes lead to a great amount of genetic variation, and ensure that no gamete will have exactly the same genetic constitution as any of the other gametes produced by the same individual.

In man the diploid chromosome number is 46. In the male there are 22 matching pairs, known as *autosomes*, and two unmatched sex chromosomes, known as *heterosomes*, the larger one of which is known as the X chromosome and the smaller one as the Y. In the female there are 23 matching pairs, both of the sex chromosomes being of the X type. In the case of the gametes, the ova always contain an X chromosome, while half of the spermatozoa contain an X and the other half contain a Y. The total possible number of chromosome arrangements due to reassortment in meiosis alone is 2^{23}, which is more than 8×10^6. Further rearrangement takes place because of crossing-over, so it is not surprising that individual zygotes from the same parents are never alike genetically.

10–3
Chromosomes
and heredity

The close parallel between the behaviour of chromosomes at meiosis and the segregation of Mendel's hereditary factors was pointed out by Sutton and others, and the suggestion was made that the genetic factors involved, the genes, were located on the chromosomes. Mendel's first series of experiments may be explained in terms of chromosome behaviour as follows.

We assume that the factors R (round peas, dominant) and r (wrinkled, recessive) are determined by a slight difference in corresponding regions of a homologous chromosome pair. In modern terminology, these alternative forms of a gene are called allelomorphs or *alleles*. Because both of the original parental strains bred true, the R gene must be on both of the matching chromosomes, and the r gene must be on both of the homologous pair concerned. All of the gametes from one

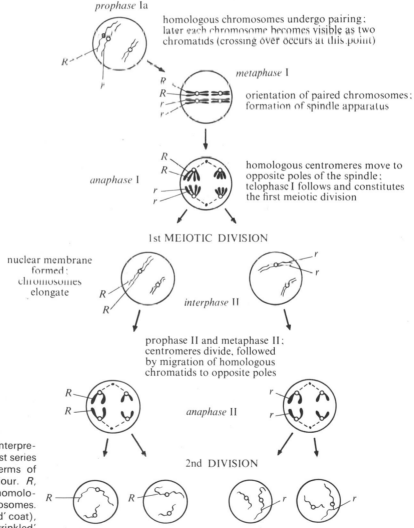

Fig. 10–8 The interpretation of Mendel's first series of experiments in terms of chromosome behaviour. *R, r* are alleles on an homologous pair of chromosomes. *R* is dominant ('round' coat), *r* is recessive ('wrinkled' coat)

parent will have the R gene, and all from the other the r gene. Thus the members of the F_1 hybrid generation must all have one chromosome with the R gene, and the other matching chromosome with the r gene.

The sequence of events leading to the formation of the gametes of the F_1 generation is illustrated in Fig. 10–8. Clearly half the gametes, whether male or female, will carry the R gene and half the r gene. The possible recombinants in the F_2 generation are as shown in Fig. 10–1.

Similarly, Mendel's work on the independent assortment of two separate pairs of factors may be shown as in Fig. 10–9, (a), (b), and (c), assuming that the genes determining these factors were located on two different chromosome pairs.

However, as Sutton originally pointed out, in any species there are more hereditary characters than there are chromosomes, and so one chromosome must carry a number of factors. These cannot assort independently, but will remain in association; that is, they show *linkage*.

An example of linkage was observed by Bateson, Saunders, and Punnett (1905 and 1906) in their studies of sweet peas. They were studying the inheritance of two pairs of characters—purple flower, which is dominant to red flower, and long pollen grain, which is dominant to round pollen grain. The parents were purple long and red round. Out of 2,132 plants of the F_2 generation, the numbers in the four possible groups were found to be as follows:

Purple long	Purple round	Red long	Red round
1,528	106	117	381

This was very different from the $9:3:3:1$ ratio to be expected

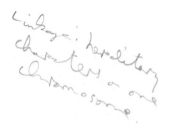

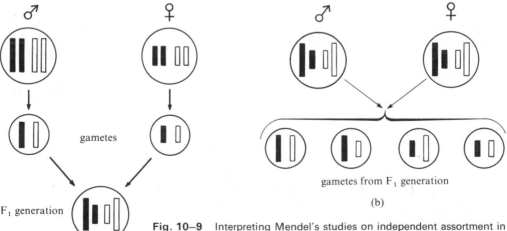

gametes from F_1 generation

(b)

Fig. 10–9 Interpreting Mendel's studies on independent assortment in terms of chromosome behaviour: **(a)** formation of the F_1 generation; **(b)** formation of gametes from the F_1 generation. Both male and female produce the four types of gametes shown; **(c)** *opposite:* formation of the F_2 generation. R and r, and Y and y are on homologous pairs of chromosomes, the size difference is used for diagrammatic purposes only to indicate recessiveness of r and y

9 round yellow
3 round green
3 wrinkled yellow
1 wrinkled green

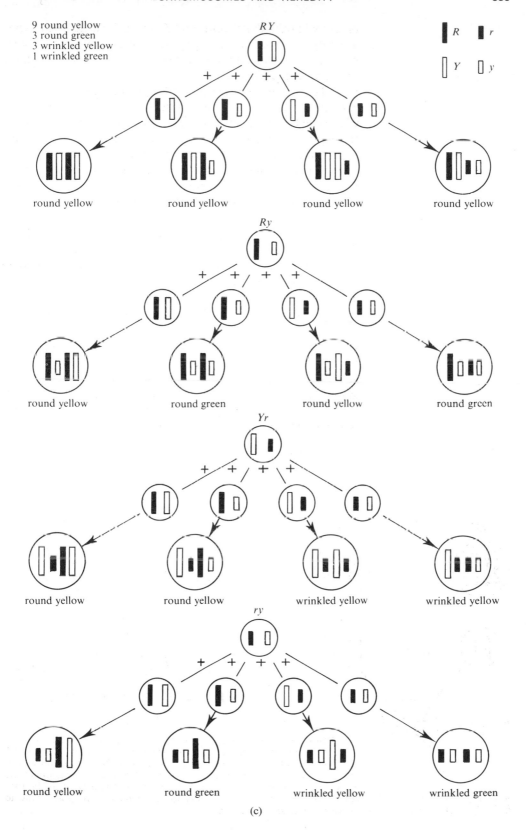

(c)

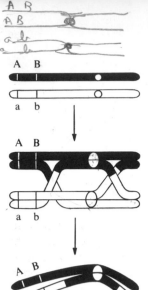

from independent assortment, which would give approximately

Purple long	Purple round	Red long	Red round
1,199	400	400	133

That is, there were more plants with the original parental combinations and fewer recombinants (purple round and red long) than expected. So the characters of flower colour and pollen grain shape had remained linked in some way during cell division. This linkage, however, could not have been complete, otherwise there would have been no recombinants in the F_2 generation, and purple long and red round

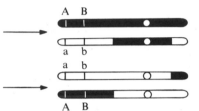

Fig. 10–10 Suggested scheme of crossing-over during meiosis. A and B, and a and b remain linked as no crossing-over takes place between them

plants would have occurred in the proportion of 3:1. The figures obtained are what would be expected if *partial linkage* were to occur. This partial association of characters was found to take place if the cross was made between purple round and red long parents. Again, there was an excess in the parental classes and a deficiency in the recombinant classes in the F_2 generation.

Janssen's observations in 1909 on chiasma formation during meiosis provided a possible explanation of how partial linkage could occur. Whether or not two genes on the same chromosome remain linked during meiosis depends upon the exact position of crossing over of the chromatids (see Fig. 10–10). Morgan realized the implications of Janssen's theory, and his achievements (1911 to 1915) in locating the relative positions of genes on chromosomes in the fruit fly *Drosophila melanogaster* resulted directly from this understanding.

10–4
Gene mapping
with *Drosophila*

As Morgan pointed out, the closer together two genes lie on a particular chromosome, the less likely is crossing-over to occur between them, and the fewer the recombinants occurring in the progeny of crossing experiments. The recombination frequency is in fact 50 per cent when independent assortment of individual chromosomes is operating. (See Figs 10–2 and 10–9b also Section 10–1. Here the parental combinations of RY and ry are present in 50 per cent of the gametes of the F_1 generation and so in 50 per cent of the zygotes of the F_2 generation, and the recombinants Ry and rY also occur with a frequency of 50 per cent.) So an observed frequency of recombinants of less than 50 per cent indicates that linkage is occurring and that the genes involved must lie on the same

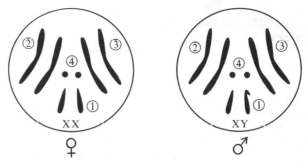

Fig. 10–11 Chromosomes of *Drosophila melanogaster:* *left,* female; *right,* male

chromosome. Further, since the fewer the recombinants the closer together the genes are situated, it follows that the observed recombination frequency may be used as a measure of the distance between any two linked genes on a chromosome. Using this method, Morgan and his associates were able to determine, for many *Drosophila* genes, which were linked, and their relative positions on the various chromosomes.

The choice of *Drosophila* for genetic experiments has several advantages; first, the breeding time from egg to maturity is a matter of a few days only, whereas plant experiments take many months; second, there are many variations of external appearance which are easy to recognize; third, and all-important to a geneticist, there are only four pairs of chromosomes in *Drosophila.* Three of these pairs, or bivalents, are autosomes, two of them long and one very short (Fig. 10–11), and the fourth pair, the sex chromosomes, consists of two straight medium-length X chromosomes in the female, and one X and one hook-shaped Y chromosome in the male. The appearance of the giant chromosomes of *Drosophila* salivary glands is shown in Fig. 10–12.

The method Morgan used was to look for a fly of abnormal appearance in his stock of *Drosophila.* If he found after certain breeding tests that the abnormality was inherited, he concluded that a *mutation* had occurred—in other words, that the gene responsible for this factor had altered. Morgan had to rely mainly on spontaneously occurring mutations, although he attempted to induce mutations artificially by means of chemicals, X-rays, and so forth. Characters such as body colour, wing shape and size, etc., were used to identify changes in the normal or wild-type appearance which had occurred due to the presence of mutant genes.

The first mutant form observed by Morgan was a male fly with white eyes which appeared among large numbers of normal red-eyed males and females. Breeding experiments showed that the gene for white eyes was connected with sex determination—that is, was *sex-linked.* Morgan's results made sense, as he pointed out, if this mutant gene could be carried on the X chromosome but never on the Y chromosome. In this case, white-eyed females would only be observed if they were homozygous for this recessive gene, otherwise the

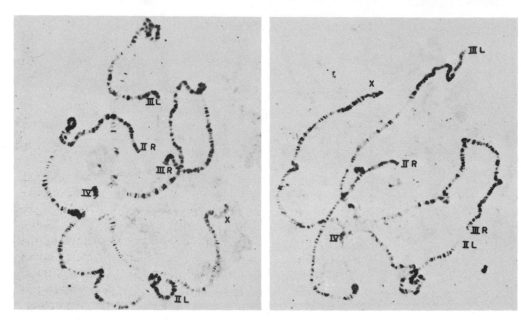

Fig. 10–12 The salivary gland complement of *Drosophila melanogaster* showing the male (*left*) and female (*right*) sets. Each complement consists of five main arms—one for the sex-bivalent, two for each of the V-shaped autosomes, and a small rod corresponding to the IVth or 'dot' bivalent. The conventional nomenclature of the various arms is indicated, and is based on their characteristic banding. The sex-bivalent in the female (XX) is thicker and denser than that of the male (XY), since the Y is mainly embedded in the chromocentre (the region of association of the centromeres of the chrosomes). Magnification ×1,000. (With permission from O. G. Fahmy and M. Fahmy.)

dominant wild-type gene for red eyes on one X chromosome would mask the mutant on the other X chromosome. The Y chromosome, however, since it carried no dominant gene for red eyes, did not mask the effect of the mutant gene when it occurred on the X chromosome in males. For the first time, experimental work indicated that a specific gene was associated with a specific chromosome which could be identified in microscopic preparations. More than 100 genes have now been identified as being carried on the X chromosome in *Drosophila*.

The method used by Morgan and his associates was that of the three-point cross. If one considers three genes *A*, *B*, and *C* located on the same chromosome, and their alleles *a*, *b*, and *c* on the homologous chromosome, as shown below

then the possible recombinants resulting from crossing-over are

| *A* | *b* | *c* |
| *a* | *B* | *C* |

from a crossing-over occurring between *A* and *B*

A	*B*	*c*
a	*b*	*C*

from a crossing-over between *B* and *C*

A	*b*	*C*
a	*B*	*c*

from a crossing-over between *A* and *B*, and between *B* and *C*.

Morgan crossed mutant forms of *Drosophila* possessing three variant characters—that is, carrying three recessive genes—with the normal wild-type, and then back-crossed the offspring with flies carrying the recessive genes. By careful examination of the offspring he was able to determine the percentages of recombinants of the different types, and so the percentages of crossing-over between the three pairs of genes, *A* and *B*, *B* and *C*, and *A* and *C*. The types of characters which he studied were body colour, wing shape, eye colour, absence of bristles, and so on; some of these are shown diagrammatically in Fig. 10 13a. The symbol given to the mutant gene refers to the appearance of the mutant form (for instance, *vg*—vestigial wings) and the normal wild-type gene is written as *vg*¹, or sometimes as + alone where there is no ambiguity.

Morgan showed that the proportion of crossing-over between a particular pair of alleles of two different genes (e.g. *A* and *B*) was of the same order, irrespective of whether they

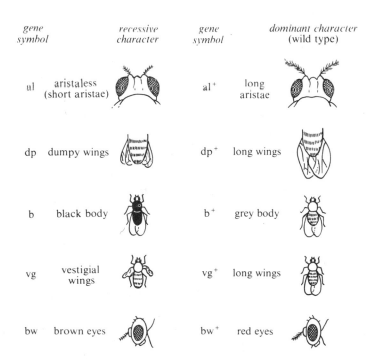

gene symbol	recessive character	gene symbol	dominant character (wild type)
al	aristaless (short aristae)	al⁺	long aristae
dp	dumpy wings	dp⁺	long wings
b	black body	b⁺	grey body
vg	vestigial wings	vg⁺	long wings
bw	brown eyes	bw⁺	red eyes

Fig. 10–13 (a) Some of the recessive and dominant characters studied by Morgan in *Drosophila* with their corresponding symbols

were both initially on the same chromosome (in *coupling*) or on different chromosomes (in *repulsion*). Morgan also found that the percentage of crossings-over between two of the factors of a three-point cross as obtained from the observed percentage of recombinants (when corrected for double cross-overs, resulting in recombinants of the type *AbC* and *aBc*), was approximately the sum of the percentages of crossings-over between the other two possible pairs of factors. That is, if there was found to be 30 per cent crossing-over between *A* and *C*, and 4 per cent between *C* and *B*, then there was approximately 34 per cent between *A* and *B*. On the basis of these

Fig. 10–13 (b) Linear genetic map of *Drosophila* showing the four linkage groups corresponding to the four chromosomes

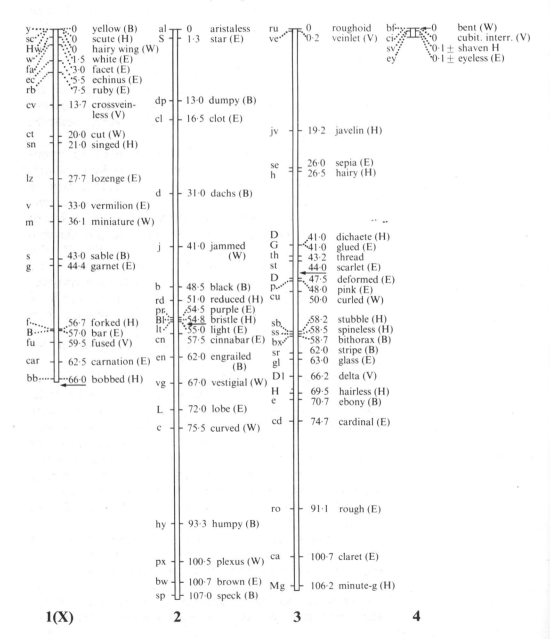

results, Sturtevant, one of Morgan's colleagues, suggested in 1913 that the percentage of crossings-over could be used as an index of the distance between any two genes. That is, in the example given above the genes must be arranged on the chromosome in the order $A—C—B$, with a much shorter distance between B and C than between A and C:

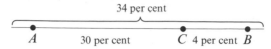

Since the crossing-over percentages were found to be additive, then the only arrangement of gene loci which would fit these results was that of a linear array. So gene positions could be plotted on a *linear map* (Fig. 10–13b). Sturtevant suggested that the unit of distance on the map, known as a centiMorgan, should be defined as the distance in which there was 1 per cent recombination frequency—that is, one crossing-over for every 100 gametes. Thus, if the recombination frequency between two genes is 6 per cent, then the genes would be separated by six map units.

After analysing the progeny of many breeding experiments using flies bearing various mutant genes Morgan, with Sturtevant, Bridges, and Muller, was able by 1915 to map the position of about 50 *Drosophila* genes. Many more genes have since been mapped, over 500 having been mapped on one of the longer chromosomes alone. A simplified map of chromosome X and the three autosomes is shown in Fig. 10–13b. The fact that gene position could be represented by a linear map was a new argument in support of the chromosome theory, since chromosomes themselves were known to be of thread-like proportions. Support was also obtained from Morgan's experiments indicating that the genes fell into four linkage groups, which corresponded with the haploid number of four chromosomes.

Sturtevant pointed out that the distances between linked genes shown on genetic maps might not be proportional to the actual spatial distances, as certain parts of chromosomes were perhaps more liable to exchange material than others. It is also assumed in gene mapping that the existence of one cross-over does not affect the probability of others occurring. In practice, however, *interference* takes place, which generally decreases (Muller 1916), or occasionally may increase, the probability that a second cross-over will occur adjacent to an existing one.

Sturtevant also showed that the additive properties of cross-over frequencies did not hold if the frequencies were large, and were only accurate if the genes concerned were spaced closely together. This is due to the fact that if the genes are far apart, multiple cross-overs can occur. However far apart two genes are, even at the ends of a long chromosome, they will never show less than 50 per cent linkage due to multiple cross-overs; if there are 2,4,6—up to any even number of cross-overs between them, they will assort together, but if there are 1,3,5—up to any odd number of cross-overs, then they will be separated. So genes placed far apart on a chromosome will tend to assort almost at random in gamete formation, due to frequent cross-overs, which explains how the number of apparently freely assorting characters can exceed the haploid number. In fact it is often impossible to tell in beginning a new genetic map whether two genes are situated at opposite ends of a chromosome or on different chromosomes. This also accounts for the fact that maps of up to 107 units (Fig. 10–13b) are obtained, since genes placed far apart will always tend to approach the recombination frequency of 50 per cent—that of unlinked genes—and the accuracy of the method decreases.

In 1933, Painter began a study of the giant chromosomes of the salivary glands of *Drosophila*. Because he could recognize certain characteristic chromosomes and compare the changes in their appearance with the genetic changes in the *Drosophila*, he could construct chromosome maps of the exact physical position of the genes. He confirmed that the order of the genes on the statistical maps was correct, although the distances between genes varied due to the variation in probability of cross-overs occurring along the length of the chromosome. It was also possible to confirm that the four linkage groups had been correctly assigned to the four chromosome pairs.

As a result of the work of Morgan and his colleagues, genetics was firmly placed on a numerical basis, and it was possible to predict with considerable accuracy what would occur in given situations. Genetic maps have now been constructed for many organisms, and these are often a good reflection of the actual physical map of the chromosome.

**10–5
Gene mapping
with
micro-organisms**

Considerable time and labour were involved in searching through large numbers of *Drosophila* mutants to detect individuals with variations in appearance, particularly when the gene sites concerned were situated close together and the number of recombinants was very small. Since the mid-1940s, certain micro-organisms—moulds, bacteria and viruses—have become the material used for genetic studies. The great advantages of using micro-organisms are that the time interval between generations is comparatively short, and that they are easily grown under controllable conditions in the laboratory. The disadvantage is the apparent lack of criteria for assessing genetic change, since there are no easily recognizable morphological features. It was not until biochemists had shown that micro-organisms could be classified in terms of their nutritional requirements that their potential as material for

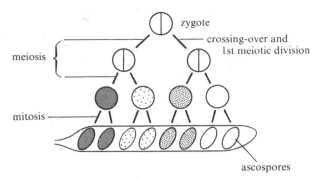

Fig. 10–14 How the ascospores are ordered in the ascus of *Neurospora*

biochemical genetic studies came to be realized. It then became possible to deal with millions of individuals (whereas with *Drosophila* it was difficult to look at more than 50,000 progeny) and at the same time, by the use of selective media, to reduce the labour involved.

The first studies on biochemical genetics were those of Beadle, in the early 1940s, on the bread mould *Neurospora*. *Neurospora* can reproduce asexually or sexually. In the latter case, the fusion of the two parental nuclei results in a diploid zygote, which, however, only exists for a short time. It undergoes meiosis rapidly to give four haploid daughter nuclei, which then undergo a further mitosis resulting in a total of eight nuclei (Fig. 10–14). These develop into spores, which are regularly arranged in a spore case known as an ascus. There are therefore four pairs of spores, the members of each pair being identical genetically, and their arrangement in the ascus reflects exactly the order in which nuclear division took place (see Fig. 10–14).

The advantage of using *Neurospora* was that the ascospores could easily be dissected out individually in order and then allowed to germinate separately in different media. Also, since the nuclei are haploid, there was no possibility of a recessive mutation being masked by a dominant wild-type allele. By varying the composition of the medium it was possible to determine if a mutation resulting in the loss of ability to biosynthesize a particular chemical compound, which was metabolically essential, had occurred, and if so what this compound was. For instance, one mutant strain of *Neurospora* was found to require the amino acid arginine (Fig. 4–1a) for growth, whereas the wild type had been able to grow in a 'minimal' artificial medium containing no arginine. This was an example of what Garrod, in 1908, speaking of a rare hereditary disease in humans, termed an 'inborn error of metabolism'.

Beadle and Tatum published their first results in 1941 and 1942. They soon had dozens of mutant moulds with different biochemical deficiencies. By crossing with the wild-type and obtaining the expected distribution of progeny —that is four mutant and four non-mutant spores in each ascus—they checked that each induced deficiency which they observed was due to a mutation in only one gene. Since it appeared from their work that one gene controlled a specific biochemical synthesis, and since each synthesis generally requires a specific enzyme, Beadle and Tatum reasoned that one gene probably controlled a specific enzyme. This later became known as the 'one gene, one enzyme' concept.

Early geneticists had to rely on spontaneous mutations for their studies. However, the work of Muller in 1927 showed that irradiation of sex cells with X-rays caused mutations, and later ultraviolet radiation was also shown to be mutagenic. These were the means which Beadle and Tatum used to obtain their *Neurospora* mutants. Now, certain specific chemicals are more often used as mutagens, as the number of mutated genes which they produce is considerably higher.

10–6
Bacterial studies

Similar studies to those outlined above on *Neurospora*, involving mutations which affected the ability to synthesize certain essential compounds, were carried out on bacteria in 1944. Other bacterial mutations studied made use of strains which showed resistance to various antibiotics and those which were resistant to bacteriophage infection (see below).

A major step forward occurred in 1946 when genetic recombination in *Escherichia coli*, implying some form of sexual reproduction, was reported by Lederberg and Tatum. The method they used was to take two parental strains of bacteria, both of which had specific, but different, growth requirements. The strains were mixed and placed in a medium from which the necessary growth factors were missing, so that neither strain could grow by itself under these conditions. It was found that the majority of cells died as expected but a very small number were able to grow and produce colonies. This meant that these cells had acquired, in place of their mutant genes, copies of the original functional genes.

This implied that there must in some cases have been a sexual phase, during which genetic material was exchanged, resulting in the formation of one genome with completely functional copies of all of its genes. (As explained in Chapter 14, the term chromosome does not really apply to the genetic material of viruses and bacteria; *genome* is a more appropriate expression.) The possibility of the observed results being due to a reverse mutation, from the mutant to the wild-type form, was excluded by the use of strains of bacteria which lacked the ability to synthesize two specific growth factors. The probability of two reverse mutations in the same bacterium, which was of the order of 1 in 10^{14}, was judged by Lederberg and Tatum to be small enough to be ignored.

Lederberg studied the recombination frequencies of 15 pairs of mutant characters in *E. coli*, found that they all belonged to the same linkage group, and concluded that

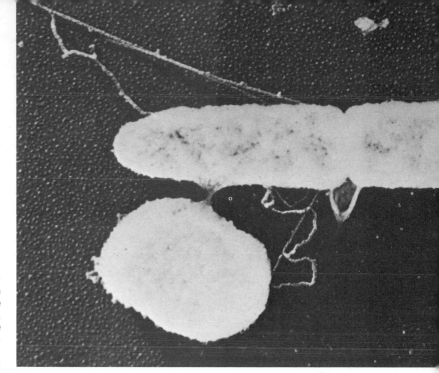

bacteria must contain a single strand of genetic material, a conclusion later found to be correct. Also correct was his assumption that bacteria must be haploid, since he could see no signs of dominance or recessiveness. The first genetic map of *E. coli*, with eight loci, was published by Lederberg in 1947.

The actual sexual process, or conjugation, was found to have certain very distinctive features. Instead of the fusion of two cells, followed by segregation during meiosis, it was found in the early 1950s by using electron microscopy that genetic material moves from the male cell into the female by means of a narrow interconnecting bridge. The electron micrograph illustrating this (Fig. 10–15) was taken by Anderson, one of the investigators in the field. The male bacterium, F^+, differs from the female, F^-, in containing the *F* (fertility) genetic factor. The *F* factor may exist as a very small separate genome, now thought to be circular as the bacterial genome itself has been shown to be, or it may be part of the normal bacterial genetic material (Fig. 10–16). This type of genetic particle is known as an *episome*.

The *F* factor may become integrated into the bacterial genome by means of a crossing-over process. In cells where

Fig. 10–16 Conjugation in *E. coli* bacteria

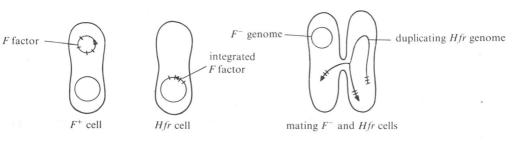

F factor

F^+ cell

Hfr cell

F^- genome

integrated F factor

duplicating *Hfr* genome

mating F^- and *Hfr* cells

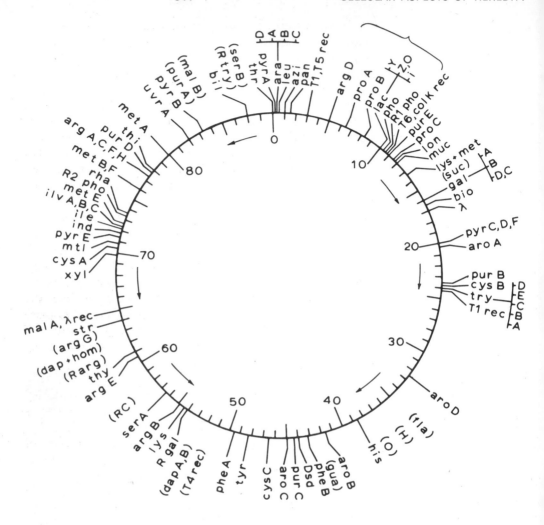

Fig. 10–17 Genetic map of *E. coli*. The symbols mark the location of genes. When locations are only approximately known they are shown in parentheses. The numbers divide the map into time intervals of minutes. (From A. L. Taylor and M. S. Thoman, The genetic map of *Escherichia coli* K-12. *Genetics*, **50**, 659, 1964.)

the *F* factor is free, transfer of genetic material has been found to be a relatively rare event, whereas when the *F* factor is part of the bacterial genome the recombination frequency with *F⁻* cells is much higher. Bacteria with the factor integrated in the genetic material are thus known as *Hfr* (high frequency of recombination).

In conjugation the male genome always breaks at the point where the *F* factor is integrated. This is now thought to be connected with genetic duplication, so that, as breakage occurs, replication begins (Fig. 10–16). One of the daughter strands of DNA then moves gradually into the female cell. The time required for complete transfer of genetic material in *E. coli* is about 90 minutes, but transfer is usually incomplete. It is not yet known what are the forces involved in the process. Crossing-over then takes place between the partial male genome and the female genome, followed by a segregation process which results in the formation of a haploid cell.

It is possible to interrupt the transfer of genetic material at any time, and to tear apart the conjugating bacteria—for example, by means of a Waring blender. Since a particular end of the male genome always enters the female cell first, and since the genes enter in a definite order, the frequency of occurrence of male genes in the recombinant genome affords a measure of how close they were to the entering end of the male genetic material. Thus by using parental strains with differing metabolic requirements—resistance to antibiotics and so on—and by stopping genetic transfer at different times and analysing the recombinants, a genetic map of the *E. coli* genome could be built up.

If there had only been one strain of *Hfr* bacterium it would have been very difficult to assign positions to genes near the end of the male genome that enters last, as they would enter the female bacterium only rarely. Fortunately, however, there exist various *Hfr* strains in which the *F* factor is situated in each case at a different point in the bacterial genome. So when conjugation occurs, breakage takes place at a different point for each *Hfr* strain, and the order of transfer of genetic material also differs. By combining the results obtained with various *Hfr* strains, it has been possible to build up a very complete genetic map of *E. coli* (Fig. 10–17).

The most interesting feature of the results is that they indicated that the bacterial genome is circular. The arrows on the map indicate the points at which the various *Hfr* genomes break before genetic transfer, and the direction of the arrows indicates the direction of transfer of the genetic material. The symbols refer to the character of the mutant gene concerned. A list is given in the table (Fig. 10–18) of the genes transferring between about 5 minutes and about 11

Fig. 10–18 Genes transferring from *Hfr* to *F⁻* cell in *E. coli* bacteria between 5 and 11 minutes after conjugation commences

Approximate time of entry (mins)	Symbol on map	Enzyme or reaction affected	Mutant character
5	arg D	ornithine transcarbamoylase	requires arginine for growth
7	pro A		requires proline
9	pro B		requires proline
10	lac Y	galactoside permease	unable to concentrate β-galactosides
10	lac Z	β-galactosidase	unable to utilise lactose as a carbon source
10	lac O	defective operator gene	constitutive synthesis (maximum rate) of lactose operon proteins
11	pho	alkaline phosphatase	cannot use phosphate esters
11	R·1 pho	alkaline phosphatase repressor	constitutive synthesis of phosphatase

minutes after entry, together with the enzyme or reaction affected, when known.

There are two other ways in which genetic recombination may take place in bacteria and which have proved of use as additional methods of gene mapping. In the first, genetic material is carried from one bacterium to another by an infective virus, a process known as *transduction*. Bacterial viruses or *bacteriophages* were discovered in 1914 and some of the most important work on the nature of the gene itself has been carried out on T4 phage. In transduction, an infective viral particle carries over a small part, about 1 per cent, of a host bacterial genome and then injects it into another bacterium. Here crossing-over may take place, resulting in a genetically altered bacterium if different strains of bacteria were used at the two stages. Transduction is a rare occurrence, but has proved extremely useful in determining whether two genes are placed close together, since only a small number of genes can be carried by a single transducing particle.

The other method of genetic recombination is that of *transformation*. This method involves the introduction of purified DNA from one strain of bacterium to a genetically different strain. It may then be shown, as Griffiths first showed in 1928 using strains of pneumococci, that some of the characteristics of the donor cells have been acquired by the recipients. Transformation is a very inefficient process but has proved useful for gene mapping in bacteria where a suitable transducing phage is not known, and particularly in determining which genes lie close together, since again only small fragments of genetic material involved.

10–7
Phage genetics

In the late 1930s, geneticists began to investigate the various types of phage which attack *E. coli*. The best known of these are the similar strains T2, T4, and T6 (Fig. 10–19). The head of a typical phage contains DNA surrounded by a protective protein coat. The tail is a hollow core, surrounded by an outer contractile sheath, through which the DNA is injected into a

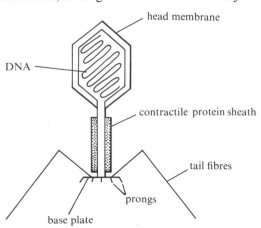

Fig. 10–19 Structure of T4 bacteriophage

head membrane

DNA

contractile protein sheath

tail fibres

prongs

base plate

bacterium. There is a base plate to the tail, carrying fine fibres, the function of which is to attach to the bacterial coat. The fibres are used to place the virus in the correct position, and the prongs on the base plate anchor it there. A small hole is then digested in the bacterial cell wall by means of the enzyme lysozyme which is present in the tail of the phage. The viral DNA is injected through this hole, and generally it rapidly begins to reproduce inside the bacterial cell (Fig. 10 20). Between 100 and 1,000 new copies of viral DNA are produced which then acquire a newly synthesized protein coat and a tail, and finally form complete phage particles. The entire life cycle, which takes only a few minutes, finishes when the bacterium lyses, that is its cell wall breaks down, releasing the contents and freeing the many phage progeny. This process is known as a *virulent* infection and the phage which causes it is known as a *lytic* virus.

There is another process, which may occur in certain cases, known as a *temperate* infection. Some bacterial viruses, known as *lysogenic* viruses, do not always cause lysis in the usual way. Instead, the viral DNA becomes attached to the host cell genome, that is, the viral DNA can behave as an episome (Fig. 10–20), either existing autonomously or as part of the host genome. When it is attached to the bacterial genome it is known as the *prophage*, and appears to behave as part of the bacterial genetic material, duplicating once every cell generation. The prophage may be carried on the bacterial genome indefinitely, without destroying the host. Very rarely it becomes released from the host bacterium and multiplies as in a virulent infection, with resultant lysis.

Fig. 10–20 Life cycle of a lysogenic bacterial virus, showing virulent infection and temperate infection

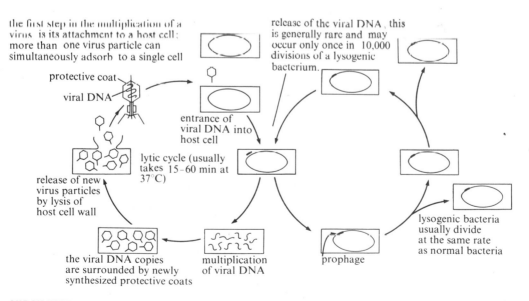

the first step in the multiplication of a virus is its attachment to a host cell; more than one virus particle can simultaneously adsorb to a single cell

release of the viral DNA, this is generally rare and may occur only once in 10,000 divisions of a lysogenic bacterium.

protective coat

viral DNA

entrance of viral DNA into host cell

lytic cycle (usually takes 15–60 min at 37°C)

release of new virus particles by lysis of host cell wall

the viral DNA copies are surrounded by newly synthesized protective coats

multiplication of viral DNA

prophage

lysogenic bacteria usually divide at the same rate as normal bacteria

VIRULENT
INFECTION

TEMPERATE
INFECTION

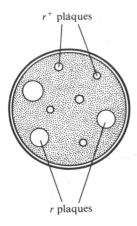

r⁺ plaques

r plaques

Fig. 10–21 Plaques caused by phage multiplication in a confluent culture of bacteria. The small plaques are caused by the normal wild-type phage (*r*⁺) and the larger ones by a rapidly growing mutant (*r*). Intermediate plaques caused by the growth of both types also occur

The method of studying virulent phage infections of bacteria in the laboratory is to inoculate confluent cultures, or 'lawns', of bacteria with a few virus particles. Each virus infects a cell and multiplies. On cell lysis the progeny are released, and infect neighbouring cells. In a few hours a circular *plaque*, consisting of a clear region of killed bacterial cells, is clearly visible in the otherwise healthy lawn of bacteria (Fig. 10–21). Different mutant strains of bacteria can be recognized by the variations in plaque morphology which they produce; some result in large plaques, some in very small ones, some cause plaques with clear-cut outlines and others plaques with blurred and fuzzy edges. These differences are easy to recognize and count (Fig. 10–21) and a large number of plaque-type mutants have been found.

Other mutant strains of phage may be recognized by the fact that they attack some strains of bacteria but not others. This is probably due to a changed ability to adsorb to the bacterial coat because of an alteration in the phage tail fibres. Yet another large class of mutants is a temperature-sensitive group. Mutant phages of this type can multiply at 25°C but not at 42°C, possibly because of the denaturation of proteins required for their protective coat. This type of mutant is known as a *conditional lethal*. By using these various characteristics it has been possible since 1945 to find mutations in many of the genes of the T4 phage.

If an inoculum of one strain of phage only is given to a bacterial colony, then all of the progeny will be identical except for random mutants, which occur at the rate of approximately one mutation per gene for 10^5 to 10^6 phages. However, if several strains of phage are inoculated simultaneously, since a bacterial cell may be infected by more than one particle at once, several kinds of viral DNA can be present and duplicate inside the host cell at the same time. It has been shown by means of the distinguishing methods mentioned above that under these conditions crossing-over occurs between similar viral genomes, and genetic recombinants occur among the viral progeny.

Crossing-over must occur several times during the period when viral genomes exist free inside the bacterial cell. For instance, if a bacterium is infected with three distinct phage particles, single viruses occur among the progeny which carry characteristics from all three types. However, this does not affect the argument that if genes are situated close together on a genome they recombine only rarely, and vice versa. Thus, by counting the frequency of recombinants obtained using various types of mutants, it has been possible to build up a very complete genetic map of T4 phage (Fig. 10–22), which is the best-known of the phage maps. Like that of *E. coli* it has been found to be circular, although it is not yet known if this is so for all viruses. The genes which control

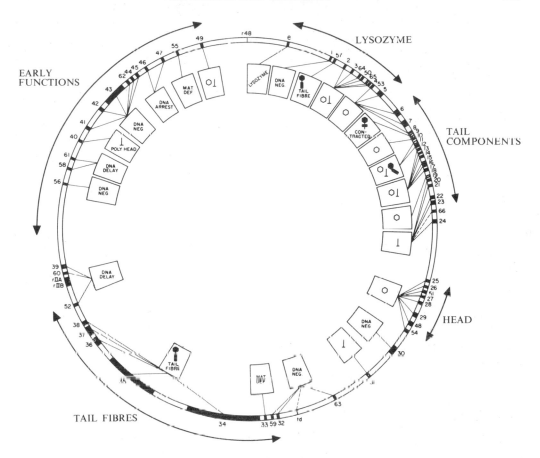

Fig. 10–22 Genetic map of T4 bacteriophage. Minimal lengths of various genes are indicated by filled-in areas, and symbols show phage components present in lysates of mutant-infected bacteria. The first half of the life cycle is controlled by genes in the upper left of the circle, and the second half by the remaining genes. (With permission from R. S. Edgar and W. B. Wood.)

synthesis and function in the early part of the life cycle are in the upper left of the circle in Fig. 10–22 and the other genes control the subsequent development and function of the head, tail, tail fibres, and so on.

10–8
Mutations and
mutagens

Spontaneous mutations occur in both plant and animal cells. A mutation may take place in the gamete, or at any stage in the development of an organism. In the latter case the number of affected cells in an individual will be less the later during development a mutation has occurred. A mutation may also occur in a single somatic cell; one of the hypotheses of carcinogenesis is that tumour cells arise by somatic mutation.

The term mutation covers any process of stable variation in a chromosome and the result of this is called a mutant. Thus a mutation may vary in magnitude from a very large effect to one so minute that very refined methods would be necessary

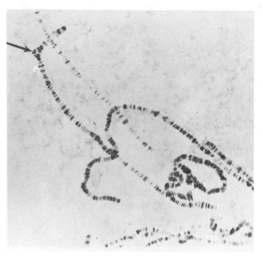 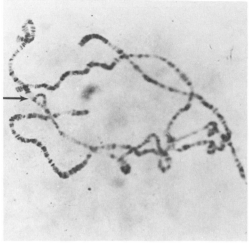

Fig. 10–23 Chromosome aberrations detectable in the salivary gland chromosomes of *Drosophila melanogaster*. *Left:* Deficiency (or small deletion) in the X chromosome. *Right:* Large deletion in IIIR (right arm of 3rd chromosome). (With permission from O. G. Fahmy and M. Fahmy.)

to detect it. For instance, at the upper end of the scale there may be loss (deletion) of part of or even the whole of a chromosome; such a large-scale loss as this is often lethal to an organism. Conversely, there may be the addition of a whole chromosome; an example is Down's syndrome (mongolism) in man, in which there are three copies (trisomy) instead of two of chromosome 21. Other major changes may arise from the turning of part of a chromosome back to front (inversion), or the exchange of one broken-off bit of chromosome for the broken-off bit of another (translocation).

At the lower end of the scale, a single nucleotide base pair may be involved, either by the addition or deletion of a base pair in the chromosome, or by the substitution of a different base pair for one of the normal pairs. An example is the alteration of the haemoglobin molecule in the human disease known as sickle-cell anaemia. Here the only change in the protein of the haemoglobin is the replacement of one glutamic acid residue by a valine residue (see Fig. 4–1a), which must result from a change in the three nucleotide base pairs of the DNA which code normally for glutamic acid. Possibly a change occurs in only one base pair (see Fig. 4–7). When changes involve one nucleotide pair, a back mutation to the normal arrangement can occur. However, with large changes it is clearly impossible for a reverse mutation to the normal situation to occur.

The frequency of spontaneous mutations is low—of the order of 1 per 10^6 cell divisions. The rate can be greatly increased by the use of *mutagenic agents*, which affect DNA either directly or indirectly, and act apparently indiscriminately along the length of the molecule. Examples of mutations induced in *Drosophila* chromosomes are shown in Fig. 10–23. X-rays and γ-rays may cause the rupture of various phosphodiester linkages in DNA by the action of free radicals, which are formed when the radiation is absorbed by water.

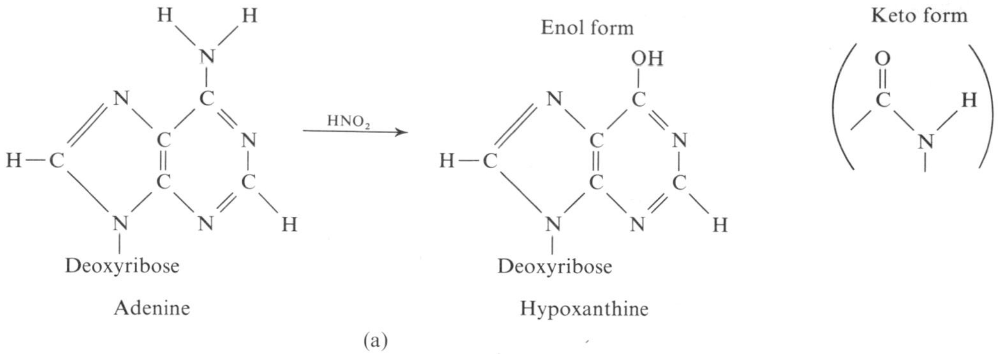

Enol form   Keto form

Deoxyribose
Adenine

Deoxyribose
Hypoxanthine

(a)

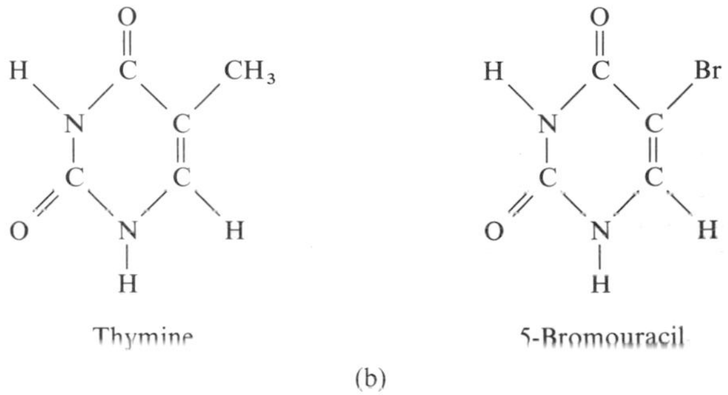

Fig. 10–24 Some chemical mutagens: (a) the mutagenic effect of HNO_2; (b) 5-bromouracil, a base analogue mutagen

Thymine   5-Bromouracil

(b)

Only if breaks occur opposite each other in the two complementary chains will the molecule fall apart; otherwise hydrogen bonding will tend to retain the helical structure. Ionizing radiation may also cause breakage of hydrogen bonds, again destroying the molecular structure of DNA in that region.

An example of a powerful chemical mutagen is nitrous acid, HNO_2, which converts one normal DNA base into another base by deaminating the amino groups. Figure 10–24a shows the deamination of adenine to the keto–enol derivative hypoxanthine. This no longer pairs with thymine, but with cytosine, and so modifies the hydrogen bonding.

Another type of chemical mutagen is an analogue of a normal DNA base which substitutes for the base in the DNA molecule. The analogue bases often do not pair as accurately as the normal bases. An example of this class is the powerful mutagen 5-bromouracil, which is an analogue of thymine (Fig. 10–24b).

Another group of chemical mutagens causes an extra base to be inserted in the DNA molecule. An example of this group is the acridine dye proflavine which was used by Crick to produce the mutants which provided evidence for the triplet nature of the genetic code.

Alkylating agents such as the nitrogen mustards, $RN(CH_2—CH_2—Cl)_2$, bring about mutations by interalkylating aligned bases on the two strands of the DNA molecule.

This cross-linking effect prevents the separation of the double helix during replication. Alkylation of bases on one strand of the DNA probably causes changes in hydrogen bonding.

10–9
What is a gene?

The idea of a particle of heredity was implicit in the studies of classical genetics of the 19th century; Mendel, for example, spoke of 'elements'. The expression 'gene' was used much later, but the concept still tended to be that of particulate units which were arranged on a thread-like chromosome.

Morgan's work on *Drosophila* led to the definition of a gene as a unit of recombination—that is, the shortest segment of chromosome which can be separated from its neighbouring segments by crossing-over (Fig. 10–25b). As more was discovered about mutations and mutagenic agents, another definition of the gene was formulated—that of the shortest chromosomal unit which can undergo mutation (Fig. 10–25a). Since the activity of a gene can only be detected by observing its effects, a third possibility was to define it as a unit of function, each unit determining a particular product (Fig. 10–25c). However, this third concept could not be clearly defined. It was known from *Drosophila* studies that one gene can affect more than one part of a phenotype, and conversely that a number of genes may affect one particular characteristic, such as eye colour. It was also realized that the functional effectiveness of a gene depends upon its relation to other neighbouring genes, and it was suggested that there might be overlapping regions of gene function.

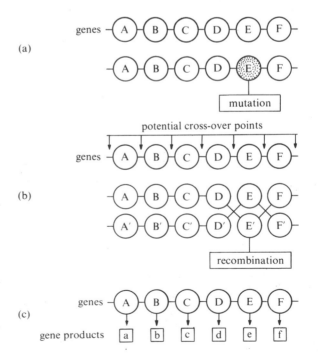

Fig. 10–25 The gene as **(a)** a unit of mutation, **(b)** a unit of recombination, **(c)** a unit of function

The three definitions of the gene mentioned above could certainly not be identical. It was apparent from studies on mutation and mutagenic agents that the unit of mutation could be much smaller than the functional unit, and it appeared to be also smaller than the unit of recombination.

As a result of all of these conflicting ideas, the concept changed to that of a 'gene complex' rather than that of an autonomous unit. This rather vague definition was replaced by the 'one gene, one enzyme' concept, which was a considerable simplification, and more in keeping with the growing knowledge of cellular biochemistry.

The work of Pontecorvo (1952) on mutant forms of the mould *Aspergillus nidulans* suggested that crossing-over could take place within one functional unit. He studied three closely situated mutant sites, all connected with the ability to synthesize biotin. This was the beginning of genetic analysis *within* a gene, known as fine genetic mapping.

Benzer (1955) carried out fine genetic mapping of T4 phage, using a class of mutants known as *r*II. These may be recognized by their plaque morphology, and by the fact that they will grow in the B strain of *E. coli* but not in the K strain. By infecting a culture of K bacteria with a mixture of two of these *r*II mutants one of two very clear-cut results was obtained. The first was the expected result of no, or very few, lytic plaques, implying little or no phage growth. The second and surprising result was a large number of plaques over the whole bacterial plate. Thus in the latter case, two mutants which could not grow individually on K bacteria were able, when mixed together, to survive and function on the bacteria in the same way as a wild-type phage. When this result was observed, the two mutants concerned were termed *complementary*, because each could complement the function of the other, with the result that they both survived and multiplied as neither could individually.

By crossing a large number of pairs of mutants, determining their recombination frequencies and plotting the mutant loci on a genetic map, it was found that the *r*II region of the T4 genome could be subdivided into two distinct regions as far as the *complementation test* was concerned. Any phage particle with a mutant occurring in one of these regions could complement any particle with a mutant in the other region, but could not complement a mutant in the same region as itself.

This is illustrated in Fig. 10–26. In (a), both mutations occur in the *A* region of the genome. Thus, except for the very rare occurrence of crossing-over and recombination between the two closely placed mutation sites, there is no fully functional *A* region. The result is that neither phage particle survives and multiplies and with rare exceptions no plaques are observed. Similarly, in (b), because both mutations are

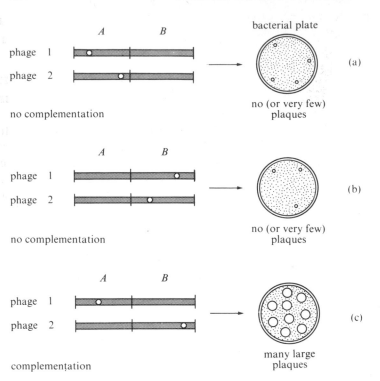

Fig. 10–26 Schematic representation of the complementation test. ○ represents a mutation locus.

situated in region *B*, there is no fully functional *B* region of the genome and the same result is observed. In (c), however, one mutant locus is in the *A* region, and the other in the *B* region. In phage 1 the *B* region is fully functional, and in phage 2 the *A* region is fully functional. These two phage particles can complement each other, each supplying the defective region of the other. They can therefore survive together, forming many large lytic plaques on the bacterial plate in the same way as the wild-type phage, although neither phage particle could grow by itself on K bacteria.

These observations suggested to Benzer a new definition of the gene in terms of a functional unit. He called the region of a chromosome within which a mutant cannot complement another mutant in the same region, but can complement a mutant in a different region, a *cistron*. The word cistron was derived from the observed *cis–trans* effect in diploid organisms. If two recessive mutant sites are located on the same chromosome (*cis* arrangement) in a heterozygote, one chromosome will be fully functional, and the appearance will be wild-type (Fig. 10–27). But if the same mutant sites are on opposite chromosomes (*trans*), both chromosomes are defective, and the *trans* form will be mutant in appearance.

It now seems probable that a cistron is the region of the DNA molecule which codes for one polypeptide chain. It is certainly a much larger unit than either that of recombination or that of mutation.

By means of finer and finer mapping of the *r*II region of

Fig. 10–27 The *cis-trans* effect in diploid organisms. **(a)** *Cis* arrangement of mutant sites; one normal wild-type chromosome. The heterozygote is wild-type in appearance. **(b)** *Trans* arrangement of the same mutant sites; both chromosomes are defective so the heterozygote will be mutant in appearance

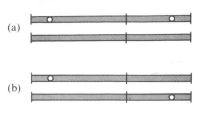

(a)

(b)

T4 phage, Benzer obtained an estimated value for the smallest unit of recombination. He first determined the approximate region in which a mutation was located by the use of *deletion mapping*. Certain mutations arising in the *r*II region result in the deletion of genetic material, which varies from loss of small parts only of the *A* or *B* cistrons to complete loss of both. A large number of these deletion mutants have been characterized and crossing-over experiments carried out between a phage particle carrying a mutation at an unknown locus and a series of mutants with known deletions. A functional recombinant will only be obtained in this type of crossing-over if the mutation is not in a region affected by the deletion (Fig. 10–28). This determines the approximate locus of the mutation, and further fine mapping using a large number of mutants may then be used to define its position completely.

Benzer used hundreds of mutants to obtain very fine mapping of the *r*II region, and to push the technique to its ultimate limits of resolution. Although the method is theoretically capable of detecting recombination frequencies of 1 in 10^6, the smallest observed frequency was in fact about 2 in 10^4, which indicated that this was the real lower limit of recombination frequency. Benzer made an estimate of the length, in nucleotide pairs, of the DNA molecule carrying the genetic information in T4 phage. By summing all of the short distances obtained from the mapping experiments, he was able to give an approximate value to the number of map units (each unit corresponding to 1 per cent recombination) involved in his complete linkage map. From this he could calculate an approximate value for the number of nucleotide pairs involved in the smallest observed length of DNA which had functioned

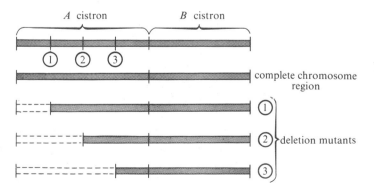

A cistron   *B* cistron

complete chromosome region

deletion mutants

Fig. 10–28 Diagram of deletion mapping. Broken lines indicate region deleted.

as a recombinant unit. This value was about one to three nucleotide pairs, which is remarkably close to the minimum theoretical value of one base pair. So it seems probable that the smallest unit of recombination (the *recon*) has in fact the same dimensions as the smallest unit of mutation (the *muton*), that is a single nucleotide base pair in the DNA molecule.

The present concept of a gene is that of a unit of function, the cistron. Each cistron codes for a complete polypeptide chain, which may be a biologically active protein on its own, or which may aggregate with other polypeptide chains to form a functional protein. Present-day ideas may be summarized as 'one cistron, one polypeptide chain'.

Bibliography

Beadle, G. W. and E. L. Tatum, Genetic control of biochemical reactions in *Neurospora. Proc. Nat. Acad. Sci.* **27,** 499, 1941.

Benzer, S., The fine structure of the gene. *Scientific American,* January, 1962.

Haggis, G. H. (Ed.), *Introduction to Molecular Biology.* Longmans Green, London, 1964.

Herskowitz, I. H., *Basic Principles of Molecular Genetics.* Little, Brown and Company, Boston, Mass., 1967, and Nelson, London, 1968.

Jacob, F. and E. L. Wollman, Viruses and genes. *Scientific American,* June, 1961.

Lederberg, J., Recombination mechanisms in bacteria. *J. Cell. Comp. Physiol.,* **45,** suppl. 2, 75, 1955.

Ramsay, J. A., *The Experimental Basis of Modern Biology.* Cambridge University Press, London, 1965.

Sinnott, E. W., L. C. Dunn, and T. Dobzhansky, *Principles of Genetics.* McGraw-Hill, New York, 1958.

Sturtevant, A. H. and G. W. Beadle, *An Introduction to Genetics.* Dover Publications, New York, 1962.

Watson, J. D., *Molecular Biology of the Gene,* W. A. Benjamin, New York, 1965.

Whitehouse, H. L. K., *Towards an Understanding of the Mechanism of Heredity,* 2nd edn. Edward Arnold, London, 1969.

PART 4

Cellular dynamics

We now consider the more dynamic aspects of cell behaviour, notably those involving cell movements, cellular interactions, and control mechanisms. Biologists have long recognized that coordinated movement, particularly in response to an external stimulus, is one of the most striking characteristics of living cells. In unicellular organisms, both plant and animal, such movements are commonly observed. In angiosperms and other multicellular plants they are generally restricted to movements within the cytoplasm, while in vertebrates highly specialized smooth muscle and striated muscle cells generate coordinated movements of limbs and tissues.

In Chapters 11 and 12 we examine the behaviour of cells during embryonic development and cellular control mechanisms. Because the embryo develops from a single fertilized ovum, this study illustrates the behaviour of small groups of cells as they grow and develop, by cell differentiation and morphogenesis, into the adult organism. A stepping stone is provided here between cell biology and general biology; we have therefore tried to show the relevance and importance of studies in cell biology as a basis for all the aspects of general biology.

In Chapter 13 these problems are considered in terms of cell biology at the molecular level, including intracellular mechanisms which regulate gene function, and extracellular regulators including interactions between cells on contact.

11 Cellular movements

The movement of living cells has fascinated biologists ever since van Leeuwenhoek first observed a new world of moving Protozoa under his simple microscope. Later, developments in fixing and staining cells unfortunately led many cytologists to concentrate their efforts on cell structure and chemistry, the revival of interest in living cells being comparatively recent. One of the pioneers in this field was Sir James Gray who worked on the movements of sperm tails and cilia in the 1920s. As with studies of mitosis, the development of time-lapse and other recording methods has been of immense help in the study of the behaviour of moving cells. In particular, Kamiya and other Japanese botanists have made detailed studies of cytoplasmic movements in plant cells, while Goldacre, Allen, and others have studied the movements of the giant amoeba *Chaos chaos*, and *Amoeba proteus*. Abercrombie, Ambrose, and Easty have studied the movements of mammalian tissue cells, both normal and malignant, using interference microscopy combined with time-lapse methods.

In 1820, the Scottish botanist Robert Brown examined the movements of pollen grains suspended in water under a high power microscope and noticed that the particles were in continuous motion. Their irregular movements, illustrated in Fig. 11–1 a, are the results of random bombardment by water molecules. The kinetic energy of molecules bombarding the particle fluctuates and is at times greater on one side than on the other, so that the particle is displaced (Fig. 11–1b). This type of movement occurs in any solution and depends only on temperature; it does not require an external energy supply for its maintenance. Physicists who have studied biology have sometimes attempted to interpret movements when seen inside living cells as a consequence of a random type of Brownian movement. Such movements do in fact occur on a small scale in all cells and they increase in dying cells. But the characteristic movements of living cells are different; they are coordinated and require a continuous source of chemical energy. Recent studies at the molecular level now suggest that cell movements depend on important and specific properties of the macromolecules and organelles in the cell cytoplasm. Electron microscopy has shown that many cells contain

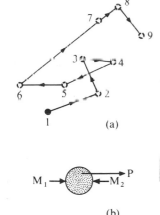

Fig. 11–1   (a) Diagram of Brownian movement. 1–9: successive positions of a particle observed at equal time intervals. (b) A particle in suspension is bombarded by molecules of unequal kinetic energy. If M_1 is greater than M_2, then the particle will move in the direction P

assemblies of fibres or tubules. In muscle cells the fibres are arranged in a highly organized structure. In other cells, the connection of the fibres with cell movements is less clearly demonstrated, but it is likely that fibres and tubules also play some role in the movements of the cell.

11–1 Structure of cytoplasm

The structural elements of cellular cytoplasm include mitochondria, ribosomes, lysosomes, Golgi bodies, endoplasmic reticulum, granules of lipid, and glycogen, as described in Chapter 5. But little has been said so far about the ground substance of the cytoplasm. One of the most characteristic properties of cytoplasm is its ability to undergo a sol–gel transformation—the role of such transformations in amoeboid locomotion was briefly mentioned in Chapter 1. In living cells the nature of this sol–gel transformation has been studied by a number of refined techniques. The viscosity has been measured by drawing steel spheres through the cytoplasm with magnets, and by other methods. The relative viscosity of the plasma sol is quite low, being about two to ten times greater than that of water; on the other hand, the gel is a moderately rigid structure which shows elasticity and breaks on application of a critical force.

The plasma gel is located near the plasma membrane; it is generally free from granules and other inclusions. The low-viscosity plasma sol is located in the interior of the cytoplasm. Marsland showed that the plasma gel can be liquefied under high hydrostatic pressure. An elegant technique for the study of the cytoplasm has been used by Harvey. A microcentrifuge is mounted on the microscope so that cells can be observed while under strong centrifugal force. When a sea-urchin egg is subjected to such centrifugation it becomes stratified, with a layer of fat droplets in the centripetal region, a clear central zone containing the nucleus, and a layer containing mitochondria and granules of pigment at the centrifugal pole. The plasma gel is not displaced because of its gel structure. Calcium ions are necessary for the preservation of the gel structure, which becomes liquefied in the absence of calcium.

The structure of microtubules as found in the mitotic spindle was described in Chapter 9; similar structures also occur in non-dividing cells. More recently, microtubules have been isolated from the fibres of cilia and flagella; they are formed from globular units 40 Å in diameter which have a molecular weight of 60,000 and an amino acid composition similar to actin, one of the components of actinomyosin, the muscle protein. One mole of the nucleotide guanine is found per mole of protein sub-unit. Microtubules can be disaggregated in a detergent solvent containing guanosine triphosphate. On dilution with water, ribbon-like fibres are re-formed.

Actin-like proteins have now been isolated from a number of cells, including amoebae and slime moulds.

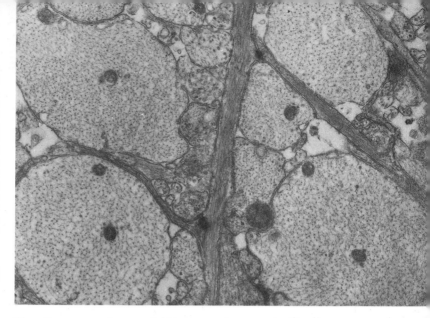

Fig. 11-2 Electron micrograph of the central nervous system of a larval lamprey, *Petromyzon fluviatilis*. The axoplasm of nerves contains filaments running parallel to the long axis of the cell process, and appearing in transverse section as dots of uniform size fairly evenly distributed throughout the cross section; commonly called neurofilaments, they appear to be fundamentally similar to the tonofibrils of epithelial cells. Magnification ×18,000. (With permission from D. W. Fawcett (ed), *The Cell, An Atlas of Fine Structure*. W. B. Saunders Co, Philadelphia, 1966.)

Usually a complex, probably an actinomyosin-like material, has been identified; the myosin component shows an ATPase activity. This enzyme is necessary for the release of chemical energy in muscular contraction. Fibres about 40 Å in diameter are revealed in electron micrographs of a number of cells, either located near the membrane and lying parallel to the surface, or more generally distributed. They are present in the axons of neurones as shown in Fig. 11-2. The fibrous aggregates seen in epithelial cells are known as tonofibrils.

11–2	**Cyclosis in plant cells.** Much of the work on cytoplasmic

**11–2
Cytoplasmic
movements**

Cyclosis in plant cells. Much of the work on cytoplasmic movements has been carried out on amoebae, but results so far have proved difficult to interpret, because of the simultaneous locomotion of the whole cell. The situation is simpler in plant cells where movements occur by a continuous and uniform flow of cytoplasm, generally in circular pathways. In describing cellular movements we shall examine only a restricted number of such cell types. Cells which have been widely used, particularly by Japanese botanists, are the freshwater algae, *Nitella* and *Chara*. *Chara* cells grow to an enormous size, reaching 0·5–2 cm in length. The streaming in these cells is illustrated in Fig. 11-3. Since the cytoplasm contains numerous granules, the flow can be readily observed under the microscope. In these cells the chloroplasts are not located freely in the cytoplasm, as in leaf cells of higher plants, but are attached to the inner wall as shown.

Interesting results have been obtained by Hayashi, who finds that centrifuging the cells at 400 times the force of gravity causes chloroplasts in some regions to become detached (Fig. 11-4). Cytoplasm is not driven along the cell in these regions but accumulates in the manner shown; eventually chloroplasts become redeposited and streaming of cytoplasm in these regions is resumed. The simplest interpretation of these results is that the chloroplast surface provides the motive force

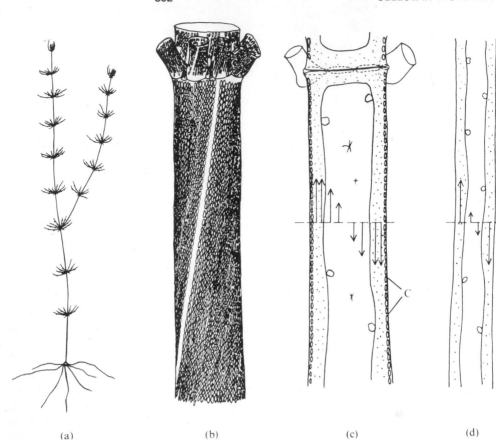

(a)   (b)   (c)   (d)

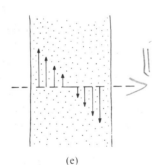

(e)

Fig. 11–3 Streaming in the cells of *Chara braunii*. **(a)** The whole plant; the internodal cells are about 0·5–2 cm long. **(b)** Close-up of the upper end of an internodal cell. **(c)** Longitudinal section of **(b)**. **(d)** Longitudinal section of a part of the rhizoid. **(e)** Streaming in a plasmolysed cell. C   chloroplasts; arrows indicate the direction and relative rates of streaming in the cell. (With permission from T. Hayashi in *Primitive Motile Systems in Cell Biology*, R. D. Allen and N. Kamiya, (eds). Academic Press, New York 1964.)

for streaming; but the rhizoid cells do not contain chloroplasts and streaming also occurs in these cells, as shown in Fig. 11–3d. Rhizoid cells can be plasmolysed with 0·2 M calcium chloride solution, whence the tonoplast disappears (Fig. 11–3e). The cytoplasm then streams at its maximum rate along the inner side of the plasma gel, which is approximately 1 μ thick. When the plasma gel is absent, streaming does not occur. These observations support the view that the motive force for streaming occurs at the inner surface of the plasma gel. In the internodal cells the chloroplasts are coated with a thin film of plasma gel and it is at this surface that the motive force appears to reside. Supporting evidence for this was obtained from Marsland's observation that high hydrostatic pressure which transforms the cortical gel into a sol also arrests cytoplasmic streaming in *Elodea*.

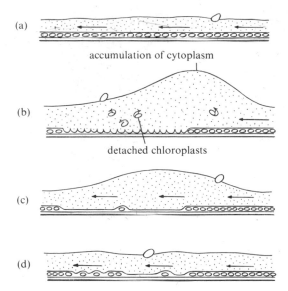

(a)

accumulation of cytoplasm

(b)

detached chloroplasts

(c)

(d)

Fig. 11–4 Centrifugation brings about the detachment of chloroplasts and cessation of cytoplasmic streaming: **(a)** before centrifugation; **(b)** 5–10 hours after centrifugation; **(c)** 3–5 days later; **(d)** about 2 weeks later. (After T. Hayashi, 1963.)

Some elegant experiments on cyclosis have been conducted by Kuroda (Fig. 11–5) on *Nitella* cells. The cells are ruptured at the base under conditions in which a negative hydrostatic pressure is maintained (Fig. 11–5). The plasma sol continues to stream and flow round the wall of the cell, leading to the deposition of a naked mass of cytoplasm on the glass plate below. The pools of cytoplasm were studied particularly by Jarosch and Kuwada, who observed most interesting movements:

> Isolated chloroplasts spin round rapidly on their own axes. Nuclei spin round more slowly on their axes. Individual particles occasionally make jumping (saltatory) movements. Circular fibres or bundles of fibres spin continuously within the cytoplasm. These later change into polygons which can be picked up on the end of a micromanipulator needle, when they continue to show the same type of movement (Fig. 11–6).

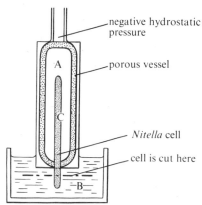

negative hydrostatic pressure

porous vessel

Nitella cell

cell is cut here

A

C

B

Fig. 11–5 The equipment used for amputating one end of a *Nitella* cell. A negative hydrostatic pressure is applied in the chamber, A, and the cell, C, is cut in the cuvette, B, at the site indicated with the broken line. (After K. Kuroda.)

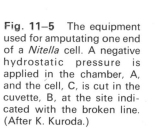

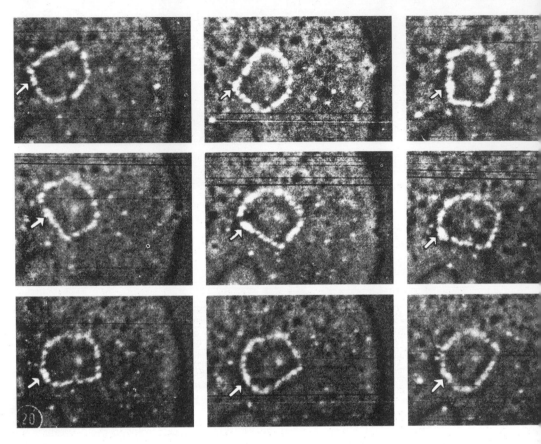

Fig. 11–6 Frames from a time-lapse ciné film showing wavelike propagation of a corner in a ring of fibril bundles; a particle embedded in the ring (arrowed) scarcely changes its position. The nine ciné pictures were taken at intervals of 0·25 second. (With permission from R. Jarosch in *Primitive Motile Systems in Cell Biology*, R. D. Allen and N. Kamiya (eds). Academic Press, New York, 1964.)

How relevant these various movements are to the general problem of cytoplasmic streaming is discussed below.

In some plant cells the cytoplasm is divided into strands linked to form a group of interconnecting channels. The tonoplast is crossed by these channels, and each section of the vacuole must be surrounded by a membrane. This type of movement is observed in the staminal hairs of *Tradescantia*. The terminal or apical cell is the dividing cell of these hairs and does not show streaming. The second cell adjacent to the apical cell continues to grow, and strands of cytoplasm form across the cell vacuole as shown in Fig. 11–7.

Streaming in slime moulds.   The slime mould *Physarum polycephalum* shows the characteristic feature when grown on agar of streaming in channels in the plasma gel. Fragments can be isolated in the form of threads (Figs 11–8 and 11–10a) which reveal an interesting streaming of the cytoplasm.

Fig. 11–7 Interconnecting channels of cytoplasm running across the vacuole, as observed in *Tradescantia*; the terminal cell is the dividing cell and does not show streaming.

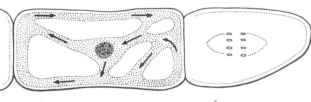

Tonoplast – single layer membrane around vacuole in plant cells.

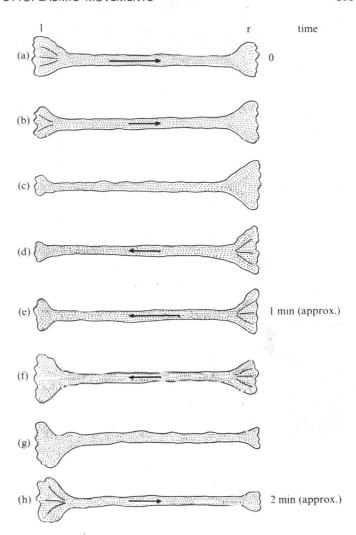

Fig 11–8   Threads of the slime mould *Physarum poly-cephalum*, showing periodic reversal of cytoplasmic streaming. **(a)** and **(b)** show streaming from left to right, **(c)** arrest, **(d)**, **(e)**, and **(f)** streaming right to left, **(g)** arrest, **(h)** streaming left to right. Length of arrows indicates rate of streaming

The series shown in Fig. 11–8 indicates that the direction of streaming reverses periodically. The diagram shows a strand with two terminal gel-like regions 1 and r. These contain many fine channels, not unlike the tributaries of a river, which converge to form a central channel along which the cytoplasm flows in a uniform stream. The cytoplasm flows progressively from left to right, the gel-like state being converted into a sol (b) until it is reduced to a small volume as shown in (c). At this point streaming stops and begins again in reverse. The period of reversal is about two minutes but is not exactly periodic.

The streaming mass of cytoplasm in the central channel displays flow birefringence which must arise from elongated structures of some kind, probably protein molecules or fibres which become orientated. This is well illustrated in the polarized-light picture of Fig. 11–9. Unlike the streaming in *Chara* and *Nitella*, streaming in *Physarum* depends on a considerable pressure gradient. Kamiya has demonstrated that

Biology:
What function does
this have?

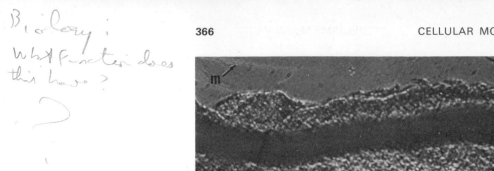

Fig. 11–9 Polarized light picture of birefringence in a plasmodial strand: **(a)** and **(b)** are taken at opposite compensator settings to show reversal of contrast in the two pictures; m mucus secreted by plasmodia; scale, 10μ interval. (With permission from H. Nakajima in *Primitive Motile Systems in Cell Biology*, R. D. Allen and N. Kamiya (eds). Academic Press, New York, 1964.)

|ııııı
50μ

a water pressure of 15 cm is required to reverse the direction of streaming, this rate of streaming being far higher than the fastest in *Chara* or *Nitella*. He has shown that the flow is like the flow of water through a pipe under hydrostatic pressure, the maximum velocity being reached in the middle of the channel, not at the walls as in *Nitella*. Wohlfarth-Bottermann and Kamiya have shown that there are fibres in electron micrographs of sections of the gel region. Wohlfarth-Bottermann has examined the location of the fibres in the gel regions l and r of suspended strands and found many more fibres along the walls of channels in a region of efflux than in the region of influx (Figs 11–10a and b). Electrical potential differences between the ends of the strands are also associated with the reversal of streaming. The potential difference between l and r of Fig. 11–8 a is reversed on reversal of flow in Fig. 11–8 d, but the two do not occur simultaneously (Kamiya). Anderson has also found a difference of ion concentration between the

Fig. 11–10 *Opposite* (a) Drawings of the functional arrangement of fibrils in a hanging loop of a protoplasmic strand of *Physarum*. A efflux of cytoplasm, showing many fibrous regions; B influx of cytoplasm. (b) Electron micrograph of a fibrous region of A above. Magnification ×26,000. (With permission from K. E. Wohlfarth-Botterman in *Primitive Motile Systems in Cell Biology*, R. D. Allen and N. Kamiya (eds). Academic Press, New York, 1964.)

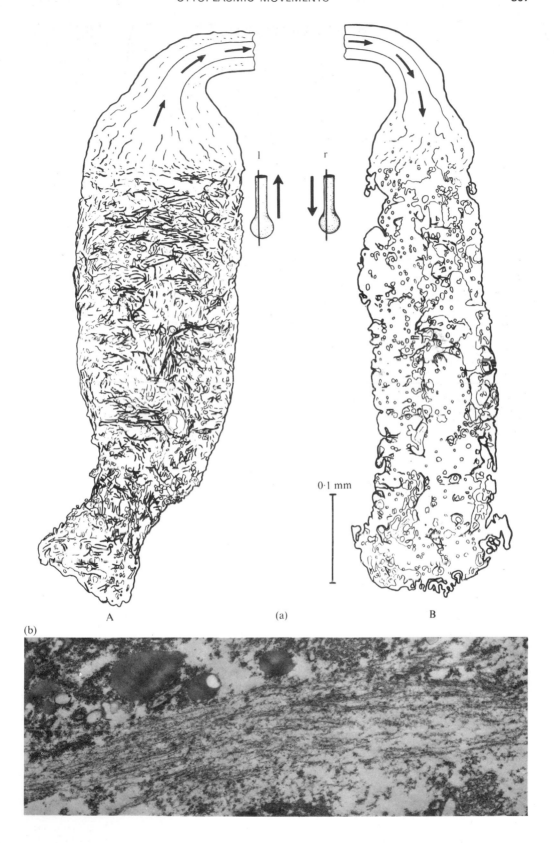

A (a) B

0·1 mm

(b)

advancing non-channelled region and the posterior channelled region. The potassium concentration is 30 mequiv/kg in the advancing region, and 20 mequiv/kg in the posterior. Sodium concentrations are also reversed, being 3–8 mequiv/kg in the advancing region and 12–15 mequiv/kg in the posterior region. These changes probably have some bearing on the electrical potentials generated.

Cytoplasmic movements in animal cells. *Behaviour of mammalian tissue cells.* Although most of the studies of cytoplasmic movements in animal cells have been carried out on

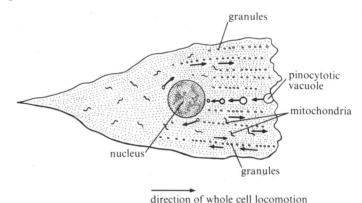

Fig. 11–11 Representation of cytoplasmic movements in a mammalian fibroblast cell

large amoebae, the movements seen in the cells of some animals and plants are less complex and will therefore be described first. Figure 11–11 shows a mammalian fibroblast as seen in tissue culture. This is the most characteristic form seen when the cells are growing in a culture of low density, the fan-like leading edge being typical of these cells. Under the high power of the light microscope, using time-lapse filming, the following types of movement can be observed:

General irregular movements of individual granules, due partly to Brownian movement.

Autonomous movement of individual organelles, particularly of mitochondria.

Occasional rotation of the nucleus for short periods (Pomerat, Abercrombie, Ambrose, and Easty).

Streaming of cytoplasm in limited regions, possibly formed by channels in the cell structure and by sheets of endoplasmic reticulum. Two adjacent channels are sometimes seen in which streaming occurs in *opposite directions*, that is, one towards and one away from the nucleus.

On the leading edge of the cell, drops of medium are occasionally enclosed by the cell membrane and drawn into the cytoplasm, a process known as pinocytosis.

Pinocytosis is related to phagocytosis, the means by which protozoans and other cells ingest particles of food. The pinocytotic vacuole, once it enters the cytoplasm, is drawn by cytoplasmic streaming towards the cell nucleus as shown in Fig. 11–11. It slowly decreases in volume during this process. Note that in tissue cells, although the cells can migrate on the glass surface as shown in Fig. 11–11, there is no predominant direction of cytoplasmic streaming, such as occurs in amoebae, and most of the streaming in mammalian tissue cells occurs near to cell membranes. As in cyclosis in plant cells, the motive force appears to be localized at or near the interface between plasma sol and plasma gel. In organelles such as nuclei and mitochondria the force again seems to be localized at or near their surface.

Cytoplasmic movements in amoebae. The type of locomotion seen in the amoeba *Amoeba proteus* is accompanied by continuous streaming of the cytoplasm in the forward direction. Clearly visible under the light microscope, this movement

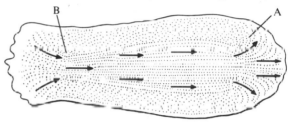

Fig. 11–12 Cytoplasmic streaming in an amoeba: A region of conversion of plasma sol to plasma gel; B region of conversion of gel to sol

is accompanied by locomotion of the whole cell. In a technique used by Goldacre, amoebae can be suspended in a dextran solution of the same density as the cells and the freely suspended cells continued to stream uniformly. The most striking feature of the flow is the sol-gel transformation, first described by Mast (1926). In region A (Fig. 11–12) the central plasma sol is converted into a plasma gel which remains attached to the inner wall of the plasma membrane but at point B the gel is reconverted into a sol. In *Amoeba proteus* the cytoplasm has numerous channels of plasma sol within the plasma gel, resembling the channels within the gel region of *Physarum polycephalum* (Fig. 11–10a).

Where is the motive force in these cells? According to Goldacre, it is located in the tail region B. In his model, the protein molecules of the cytoplasm are extended in the gel to form a network, while in the sol region they are coiled up to form soluble compact molecules. The active process is located in the tail region, where ATP causes contraction of the gel to form a sol. Evidence in favour of this view was obtained by injecting ATP locally into the cytoplasm as shown in Fig. 11–13. Injection into the tail region causes increased contraction, while injection into the head region causes an actual *reversal* of the head-to-tail polarity. Goldacre has described other experiments in support of his theory.

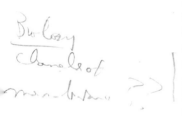

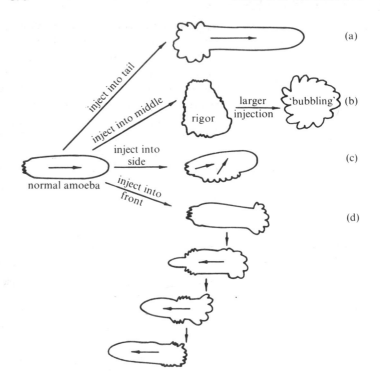

Fig. 11–13 Effect of micro-injection of ATP into various parts of an amoeba. Vigorous contraction of plasma gel occurred immediately at the site of injection, resulting in wrinkling of the cell membrane. Reversal of streaming occurred after injection into the front end, which then became a tail region. (With permission from R. J. Goldacre in *Primitive Motile Systems in Cell Biology*, R. D. Allen and N. Kamiya (eds), Academic Press, New York, 1964.)

According to Allen, however, the motive force is localized in the front region A. This is the fountain zone theory, according to which the molecules of cytoplasm in the plasma sol are extended. In the fountain zone region they are actively condensed into a plasma gel. This process draws the cytoplasm forwards towards the front of the cell. Allen has shown that, if an amoeba is placed in a fine capillary tube and the tail region is ruptured, cytoplasm continues to flow forwards for a time, which he attributes to the continued activity of the motive force in the fountain zone region. Note that the pressure difference responsible for flow is extremely small in *Amoeba proteus*. Kamiya finds that less than 0·5 mm of water pressure will reverse the direction of flow, so the forces responsible for streaming must be extremely delicately balanced.

Under natural conditions, when amoebae lie at the bottom of a pond on a solid surface, the cytoplasmic movements lead to cellular locomotion. This movement is closely related to cellular locomotion in other types of cell and involves contact between the cell *surface* and the solid substrate (see Section 11–4).

11–3 Muscular contraction

Cytoplasmic contractions are highly organized in the muscle tissues of animals. There are two main types of muscle tissue, smooth muscle and striated muscle. Striated muscle has a highly regular structure and so can be investigated by X-ray diffraction, electron microscopy, and other methods. Figure 1–12, an electron micrograph of sheep muscle, shows the type of

structure in tissue which undergoes rapid contraction and
relaxation, as, for example, in limbs of vertebrates and in the
flight muscle of insects. In the polarizing microscope the
muscle fibres show banding; some bands are anisotropic (A
bands) and others are isotropic (I bands). (Anisotropic—shows
birefringence, isotropic—non birefringent.)

Figure 1–12, giving the structure in more detail, reveals a
less dense region in the middle of the A band; this is the H
region. In the centre of the I region is a narrow dense band, the
Z band. In striated muscle all the myofibrils are in register, all
A bands and corresponding I bands being in line. This is why
the structure can be seen clearly in the light microscope. The
detailed electron microscopy has been worked out particularly by Huxley and Hanson, although the earliest work was
done by Hall, Jakes, Schmitt and others. The individual myofibrils have a diameter of about 1 μ. In fully relaxed muscle, the
length of the A bands is commonly about 1·3 μ and the length
of the I bands about 0·8 μ.

The myofibril is itself composed of fine filaments. In the
more dense regions, the A regions, there are two sets of filaments; first, the large filaments 100 Å in diameter and 1·5 μ
long (in rabbit muscle), which extend only across the length of
the A band, and second, fine filaments which are about 60 Å in
diameter and 2 μ long. The fine filaments also extend into the I
bands, which consist only of fine filaments. There is now
considerable evidence that muscle contracts by a sliding of the
fine filaments within the larger filaments (Fig. 11–26). The
coarse filaments are myosin, a fibrous protein, while the fine
filaments are a globular protein, actin. In myosin the long-chain
molecules are folded up in an α-helix (Figs 4–15a and 4–15b)
and lie parallel to each other. It is this structure which gives rise
to filaments. In actin the chains are folded up into a globular
structure and the 60 Å diameter filaments are an aggregate of
these particles. The complex formed between actin and
myosin, in muscle, is called *actomysin*.

In smooth muscle, which may contract extremely slowly,
there is no organization into A and I bands. The basic molecular structure appears to be rather similar, but the lining up
of the myosin fibrils does not form a banded structure.
The bands may be distorted so that there is an overlap.
(Hermann and Zebe). The contractile mechanism appears to
be basically similar to that of striated muscle (Section 11–5).

An important component of the muscle cells which plays
a major role in the initiation of muscular contraction is the
sarcolemma. The sarcolemma consists of an intercommunicating arrangement of membrane vesicles which penetrates
between the assemblies of actin and myosin within the muscle
cells. The motor neurones operate on this membrane system,
causing a sudden release of calcium ions into the cell cytoplasm,
which initiates contraction, as described in Section 11–5.

**11–4
Cellular
locomotion**

Movements which affect the whole cell and lead to cellular locomotion are of two main kinds—those connected with special cell surface structures (flagella and cilia) and those involving cytoplasmic movements. Bacteria carry flagella while some of the larger unicellular organisms move mainly by amoeboid movement, although cilia, and flagella, are found on some of the largest protozoans.

Flagella and cilia. These are similar types of whip-like projection on cell surfaces. When only one or two are present they are called flagella (from the Latin *flagellum*, meaning a whip), while when they are numerous and beat in a co-

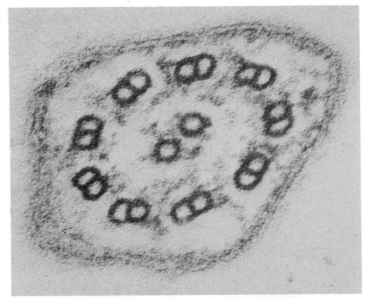

Fig. 11–14 Electron micrograph of a transverse section of a flagellum of *Chlamydomonas reinhardii*. Magnification ×259,200. The nine pairs of tubular filaments or fibrils and the two central ones are clearly seen. (With permission from J. R. Warr *et al. Genetics Research*, **7**, 3, 1966.)

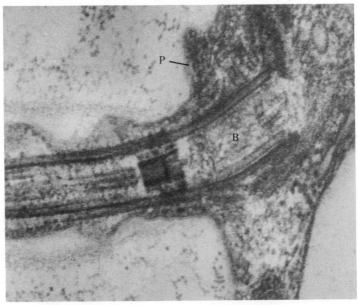

Fig. 11–15 Electron micrograph of a longitudinal section of a flagellum and basal body of *Chlamydomonas reinhardii*. Magnification ×124,000. B basal body; P plasma membrane. (With permission from J. Randall *et al. Developmental Biology*, Supp. 1, 1967.)

ordinated rhythm they are called cilia (Latin, *cilium*, an eyelash). Flagella can be as much as 150 μ long, while cilia are generally shorter, only 5–10 μ. Sperm tails are also built on the same basic macromolecular structure as flagella and cilia, and their movements also involve a rhythmic beating of the fibre. This is clearly shown in electron micrographs of cross sections through the structures (Fig. 11–14), the hollow cylinder being composed of nine pairs of tubular filaments arranged in rather the same way as the centrioles described in Chapter 9.

In ciliated epithelial cells which are not dividing, centrioles exist which eventually form the basal bodies of the cilia. In the longitudinal section of a flagellum shown in Fig. 11–15, the basal body can be seen attached to the flagellum. It lies just below the plasma membrane. The plasma membrane is continued over the shaft of the flagellum and forms a continuous sheath over the structure. The transverse section of the main structure indicates that besides the nine outer filaments there are two central filaments, making a total of eleven. This is the basic pattern for the structure of cilia, flagella, and sperm tails. Note that the filaments have the form of microtubles. These fibres contain an actin-like protein, similar to the actin of striated muscle. The outer filaments are associated in pairs of subfibrils (Fig. 11–14).

Chemical analysis reveals ATPase in flagella and cilia, and calculations indicate that ATP can diffuse up the shaft from the cytoplasm quickly enough to provide the energy needed for motion within the shaft during the beating cycle.

Flagella, cilia, and sperm tails can beat in various ways (Brockshaw, Sleigh, and others). The simplest type of movement is shown by the sperm tails of lower organisms (Fig. 11–16). It is in the form of a planar wave shown by Sir James

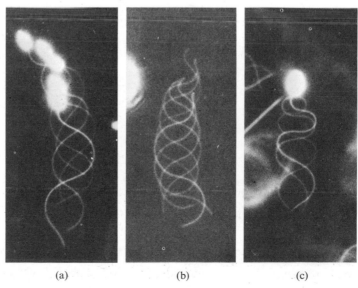

Fig. 11–16 Photomicrographs of the movement of sperm flagella. Each photograph is a multiple exposure taken with a series of flashes, using dark field illumination; they are reproduced as negative prints. **(a)** Normal movements of spermatozoa from a sea urchin. **(b)** Movement of a headless sea-urchin sperm. **(c)** Movement of a sea-urchin sperm attached to the slide by its head. Magnification × 1,050. (With permission from C. J. Brokaw.)

(a)   (b)   (c)

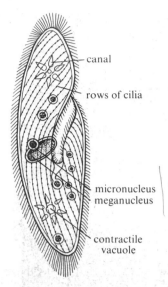

canal

rows of cilia

micronucleus
meganucleus

contractile
vacuole

Fig. 11–17 Diagram of *Paramecium caudatum* from the ventral side (after Doflein)

Gray (1926) to be sinusoidal. Figure 11–16 indicates that regions of the tail are either straight or have a constant curvature. During the movement of the wave along the tail there is a progressive sharp transition from the straight position to constant curvature and back again. The beating of both flagella and sperm tails can involve a more complex movement in which a helical motion of the rod is superimposed on the planar wave. This must also involve some twisting of the elements in the structure.

Beating of cilia involves the coordinated movement of groups or rows of individual surface extensions (Fig. 11–17). There is a 'rowing' action, the beat being transmitted progressively along the row. This movement, which occurs as a metachronal rhythm, has been studied by Sleigh using a high-speed ciné-camera technique. His analysis reveals that the beating of cilia on the gills of molluscs and in protozoans is basically similar, but form and magnitude differ in different species. The beat in the protozoan *Stentor polymorphus* is analysed in Fig. 11–18, where numbered outlines portray the various positions in one cycle of the beat. The cycle can be divided into the following phases:

Rest position.
Effective stroke.
Recovery phase.

As we might expect, bacterial flagella (Fig. 11–19) have a simpler structure than the flagella of flagellates. By agitation or ultrasonic treatment, the flagella can be broken off from the

Fig. 11–18 Diagram of the beat of a cilium in *Stentor polymorphus*, filmed at 400 frames per second at 22°C. (a) Numbered outlines show the times of the successive positions in one cycle of the beat. (b) Times on the graph correspond to those on the profiles in (a). (With permission from M. A. Sleigh in *Aspects of Cell Motility*, Symposium of the Society for Experimental Biology, 1968, Cambridge University Press, London.)

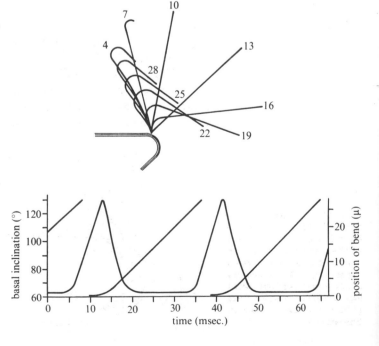

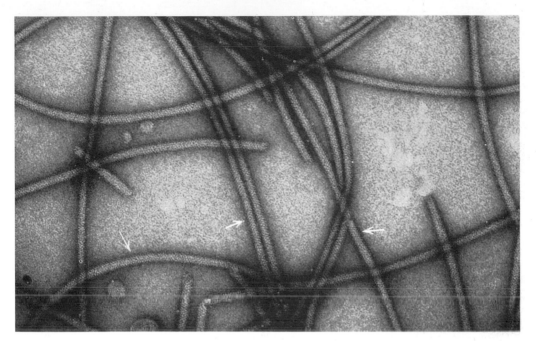

Fig. 11–19 Flagella of *Salmonella typhimurium* negatively stained with uranyl acetate. Arrows point to zones where globular units are arranged along rows which make an oblique angle with the flagellar axis. Magnification ×125,000. (With permission from J. Lowy and M. Spencer.)

surface of the bacteria. Analysis of a purified preparation reveals that they are composed of a single protein, flagellin. This has an amino acid composition similar to the actin of striated muscle. Flagella can also be deposited from the suspension as a bundle of parallel filaments and set up for X-ray diffraction. Astbury and others found that the pattern obtained is the α-type, similar to that of muscle and α-keratin. Along the axis there is a spacing corresponding to a repeating pattern of 5·1 Å, and along the meridian there is a 9·8 Å spacing. The α-pattern is produced by the α-helix. Astbury, Beighton, and Weibull called bacterial flagella 'monomolecular muscles'. Unlike other flagella and cilia, bacterial flagella do not contain an ATPase.

Electron microscopy indicates that bacterial flagella are not made up of 9 + 2 filaments but are simply cylinders, composed of globular molecules 40 Å in diameter, arranged in a hexagonal packing with a helical twist (Fig. 11–20). The

Fig. 11–20 Structure of bacterial flagella. The motive force is located at the base of the narrow shaft, which is composed of globular molecules, 40 Å in diameter, arranged in hexagonal packing with a helical twist. (After J. Lowy and M. Spencer.)

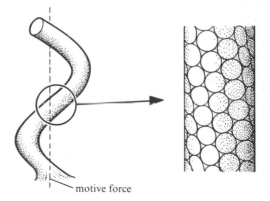

motive force

flagellum consists of three portions, a terminal hook, a main shaft, and a basal region associated with the cell membrane of the bacterium. According to Lowy and Spencer, flagella beat like whips, as their name implies. The motion is brought about by a rotating movement of the base (Fig. 11–21); in other words, the motive force is located at the base. It is difficult to see how ATP can diffuse readily along the extremely narrow shaft of bacterial flagella to provide the motive force within the shaft, as appears to be the case with other flagella and cilia. A rotating or oscillating movement of the base could possibly be transmitted along the hollow helix shown in Fig. 11–20 by a progressive change in packing of the sub-units.

A diametrically opposite situation obtains in the sperm tails of mammals. These are much larger than invertebrate sperm tails and have an outer set of fibres in addition to the $9 + 2$ filaments. In these large structures, the ATP is generated 'on site' by mitochondria which are wrapped around the outer filaments.

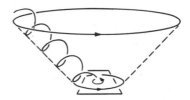

Fig. 11–21   Model for flagellar motion: a rigid helical rod, inserted at right angles to the plane of the base. There is no rotation of the rod relative to its base; the apparent motion results from the progressive transference of a small contraction from one sub-unit helix to the neighbouring ones. (With permission from J. Lowy and M. Spencer, in *Aspects of Cell Motility,* Symposium of the Society for Experimental Biology, 1968. Cambridge University Press, London.)

Locomotion of mammalian tissue cells.   This appears to depend on a mechanism intermediate between that observed in cells with highly organized processes, such as flagella and cilia, and the streaming type of motion seen in amoebae. Studies of these movements with time-lapse films in monolayer tissue cultures (Abercrombie, Ambrose, and Easty) have shown that these cells develop transient surface protrusions which appear to play a part in cellular locomotion. The protrusions take the form of ruffles or undulations of the membrane. They are shown diagrammatically in Fig. 11–22 which represents a

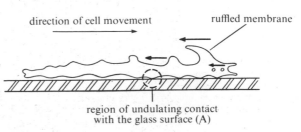

direction of cell movement   ruffled membrane

region of undulating contact
with the glass surface (A)

Fig. 11–22   Vertical section through a fibroblast moving on a glass surface.

cross-section through a moving fibroblast. The form of the ruffles is clearly seen in the Stereoscan picture of an instantaneously frozen membrane (Fig. 11–23). A surface contact microscope, which enables the contacts between the cell and the surface on which it is moving to be visualized directly, reveals intermittent contacts with the glass apparently produced by these wave-like movements of the membrane. The Stereoscan allows us to see these contacts directly from the side of the cell. The direct contact between the cell membrane and the glass is only at position A in Fig. 11–24. This corresponds to a region of contact as shown in Fig. 11–22.

In other words, tissue cells move as a result of peristaltic waves on the cell surface, not unlike the peristaltic locomotion demonstrated in earthworms by Sir James Gray. It is produced by a combination of a transverse wave and a compressional wave at the cell surface. The points of attachment (A) are nodes that is, points of no movement; extension and moving forward of the membrane occur instantaneously in the unattached regions. This type of locomotion does not require a forward flow of cytoplasm as in amoebae. The amoeboid type of streaming is not observed in fibroblasts or epithelial cells. Note that it is also possible for groups of cells when attached together in the form of a sheet to migrate in a coordinated manner. This is observed in sheets of epithelium growing on glass. The ruffled membranes are particularly pronounced on the edges of the leading cells.

The ruffled membranes of tissue cells are also of major importance in the dynamics of cell-to-cell interactions and in the control of cell movements by contact with other cells. This problem will be further discussed in Chapter 14.

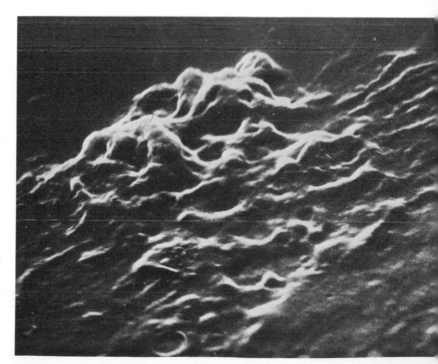

Fig. 11–23 Stereoscan microscope picture of an instantaneously frozen cell, showing the ruffled membrane structure. Magnification × 5,700.

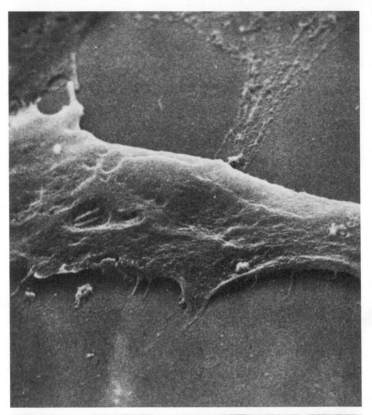

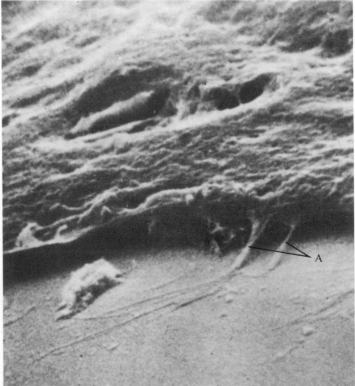

Fig. 11–24 *Above:* Stereoscan microscope picture of contacts between the underside of a fibroblast and the glass surface on which it is moving. Magnification ×2,250. *Below:* the same at higher magnification, showing the undulating membrane making local contact with the glass. (A in Fig. 11–22.) Magnification × 7,500

Amoeboid locomotion. When the amoebae are lying on a solid surface, streaming in the cytoplasm produces the characteristic amoeboid locomotion. Amoebae vary greatly in the number of advancing pseudopodia they produce. A few species are monopodal. *Amoeba proteus* produces a pseudopodium which appears to 'break out' from the main cell mass. For a time this pseudopodium is dominant in controlling forward movement. A new pseudopodium is then formed and becomes dominant while the leading pseudopodium is slowly withdrawn. Other species, of the Foraminifera and Radiolaria, may produce numerous pseudopodia at any one time. Numerous pseudopodia are produced by other cells as well as amoebae, an extreme case being the formation of microvilli by tissue cells in culture. The microvilli are fine extrusions of the cytoplasm 1,000–2,000 Å in diameter which are suddenly produced and then withdrawn into the cell surface.

According to the mechanisms for amoeboid locomotion proposed by Goldacre (tail contraction) and by Allen (fountain zone) the leading pseudopodium is formed as the result of pressure from the cytoplasm. It is driven forwards like an advancing drop of water on a greasy surface (Fig. 11–12). That is, a new membrane is being formed in the front, while the membrane is drawn up in the rear of the cell or tail region. Examination of a moving amoeba under the microscope certainly suggests that this is the case, because the hyaline cap, on the leading edge—a region free from granules, seems to press forwards, as though under pressure from behind. One problem here is that the turnover of the membrane surface seems slower than we would expect; for example, Wolpert, Thompson, and O'Neill prepared a fluorescent antiscrum to the amoeba's surface material and found that the attached stain remained on the surface for many hours. A possible explanation for this low turnover is that the stain is taken up by an outer coat or sheath under which the amoeba moves.

Electrical potential differences between parts of moving amoebae have also been demonstrated (Bell, Kamiya, and others) although as with *Plasmodium* the relationship between these potentials and locomotion is not exactly defined.

Other types of cellular locomotion. It has been claimed by Piper that bacteria can still move after their flagella have been removed. The residual locomotion appears to arise from an undulating movement of the whole cell, a type of movement that is also observed in the alga *Oscillatoria*, which in the surface-contact microscope can be seen to move forwards by making oscillating motions of the thread of cells. These movements are probably similar to those already described for the locomotion of mammalian tissue cells but they involve a general, coordinated undulation of the more rigid cell wall. Some cells apparently move by combining several activities:

for example, mammalian macrophages exhibit highly active membrane undulations but can also put out pseudopodia like amoebae, while lymphocytes produce numerous pseudopodia simultaneously and appear to 'dance' under the microscope. Many cells can change their type of locomotion in different media. The most dramatic example is *Naegleria gruberi*, studied by Willmer, which can change from a flagellate to an amoeboid type of locomotion. Positive ions favour the amoeboid form and negative ions the flagellate form. Concentrated medium favours the amoeboid form, dilute medium the flagellate form.

11–5
Mechanisms of
cell movement

How can we explain these complex movements of cytoplasm and of whole cells as observed in bacteria, fungi, plants and animals? Coordinated movements in living organisms are harder to explain in terms of molecular and cell structure than any of the properties of cells considered so far. The wonderful nature of the cell becomes most apparent with the study of cellular dynamics. Various simple physical explanations for cell movements have had to be abandoned one by one as the structure of the cytoplasm has become revealed by electron microscopy and the behaviour of living cells examined by more refined physical methods. We now consider the various working hypotheses about cell movements currently in favour.

Biochemistry, molecular biology, and cytogenetics have demonstrated that the basic structures and functions of cells are remarkably similar in many types of organism. Specialized functions generally arise as refinements of already existing functions. This is the rationale for seeking unifying principles relating to cell movements. A starting point for such a search is to study two particular types of cell which represent the extreme forms of behaviour seen in cell movements, each showing essentially one particular type of movement:

The cell of mammalian striated muscle in which an extremely regular contractile process occurs within a highly organized macromolecular structure.

The freshwater alga *Nitella* in which movement is restricted to cyclosis, with the motive force localized at or near the surface of the plasma gel.

Huxley has shown that muscular contraction is connected with a sliding filament movement. Striated muscle consists mainly of the two proteins myosin and actin. Myosin consists of rod-like molecules with a massive head region (Fig. 11–25) and the myosin molecules pack together as bundles in a helical array. Shortening of the muscle involves sliding of the actin filaments between the myosin filaments (Fig. 11–26) and the heads of the myosin molecules attach to the actin at certain points to form molecular bridges (Fig. 11–27). The bridges are probably made and broken continuously

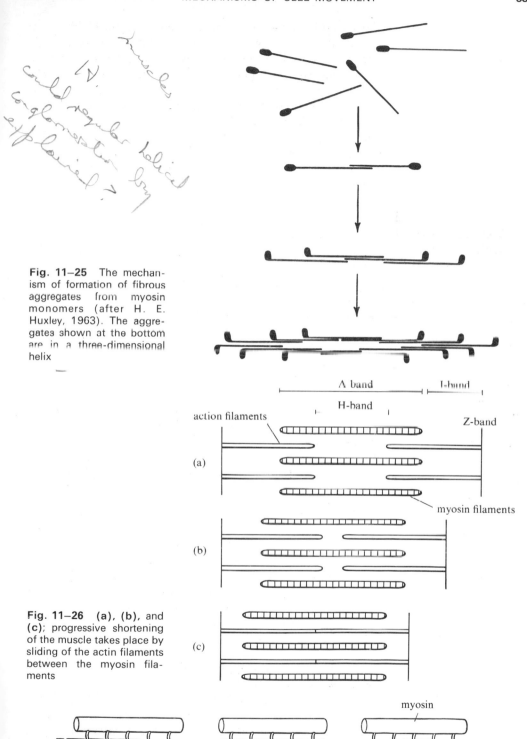

Fig. 11–25 The mechanism of formation of fibrous aggregates from myosin monomers (after H. E. Huxley, 1963). The aggregates shown at the bottom are in a three-dimensional helix

Fig. 11–26 (a), (b), and (c); progressive shortening of the muscle takes place by sliding of the actin filaments between the myosin filaments

A band

I-band

H-band

action filaments

Z-band

(a)

myosin filaments

(b)

(c)

myosin

actin

Fig. 11–27 Suggested ratchet action in striated muscle. *Left:* thick and thin filaments of relaxed muscle; *centre:* as muscle contracts it reaches maximum tension; *right:* further contraction will lower the tension because fewer ratchets are engaged. (After T. Hayashi.)

like a ratchet, but it is essential that not all bridges should
form and be broken at the same time. A person climbing a
ladder must keep hold at least with one hand or foot, and the
same principle appears to apply in muscular contraction.
Individual heads seem to form and break bridges in a cyclic
fashion during sliding of the filaments; in insect flight muscles,
however, where an extremely rapid making and breaking of
contacts is required, the bridges make and break synchronously
(Pringle).

As in most other cellular processes requiring energy,
muscular contraction uses up ATP. The heads of the myosin
molecules carry ATPase, which like other ATPases is activated
by calcium ions (Ca^{2+}), though not by Mg^{2+}. In association
with actin, however, it becomes Mg^{2+}-dependent. In muscle
the Mg^{2+} concentration stays within $5–10 \times 10^{-3}$ M. But the
system is extremely sensitive to a triggering action by calcium
ions. Changes in the concentration of between 10^{-5} and
10^{-7} M can change ATP hydrolysis from a high rate (con-
tracting muscle) to a low rate (resting muscle). In striated
muscle, the calcium ions are released into the muscle cell
through the membrane system or sarcolemma, within which a
reserve of calcium ions is maintained. This release leads to
hydrolysis of ATP via the myosin head protein in association
with the actin filaments.

The sliding filament model involves a highly organized
molecular arrangement within the muscle fibres. Some earlier
views of muscular contraction were rather similar to those
indicated in Goldacre's model for amoeboid movement. In the
resting muscle the molecules were thought to be extended as
long chains, while the contraction involved a coiling up of the
molecules, as a random coil. The reverse process occurs when a
piece of rubber is stretched: before stretching the molecules
lie in random coils, while after stretching they become arranged
as parallel extended chains. In thermodynamic terms the
random coil is in a state of low free energy (high entropy) while
the parallel array is in a more ordered state of high free energy
(low entropy). According to this model contracted muscle
should have a lower free energy than resting muscle but recent
studies by physiologists have shown that this is not the case.
Contraction of muscle is not associated with a free-energy
change and in the contracted state the molecules are as highly
orientated as in the resting state. This is to be expected in the
sliding filament model (Fig. 11–26).

What relevance have striated mammalian and insect
muscle to the extremely labile movements seen during cyclosis
in *Nitella*? Evidence that there may be some connection has
come from chemical studies. The two proteins of striated
muscle, actin and myosin, have been well characterized in
terms of their amino acid composition. As mentioned above,
actin fibres are formed from globular sub-units, about 60 Å in

diameter, which join together like a string of beads. Fibres in other cell types are known to have a similar constitution. Even the fibres of microtubules are formed in much the same way from similar units, as is the flagellin of bacterial flagella. In some cases, a protein complex with many of the properties of the actomyosin of striated muscle has been isolated. A particularly important property is superprecipitation. A gel of actomyosin to which ATP is added in the presence of Mg^{2+} ions contracts into an extremely small volume. It is probably related in some way to the contraction of the macromolecular units in living muscle, but because of the disorganized structure of the isolated actomyosin gel can only occur as a gross contraction of the gel network. A protein gel which behaves in the same way has been isolated from *Plasmodium*, amoebae and other cells. Probably filaments are embedded at the surface of the plasma gel in *Plasmodium*, particularly where active protoplasmic streaming is occurring (Fig. 11–10).

The rotating filaments which are found in isolated *Nitella* cytoplasm (Fig. 11–6) probably break off from the inner surface of the plasma gel and a protein containing ATPase is certainly present in the cytoplasm of these cells. The work of Wohlfarth-Bottermann indicates that fibres are present in the region of the plasma gel where the most active streaming is occurring (Fig. 11–10); this has been confirmed by the observations of Kamiya on the same material. Protoplasmic streaming is usually in a definite direction, at least for a time. Here again the properties of actin filaments may provide a clue. Electron microscopy indicates that the bridges which form between myosin heads and actin can have an arrowhead form which implies some head-to-tail polarity in the structure of the strands. Such polarity has already been mentioned in the case of nucleic acid chains where the phosphate ester links provide a head-to-tail polarity and determine the direction of synthesis of the chains. In a sliding filament model some kind of polarity of the molecules is required to account for the direction of contraction on activation with Ca^{2+} ions.

In cyclosis, Pringle has suggested that the water molecules may be caused to stream by adsorption on protein molecules— that is, by hydration of protein molecules. He suggests that the carrier protein could be a myosin-like protein located in the plasma sol similar to the heads of Fig. 11–25. This is carried along the surface of an actin filament embedded at the plasma gel surface, and operates by a mechanical ratchet type of motion similar to muscle. On the other hand there is much to be said for the view that electrical forces are more likely to predominate in this situation. In the phenomenon of electro-osmosis, liquids stream through narrow tubes when an electric field is applied. This is due to hydration of *ions*: as the ions move, the shell of water molecules is also carried along with the ion. The progressive movement of a *micro* surface potential along fibres at the surface of the plasma gel could be generated by progressive Ca^{2+} ion adsorption followed by ATP hydrolysis in the neighbourhood of myosin head protein, which could carry the shell of water molecules along the surface. But these are only speculations; experimental methods for testing such hypotheses are likely to be hard to develop.

It should be mentioned that the sliding filament theory also appears to fit the behaviour of flagella and cilia. Sleigh has suggested that transmission of the wave along the filament involves a progressive forward and backward relative movement of neighbouring filaments away from an equilibrium position.

Bibliography   Allen, R. D., Amoeboid movement, in J. Brachet and A. E. Mirsky (eds), *The Cell,* vol. 2. Academic Press, New York, 1961.

Allen R. D. and N. Kamiya (eds), *Primitive Motile Systems in Cell Biology.* Academic Press, New York, 1964.

Fawcett, D. W., Cilia and flagella, in J. Brachet and A. E. Mirsky (eds), *The Cell,* vol. 2. Academic Press, New York, 1961.

Gray, J., *Ciliary Movement.* Cambridge University Press, London, 1928.

Hayashi, T., How cells move, in *The Living Cell.* W. H. Freeman and Company, San Francisco, 1961.

Huxley, H. E., Muscle structure and theories of contraction. *Prog. Biophys. Biophys. Chem.* **7,** 255, 1957.

Miller, P. L. (ed.), *Aspects of Cell Motility,* Symposia of the Society for Experimental Biology, vol 22. Cambridge University Press, London, 1968.

12 Cell biology of development

**12–1
Historical
introduction**

The development of organisms has fascinated man since early times. Aristotle, for example, in the fourth century B.C., observed chick embryos and suggested with remarkable foresight that the development of organisms is brought about by a causal interaction between their parts—that is, by epigenesis.

For many years embryology remained a descriptive science, although the introduction of the microscope enabled biologists in the 17th and 18th centuries to make observations on a new level, particularly of sperm cells. Many imagined that they could see preformed organisms (animalcules) in the sperm (Fig. 12–1). According to the preformationists, as followers of this school of thought were known, the sperm or ova contained replicas of the adult which unfolded and grew like a bud. With better methods of observation Caspar Friedrich Wolff (1738–1796) studied chick embryos but could see little evidence for preformed structures; he favoured the earlier epigenetic view of Aristotle.

One of the first systematic studies of development was made in 1877 by the American biologist Louis Agassis who studied starfish embryos. Besides the two rival schools of preformationists and epigeneticists an additional unifying concept was the recapitulation theory as put forward by Meckel in 1810, which came much into favour after Darwin's work on the theory of evolution. Meckel suggested that the developing embryo of an animal such as a bird or a mammal passes through a series of stages which resemble the adults of the forms ancestral to it; for example, the embryo bird has gill slits at a certain stage of development, which have an obvious function in the adult fish but disappear in the bird before the adult stage is reached. Von Baer observed these structures with an improved form of microscope and he was able to show that the gill slits in the bird embryo are like those of a fish embryo but only remotely resemble those of an adult fish. Haeckel became the chief exponent of Meckel's theory of recapitulation, which became the guiding principle of embryology for many years.

Many years later de Beer and other embryologists showed that von Baer's views were much more realistic. We

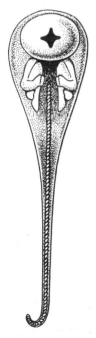

Fig. 12–1 An animalcule as visualized by a preformationist (after Hartsoecker, ca 1700)

now recognize that it is the *later* stages of embryonic development, which involve minor changes in structure, that are likely to be modified during evolutionary change. In the intermediate stages embryos can develop in a limited number of ways which are controlled in part by structural and geometrical limitations, so that most embryos tend to be rather similar at those stages. However, the early stages of development differ considerably in different groups.

Towards the end of the 19th century embryology entered a new phase; biologists began to look for mechanisms underlying the changes seen during development, the pioneers in this field being the two German embryologists, Wilhelm Roux and Hans Driesch.

Roux called his subject *Entwicklungsmechanik*, developmental mechanics. In one of his experiments, he made a hole with a hot needle in one of the cells of an amphibian embryo at the two-cell stage, and found that the remaining cell gave rise to a half embryo. On the basis of these studies Weismann in 1892 put forward the theory that the course of development is laid down in the fertilized egg, and that every cell is endowed with a definite fate in future development, with the result that the young embryo can be looked upon as a mosaic. This view had to be modified when later experiments were carried out by Endres (1895) and Herlitzka (1897). If the two cells are separated in a more delicate manner with a hair loop (a method later improved by Spemann) two normal embryos develop as shown in Fig. 12–2.

Roux's earlier experiment must have caused some damage to the untreated cell. Driesch carried out a number of experiments with sea urchins in which individual cells, when separated from each other, develop into normal embryos. He concluded that the egg is a harmonious equipotential system and that by entelechy (from the Greek *entelekhia*, of perfection), normal development is restored to a manipulated embryo.

Great advances in the field of experimental embryology were made by Spemann and his school at the beginning of the 20th century. He showed that grafting tissues from one part of an embryo to another could modify development. He provided ample proof of Aristotle's original postulate that embryos develop through a causal interaction between the parts. Other experiments of this kind, which are of continuing interest to embryologists, are described below.

Embryology is the meeting place between cell biology and general biology. Although embryology cannot here be considered in detail (fuller treatments are cited in the Bibliography) earlier chapters on the structure and function of cellular components, on the life of cells, and on cellular dynamics, have provided a fitting context for considering developmental biology from the point of view of cell differentiation, molecular biology and general cell biology.

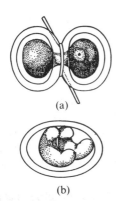

(a)

(b)

Fig. 12–2 (a) Constriction of the two-cell stage of a newt embryo with a hair loop. (b) Two normal embryos develop if the ligature lies along the presumptive median plane. (After Spemann, 1936. With permission from L. Saxén and S. Toivonen, *Primary Embryonic Induction*. Logos Press, London, 1962.)

A complete understanding of the structure and function of whole organisms, including man, can only be achieved by an understanding of how individual cells arise, develop, and interact with one another. In this sense developmental biology provides a good stepping stone between cell biology and general biology.

In the account that follows we preface cellular aspects of development by descriptions of how the whole embryo develops so as to provide the necessary frame of reference. The embryos described are three that are often used for experimental work: the sea urchin, amphibian, and bird embryos. Because of the importance of the subject a brief account is given of development in mammalian embryos. Chapter 13 describes the biochemical work that has been carried out on embryos, such as studies of DNA, RNA, and protein synthesis, respiration and energy requirements, and so forth.

In earlier chapters we assumed that the daughter cells arising from mitosis exhibit the same character and biological behaviour as the parent cell. In embryonic development, however, there is a progressing divergence to produce different cell types such as muscle, connective tissue cells, and renal cells during the growth of the embryo. The stages leading to the appearance of new cell types have been beautifully demonstrated by Vogt and other workers. It is possible to find out which particular regions of the embryo will develop *later* into nerve, epidermis, and other tissues, and to prepare maps of early embryos in which these regions are designated. Such maps, called presumptive, prospective, or fate maps, are described below. Experiments of the type performed by Roux and Driesch and by Spemann, which involve grafting tissues from other embryos into a growing embryo, have shown the extent to which the fate of tissues in early stages has already been determined or can be modified in a different environment. These experiments show that some cells at certain stages are *pluripotent*, that is, they are able to develop into several different sorts of cells, whereas other cells are already determined in their future development (*unipotent*). The direction in which a particular pluripotent cell develops can be affected by neighbouring cells and tissues. This is an important phenomenon known as *embryonic induction*. The interesting results of such grafting experiments are described in the next section.

Finally we come to consider the role of the gene in development, 'the strategy of the gene' as it has been aptly called by Waddington. Each cell derived from the original fertilized egg contains the *diploid* chromosome complement derived from the male and female gametes. But in some cells protein synthesis is restricted mainly to one or two proteins; for example, reticulocytes produce haemoglobin and the protein synthesis of muscle cells is largely restricted to actin and myosin. Clearly the whole genome is not utilized by every cell in the embryo or by the same cells for more than a limited period. The control of gene function in development, both at the biological and at the molecular level, is a problem that is relevant not only to normal growth, but also to aspects of cancer and certain diseases that arise from inborn errors of metabolism. Gene control is further discussed in Chapter 13.

12-2
The gametes

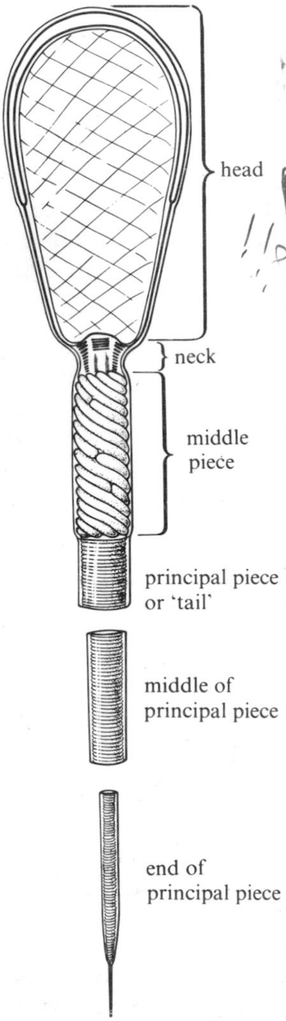

The formation of the gametes from the germ line cells is important because the early stages of development depend heavily on materials that are synthesized before fertilization. Figure 10-7 shows the stages of spermatogenesis in animals which lead to the formation of four haploid cells, known as the spermatids, from the second meiotic division. These cells develop into mature sperm with dense heads containing a haploid complement of DNA and motile tails. The sperm head contains densely packed DNA associated with protamine, a comparatively simple protein with a higher proportion of basic amino acids than the histones which are associated with DNA in somatic cells. A bull sperm head is drawn in Fig. 12-3. On the front is an acrosome which contains lysosomes and associated enzymes. The shape of the head varies considerably in different species (Fig. 12-4). The closely packed mitochondria are located in the middle segment and provide the energy, in the form of ATP, for locomotion.

Whereas spermatogensis produces four haploid cells, oögenesis gives rise to only one ovum (Fig. 10-7). The oöcyte remains in the stage before the first meiotic division for a considerable time. The nucleus is in most cases arrested at meiotic prophase, at which time the well-known 'lampbrush' chromosomes appear in amphibia. In other phyla the structure of the oöcyte nucleus is less apparent, but there is a gross enlargement and it becomes filled with nuclear sap (Fig. 12-5). In this state the nucleus is called a *germinal vesicle*.

The development of lateral loops on the chrommomeres of lampbrush chromosomes is associated with a high rate of RNA synthesis in the cell nucleus and of protein synthesis in the cytoplasm. Most of the genome is active. As mentioned in Chapter 3, an excessive replication of what is known as slave DNA favours the active synthesis of messenger RNA. The net effect is to produce a store of gene products within the cytoplasm of the oöcyte. In many species the oöcyte is surrounded by cells, known as nurse cells, which form cytoplasmic bridges with the oöcyte cytoplasm, enabling them to feed material into the oöcyte cytoplasm.

An important material that is produced at this time is the egg yolk. Egg yolk consists of protein and lipid and is secreted in the form of granules. It is basically a food material, which is broken down during development to provide a source of food

head

neck

middle piece

principal piece or 'tail'

middle of principal piece

end of principal piece

Fig. 12–3 Diagram of a bull sperm. (From *Interacting Systems in Development* by James D. Ebert. Copyright © 1965 by Holt, Rinehart, and Winston, Inc. Reprinted by permission of Holt, Rinehart, and Winston, Inc., New York.)

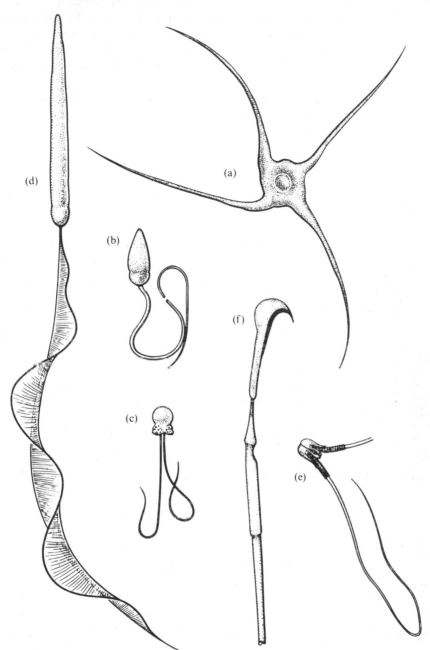

Fig. 12–4 The morphology of various types of sperm: (a) crayfish; (b) sea urchin; (c) toadfish; (d) toad; (e) opossum; (f) guinea pig. (From *Interacting Systems in Development* by James D. Ebert. Copyright © 1965 by Holt, Rinehart, and Winston, Inc. Reprinted by permission of Holt, Rinehart, and Winston, Inc., New York.)

for the growing embryo. The quantity of yolk found in the ovum depends upon the amount required for development. This embryos of sea urchins do not increase appreciably in mass during development, bird eggs contain the whole reserve for embryonic growth, while with species such as mammals the embryo depends on the mother during intra-uterine life and does not require a large reserve of food. Ova with a small quantity of yolk uniformly distributed within the ovum are called homolecithal echinoderms), those with intermediate

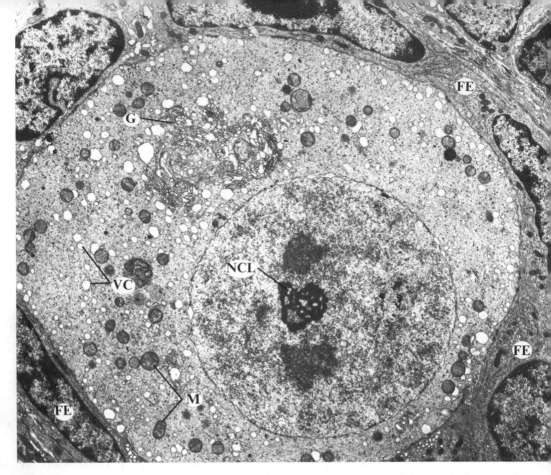

Fig. 12–5   Electron micrograph of part of a guinea pig ovary showing an oögonium surrounded by follicular epithelial cells (FE). NCL nucleolus; G Golgi complex; M mitochondria; VC vesicular component. Magnification ×9,300. (With permission from Everett Anderson.)

amounts are called mesolecithal (amphibia), and those with large amounts of yolk which gravitates towards the lower part of the ovum are called telolecithal (elasmobranchs, reptiles, and birds). The oöcyte acquires other components which can be separated by centrifugation (Fig. 12–6).

In some species the meiotic divisions of the nucleus occur before fertilization and give rise to an ovum; in other cases the divisions are delayed until after fertilization. Division of the oöcyte nucleus leads to *maturation* and appearance of a mature ovum. Of the four haploid nuclei formed in the ovum only one remains as the egg nucleus and the remaining three degenerate into condensed *polar bodies* (Fig. 10–7), with the result that only one ovum is formed from each oöcyte in contradistinction to the four sperm formed from each spermatocyte.

The part called the ovum is the living part of the egg. In most species the formation of the mature egg involves the deposition of successive layers of protective coverings over the living ovum. The membrane which surrounds the living ovum is called the vitelline membrane, but its origin may vary with different species; it consists of a proteinaceous sheath (Bellairs) but is distinct from the plasma membrane. Surrounding the vitelline membrane is a layer of jelly which in mammals also contains a layer of nurse or follicle cells.

dictyosome: Plant equivalent of Golgi App. but with no orientation toward nucleus, Produces ER [handwritten annotation]

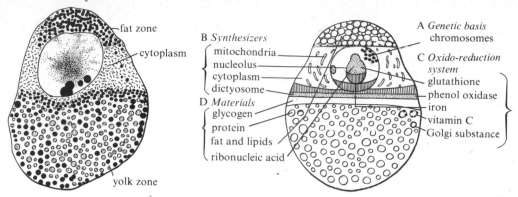

Fig. 12–6 The effect of centrifugation on the oöcyte of the snail *Limnea*. *Left:* drawing of actual section. *Right:* distribution of the various components shown diagrammatically. (After Raven, 1948, and Bretschneider and Raven, 1951.)

Labels in figure:
- fat zone
- cytoplasm
- yolk zone
- B *Synthesizers*
  - mitochondria
  - nucleolus
  - cytoplasm
  - dictyosome
- D *Materials*
  - glycogen
  - protein
  - fat and lipids
  - ribonucleic acid
- A *Genetic basis* chromosomes
- C *Oxido-reduction system*
  - glutathione
  - phenol oxidase
  - iron
  - vitamin C
  - Golgi substance

These cells secrete hormones which play a role in the oestrus cycle. In mammals and birds additional membranes known as the *chorion* and *amnion* are formed. In birds and reptiles an outer sheath consisting of albumin and a membrane or hard shell is deposited. A longitudinal section through a hen's egg as it appears at the time of laying is shown in Fig. 12–17.

According to Needham the evolution of species with protective membranes was one of the most crucial steps in the colonization of dry land by animals. The leathery membrane of reptiles helps to prevent desiccation but the egg must be laid in moist sand so that loss of water is retarded. In birds, the hard shell makes the egg less susceptible to desiccation although equipment for dealing with waste products produced by the growing embryo must be present. In the mammals there is a return to a much simpler membrane structure, the egg being protected within the uterus.

12–3 Fertilization

The egg is fertilized by the sperm in various ways, depending on the species. With aquatic or marine animals, such as the sea urchin, fertilization generally takes place after the eggs have been laid. In birds fertilization occurs before laying and before the deposition of the outer membranes in the oviduct. In mammals fertilization also takes place within the female and the embryo develops within the uterus. The various stages in fertilization are as follows. (Figs 12–7 and 12–8.)

1. Activation of the sperm. In the oviduct of mammals sperm are partly inactive because of the high concentration of CO_2. A decapacitating factor is deposited on the sperm from the seminal fluid which is removed enzymatically when the sperm enter the uterus.

2. In eggs which are surrounded by jelly and nurse cells the sperm must penetrate to reach the vitelline membrane. The sperm acrosome secretes enzymes (including hyaluronidase) which assist penetration.

3. In some species of echinoderm, the egg itself takes an active part and sends out conical projections from its surface which trap the approaching sperm when the acrosome makes contact with the plasma membrane.

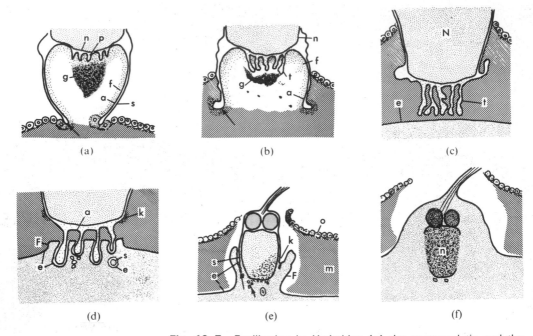

(a)   (b)   (c)

(d)   (e)   (f)

Fig. 12–7 Fertilization in *Hydroides*: **(a)** the acrosomal tip and the sperm plasma membrane become continuous; **(b)** and **(c)** the acrosomal wall is everted; **(d)** the sperm and egg plasma membranes meet; the fertilization cone rises; **(e)** the sperm parts still protrude into the egg envelope; **(f)** in the zygote the internal sperm parts mingle with the egg cytoplasm. a acrosomal membrane; e egg plasma membrane; F fertilization cone; f fine granular material lining acrosomal membrane; g acrosomal granule; k acrosomal remnant, which serves as a marker for identifying sperm plasma membrane; m middle layer of egg envelope (vitelline membrane); N nucleus; n nuclear membrane or components; o outer layer of vitelline membrane; p periacrosomal material; s sperm plasma membrane, or part in zygote derived from it; t acrosomal tubule. (With permission from A. L. Colwin and L. H. Colwin. *J. Biophys. Biochem. Cytol.,* **10**(2), 1961.)

4. In the annelids, the acrosome, once it has broken down, sends out filamentous tubules (in Fig. 12–7c) that touch the plasma membrane of the ovum (d) and fuse with it. In the case of the sea urchin, according to Rothschild and Swann, the plasma gel of the ovum changes in optical appearance and becomes rigid, which tends to prevent the penetration of further sperm. Before fertilization the plasma membrane of the ovum is highly convoluted and numerous transparent cortical granules (probably membrane vesicles) are associated with it (Fig. 12–9). When the sperm penetrates, there is a local expansion produced by the explosion of these vesicles which fuse to form a membrane in association with the vitelline membrane and give rise to the elevated fertilization membrane.

The role of vesicles which can fuse and give rise to new membrane structures was mentioned in Chapter 5 in

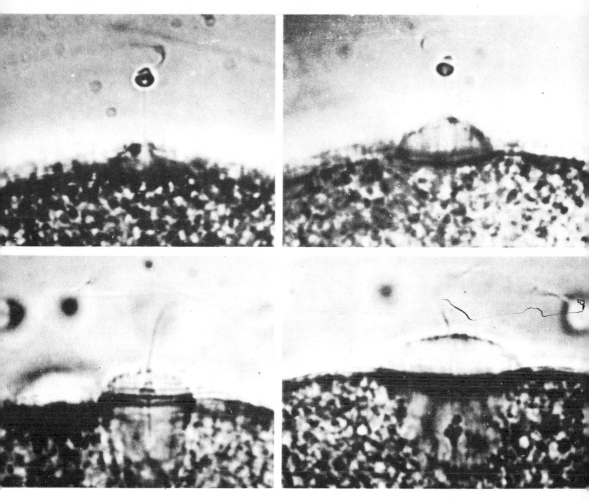

Fig. 12–8 Stages of fertilization as seen in living cells by phase-contrast microscopy. (With permission from A. L. and L. H. Colwin.)

connection with other cellular membranes. This may provide a rather general mechanism for formation of new membrane structures, both at cell surfaces and within the cell cytoplasm. The striking changes seen in sea urchins do not always occur on fertilization of the ovum. In birds, several sperm may enter the egg but only

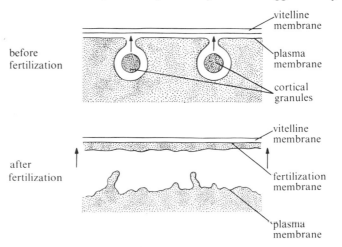

vitelline
membrane

before
fertilization

plasma
membrane

cortical
granules

vitelline
membrane

after
fertilization

fertilization
membrane

plasma
membrane

Fig. 12–9 The formation of the fertilization membrane in the sea urchin

Handwritten margin notes: "How male aster is attracted to female", "H.", "H.", "forced attraction", "Bilateral symmetry"

Fig. 12–10 Fertilization in the annelid *Urechis*. (a) Sperm entering the egg. (b) Fertilization cone has formed and the germinal vesicle (nuclear region) is breaking down. (c) Second polar body division and sperm aster. (d) Egg and sperm nuclei are approaching one another; the polar bodies are at the animal pole. (e) Union or conjugation of male and female nuclei. (f) First cleavage division. (After Bêlăr; with permission from C. H. Waddington.)

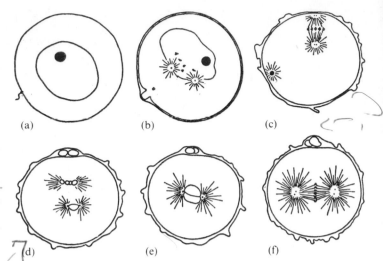

(a)　　　　　　　(b)　　　　　　　(c)

(d)　　　　　　　(e)　　　　　　　(f)

one conjugates with the egg nucleus, the remainder becoming associated with yolk granules. As indicated in Fig. 12–7f, the sperm nucleus and the associated centrioles now enter the cytoplasm of the ovum.

5. Conjugation. If the meiotic divisions (maturation) of the ovum have not already taken place, these now begin as shown in Fig. 12–10. The germinal vesicle breaks down and the divisions occur while the centrioles of the sperm nucleus begin to form asters. Conjugation or union of the male and female nuclei follows and first cleavage begins.

12–4 Early development

The diagrams of the embryos described in this section are referred to by the conventions given in Fig. 12–11. A sphere is shown with a centre 0 and three axes $0x$, $0y$, and $0z$. The $0z$ axis is vertical, meeting the upper surface at the *animal pole*, A, and the lower surface at the *vegetal pole*, V. In later stages of development embryos have a *bilateral plane of symmetry*. In the human body this vertical plane passes from back to front through the spinal cord. Right and left arms, legs, eyes, ears, and so on are all arranged symmetrically about this plane. Only a few organs, such as heart, spleen and liver, are not arranged symmetrically about the plane. This plane, called the *sagittal* plane by embryologists, is represented in Fig. 12–11 as the plane containing the x and z axes. The cephalic or head region lies towards pole A in this plane, while the caudal or tail region lies towards pole V in amphibia.

The sagittal plane is indicated by the great circle (like a line of longitude) where the plane meets the sphere. In diagrams, sections of embryos are generally shown in this plane. Sometimes the vertical plane at right angles to this, the transverse plane, is used. The plane containing the x and y axes is horizontal and meets the sphere in the equatorial great circle (equator).

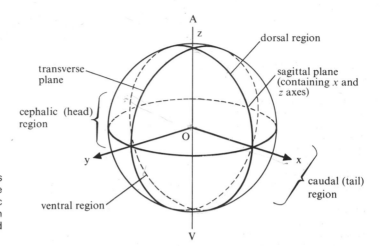

Fig. 12–11 Coordinates used to show the relative positions of embryonic development as shown in Figs 12–12, 12–15, and 12–18

Sea urchin development. The sea urchin egg is fertilized in sea water and appears to be radially symmetrical as shown in Fig. 12–12. There is however a vertical gradient, indicated by shading between the animal and vegetal pole, the nature of which will be discussed below.

Cleavage. The first mitoses in a fertilized egg are known as *cleavage*, because the egg is split in a vertical or horizontal plane across the circumference of the egg. The direction of the plane is determined by the division of the nucleus and lies at right angles to the mitotic spindle. The development of the spindle and separation of chromosomes occurs in much the same way as in other cells.

Cleavage probably occurs in the same way as does cell division in other animal cells, but the largeness of the egg makes the sea urchin embryo of particular interest. Two theories of cleavage are illustrated in Fig. 12–13. According to the model of Mitchison and Swann (1952), cleavage is produced by membrane expansion—that is, by an unfolding of contracted proteins in the plasma gel region because of secretion of an unknown substance which first reaches the polar region. Changes in birefringence of the plasma gel begin at the poles (Fig. 12–13a), and spread progressively towards the furrow region. According to this model progressive expansion of the membrane drives it towards the furrow, which is forced inwards by movement initiated at the poles. Wolpert (1960) has put forward an alternative theory according to which a substance which is secreted from the asters makes the polar regions plastic. These regions are then stretched passively by the active contraction within the region of the furrow, probably in a dense layer below the plasma membrane, not in the plasma membrane itself.

The second cleavage plane is also vertical, and the next horizontal, giving rise to the eight-cell stage (Fig. 12–12d). During these mitoses cell growth does not occur in the same

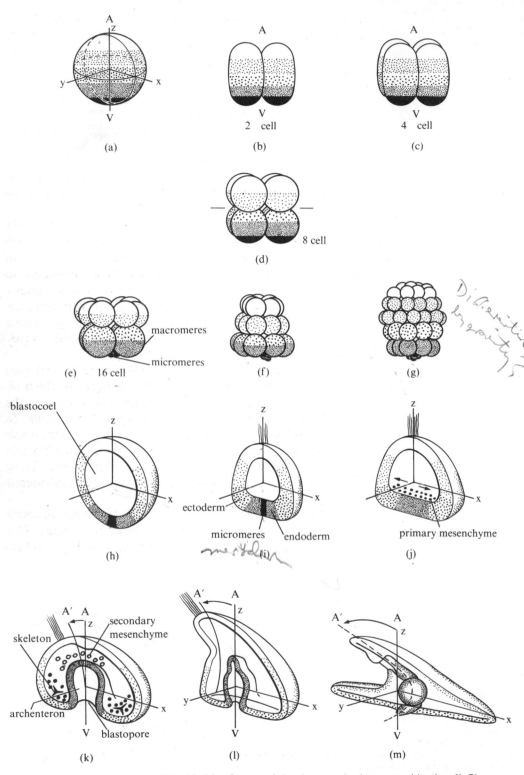

Fig. 12–12 Stages of development in the sea urchin. **(a–d)** Cleavage. **(e–h)** Formation of blastula. **(h–k)** Gastrulation. **(l–m)** Formation of mature larva (pluteus)

way as in the growing cells described in Chapter 9. At the eight-cell stage all cells are the same size; at the next stage the upper cells form a ring of eight *animal* cells. The lower give rise to a ring of four very large cells (macromeres) and four small cells (micromeres); in other words, an *unequal* division of cytoplasm takes place. Further divisions now give rise to a hollow sphere, the *blastula*, in which a fluid fills the centre.

The Blastula. Divisions continue, the cells becoming pro gressively smaller during the blastula stage. Figure 12–12h shows a section through the sagittal (*x–z*) plane. Fairly early in the development of the blastula, beating cilia form on the outer surface of the embryo, which now escapes from its enclosing membranes. Extra large cilia then form an *apical* tuft at the animal pole, while flattening of the vegetal pole occurs (Fig. 12–12i). The gradation of tissue types as they have arisen from the upper animal and lower vegetal parts of the fertilized egg are shown by shading. These are destined to become *ectoderm* (upper region), *endoderm* (basal region), and *mesoderm* (micrometres). Ectoderm will later form the outer regions of the embryo, and mesoderm will produce a middle layer of muscle, calcareous skeleton, and other tissues. The *endoderm* will give rise to the gut.

In the sea urchin, the mesoderm is formed in a different way from mesoderm of birds and amphibians in which all the mesoderm is derived from tissue which follows the endoderm in penetrating into the cavity of the blastula. In sea urchins, cells derived from the micromeres (black) break away from the central part of the inner wall of the blastula (Fig. 12 12j) and come to lie along the inner surface. These cells form *primary* mesenchyme and give rise to the calcareous skeleton.

Gastrulation Two changes now take place, the first of which involves the bending inwards of the base of the blastula. This invagination, called *gastrulation*, is shown in Fig. 12 12k.

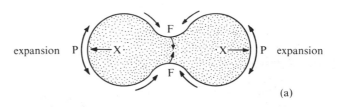

expansion P ←X· F X→ P expansion

(a)

Fig. 12–13 Suggested mechanisms of cleavage. **(a)** Polar expansion model. **(b)** Contraction model. P polar region; F furrow; C con- traction; X changing bire- fringence

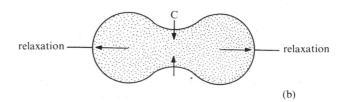

relaxation — ← C → relaxation

(b)

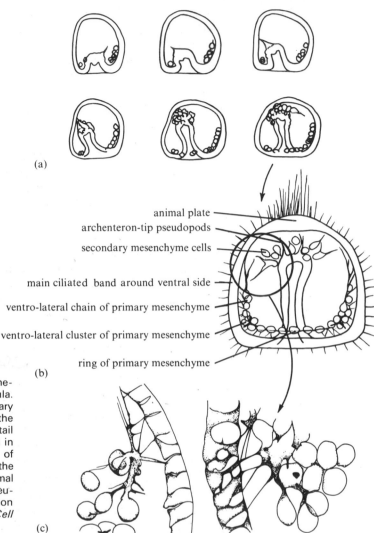

(a)

animal plate
archenteron-tip pseudopods
secondary mesenchyme cells

main ciliated band around ventral side

ventro-lateral chain of primary mesenchyme

ventro-lateral cluster of primary mesenchyme

ring of primary mesenchyme

(b)

Fig. 12–14 Morphogenesis in the sea urchin gastrula. **(a)** Migration of secondary mesoderm cells from the archenteron tip. **(b)** Detail of the structures involved in **(a)**. **(c)** Establishment of contact with the wall of the gastrula by mesenchymal cells with elongated pseudopodia. (With permission from T. Gustafson. *Exp. Cell Res.*, **32**, 576, 1963.)

(c)

Invagination of the vegetative region produces a primitive gut or archenteron. This cup of tissue has a hollow cavity, which communicates with the exterior by the blastopore. From the inner surface of the top of the archenteron more cells are released, which form the *secondary* mesenchyme. This region produces skeleton and muscle.

Previous stages in development of the embryo mainly involve cell division and the appearance of cells of differing size. During gastrulation cell movement is much more important. Consider the migration of the secondary mesenchyme (Fig. 12–14). These cells put out extremely elongated pseudopodia which attach to the inner wall of the blastocoel cavity and draw the cells towards the wall. (Details of movements of this type are discussed in Chapter 13.)

During gastrulation another change of shape occurs, the

animal pole region bending in the sagittal plane, so that the embryo is no longer radially symmetrical. This process continues as A moves to A′ (Figs 12–12k, l, and m) until the characteristic conical form of the mature larva is produced. These changes in the sagittal plane also occur in amphibian and bird embryos, resulting in the characteristic elongation from head to tail, the embryo being symmetrical about the sagittal plane.

Larval stage.   The upper part of the archenteron becomes extended until it touches the ectoderm, when the two fuse to form a channel. Consequently there are now two openings to the exterior, a mouth and an anus, between which is formed a primitive digestive tract. Further changes eventually produce the mature larva (pluteus).

Amphibian development.   Fully grown frog or newt eggs are quite large, reaching 2–3 mm in diameter, and they provide useful material for embryological work. The eggs contain considerable quantities of yolk. Before fertilization the egg appears to be radially symmetrical like the sea urchin egg. The position of entry of the sperm apparently introduces some asymmetry and a so-called grey crescent is formed, whose position is related to the point of entry. The grey crescent is a horizontal crescent of pigmented material which forms where the yolky vegetative material meets the clear animal region. Once the position of the grey crescent has been fixed, it determines the dorso-ventral plane of symmetry (containing the axes $0x$, $0z$, Fig. 12–15a).

Cleavage.   The early stages of cleavage in amphibian eggs resemble those of the sea urchin. However, the concentration of yolk in the vegetative region affects the size of the cells; mitosis occurs more rapidly in the animal region and large yolky cells are formed in the vegetative region.

Blastula.   A blastula consisting of a hollow sphere with a blastocoel cavity is formed as in the sea urchin. Figure 12–15d is a cross-section showing the large vegetative cells at the bottom.

Gastrulation.   As in the sea urchin, gastrulation occurs by invasion of the spherical blastula by the inward movement of tissues (Fig. 12–15e). But unlike the sea urchin, much of the mesoderm is derived from tissue which moves into the blasto-pore cavity round the dorsal lip (Fig. 12–15e). Even the most experienced observer who sees the timing and coordination of these movements, in time-lapse films of the early stages of development, is filled with wonder to see the precision with which they are controlled. Once the mesoderm has been drawn into the archenteron, the cells at the base move forwards to form a yolk plug (Fig. 12–15f).

Neurula.   Changes now occur in the region of the gastrula surface immediately above the blastopore. This region is fated

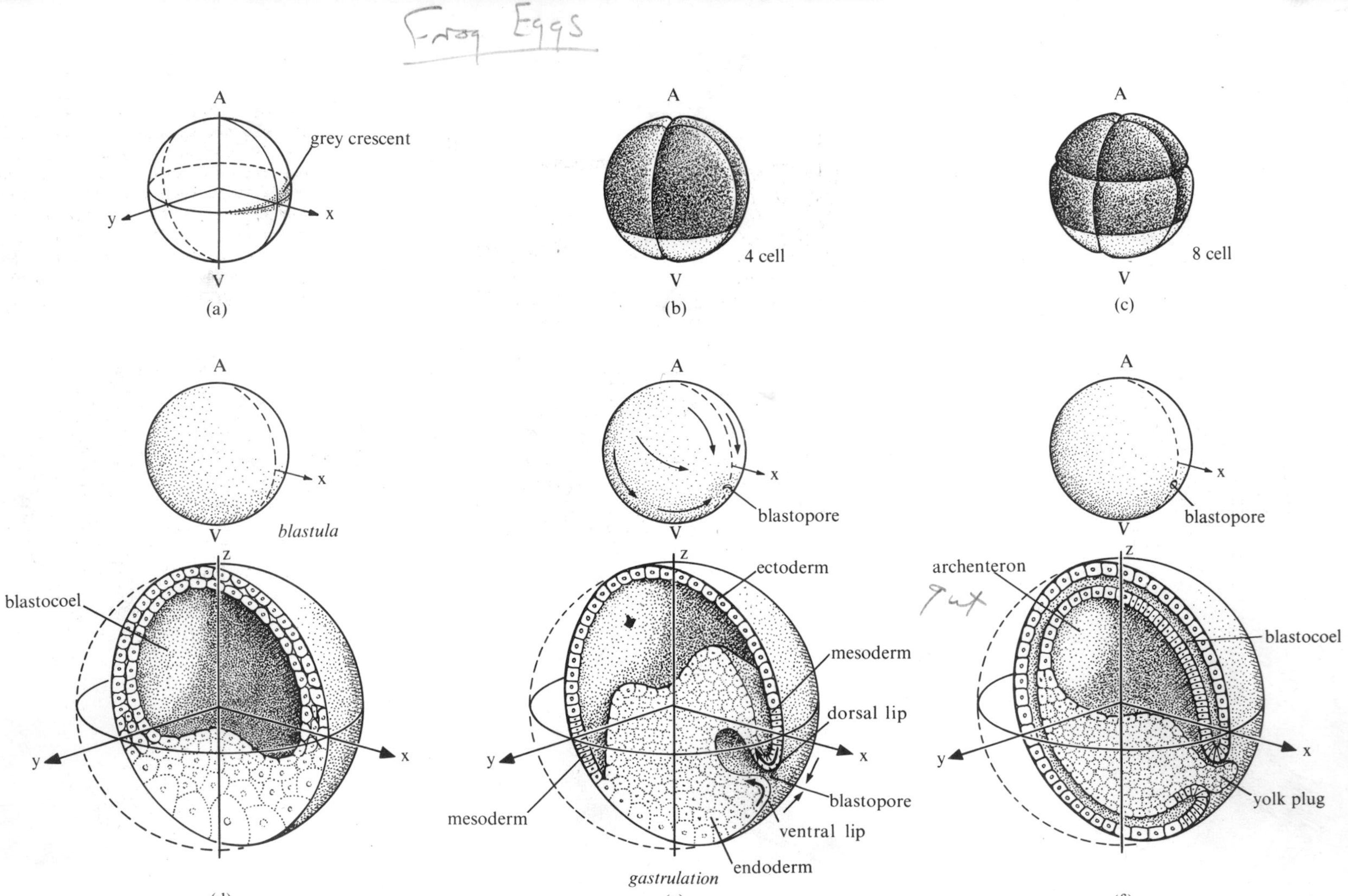

(a) grey crescent

(b) 4 cell

(c) 8 cell

(d) blastula — blastocoel

(e) gastrulation — ectoderm, mesoderm, dorsal lip, blastopore, ventral lip, endoderm, mesoderm

(f) blastocoel, yolk plug, blastopore, archenteron, gut

Frog Eggs

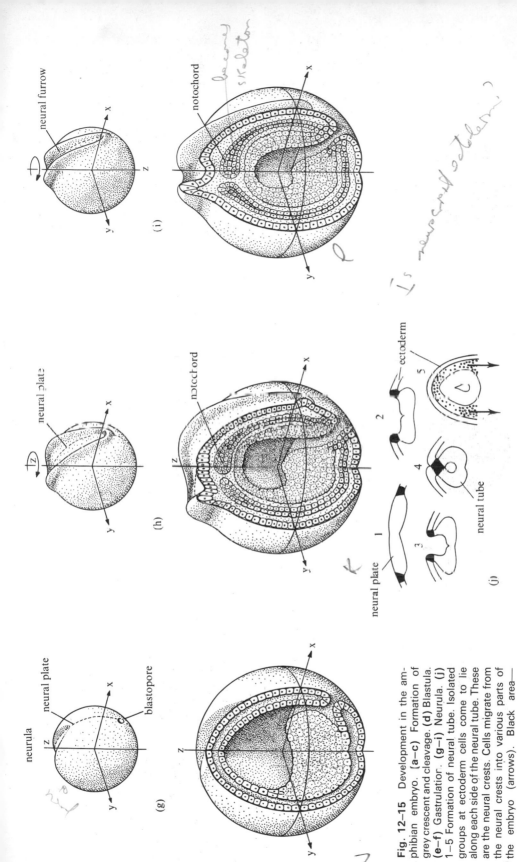

Fig. 12–15 Development in the amphibian embryo. **(a–c)** Formation of grey crescent and cleavage. **(d)** Blastula. **(e–f)** Gastrulation. **(g–i)** Neurula. **(i)** 1–5 Formation of neural tube. Isolated groups at ectoderm cells come to lie along each side of the neural tube. These are the neural crests. Cells migrate from the neural crests into various parts of the embryo (arrows). Black area— neural crest.

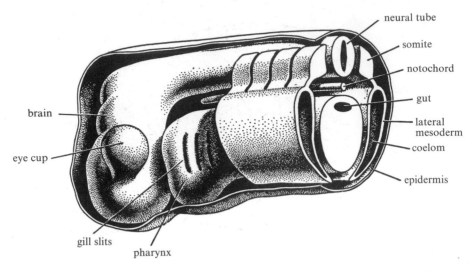

brain

eye cup

gill slits

pharynx

neural tube

somite

notochord

gut

lateral mesoderm

coelom

epidermis

Fig. 12–16 Generalized basic structure of a vertebrate embryo. (With permission from C. H. Waddington.)

to become neural tissue. A groove forms along the sagittal plane with a flattened region, the *neural plate*, bordered on each side by two ridges or neural crests (Figs 12–15g and h). Sections of the embryo at this stage, partly along the sagittal and partly along the transverse plane, are shown in Figs 12–15g, and i.

Of particular importance are the cells of the neural crest (Fig. 12–15j(5)), which behave in an extraordinary way as compared with the remaining cells of the neural plate. They migrate very extensively throughout the embryo. Those from the head region of the fold are destined to become cartilage; those from the trunk region give spinal ganglia and sensory nerves. Those in the middle region become sheath cells (Schwann cells) which later will cloak the nerve axon with myelin; a stream of cells also moves within the ectoderm and will later produce pigment cells.

As the groove deepens, the ridges eventually move together and fuse. The *neural tube* will eventually form the central nervous system.

A somewhat similar change takes place in the archenteron roof immediately below the neural tube. A folding of the mesoderm in this region forms a rod of tissue, the *notochord*. On each side of the notochord grow thickened regions of mesoderm, which later break up into blocks of tissue along the cephalo-caudal axis. These regions of mesoderm are called *somites*. Those in the cephalic (head) region are formed first.

The later stages of development may be briefly summarized as follows:

The neural tube will become the brain, at its anterior end, and other parts of the nervous system.

The somites develop into the segmental muscles of the trunk, and the inner layers (connective tissue) of the skin.

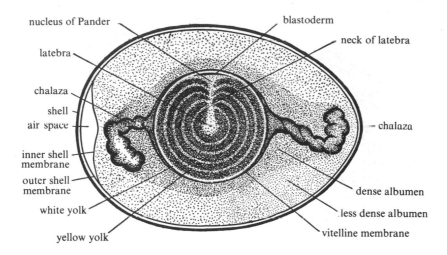

nucleus of Pander

blastoderm

latebra

neck of latebra

chalaza

shell
air space

chalaza

inner shell
membrane

outer shell
membrane

white yolk

dense albumen

less dense albumen

yellow yolk

vitelline membrane

Fig. 12–17 The structure of a hen's egg at the time of laying. (After Lillie. With permission from B. M. Patton, *Early Embryology of the Chick.* McGraw Hill, New York, 1951.)

The mesoderm farther removed from the sagittal plane (lateral mesoderm) will form the nephros or kidneys. The still more lateral mesoderm will give rise to limbs, the more lateral muscle, and the *subepidermal* tissues.

The endoderm becomes fused in the anterior region with the ectoderm so that an opening is formed to the exterior. This becomes the mouth, leading into the oesophagus and the *gastro-intestinal* tract.

The notochord is the first element of the skeleton to appear, and it quickly becomes rigid. At first this is due to absorption of water so that the structure becomes turgid. Skeletal structures are then laid down.

The beginning of these changes is indicated in Fig. 12 16.

Bird embryos. The development of bird embryos resembles that of mammalian embryos, including human embryos. Because the chick embryo provides a convenient experimental material we shall describe it in some detail. The mature ovum of the bird is enclosed within a membrane, the vitelline membrane, which is a layer of protein (Fig. 12–17). Fertilization occurs before laying while the ovum is still in this condition. Additional membranes are later deposited round the ovum as it passes through the oviduct; these ensure that the embryo can develop under stable conditions in aqueous surroundings, isolated from the environment. Most of the egg is filled with yolk and the embryo proper develops in an extremely small clear region at the top, as shown. Cleavage of this region produces a sheet of cells or blastoderm (Figs 12–18a, b, and c). Cleavage is vertical up to the four-cell stage, but subsequently becomes horizontal. The *blastoderm* so produced is shown in Fig. 12–18d. A section through the (*x–z*) or future *sagittal*

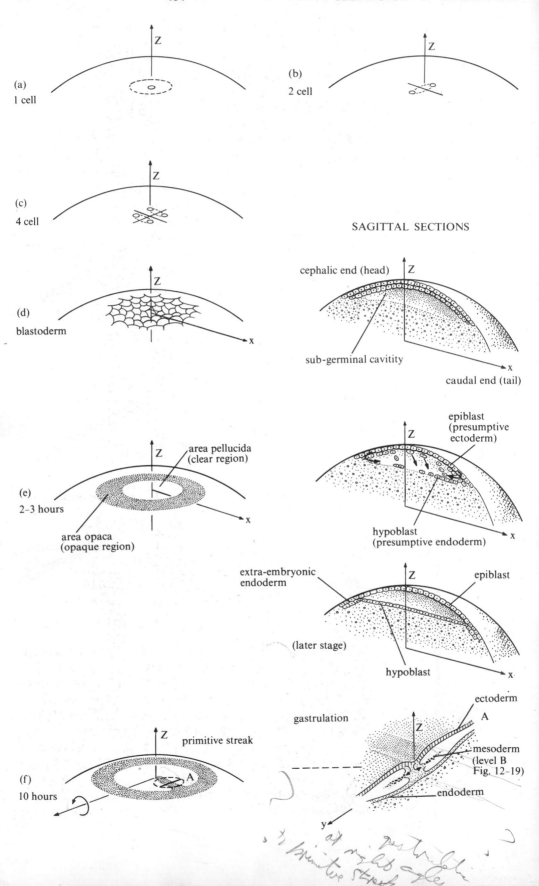

(a)
1 cell

(b)
2 cell

(c)
4 cell

SAGITTAL SECTIONS

(d)
blastoderm

cephalic end (head)

sub-germinal cavitity

caudal end (tail)

epiblast
(presumptive
ectoderm)

(e)
2-3 hours

area pellucida
(clear region)

area opaca
(opaque region)

hypoblast
(presumptive endoderm)

extra-embryonic
endoderm

epiblast

(later stage)

hypoblast

gastrulation

ectoderm
A

mesoderm
(level B
Fig. 12-19)

endoderm

(f)
10 hours

primitive streak

A

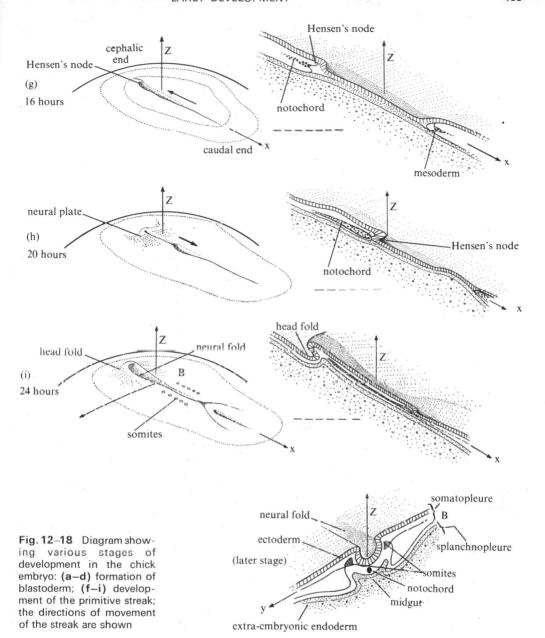

Fig. 12–18 Diagram showing various stages of development in the chick embryo: **(a–d)** formation of blastoderm; **(f–i)** development of the primitive streak; the directions of movement of the streak are shown

plane is also shown together with the region where the head and tail will later develop. A clear space, the sub-germinal cavity, appears below the layer. When seen from above, the disk-shaped embryo contains extremely yolky cells on the outside which merge into the remaining yolk at the periphery. This is called the *area opaca* (thickened or opaque region).

The central region is called the *area pellucida* (clear or transparent region). The upper layer of cells is called the epiblast. The hypoblast is initially formed by movement of cells downwards (delamination) from the upper layer (Fig. 12–18e); later during the formation of the primitive streak

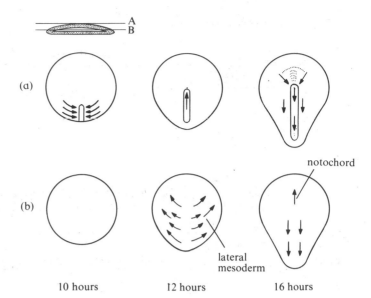

Fig. 12–19 The migration of cells during the primitive streak stage, in upper (A) and lower (B) regions in top figure: **(a)** at the surface of the embryo, (A); **(b)** by tissues which have passed into the lower layers during gastrulation (B)

(see below) migrations begin in the caudal region and contribute to the hypoblast.

Gastrulation. The beginning of gastrulation is the development of a so-called *primitive streak*, an elongated region at the border between the *area opaca* and the *area pellucida*.

Figure 12–19 portrays the development of the chick embryo in relation to the coordinates in Fig. 12–11, Fig. 12–12 (sea urchin), and Fig. 12–15 (amphibian). To understand the relationship of the gastrulation movements in the chick to those in the sea urchin and amphibian embryos, imagine that at 10 hours after fertilization (Fig. 12–18f) the embryo has been rotated about the y axis as shown by the circular arrow, so that the gastrulation movements are not at the bottom, as in the sea urchin, or towards the base of the sphere, as in the amphibian embryos, but from the top downwards. Because of yolk, the amphibian blastula cannot gastrulate by invagination as does the sea urchin blastula. The bird embryo, with an enormous amount of yolk, must undergo a special gastrulation, in which the yolky cells do not participate.

The movements begin in the primitive streak region as shown in Fig. 12–18f. This is a *transverse* section in the 0y, 0z plane. Cells at and near the central region of the groove in the primitive streak move downwards and laterally outwards, until they lie between the ectoderm and endoderm as shown. The primitive streak now elongates progressively (Fig. 12–18g). A thickening in the surface of the blastoderm forms in the extending regions; this is known as Hensen's node. Here cells are migrating *downwards* and *forwards* to form a strip of tissue which will contribute to the notochord and is known as the head process. The primitive streak lengthens until it extends beyond the centre of the *area pellucida*, lying on the sagittal plane as shown. After 16 hours it begins to shorten, moving

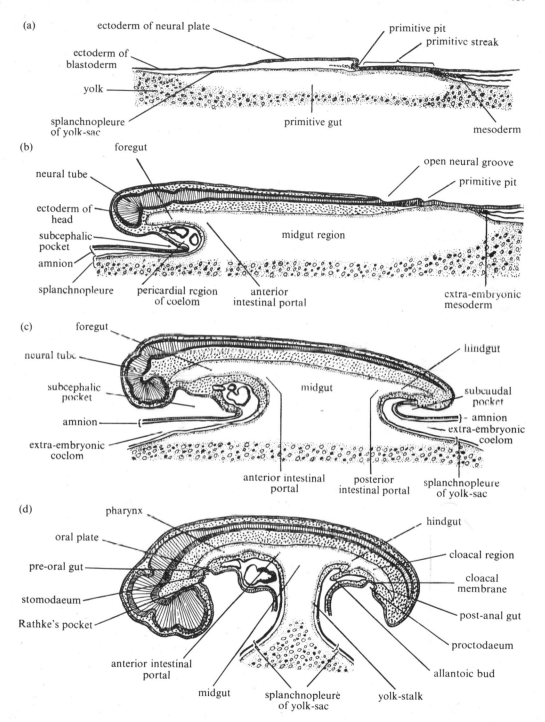

Fig. 12–20 Later stages of development of the chick embryo: **(a)** primitive streak stage (end of first day); **(b)** foregut established (end of second day; **(c)** foregut, midgut, and hindgut established (about 3 days); **(d)** foregut and hindgut increased in length at expense of midgut; yolk stalk formed. (With permission from B. M. Patten, *Early Embryology of the Chick.* McGraw-Hill, New York, 1951.)

backwards towards the caudal region (Fig. 12–18h and Fig. 12–19).

These gastrulation movements result in an arrangement of ectoderm, mesoderm, and endoderm somewhat similar to that of the sea urchin and amphibian. But the process starts from a sheet of cells which separate downwards into layers of ectoderm and endoderm with mesoderm lying between. Because development is from a sheet instead of a sphere some additional movements are required to generate the gut region, as will be described below.

Neurula. The region in front of Hensen's node now generates a neural plate as in amphibia. The neural fold elongates and comes to occupy the whole region which will eventually become the brain and spinal cord. The primitive streak is now confined to the tail region. The head fold is extending beneath the neural tissue, as shown in Fig. 12–18i. The tips of the neural fold fuse together to give a spinal cord and brain region and become covered with ectoderm. During this process the somites are beginning to form on each side of the notochord. They first develop near the cephalic region and appear ~~appear~~ progressively further along the axis towards the caudal region. The embryo grows considerably during these stages.

Later stages. The later stages of development are illustrated in the sagittal sections in Fig. 12–20. Up to 24 hours the gut region has not been closed, only the upper surface being covered by endoderm, the lower region of the sub-germinal cavity being in direct contact with the yolk. The gut is closed by the development of two pockets, a head pocket (Fig. 12–20b —sub-cephalic pocket) which develops first and a tail pocket (Fig. 12–20c—sub-caudal pocket) which develops later. An amnion and chorion are formed. The embryo lies enclosed in fluid within the amniotic cavity. A membranous *allantoic sac* forms from the allantoic bud (Fig. 12–20d) which grows between the amnion and chorion, eventually encompassing the embryo. These membranes facilitate respiration and excretion as well as protecting the embryo.

(a)

(b)

(c)

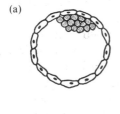

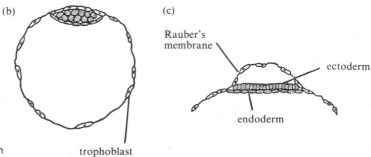

Rauber's membrane

ectoderm

endoderm

trophoblast

Fig. 12–21 Sections of an early mammalian embryo showing trophoblast and inner cell mass. (With permission from C. H. Waddington.)

The human embryo. In mammals, except for the most primitive group, the embryo develops within the uterus and is supplied continuously with nutrients from the placenta. There

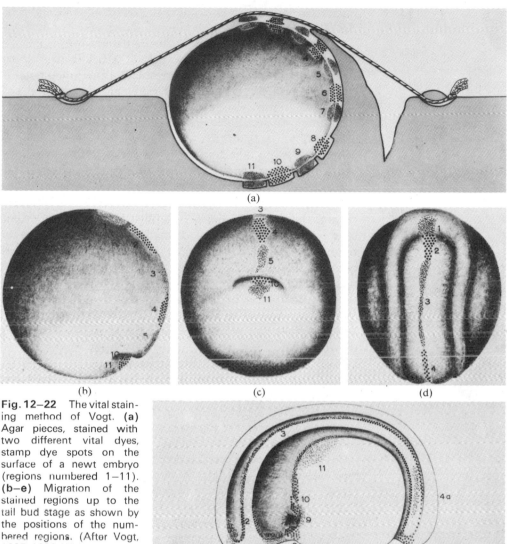

(a)

(b) (c) (d)

Fig. 12–22 The vital staining method of Vogt. **(a)** Agar pieces, stained with two different vital dyes, stamp dye spots on the surface of a newt embryo (regions numbered 1–11). **(b–e)** Migration of the stained regions up to the tail bud stage as shown by the positions of the numbered regions. (After Vogt, 1925, 1929. With permission from L. Saxén and S. Toivonen, *Primary Embryonic Induction*. Logos Press, London, 1962.)

(e)

is no need for a food store of yolk as in amphibian and bird embryos. Cleavage is once again total. The mass of cells produced develops a central cavity and is known as a *blastocyst*. This soon becomes differentiated to give rise to an inner cell mass which will form the embryo. The outer layer of cells of the sphere is called a *trophoblast* and anchors the embryo to the wall of the uterus; the cells invade the uterine tissues in some mammals. This is illustrated in Fig. 12–21, which shows a section of an early mammalian embryo. The trophoblast lying immediately on top of the inner cell mass now lifts off to form a blastocoel.

In some mammals the trophoblast disappears completely over the embryo to form an amniotic cavity. The embryo is then exposed at the surface of the sphere. This is now a *blastoderm* similar to the blastoderm of chick embryos. Development then proceeds in much the same way as in birds. In man, the primitive streak is shorter than in birds.

The application of new culture methods is proving valuable in studying the early stages of development in mammalian embryos, including fertilization of human embryos. Such methods may be relevant to problems of genetic abnormalities.

Other groups. In the molluscs (with the exception of cephalopods) and annelids there is an interesting development of the cells during the early stages. The first two cleavages are vertical, but slightly inclined to each other; the third cleavage is horizontal, but the upper cells (micromeres) are very small and slightly rotated relative to the lower layer. Such eggs are known as spirally cleaving eggs. After fertilization the egg nucleus divides; this happens a number of times without any corresponding cytoplasmic division. The nuclei then migrate to the outer layer or plasma gel where they fuse with the cytoplasm and give rise to separate cells. This outer structure is rather like the blastula of other groups but is called a blastoderm.

The details of these and other groups are given in the books on embryology listed in the Bibliography at the end of this chapter.

12–5 Mapping the embryo

Brief reference has already been made to the *fate* of particular regions in early embryos which will later become such tissues as ectoderm, mesoderm, and neural plate. The classic work of the German embryologist W. Vogt and his school enabled these regions to be precisely mapped in amphibia. Earlier work had been done on fixed tissues of sections prepared from embryos at various stages of development. Vogt's work with living embryos completely revolutionized the ideas of embryologists, showing the great value of sequential studies of living cells or tissues as compared with fixed preparations. His technique is shown in Fig. 12–22. Small pieces of agar were stained with two different colours and mounted in a hollow block as shown; a living embryo at the blastula stage was then pressed against them. When the embryo was removed from the block, various patches of the surface were stained in different colours, and the migration of these patches during subsequent development could be followed Fig. 12–23.

Vogt showed that the lower part of the embryo becomes *endoderm* and subsequently forms the gut and its annexes; the whole upper hemisphere becomes ectoderm which differentiates into the nervous system and the epidermis with its derivatives such as auditory and nasal placodes. In between lies a broad belt which becomes mesoderm. In the *dorsal* side immediately above the blastopore (dorsal lip) this tissue

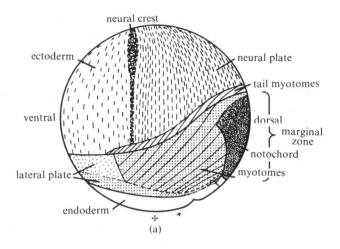

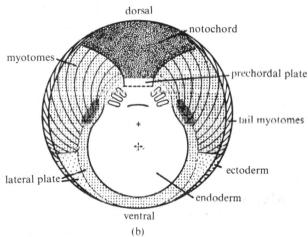

Fig. 12–23 Fate map obtained by the method shown in Fig. 12-22. (a) Left side of a newt embryo with the different presumptive areas indicated. (b) The map seen from the vegetal pole. (After Vogt 1925, 1929, and Pasteels, 1942. With permission from L. Saxén and S. Toivonen, *Primary Embryonic Induction*. Logos Press, London, 1962.)

becomes notochord; below this region, but farther to the side, lie the two areas which give rise to mesodermal somites. The maps in Fig. 12–23 were called *presumptive* maps (German *präsumtiv*); they have also been called *prospective* and *fate* maps. Despite much further work on the subject, the main outlines of Vogt's maps have remained unchanged.

The important migrations of the regions of tissues surrounding the blastopore cavity during amphibian development have already been described and are indicated in Fig. 12–15e. The relationship between the presumptive map and the gastrulation movements then becomes clear and is particularly well seen in the region of the blastopore; tissue from the hollow sphere of the blastula flows into this cavity, a little like water of a shallow river flowing into a rock fissure to form an underground stream. To begin with, the blastopore lip is surrounded by pale ectoderm; it later becomes surrounded by the darker material of the animal hemisphere as the tissues from the animal region move downwards into the blastopore.

Gastrulation movements can also be mapped, using carbon particles to mark the patches of tissue instead of vital

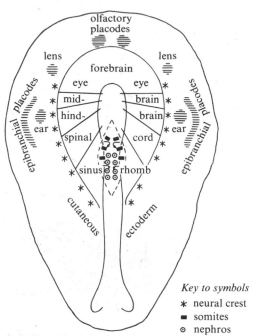

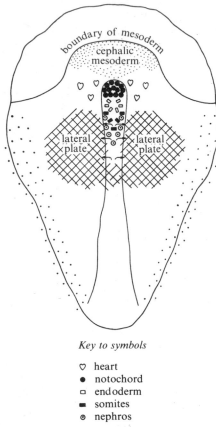

Key to symbols

* neural crest
▪ somites
⊚ nephros

Left: **Fig. 12–24** Presumptive map of the outer layer of the chick embryo in the primitive streak stage. (After D. Rudnick. With permission from B. M. Patten, *Early Embryology of the Chick.* McGraw-Hill, New York, 1951.)

Key to symbols

♡ heart
● notochord
□ endoderm
▪ somites
⊚ nephros

Right: **Fig. 12–25** Map similar to that of Fig. 12–24 but showing the lower layers of tissue which have migrated during the primitive streak stage. (After D. Rudnick. With permission from B. M. Patten, *Early Embryology of the Chick*, McGraw-Hill, New York, 1951.)

stains. Waddington, Pasteels, Spratt, and others have made presumptive maps of chick embryos, as shown in Figs 12–24 and 12–25. The gastrulation movements in the upper layers of the chick embryo have already been shown in Figs 12–19c and 12–19d. The presumptive map in Fig. 12–24 corresponds to layer A of Fig. 12–19 and Fig. 12–25 corresponds to the invaginated layers B. These maps show the presumptive regions after the elongation and regression of the primitive streak. The upper layer is fated to become nervous tissue and ectoderm while the lower layer gives rise to the notochord, somites, and so on.

12–6
Manipulative
experiments

Since the fate or presumptive maps were first prepared experimental embryologists have often tried to modify the course of events during development: for example, to convert a tissue fated to become ectoderm into neural tissue. Only a few such experiments can be described here, of which perhaps the simplest is illustrated in Fig. 12–26. When an amphibian embryo is placed in a solution containing more than the physiological concentration of salts, gastrulation cannot take place, possibly because the blastula becomes too contracted in high salt concentration. Gastrulation movements still occur but the ectoderm is now extruded outwards and is formed as

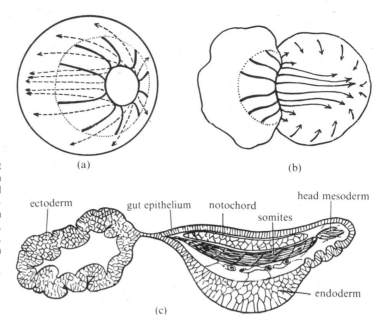

Fig. 12–26 The effect of hypertonic medium on gastrulation. **(a)** Normal gastrulation. **(b)** Exogastrulation. **(c)** Differentiation in an exogastrulated embryo. (After J. Holtfreter and V. Hamburger. With permission from B. H. Willier, P. A. Weiss, and V. Hamburger (eds), *Analysis of Development*. W. B. Saunders Co., Philadelphia, 1955.)

a hollow mass (Fig. 12–26). This change merely represents a change in relative positions of ectoderm, endoderm, and mesoderm; no nervous tissue is formed. More profound changes can be brought about by grafting experiments.

Pioneering work in this field was carried out by Spemann and his school using amphibian embryos (Fig. 12–27a). A piece of neural plate tissue was taken from an early gastrula stage of one newt embryo and implanted into another embryo of the same age. The embryo developed normally, in that the grafted region of *neural* plate tissue developed into ectoderm in its new surrounding. Clearly the fate of the grafted neural

Fig. 12–27 **(a)** The effect of removing a piece of presumptive neural plate and grafting it on to the ventral side of another embryo of the same age; no induction occurs and the neural plate fragment develops in accordance with its new surroundings. **(b)** Corresponding experiment in which the fragment has been removed from a late gastrula. The tissue now develops according to its original fate. (With permission from L. Saxén and S. Toivonen, *Primary Embryonic Induction*. Logos Press, London, 1962.)

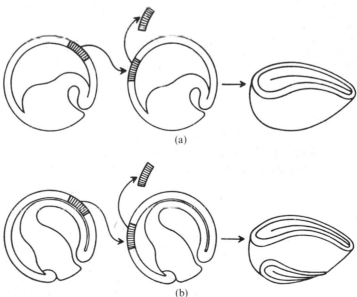

plate tissue can be modified and the cells are pluripotent at this stage and are not committed to become neural tissue.

Although this is true for neural plate tissue at the early gastrula stage, a fragment transplanted from a late gastrula does *not* develop in the same way as its new surroundings (Fig. 12–27b). It develops to form a second neural plate at the neurula stage. In other words, it is now committed. A general property of embryonic tissues is that the degree of pluripotency of a given tissue decreases progressively as the embryo develops.

How can we determine the extent to which cells are pluripotent at various stages of development and in various species? The simplest type of experiment is to separate the cells of embryos at various stages. Such experiments have been conducted by Driesch on sea urchin eggs, in which the cells can be separated from the embryo comparatively easily in a calcium-free medium; calcium ions play a role in cell aggregation. Separation at the two-cell and four-cell stages produces complete but correspondingly smaller larvae (Fig. 12–28). The decrease in size is interesting; there is little overall synthesis during sea urchin development. But it is clear that these cells are still totipotent.

This situation continues if parts of the embryo are separated in a vertical plane but not in a horizontal plane. This is beautifully illustrated in experiments by Horstadius who separated layers in a horizontal plane. At the 32- or 64-cell stage five zones could be distinguished, animal 1, animal 2, vegetative 1, vegetative 2, and micromeres as indicated (Figs 12–12e, f, and g). When isolated by horizontal planes of dissection these produce different regions of the embryo. But when recombined, the joined parts do not produce functions which are strictly additive. One is dealing with animal and vegetative potentialities of various strengths and there is a tendency for a normal equilibrium to be restored even when the two parts grafted together are not in perfect balance (Lehmann). The nature of the built-in factors that lead to stabilization is still largely unknown.

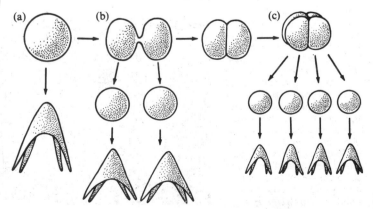

(a)   (b)   (c)

Fig. 12–28 The effect of the separation at various stages of cells of a fertilized sea urchin egg. **(a)** Normal development to form two-cell stage. **(b)** Divided vertically at two-cell stage. Two normal embryos, each half-size, now develop. **(c)** Divided at four-cell stage. Each of the four embryos is normal but one-quarter size.

An interesting approach to this problem was made in 1907 by H. V. Wilson, in the University of North Carolina, who demonstrated that sponges could be separated into individual cells by passing through a fine sieve. The dispersed cells migrated over the surface of the culture dish until they encountered one another and clumped together to form multicellular aggregates. These aggregates grew into complete new sponges with the characteristic architecture!

Such reaggregation phenomena are not encountered in vertebrates, although adult amphibia show considerable

(a)

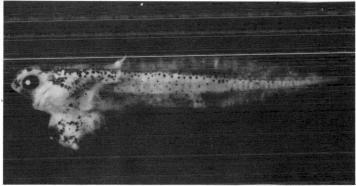

(b)

Fig. 12–29 **(a)** Implantation of the margin of the blastopore lip from an early newt gastrula (head inductor) into another of the same age causes the development of a secondary head by interaction with the neural plate region. **(b)** Similar implantation of a late gastrula blastopore lip (trunk inductor) induces a secondary trunk and tail in an early gastrula. (With permission from L. Saxén and S. Toivonen, *Primary Embryonic Induction*. Logos Press, London, 1962.)

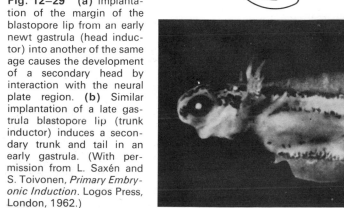

powers of tissue regeneration, the new tissues growing out from organized tissue which remains after wounding the organism.

We now turn to the important problem of *embryonic induction*, or how tissues interact during development. A classic experiment performed in 1921 by Hilde Mangold, a pupil of Spemann, is illustrated in Fig. 12–29a. A fragment of the blastopore lip of an early amphibian gastrula was implanted into the belly region of another gastrula. The result was very striking: a second neural plate was formed on the belly of the host! The experiment was then carried out using two species of *Triturus*. *T. cristatus* has pale eggs while *T. vulgaris* has darker eggs. Only a small part of the secondary neural plate formed from the *graft*, the main portion originating in host cells which differentiated in a new way under the influence of the graft. Clearly the blastopore lip acts as an inductor of neural plate tissue. We might expect this, because it comes to lie later at the archenteron roof, immediately below the presumptive neural plate.

For several years it looked as though the problem of embryonic induction—the capacity of a particular tissue to induce the development of specific properties in another tissue—had been solved. But, as so often happens in biology, further experiments showed that this simple explanation is not the answer. Ten years later it was found that blastopore lip tissue which had been killed by crushing, freezing, heat treatment, or treatment with solvents such as alcohol, ether, and chloroform, was also effective! Since then attempts have been made to characterize the inducing agent in molecular terms; this problem has still not been solved.

When dorsal lip material isolated from a late gastrula is implanted into the blastocoel cavity of another embryo, a secondary trunk and tail are induced (Fig. 12–29b). If dorsal lip material is implanted from an early gastrula, the result is a secondary head. The reason for this (see Fig. 12–15e and f) is that at stage e the blastopore lip consists of tissue which has come to lie in the archenteron roof near the upper part of the hollow sphere—that is, below presumptive neural tissue which will later become head tissue. In Fig. 12–15f the blastopore lip consists of tissue which will later come to lie below the lower portion of the sphere in the region of neural plate where the tail region will develop. Thus early blastopore lip is head inductor; late blastopore lip is tail inductor.

Experiments at other stages of development in various tissues have shown that one tissue can respond to another (the inductor) at a particular stage of embryonic development but will not respond at later stages. The capacity of a given tissue to respond to the stimulus of embryonic induction has been called by Waddington a state of competence; for example, presumptive neural plate tissue in the early blastula is in a

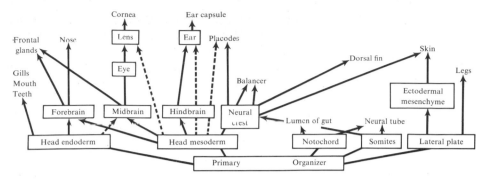

Fig. 12–30 Diagram of secondary organizers in the newt. (After Holtfreter.)

state of competence. In the late blastula its future fate has already been determined as neural tissue.

The induction of neural plate tissue by the dorsal lip of the blastopore has been called *primary embryonic induction*. All subsequent development of the embryo is controlled by this event. Spemann showed that newt embryos could be dissected in various ways; provided that the blastopore lip is left intact, normal development proceeds. This first major step of primary induction in development is succeeded by others like the head and tail regions of neural plate. These are stepwise changes in which the various tissues become canalized and more specialized in character

Our treatment of embryonic development in this chapter has mainly been concerned with the formation of tissues, morphogenesis, and tissue differentiation, whereas in previous chapters we have concentrated on cellular properties. But of course the development of each tissue also involves problems of cellular differentiation. The nucleus of each cell in the embryo as it develops from the fertilized ovum contains a full complement of genetic material in the chromosomes. One of the main factors affecting the change from totipotent cells to cells of limited potency lies in the development within the nucleus of control mechanisms which regulate, by activation or repression, the function of particular genes. We consider cell differentiation and the various extracellular and intra-cellular control mechanisms in Chapter 13.

Figure 12–30 indicates some of the interactions between tissues which are involved in cell differentiation during development. Note, however, that development involves morphogenesis and organ differentiation as well as cell differentiation. These more general aspects of embryology are described in textbooks of embryology. The arrows in Fig. 12–30 do not represent the pathways of development in the sense shown in the fate maps of Figs 12–23, 12–24, and 12–25. They indicate the interactions of a type similar to that between blastopore lip and ectoderm which is necessary for primary induction. From a single fertilized ovum cell differentiation during development progressively reduces the potency of cells by masking some genes and unmasking others, with the

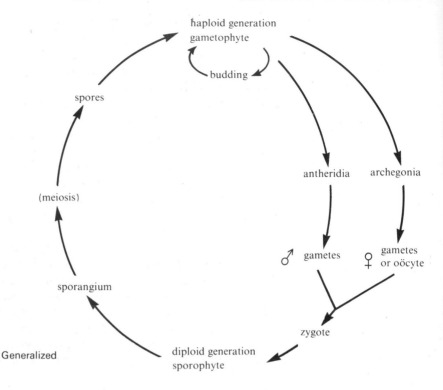

Fig. 12–31 Generalized plant life-cycle

result that cells of all the diverse types found in the adult body have emerged in a step-wise fashion.

The Development of Plants involves the same basic phenomena as described above for the development of animals. The life cycle of the green alga Chlamydomonas is shown in Fig. 1–7; it demonstrates the main stages in the life cycle of many unicellular and multicellular aquatic plants and plants living in moist conditions.

A large part of the life cycle may be spent in the haploid state. This illustrated by the bryophytes (mosses, liverworts) and the pteridophytes (ferns, club mosses) where the haploid (or gametophyte) generation usually leads an independent existence. Reproduction can take place asexually by budding, or sexually by formation of male and female reproductive organs (antheridia and archegonia). The male and female gametes are shed and fusion takes place in the water, or the male gamete swims, in a film of water covering the thallus surface, to the archegonium where fertilization of the oöcyte occurs. The resulting zygote develops into a diploid generation or sporophyte which may bear chlorophyll and have a completely independent existence from the gametophyte. The sporophyte eventually produces haploid spores by reduction division, which are shed and germinate to produce the gametophyte once more.

Usually the two forms of the plant, namely the gametophyte (haploid) and sporophyte (diploid) are dissimilar. In the liverwort *Marchantia*, for example, the gametophyte thallus is the dominant, photosynthesizing plant, and the zygote develops into a sporophyte which remains embedded in the parental gametophyte from which it continues to get nourishment. By contrast, the complex fern plant is the diploid sporophyte while the gametophyte, though free, is a tiny thallus of simple structure.

In gymnosperms (conifers) and angiosperms (flowering plants) the subordination of the gametophyte is complete and there is no independent haploid generation. In gymnosperms for example, two kinds of spores are produced, *microspores* and *megaspores*, which germinate to produce rudimentary male and female gametophyes. The microspores mature as pollen grains and are windborn to the oöcyte. The only trace of the female gametophyte is the embryo sac. After fertilization the egg develops into the embryo, within the gametophytic tissue which now forms the endosperm or food store.

The success of the flowering plants is due in part to the evolution of delicate mechanisms for ensuring fertilization (flowers) and protecting and dispersing the fertilized egg in seeds and fruits. The development of angiosperm plant embryos shows the following main features not seen in animal cells:

There is comparatively little cell movement involved.

Unequal cell division plays a major role, e.g. in the division of the zygote nucleus, and of the cell which divides to form a phloem cell, and its companion cell.

The fascinating problem of development in plants is discussed in detail in the excellent books given in the Bibliography.

Bibliography

Bonner, J., Molecular botany, in W. A. Jensen and L. G. Kavaljian (eds), *Plant Biology Today*. Wadsworth Publishing Company, Belmont, Calif., 1963.

Bonner., *The Cellular Slime Moulds*. Princeton University Press, Princeton, N.J., 1959.

Dan, K., Cytoembryology of echinoderms and amphibia. *Int. Rev. Cytol.*, **9**, 164, 1960.

Fischberg, M. and A. W. Blackler, How cells specialize, in *The Living Cell*. W. H. Freeman and Company, San Francisco, 1961.

Huxley, J. S. and G. R. de Beer, *The Elements of Experimental Embryology*. Cambridge University Press, London, 1934.

Jensen, W. A., The problem of development in plants, in W. A. Jensen and L. G. Kavaljian (eds), *Plant Biology Today*. Wadsworth Publishing Company, Belmont, Calif., 1963.

Steinberg, M. S., The problem of adhesive selectivity in cellular interaction, in M. Locke (ed.), *Cellular Membranes in Development*. Academic Press, New York, 1964.

Waddington, C. H., *Principles of Embryology*, George Allen and Unwin, London, 1956.

Waddington, C. H., *New Patterns in Genetics and Development*. Columbia University Press, New York., 1962.

Weiss, P., *The Principles of Development*. Holt, Rinehart and Winston, New York, 1939.

13 Cellular mechanisms in development and differentiation

The account we gave in Chapter 12 of the early stages of embryonic development in echinoderms, amphibia, and birds was rather general and was not entirely concerned with cellular aspects. The relationship between cell function and development becomes more complex as development proceeds after fertilization and early cleavage, and clearly cellular interactions must play an important part. Contacts formed between cells, contacts formed between cells and substrates (such as intercellular substances), movements of cells, and the transfer of information between cells are all factors in determining the direction of development at various stages. In this chapter we shall attempt to interpret development in terms of reactions at the cellular level. Those aspects which are of a more general nature are well covered by the excellent books listed in the Bibliography.

The subject is considered from four main aspects:

The role of the nucleus.
The role of the cytoplasm.
Interactions between the nucleus and the cytoplasm.
Interactions between the cell and its environment.

As far as possible, control mechanisms which may operate at various stages will also be discussed.

13–1
Role of the nucleus

The nucleus has a major role in the processes involved in development and differentiation. The genotype of a multicellular organism, which is determined at the zygote stage by those of the male and female gametes which fused at fertilization, remains unaltered in all of the many different cells (apart from the germ cells) comprising the organism. It is gene *function* which changes during development, during the formation of different cell types, and during the interactions of cells to form tissues and organs which carry out very specialized functions. The different tissues in an organism represent different regions of activity of the same genotype, or, as Jacob and Monod (1963) expressed it, 'differentiation is present when cells with the same genome synthesize different proteins.'

The activation or repression of various regions of genetic

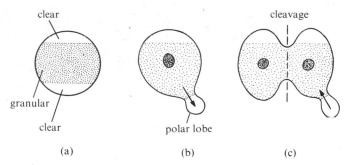

Fig. 13–1 Early development in a mollusc, showing polar lobe. **(a)** Fertilized egg showing granular region and clear basal region. **(b)** Development of polar lobe. **(c)** First cleavage; clear region flows back into right-hand cell

material in development depends, as will be discussed below, partly on intracellular processes of interaction between cytoplasm and nucleus, and partly on reactions between cells. Environmental factors also continue to operate when a cell has become fully differentiated; that is to say, certain regions of its genetic material are activated, but the extent to which this activity is expressed depends upon control messages from the environment.

13–2 Cytoplasmic factors

In the late 19th century it was thought that the cytoplasm of an egg merely contained the necessary materials for synthesis, arranged in a non-specific manner. This was later shown not to be so, and in fact different regions of the egg cytoplasm have their own characteristic specificity. It may be seen from Fig. 12-12 that the material in the cytoplasm of a sea urchin egg is arranged in a stratified manner, with less dense material in the upper part and dense granules in the lower. Division in the first cleavage takes place in a vertical plane, so that the cytoplasmic contents are equally divided between the two

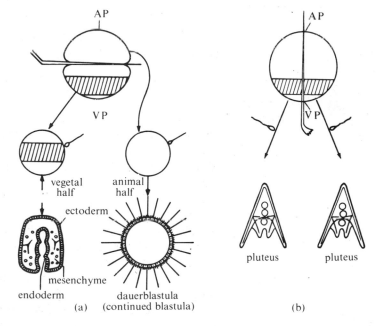

Fig. 13–2 Development of dissected sea urchin eggs. **(a)** Divided through the equator then fertilized. The animal half forms a blastula but no endoderm. The vegetal half forms an incomplete embryo. **(b)** Longitudinal dissection followed by fertilization. Each part develops into a normal embryo of half size. AP animal pole; VP vegetal pole. (From *Embryology* by L. G. Barth. Copyright © 1953 by Holt, Rinehart, and Winston, Inc. Reprinted by permission of Holt, Rinehart, and Winston, Inc., New York.)

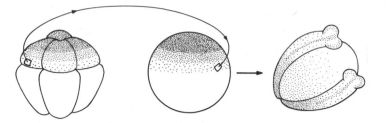

Fig. 13–3 Grey crescent grafted from a young embryo (1 to 8 cell stage) on to an uncleaved fertilized egg. A twinned embryo results. (After A. S. G. Curtis.)

daughter cells. When horizontal cleavage occurs, however (for instance at the formation of the eight-cell stage), the upper cells in Fig. 12–12 will contain fewer dense cytoplasmic granules than the lower ones.

An extreme case of the segregation of cytoplasmic material during cleavage occurs in the eggs of some marine snails and mussels. In Fig. 13–1 are shown the various stages of the first cleavage in such eggs. First, part of the cytoplasm protrudes to form a polar lobe near the vegetative pole. During the first cleavage the lobe remains attached to one blastomere; it is withdrawn into it and reappears at the next division. If the polar lobe is removed by amputation, the embryo develops in an abnormal way, lacking mesodermal structures. Consequently the polar lobe region must contain substances needed for the formation of mesoderm. Eggs of this kind, in which the cytoplasm is clearly divided into different regions required for the development of specific regions of the embryo, are known as mosaic eggs.

In contrast, when an amphibian embryo is divided at the two-cell stage by means of a ligature, as in Spemann's experiment shown in Fig. 12–2, two small but whole embryos are produced from the separated cells. The two cells must therefore have been equipotent. Eggs of this type, which have the capacity to redevelop normally after a disturbance, are known as regulation eggs.

However, an absolute distinction between mosaic and regulation eggs cannot be made. For instance, if sea urchin eggs are cut through vertically, two equipotent cells are obtained; if the dissection is horizontal, only partial development of structures from the two cells takes place (Fig. 13–2). Similarly,

Fig. 13–4 The effect of dissection of an early amphibian gastrula on subsequent development. Broken line indicates line of dissection. (a) Dissection leaves one part with blastopore lip which develops normally. Remainder does not produce embryo. (b) Similar effect with horizontal dissection. (c) Vertical dissection leaving part of blastopore lip in each half. Gives whole embryos

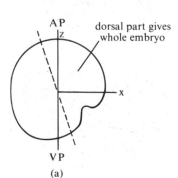

dorsal part gives whole embryo

VP

(a)

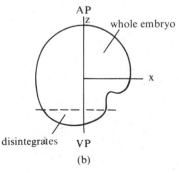

whole embryo

disintegrates   VP

(b)

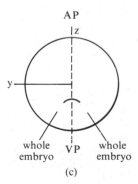

whole embryo   whole embryo

VP

(c)

when amphibian eggs are dissected, the two separated cells will develop normally only if part of the grey crescent is contained in each. The grey crescent is the region which later controls the formation of the blastopore lip.

A beautiful experiment carried out by Curtis showed that dissection of the grey crescent from one egg and grafting in another uncleaved fertilized egg leads to the formation of two blastopore lips, and eventually to the formation of a secondary twinned embryo (Fig. 13-3). The importance of the grey crescent and the blastopore region in later development is illustrated by experiments involving the dissection of embryos at the beginning of gastrulation (Fig. 13-4). If dissection is carried out as shown in (a) or (b), so that the blastopore lip is completely contained in one of the separated parts, then it is

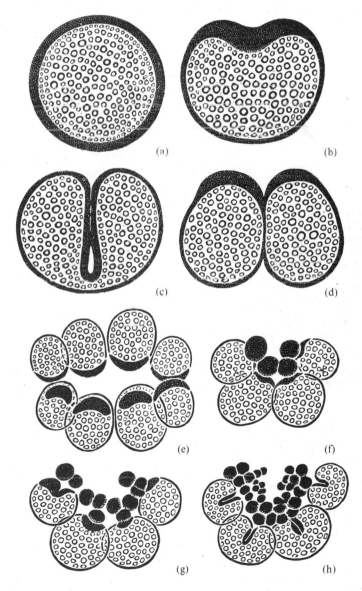

Fig. 13–5 The origin of the swimming plates in the ctenophore, *Beroë ovata.* **(a–d)** Stages of first cleavage. **(e)** 8-cell stage with accumulation of micromere material near the upper pole. **(f)** Pinching-off of the first micromeres (only half of the embryo shown). **(g–h)** If the 16-cell stage is divided into two unequal parts, one retaining five, and the other three, micromeres, each fragment develops into a larva with just as many rows of swimming plates as it contained micromeres. These plates all arose, therefore, from micromeres derived from plasmagel. (From *Interacting Systems in Development* by James D. Ebert. Copyright © 1965 by Holt, Rinehart, and Winston, Inc. Reprinted by permission of Holt, Rinehart, and Winston, Inc., New York.)

(a)   (b)

(c)   (d)

(e)   (f)

(g)   (h)

only from this part that a complete embryo is formed. If dissection is such that the blastopore lip is equally divided between the two parts, then both of these will give rise to a complete embryo.

The submembrane cortical layer of the egg cytoplasm, known as the plasmagel, plays an important role in the development of one of the ctenophores. At the fourth cleavage, as shown in Fig. 13–5, eight macromeres and eight micromeres are produced. The latter have been shown by direct observation to be formed from materials originally localized in the plasmagel region of the fertilized egg. If the embryo is dissected into two unequal parts at the 16-cell stage, one containing five and the other three micromeres, each part develops into a larva with as many rows of swimming plates as it has micromeres. So the micromere components, derived originally from the plasmagel region of the egg, determine the eventual formation of the swimming plate structures.

We see, therefore, that localized cytoplasmic structures are an important factor in determining development.

13–3
Interactions
between nucleus
and cytoplasm

Studies on dissected embryos, such as those on the sea urchin shown in Fig. 13–2, reveal that early cleavage nuclei have the same capacity as the original nucleus of the fertilized egg to develop and produce adult organisms, provided that the specific cytoplasmic components are also divided equally.

An experiment which provides further proof of this was first carried out by Spemann on fertilized newt eggs. At the two-cell stage, the egg is constricted along the cleavage furrow so that both nuclei are contained in one half (Fig. 13–6). Only a small cytoplasmic bridge connects this half to the second enucleated half. The nucleated part continues to cleave normally, and the other part does not. Finally, often as late as

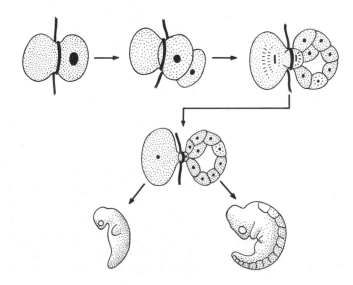

Fig. 13–6 Spemann's experiment on fertilized newt eggs. The egg is constricted along the cleavage furrow at the two-cell stage so that both nuclei are in one half. Finally a nucleus from the developing half slips back across the narrow cytoplasmic bridge; this part then develops normally

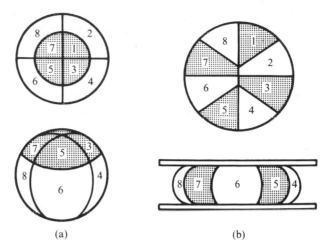

Fig. 13–7 Displacement of nuclei by compression of a developing embryo. **(a)** Normal development in un-compressed embryo. **(b)** Effect of compression between glass plates. Nuclei, not shown, come to lie in new regions of the cyto-plasm. (With permission from J. S. Huxley and G. de Beer, *The Elements of Experimental Embryology*. Cambridge University Press, London, 1934.)

(a)   (b)

the 16-cell stage, a nucleus from the developing part will pass across the narrow cytoplasmic bridge. This part will then proceed to develop normally, although there will be a time lag compared to the other embryo. So the 16-cell stage nucleus is still equivalent to the original one.

Driesch carried out an interesting experiment at the third cleavage stage of frog and sea urchin eggs. The first two cleavage planes, as mentioned earlier, are vertical, and the third horizontal. If, however, the embryo is compressed gently between glass plates after the second cleavage, then the third cleavage will take place vertically (Fig. 13 7). If the pressure is then released, the fourth cleavage is horizontal. This results in a reshuffling of nuclei (and presumably some displacement of cytoplasm) which are then located in different positions from the normal ones. In spite of this, a normal embryo finally develops, indicating that at this stage no irreversible processes concerned with subsequent development have taken place due to nuclear cytoplasmic interactions.

Much interest has centred round the problem of the changes in nuclei during development and differentiation, and how long the capacity to develop is retained in an appropriate environment. Briggs and King (1955) developed a method for carrying out such studies by transferring the nuclei of embryonic cells of the frog into enucleated egg cells. The delicate procedure is illustrated in Fig. 13-8. Nuclei taken from embryonic cells up to the late blastula stage can be transplanted undamaged, and still support normal development. Similar results were obtained by Fischberg and Blackler with toad embryos. However, when nuclei from older embryos—for instance, from endodermal cells—are transplanted, not all of the nuclear transplants develop normally.

Gurdon and Uehlinger (1966) described the transfer of nuclei from fully differentiated intestinal cells of adult frogs. A small proportion of the nuclear transplant embryos were able to develop into adult frogs, and some of these were fertile.

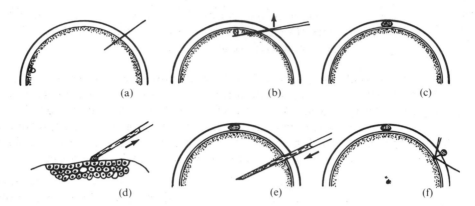

(a)   (b)   (c)

(d)   (e)   (f)

Fig. 13-8 Method for nuclear transplantation (*Rana pipiens*). **(a–c)** Egg is activated and enucleated with a clean glass needle. **(d)** Donor cell is removed from an intact blastula and drawn up into a micropipette. The pipette is smaller than the cell so the cell is broken. **(e)** The free nucleus is injected into the enucleated egg. **(f)** On withdrawing the pipette, a small canal to the exterior forms, which must be severed to prevent leakage. (From R. Briggs and T. J. King, in G. Butler (ed.), *Biological Specificity and Growth.* Princeton University Press, 1955. Reprinted by permission of Princeton University Press.)

This very important result shows that the activation or repression of genes which takes place during differentiation cannot involve irreversible loss or inactivation, although the capacity of nuclei to promote normal development (i.e. their potency) appears to decrease during differentiation. This is considered by some workers to be due to damage caused by the experimental technique. Probably nuclei from older cells are more sensitive to this type of damage.

Fig. 13–9 Fusion of a nucleated chicken erythrocyte with a cultured human tumour cell (HeLa). Inactivated Sendai virus is used to effect fusion. **(a)** Before fusion. **(b)** After fusion; erythrocyte nucleus is condensed. **(c)** Later, erythrocyte nucleus has become diffuse. DNA and RNA synthesis have begun.

Interactions between the nucleus and cytoplasm of two different types of adult differentiated cells, resulting in gene activation, have been described by Harris (1965). He has fused two or more cells by means of viruses to produce a hybrid cell (Fig. 14–11). One of the systems which he has used is the fusion of human carcinoma cells (HeLa) with chicken erythrocytes. Avian erythrocytes, unlike mammalian ones, still possess a nucleus although it is no longer functional. After fusion with the human tumour cell, however, the erythrocyte nucleus begins to resume DNA and RNA synthesis (Fig. 13–9).

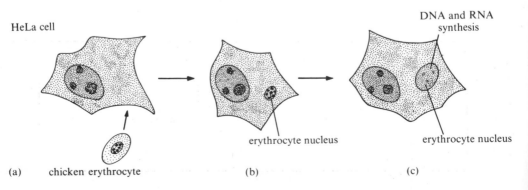

HeLa cell

DNA and RNA synthesis

(a)   chicken erythrocyte   (b)   erythrocyte nucleus   (c)   erythrocyte nucleus

Consequently the cytoplasm of the human cell must have interacted in some way with the nucleus of the chicken cell and stimulated gene activity.

13–4
Biochemical
changes during
early
development

We now consider the more important biochemical changes in the egg immediately after fertilization, before discussing the various control mechanisms which play a part in development and differentiation. Studies on synthesis during early development have been chiefly on amphibia and sea urchins. Notable workers in this field have been Brachet and Monroy.

DNA synthesis. The extremely active state of the oöcyte nucleus and the appearance of the lateral loops of amphibian lampbrush chromosomes have been described in Chapter 3. The lampbrush chromosomes contain about 16 times as much DNA as the sperm of the same species. There is also DNA present in the nuclear sap, and large amounts of DNA are stored in the cytoplasm. The latter represents reserve material which will be used for the replication of nuclear DNA after fertilization. Some DNA-containing particles fuse and accumulate in the cortex of the eggs. (Brachet, 1965). During oöcyte maturation, DNA synthesis mainly ceases in most species, although in amphibia there is still some incorporation of thymidine into the cytoplasm.

Fig. 13–10 DNA, RNA, and protein synthesis during sea urchin development

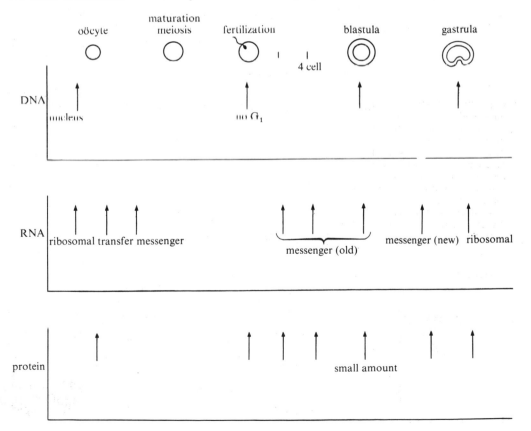

specific
with messenger
... on nuclei
only ?

At fertilization, DNA synthesis recommences rapidly. For instance in the sea urchin it starts within 15 minutes in both the egg and sperm nuclei, before they fuse. The situation is thought to be the same in amphibian eggs. This rapid synthesis of DNA, in which the G_1 phase of the mitotic cycle appears to be virtually absent, continues throughout cleavage. There is a gradual breakdown of cytoplasmic DNA, which probably provides precursors for nuclear synthesis. As shown in Fig. 13–10, the synthesis of DNA then continues through the blastula and gastrula stages.

RNA synthesis.   The growing oöcyte is the site of extensive RNA synthesis. In sea urchins and amphibians this is essentially directed towards the production of ribosomal RNA, to build up a store of ribosomes for protein synthesis after fertilization. Transfer RNA and relatively large amounts of messenger RNA are also produced.

The synthesis of RNA stops at maturation when the nuclear membrane breaks down and the nucleolus disappears. Brachet has shown that, in contrast to DNA, there is no synthesis of RNA shortly after fertilization in sea urchin eggs. This first takes place at the two to four cell stage (Fig. 13–10).

Whiteley, McCarthy, and Whiteley (1966) carried out molecular hybridization experiments to identify specific messenger RNA molecules. They found that some kinds of messenger RNA are present at all stages of development, including the unfertilized egg. This suggests that these molecules must carry messages necessary for the life of any cell—a reasonable assumption, since all cells contain a number of structures and functions in common. These studies also showed that certain messenger RNA molecules produced are specific for certain stages of development, in accordance with the idea that different regions of the genetic material are activated at different times during development. Glišin, Glišin, and Doty (1966) used the same techniques and showed that the messenger RNA produced during cleavage is identical with the messenger RNA already present in unfertilized eggs. Between the blastula and gastrula stages, however, some of this messenger RNA is replaced by a new species. Inhibitors of RNA synthesis have no effect on cleavage, but strongly inhibit gastrulation, as would be expected.

Protein synthesis.   In the oöcyte, protein synthesis occurs in both the cytoplasm and the nucleus, and unfertilized eggs already contain all of the important respiratory and hydrolytic enzymes. Monroy (1960) observed that protein synthesis is negligible after maturation in sea urchin eggs. Fertilization is followed by a marked stimulation of synthesis (Fig. 13–10). Control of protein synthesis cannot be due to the production of messenger RNA at this stage, for the reasons mentioned above, and other experiments suggest that control is at the ribosomal (translational) level.

Ribosomes are synthesized in sea urchin and amphibian oöcytes and are present in large amounts in the unfertilized eggs. In sea urchin eggs they exist in a repressed state (Monroy, Maggio, and Rinaldi, 1965), probably due to combination with a protein molecule, since they become active after trypsin treatment. Fertilization in sea urchin eggs is followed by the activation of a proteolytic enzyme; this may play a part in facilitating ribosomal function.

One would expect that inhibitors of protein synthesis would effectively stop cleavage and this is in fact so in both sea urchin and amphibian eggs. A little later the production of DNA is also inhibited, presumably due to lack of production of the necessary enzymes.

Protein synthesis continues to increase in the sea urchin egg until the end of cleavage. There is a period of low synthesis during blastula formation, then this is followed by intense synthesis until gastrulation. Little is known about the kinds of proteins synthesized during early development. Evidence indicates that, not only the enzymes required for DNA and RNA synthesis, but also the respiratory enzymes, are produced at this stage, probably all of the enzymes necessary for basic metabolism are in fact produced. At a later stage in development comes the synthesis of specific proteins in specialized cells, such as haemoglobin in erythroblasts, and actin and myosin in embryonic muscle cells. Clayton has shown that at the blastula stage of a newt embryo there are six detectable types of antigen present. Tissue-specific antigens also appear at a later stage.

Membrane composition. Little is known about the biochemistry of membranes in development, owing to the immense difficulties, mentioned in Chapter 8, encountered in the isolation of pure membrane fractions. However, there is some evidence of biochemical changes in the lipid as well as the protein composition of endoplasmic reticulum membranes (Palade and coworkers, 1966), which suggests that the structure and function may change during development.

Oxygen consumption. The level of oxidation, and possibly also of energy production, is unusually low in unfertilized sea urchin eggs. This is probably due to a limited amount of oxidizable substrate, rather than to the presence of a respiratory inhibitor. At fertilization, however, there is a rapid and dramatic increase in oxygen consumption. The NADP content of the egg increases within a few seconds. In frog eggs, however, oxygen consumption does not change after fertilization. It remains low during cleavage, which is able to

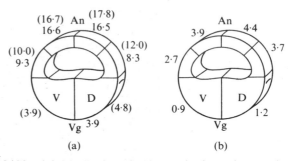

Fig. 13–11 **(a)** Distribution of glycogen in the early gastrula and, in parentheses, in the blastula. (After Heatley and Lindahl.) **(b)** Oxygen consumption in different regions of a young amphibian gastrula. (After Sze.) An animal pole; Vg vegetal pole; V ventral region; D dorsal region

proceed anaerobically, but later the uptake increases progressively. From the stage of gastrulation onwards, oxygen is essential.

Measurements have been made of glycogen distribution in the late blastula and in the gastrula stages of amphibian embryos. There is a decrease in content from the animal to the vegetative pole (Fig. 13–11). This has also been found in various other species. Similar studies have been carried out on regional oxygen consumption in amphibia. The results (Fig. 13–11) show that oxygen uptake corresponds to glycogen distribution, indicating that glycolysis is taking place aerobically. An increase in oxygen consumption in the embryo is accompanied by a corresponding increase in the levels of respiratory enzymes, particularly the cytochrome oxidases.

13–5 Nuclear-cytoplasmic feedback

The situation in cell differentiation in terms of gene function may be stated quite simply. Each cell in an organism is derived by cell division from the fertilized egg or zygote, which contains the full diploid chromosome complement. As seen at metaphase, the somatic cells of the adult also contain the diploid chromosome number. Although the genetic material contains many thousands of genes, molecular hybridization studies indicate that only 5 per cent to 10 per cent of these genes are functional at any one time in an adult differentiated cell and that different genes are active in different cell types.

From the work described earlier in this chapter it is clear that nuclear–cytoplasmic control mechanisms are important in differentiation. The work of Bonner and Huang, Paul and others, as described in Chapter 3, indicates that histones play a role in gene regulation, and that the functioning of DNA as a template for RNA synthesis in vitro is inhibited by the presence of histones. In Fig. 13–12 is shown a model made by Johns of the super-helix formed between DNA and histones according to Wilkins and other workers. It also shows a

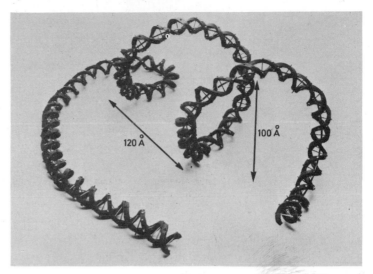

Fig. 13–12 *Upper:* model of the DNA double helix showing the approximate dimensions of the super-helix which is thought to be formed in the presence of histone. *Lower:* model of DNA, and RNA polymerase to the same scale (1 mm = 1 Å) showing how the superhelix would produce steric problems for the migration of the polymerase along the chain. (With permission from E. W. Johns.)

model of DNA and, to scale, the RNA polymerase molecule which travels along the DNA during RNA synthesis. The model suggests that the super-helix of DNA and histone may make it sterically impossible for the large polymerase molecule to move along the double helix. If histones were to combine selectively with genes, a great many highly specific histone molecules would be required. However, as mentioned in Chapter 3, the composition of histones is remarkably uniform in plant and animal cells, and it is thought that they must react with DNA in a non-specific way, while the acidic proteins of the nucleus perhaps provide the necessary specificity.

The basic biochemistry of all cells is strikingly similar, so we might reasonably suppose that the number of functional genes required for these processes must also be similar in all cells. It is surprising, therefore, that mammalian cells may contain up to 1,000 times as much DNA as bacteria. This is much too large a difference to be explained by the production of special proteins, such as haemoglobin in red blood cells, γ-globulins in plasma cells, and melanin in pigment cells. However, Britten and Waring in 1964 showed that the DNA of mice does not consist of a large number of different genes in single copies but that about one-third of it is formed of families of repeated DNA sequences which are several hundred nucleotides in length. This repetitious DNA occurs in other higher organisms, although the frequency of repetition, which varies from 100 to 1,000,000 times, and also the precision of repetition, varies greatly in different species.

How can we reconcile the presence of these repeated DNA sequences with the 'one gene, one enzyme' concept? Walker (1969) suggested that a particular kind of repetitious DNA, the so-called 'satellite' DNA of mice, may function in controlling chromosome recognition and function. Britten and Davidson (1969) further proposed that repeated DNA sequences in higher organisms may in fact be the chief mechanism by which genes are switched on or off. They postulated the existence of:

Producer genes, which produce proteins.
Receptor genes, which control producer genes and are sequences of DNA linked to them.
Activator RNA (produced by an integrator gene), which fits the receptor gene and activates it.
Sensor genes, which can bind with a messenger such as a hormone, and then activate the integrator gene or genes, so inducing the process.

The basic concept of this system is clearly derived from the Jacob and Monod theory of gene control in bacteria. The process would mean that a considerable proportion of the genes in Metazoa may not be concerned with protein synthesis, but with the coordinated control of gene function. This is an attractive idea, since we know that the interaction of genes is necessary for organ development. For instance, Waddington showed in 1940 that some 30 to 40 genes can produce abnormalities in the final wing structure of Drosophila. So all of these genes must be involved at various stages of development of the wing; that is to say, cooperative interaction between these genes must have occurred. Perhaps it is at this level that repetitious sequences of DNA may play some role. The idea that nuclear RNA may be involved in gene–gene communication is also an attractive one, since a considerable fraction of RNA is known to remain in the nucleus and never reach the cytoplasm.

Thus ways in which gene expression may be controlled in mammalian cells are beginning to emerge, and it is clear that cytoplasmic components are very important in these processes. It is of interest now to reconsider the nuclear transplants of Gurdon and Harris' heterokaryons from the point of view of control mechanisms. In both cases it was found that rapid DNA synthesis occurred in a previously inactive nucleus placed in foreign cytoplasm, accompanied by pronounced nuclear swelling. In both cases also the synthesis of RNA was induced by components in the cytoplasm. That these components were shown in both studies not to be species specific is interesting because of the similarity of histone composition in different species. Gurdon (1968) concluded that, when a nucleus is transferred to the cytoplasm of a foreign cell, it quickly assumes, in almost every respect, the nuclear activity characteristic of the host cell. Also, Harris found that, when he hybridized hen erythrocytes with rabbit macrophages (which normally synthesize RNA but not DNA), then the hen nuclei did not synthesize DNA, but only RNA. Since the nuclear transplants into unfertilized oöcytes resulted in normal embryos, Gurdon regarded the changes induced by the cytoplasm as normal, and probably the same as those in normal development.

Bell, in 1969, obtained evidence that in the cytoplasm of embryonic muscle cells there is a new class of small DNA molecules, which he called informational DNA, or *I*-DNA, present in particles which he called *I*-somes. The DNA was linear, and Bell suggested that it consisted of segments of nuclear DNA which had been transported to the cytoplasm. It was not known if this occurred only in cells which were making many copies of a restricted range of proteins (in this case muscle proteins). Later experiments by other workers, however, indicate that this cytoplasmic DNA may be an artefact produced by the method used for cell disruption.

The timing of the various control mechanisms which occur inside cells must be very precise, so that they operate at just the right stage. For instance, there is a considerable shutdown in protein synthesis in the mature unfertilized egg, which may be considered to be in a steady state, as it contains all of the components required in the early stages of development. Bell and MacKintosh suggested in 1967 that protein synthesis may be partially repressed by the regulated limitation of energy, and that repression may be maintained by the enzymic production and intracellular transport of carbon dioxide. With fertilization the energy restriction is lifted, and a rapid increase in protein synthesis begins.

Another example occurs in a mutant form of *Xenopus*, studied by Birnstiel, which is homozygous for the absence of a nucleolar organizer. Development of the fertilized egg takes place up to the stage at which the ribosomes present in the mature oöcyte are no longer sufficient to support the necessary protein synthesis, and, since no new ribosomal components can be synthesized, the embryo dies at this stage.

In the nuclear transplant studies, Diberadino and King have suggested that normal development may not occur in certain cases because the nucleus is not in the appropriate premitotic stage to enter into the cleavage cycle of the host egg. In the cellular hybridization experiments, Harris has found that it is only when nuclei are in synchronous mitosis that cell division can sometimes occur.

After reading Chapter 12 it is clear that intracellular control mechanisms alone cannot account for all the processes involved in development. The role of the blastopore lip region and the influence of various parts of the mesoderm in affecting head and tail differentiation of neural tissue indicate that many cells are pluripotent and that their differentiation is affected by the environment.

13–6 Intercellular control mechanisms

We now consider the interactions between cells which may play a part in controlling the various developmental processes; that is, in the inductive processes which lead to cell differentiation (here defined as the appearance of new tissue types), in morphogenesis, and finally in organogenesis.

It must follow, from the examples given in the last paragraph of the previous section, that there are components produced by cells which are capable of acting on other cells and initiating gene activation or suppression. However, as Grobstein (1967) states, no one has yet succeeded in isolating and chemically identifying any agent responsible for embryonic induction. So most of the work in recent years has not been concerned with early inductive events leading to differentiation, but rather with the effects of one tissue on the course of development of a second tissue. Studies have centred on the nature and the control of synthesis of intercellular materials deposited between interacting tissues undergoing development.

Kallman and Grobstein observed in 1965 that, when embryonic epithelium is separated from embryonic mesenchyme by a Millipore filter, soluble collagen synthesized by the mesenchyme interacts with polysaccharides produced by the epithelium, resulting in the deposition of collagen fibres at the surface of the epithelium. The presence of the collagen was found to facilitate the development of the characteristic branching of the epithelium. It was later established that the epithelium had a promoting effect on the synthesis of collagen by the mesenchyme.

Such studies probably yield information on the stabilization or enhancement of pre-existing patterns, rather than the acquisition of a new pattern, since Grobstein later found that embryo extract showed the same stimulating effect on the development of epithelium.

It has been suggested that molecules involved in the formation of tissues and tissue shapes must operate over fields of cells, and that there exists a gradient of the substance. Crick

(1970) has proposed that these molecules must be specific, since they should presumably not pass easily through all cell membranes, but only through those of the tissue of interest. This movement may be by means of the specialized regions of contact between adjacent cell membranes, known as tight junctions, which electron microscopists have found between cells in various tissues, or by some type of active transport mechanism. Crick agrees with Wolpert and others that embryonic fields seem to involve distances of less than 100, and sometimes less than 50, cell diameters.

Loewenstein showed in 1966 that tight junctions form diffusion pathways, as these specialized membrane structures have a low transmembrane resistance. Along such pathways substances of up to about 3,000 in molecular weight can diffuse freely. Subak-Sharpe has shown that two cell types in vitro can act in metabolic cooperation by the passage of a substance from one cell type to another, which occurs only when the cells form contacts with each other.

Grobstein (1964) has found that there is a mass effect involved in interactions between cells concerned with tissue formation. If a mouse embryonic pancreatic rudiment is grown in vitro in explant culture it will, under appropriate conditions, form functional pancreatic tissue. If, however, the rudiment is cut into eight small pieces, which are then cultivated separately, pancreatic tissue is not formed. Tissue development can be restored by growing the small fragments close together so that they fuse to form once more a single mass. Thus it appears that cooperative activity between a large number of cells is required for this process, and that small groups of cells in isolation cannot achieve the necessary condition. It is known from the work of Lash and Ellison that once cells are induced, or potentiated, to proceed along a particular pathway of differentiation, then the presence of like cells stimulates and stabilizes the process.

It should be borne in mind that in any in vitro experiments such as those described above, cells and tissues are taken out of their normal environment, and probably experience a slowing down in the functioning of a metabolic pathway which they had previously acquired. It may be that most experimental systems in vitro, which have suffered disturbances of the metabolic functions associated with development, require the presence of additional factors to restore normal functional development. Such factors may be termed 'inducers' but whether they exist or operate in the intact organism is problematical.

13–7
Cell movements
and adhesions

Gastrulation is achieved by morphogenetic movements—that is, by changes in the position and arrangement of cells. At a later stage in development similar processes must determine how sheets of epithelial cells give rise to various organ rudiments.

Gustafson and Wolpert (1962) studied by means of time-lapse cinematography the cellular mechanisms involved in the morphogenesis of sea urchin larvae, in particular in the basic processes such as changes in thickness of the blastula wall, elongation of the blastula, and attachment of cells to the hyaline layer. Before the interpretation of their work is discussed, we must consider tissue culture studies of cell movements and adhesions.

Abercrombie and Heaysman, and Abercrombie, Ambrose, Easty and Heaysman have studied the movements of both fibroblasts and epithelial cells in culture. In each case it was found that the movement of a single cell is controlled by the movement of other like cells with which it makes contact. This is illustrated in Fig. 13–13. The upper fibroblast is at first moving downwards, the process being characterized by the formation of pseudopodia on the leading edge. When the upper fibroblast touches the lower one, an adhesion is formed between the two membranes. The two cells may try to move in different directions, resulting in considerable tension in the contact, and sometimes in breakage, resulting again in freedom of movement of the two cells. As mitosis occurs in the culture, and the cell density increases, an increasing number of contacts between cells will be formed. This leads to stabilization of the culture by contacts, and the cessation of the freedom of movement of individual cells. Abercrombie has called this process *contact inhibition* of movement. A similar type of control occurs in epithelial cells. Flattened sheets of cells are formed in which individual cells are attached extremely firmly to each other by continuous contacts all round their borders.

Other studies relevant to morphogenesis concern adhesions which are formed between cells and the substrate on which they are moving. Experiments have been carried out using such surfaces as glass, plastic, collagen, and agar gel for tissue culture purposes, and considerable differences in the strength of adhesion of a given cell type to different substrates

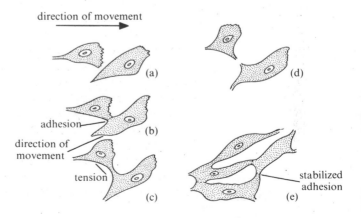

Fig. 13–13 Successive stages in the making and breaking of a contact between fibroblasts; taken directly from a time-lapse ciné film. (With permission from M. Abercrombie and E. J. Ambrose.)

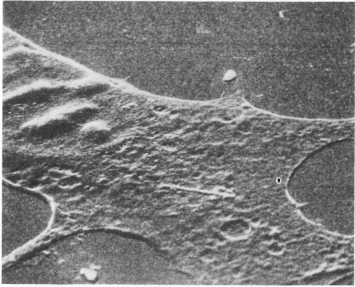

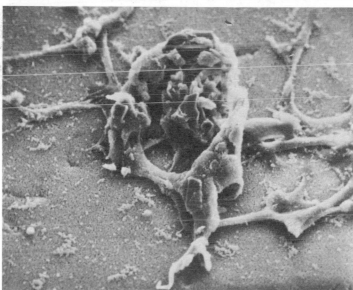

Fig. 13–14 (a) Stereo scan electron micrograph of a fibroblast similar to those in Fig. 13-13 (a) as seen when growing on a glass surface. Magnification ×1,500. (b) The same when grown on a Millipore (uneven) surface. Cells are associating to form a compact three-dimensional colony. Magnification ×300. (With permission from E. J. Ambrose and M. Ellison.)

have been found. The substrate is also of importance in controlling the direction of movement. This was first observed by Weiss in 1926; he found that, when fibroblasts are cultured on glass which has fine surface grooves (or on fish scales, which are also grooved), the cells tend to elongate and move along the grooves. This control of direction is known as *contact guidance*. In monolayer cultures of normal diploid fibroblasts it is found that the rate of mitosis decreases when the cell density increases. In a confluent layer, mitosis comes to a halt. This is not due to a deficiency in the medium; it seems to be due partly to cell–cell contacts and partly to the secretion of an inhibitor (Stoker, Clarke, and others). This phenomenon is called *density dependence of mitosis*.

Assumption: No action at a distance

Alternative explanation than material

Question: Could be adhesion; did they find the substance?

As a result of their work on sea urchin embryos, Gustafson and Wolpert suggested that the shapes of cells and tissues are largely determined by the relative strengths of cell–cell and cell–substrate adhesions and tensions. For instance, if a line of cells in contact with each other were also in contact with a membrane surface, than an increase in cell–cell adhesion with no change in cell–membrane adhesion would mean that the sheet would become concave, with the membrane on the outside. Gustafson and Wolpert found that the main changes in the external form of the larva from the blastula stage up to the time of extension of the arms could be explained in such terms. Rearrangements of ectoderm resulted in associated spaces, due to localized thickening, which caused stretching in adjacent regions, with poor cell–cell but normal cell–membrane adhesion. These spaces were frequently explored and closed by short pseudopodia from more active ectodermal cells. The authors concluded that a scale of variation in adhesive properties and in pseudopodal activity is of fundamental importance in early morphogenesis.

An interesting observation on the effect of the nature of the substrate was made by Wessels in 1964. He cultured chick embryo epidermis on a sheet of perforated Millipore, a synthetic permeable membrane. He found that restoration of DNA synthesis and orientation of the basal cells could be restored by embryo extract, but only where the basal layer was in contact with the Millipore—over the holes of the filter no DNA synthesis and no cell orientation took place. Ambrose found that fibroblasts grown on Millipore do not form a sheet as they do on a smooth glass plate (Fig. 13–14), but grow as a

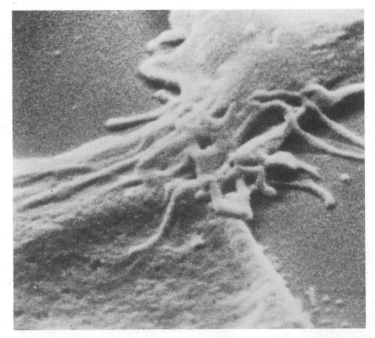

Fig. 13–15 Stereoscan electron micrograph of two fibroblasts making contact, showing association between microvilli Magnification ×15,000

epidermis + endoderm   mesoderm + endoderm   epidermis + mesoderm   epidermis + mesoderm
+ endoderm

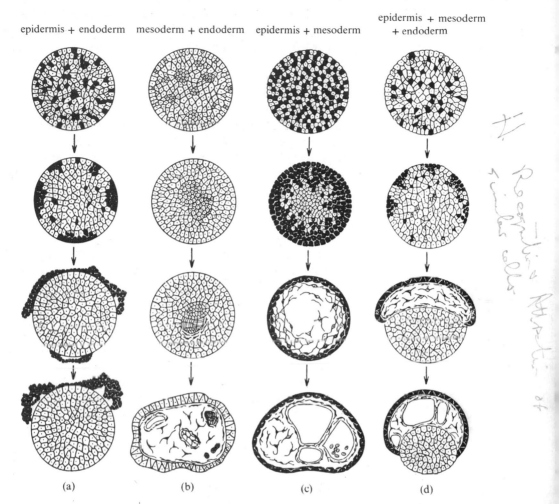

(a)   (b)   (c)   (d)

Fig. 13–16   Effect on re-aggregation of cells of an embryo separated into epidermal, mesodermal, and endodermal cells. **(a)** Combination of epidermal and endodermal cells leads to a sorting out and self isolation of homologous cells (epidermis dark, endoderm light). **(b)** Mesoderm and endoderm in combination. Mesoderm cells come to lie in centre surrounded by endoderm. **(c)** Epidermis and mesoderm in combination. Epidermis lies on outside; mesoderm eventually forms mesenchyme, coelomic cavity and blood cells. **(d)** Combination of epidermal, mesodermal and endodermal cells; cell segregation and formation of tissues. (After Townes and Holtfreter, with permission from L. Saxén and S. Toivonen, *Primary Embryonic Induction*. Logos Press, London, 1962.)

dense spherical mass of closely adhering cells, due to their inability to form strong attachments with the Millipore.

It was suggested by Bangham and Pethica (1960) that cells may adhere initially by fine projections on their surfaces, which may be the microvilli described in Chapter 8. These are the first regions to make contact between cells, as shown in the Stereoscan electron micrograph of Fig. 13–15. Bangham and Pethica pointed out that it would be easier for a fine projection to overcome the small repulsive force that it would experience

Is Time built into DNA message so different proteins are manufactured in sequence?

Fig. 13–17 (a) A confluent monolayer of fibroblasts in tissue culture showing orientation and close packing of cells. **(b)** The same after transformation to malignant cells by polyoma virus. Note overlapping of cell processes and lack of orientation due to lack of contact inhibition

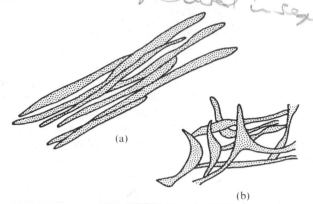

(a)

(b)

than for a large region of the cell surface to overcome a much larger forces. The formation of adhesions between microvilli may be followed by stable adhesions between much larger areas of cell membrane.

Interesting information about cell–cell adhesions has been obtained from the studies of Holtfreter, Weiss, Moscona, and others on the reaggregation of cells separated from various tissues. In Fig. 13–16 are shown the results of experiments by Townes and Holtfreter on the reaggregation of cells from germ layers of ectoderm, mesoderm, and endoderm in various combinations of two of the cell types and finally from all three types together. The diagram indicates the various types of structure obtained from the different combinations. The most striking observation is that cells first form adhesions rather indiscriminately, and contacts are made and remade until like cells eventually form contacts, which then are stable and permanent. Steinberg suggested that quantitative differences in cell adhesiveness are responsible for the sorting out. Curtis (1960) has proposed that cells are at first freely motile, but that any given cell type acquires the ability to form stable adhesions with other cells during the segregation period, and that different cell types develop this property at different times, resulting in sequential reaggregation.

It is interesting that controlled cell death is known in certain cases to play a part in morphogenesis. Saunders showed that the shaping of the chick wing, in particular the upper arm and forearm, and the removal of tissues between the digits, is partly achieved by the occurrence of localized zones of cell death. The factors involved in processes of this type are still obscure, but in some cases, such as the regression of the tadpole tail (Chapter 5), release of lysosomal enzymes is thought to play a part.

After tissue and organ formation have taken place, there must be powerful controlling mechanisms which control and stabilize their structure and function. Little, however, is known about these, except in a few cases. Hormones are known to affect the synthetic function and the size of certain organs.

the mechanism differentiation of supporting Time built into DNA

This is an example of what Weiss has called 'modulation'—that is, the change of behaviour of differentiated cells when placed in a new environment, this change being reversed when the cells are replaced in their original environment. Various examples of this have been observed in vitro; for instance, Coon and Cahn (1966) showed that pigment epithelial cells will replicate and form melanin if grown in low-molecular-weight components of embryo extract, but if they are transferred to a medium containing high-molecular-weight components of embryo extract they continue to replicate, but do not form melanin.

It has been suggested by Szent-Györgyi (1967) that there may be a growth-promoting substance (promine) and a growth-inhibiting substance (retine) present in certain tissue extracts. Variations in the relative amounts of these could then account for the growth which takes place in, for instance, regenerating liver. In mammals the greater part of the liver may be removed, and regeneration will take place until the organ has reached the approximate size of the original liver.

The process of limb or tail regeneration in amphibians is different in that it appears to involve de-differentiation of muscle, connective tissue, cartilage, and bone cells, with the emergence of cells which resemble primitive embryonic mesenchyme. Later, growth, differentiation, and morphogenesis take place resulting in the formation of a new limb or tail.

Tumour cells are cells which are no longer completely bound by the processes controlling the growth, function, and movement of normal cells. There is a superficial resemblance between some malignant cells and embryonic cells, but whether de-differentiation is involved is not known. Both contact inhibition of cell movements and density dependence of mitosis are generally deficient to varying degrees in different tumour cell types in monolayer culture (Fig. 13–17). The cells show reduced cell–cell adhesion and changes in surface properties (Section 8–7). This indicates that loss of control in malignant growth is associated with failure of surface interaction with other cells.

13–8
Summary of factors in cell differentiation

From the foregoing, we draw certain conclusions about cell differentiation during development.

1. In the oöcyte, active synthesis takes place in both nucleus and cytoplasm. This is probably connected with the synthesis of ribosomes and basic enzymes necessary for life, and as a store for future use. The nucleus is active in controlling synthesis.

2. Fertilization and early development. The nucleus plays mainly a passive role in synthesis but the chromosomes are replicated at each cleavage. The cytoplasm undergoes segregation into localized regions particularly in the region of the plasmagel and plasma membrane.

3. As a result of this, daughter cells appear with differing cytoplasm. This factor is probably sufficient to account for the early stages of cell differentiation until gastrulation.

4. These different types of cytoplasm begin to react on the nucleus at about the time of gastrulation.

5. Cells secrete and interact on contact and pass messages via the cytoplasm to the nucleus (embryonic induction).

6. A delicate interplay begins to operate, in a definite time sequence, of informational messages to the cytoplasm. At the same time changes in the cellular environment (chemical and physical interactions on the cell membrane), take place, which act eventually on the sensory genes within the nucleus.

7. Within a given type of cell population interactions occur which stabilize a pathway of cell differentiation once this has been initiated.

The subjects covered in Chapters 12 and 13 have, up to the present, been mainly studied by embryologists and by cell biologists with an interest in development. But future progress in this field is likely to be of great practical importance. It is in the field of intracellular and cellular-environmental control mechanisms that such major problems concerning human welfare as cancer, genetic abnormalities, abnormal development, and diseases of senescence will finally be understood and, we believe, brought under control.

Bibliography

Balinsky, B. I., *An Introduction to Embryology*. W. B. Saunders Company, Philadephia, 1963.

Briggs, R. and T. J. King, Nucleocytoplasmic interactions in eggs and embryos, in J. Brachet and A. E. Mirsky (eds), *The Cell*, vol 1. Academic Press, New York, 1959.

Cell Differentiation, Ciba Foundation Symposium. J. and A. Churchill, London, 1967.

Ebert, J. D., *Interacting Systems in Development*. Holt, Rinehart, and Winston, New York, 1965.

Fischberg, M. and A. W. Blackler, How cells specialize, in *The Living Cell*. W. H. Freeman and Company, San Francisco, 1961.

Grobstein, C., Differentiation of vertebrate cells, in J. Brachet and A. E. Mirsky (eds), *The Cell* vol 1. Academic Press, New York, 1959.

Gurdon, J. B. and H. R. Woodland, The cytoplasmic control of nuclear activity in animal development. *Biol. Revs.* **43**, 233, 1968.

Gustafson, T. and L. Wolpert, Cellular mechanisms in the morphogenesis of the sea urchin larva, change in cell sheets. *Exp. Cell Res.*, **27**, 260, 1962.

Moscona, A. A., How cells associate, in *The Living Cell*. W. H. Freeman and Company, San Francisco, 1961.

Waddington, C. H., *The Strategy of the Genes*. George Allen and Unwin, London, 1957.

PART 5 | Early and simple forms of life

In this final part we consider the formation of biological structures from biochemical building units, and discuss the origin of the latter.

In Chapter 14 we begin with the structure and morphology of viruses. Although they are relatively simple life forms, and may be composed of only a few biochemical sub-units, virus particles can assume quite complex shapes. Then we consider structures in bacteria, which are some of the earliest forms of life of which records still remain on the earth.

Finally, in Chapter 15, we deal with chemical evolution, which led to the formation of the complex molecules of biochemistry from much simpler molecules, and eventually the first living cells. Some process of biochemical evolution must have preceded biological evolution, and it is fascinating to review the evidence available as to the processes which may have taken place many millions of years ago.

14 Structure and organization of viruses and bacteria

14–1
Viruses Viruses cannot be classified as cells, because they have no nucleus, cytoplasm, or, except in myxoviruses, a limiting plasma membrane and cannot proliferate outside a living cell. But they are the simplest particles to possess the fundamental properties of living systems—that is, their structure and function are determined by their genetic material, and they can produce copies of themselves (by infecting suitable host cells and using the host's raw materials and metabolic machinery for their own reproduction).

Towards the end of the 19th century it was accepted that many diseases are caused by micro-organisms. In 1892, Iwanowsky, in a study of mosaic disease in tobacco plants, discovered that juice extracted from affected plants contained the infectious agent. Assuming that the agent was a bacterium, he tried to remove it by filtration but was unsuccessful even with the finest bacterial filters. Iwanowsky concluded that the agent must be smaller than any other type of bacterium then known. Other plant and animal diseases were later found to be caused by infectious agents which could pass through very fine filters and were not visible in even the highest-resolution light microscopes. The agents came to be known as *filterable viruses* or simply *viruses*, and were still considered to be very small bacteria. They could not, however, be grown on the usual media used for culturing bacteria and, unlike bacteria, they retained their infective power after precipitation from alcoholic solution. It was not until 1935 that Stanley managed to crystallize the virus that causes tobacco mosaic disease, thus proving that it could not be cellular and must differ greatly from a bacterium. The virus crystals still retained their infective capacity and caused symptoms of mosaic disease when injected into plants.

It was thought at this stage that viruses are composed entirely of protein, but we now know that they consist essentially of a *nucleic acid core* surrounded by a protein coat. In present terminology, an intact virus unit is called the *virion*, and its protein coat is known as the *capsid*; the latter is composed of a number of sub-units of a particular shape, known as capsomeres (Fig. 14–1). The capsid serves to protect the nucleic acid core against attack by nuclease enzymes.

Fig. 14–1   The generalized features of virus (virion) structure including the outer envelope; the capsid is shown in this case as having cubic symmetry

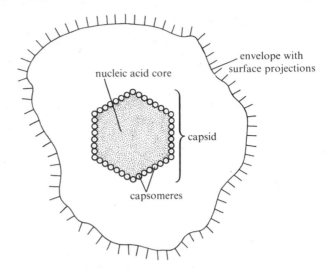

nucleic acid core

envelope with surface projections

capsid

capsomeres

Fig. 14–2   Relative sizes of viruses referred to 1μ scale at base of diagram. Types of symmetry shown are: *cubic,* shown by the five polyhedral viruses, herpes, tipula iridescent, adenovirus, polyoma, and poliomyelitis; *helical,* or screw axis, shown by tobacco mosaic and the internal components of the mumps and influenza viruses; *complex symmetry* shown by vaccinia, orf, and T-even bacteriophage. (From R. W. Horne, The structure of viruses, Copyright © 1963 by Scientific American, Inc. All rights reserved.)

Viruses can infect bacteria (bacteriophages, Section 10–7), plants, and animals. In viruses which attack plants the viral nucleic acid is RNA, but in most bacteriophages and animal viruses it is DNA (exceptions are poliomyelitis and influenza viruses, which contain RNA).

Most viruses are between 100 Å and 3,000 Å in size, the largest being the size of small bacteria. Many of them have now been examined by means of the electron microscope. Figure 14–2 indicates that the shapes as well as the sizes of viruses vary enormously. The structure of plant and animal viruses is usually simpler than that of bacteriophages; the former are rod-shaped (such as tobacco mosaic) or polyhedral (such as adenovirus) and lack the tail of the bacteriophage (Fig. 10–19), which means that they cannot inject their nucleic

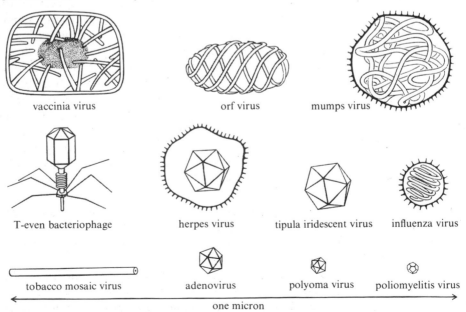

vaccinia virus

orf virus

mumps virus

T-even bacteriophage

herpes virus

tipula iridescent virus

influenza virus

tobacco mosaic virus

adenovirus

polyoma virus

poliomyelitis virus

one micron

acid core into the cells which they infect. Some viruses, such as the influenza and mumps viruses, which are members of the group of myxoviruses (Figs 14–1 and 14–2), possess an outer membrane known as the *envelope* which consists of lipid and protein and surrounds the capsid. Herpes also sometimes has an envelope. It has been reported (1969) that there is a plant virus, the potato spindle tuber virus, which consists only of RNA, without any protein coat; although it would not be the first virus studied which cannot make its own coat protein (the Rous virus, which is an avian tumour-producing virus, cannot do this), it would be the first observed to be transmissible and infectious without it.

14–2
Action of viruses

The specificity of viruses appears to vary; some may attack only one specific cell type, while others may attack a number of different cells. They all possess the property, however, that they cannot be grown on non-living material. Viruses grown in the laboratory (for production of vaccines, for example, and for research purposes) are usually cultured in fertilized hen's eggs or on cells in tissue culture. Viruses may remain inactive for long periods in the absence of living material and still retain the ability to infect the appropriate type of cell. They have been described as genetic material in search of a cell in which to reproduce.

As has been described (Section 10–7 and Fig. 10–20), bacteriophage T4 attaches to a bacterial cell wall, digests a small hole by means of enzymes in its tail, injects its DNA core into the cell and then makes copies of itself inside the host cell. We do not yet know exactly how other viruses, of simpler construction than phages, enter the plant and animal cells they infect. From the late 1940s onwards, electron microscopic studies, with very few exceptions, supported the hypothesis that intact viruses enter cells by means of a phagocytic process. However, Councilman Morgan (1968) showed that with herpes virus, which possesses an envelope, phagocytosis occurs only rarely. The usual method of penetration appears to be fusion of the viral envelope to the cell membrane, followed by digestion of the membrane at that point and passage of the capsid and core into the cytoplasm. Finally the capsid is removed and the nucleic acid core released. Morgan obtained similar results with influenza virus. The envelopes of this type of virus have regularly spaced surface projections (Figs 14–2 and 14–3) which are probably connected with attachment and entry, possibly by enzymic action on some structure present on the host cell surface.

We do not yet know whether viruses without an envelope enter the cell intact, with subsequent release of the nucleic acid core, or whether the protein coat remains at the cell surface while the nucleic acid enters the cytoplasm alone. One of the problems is to account for the extreme rapidity of

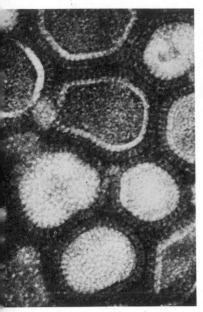

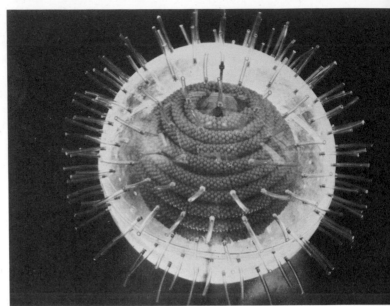

Fig. 14–3 (a) Electron micrograph of influenza virus, a member of the myxovirus family. The virus particles are irregular in size and shape, but they appear to bristle with regularly spaced surface projections. Magnification ×260,000. (With permission from R. W. Horne.) **(b)** A possible model for a typical myxovirus. (With permission from R. W. Horne.)

the process, which takes place in a matter of only a few minutes at 37°C.

Once the viral nucleic acid has entered a cell, it takes over both the raw materials and the metabolic machinery of the host, including the energy-producing systems, in order to make copies of its own molecules and of its coat proteins. Since viruses contain either DNA or RNA (but not both), this must mean that RNA is capable of carrying genetic specificity as well as DNA. Most DNA viruses, such as smallpox, polyoma, and the T-even phages, carry double-stranded helical DNA, while most of the RNA viruses, such as tobacco mosaic, influenza, and poliomyelitis, carry single-stranded RNA, as is the situation in cells. However, there are some exceptions to this; certain bacterial viruses contain single-stranded DNA, and the Reo group of RNA viruses contains double-stranded helical RNA. It is not known why the genetic material is sometimes single-stranded, but it does not appear to be important, as a complementary strand is rapidly formed after entry into the host cell.

In a *double-helical DNA virus*, the double strands of DNA come apart and replicate in the usual way; the nucleotide building units of the host cell are used to form new complementary strands, catalysed presumably by a host cell enzyme (Fig. 14–4a); the latter is necessary since viruses frequently contain no enzymatic proteins except for those sometimes involved in attaching to and penetrating the host cell. Virus-specific mRNA molecules are then made on one strand of the viral DNA in the usual way, and again the process must be catalysed by the host RNA polymerase. After this, virus-specific proteins are synthesized using host cell ribosomes and RNA. Finally the coat proteins assemble round newly

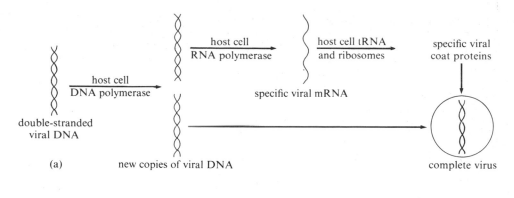

(a)   new copies of viral DNA   complete virus

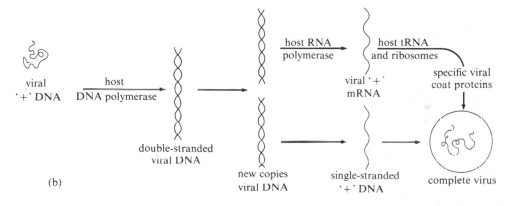

(b)

Fig. 14–4 The replication of DNA viruses inside host cells. **(a)** Virus containing double-stranded DNA. **(b)** Virus containing single-stranded DNA

synthesized DNA to form complete viral particles. In many (and possibly all) cases, no enzymes are required for the assembly of the mature virus because covalent bonds are not involved; the coat proteins are bound to the nucleic acid core by weak van der Waals forces or hydrogen bonds (see Section 14–3, tobacco mosaic virus).

With bacterial viruses containing *single stranded DNA*, the first process on entry into the host cell is formation of a strand of DNA complementary to the single viral strand, the new strand being known as the minus (−) strand and the original DNA as the plus (+) strand (Fig. 14–4b). Further copies of the double-helical form are produced, and the '+' strand is used to code in the usual way for virus-specific mRNA and coat proteins, as well as being used in the assembly of new virus particles. Some specific device is probably required to ensure that it is only the '+' strand which is used for these purposes.

In certain RNA viruses the RNA may act as a template for the formation of DNA. There is evidence for the production of a complementary DNA strand when the Rous sarcoma virus infects cells. Otherwise, the RNA appears to act as a self-replicating model. There is no evidence that either DNA synthesis on a single strand, or DNA synthesis on RNA as a template, occurs in bacterial, plant, or animal cells; as far as

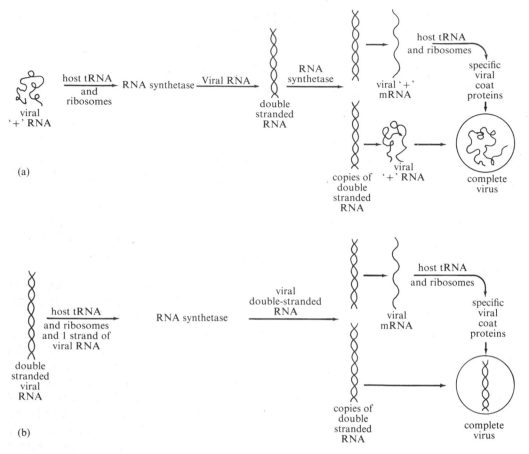

Fig. 14–5 The replication of RNA viruses inside host cells. **(a)** Virus containing single-stranded RNA. **(b)** Virus containing double-stranded RNA

we know, in uninfected cells, all cellular RNA is synthesized on DNA as a template, catalysed by the enzyme RNA poly-merase. During self-replication of a single-stranded RNA virus, a new enzyme, *RNA synthetase*, is formed on the host ribosomes using the single strands of viral RNA entering the cell, which act as mRNA (Fig. 14–5a). RNA synthetase then catalyses the formation of a complementary strand of RNA to form a double helix, which serves as a template for more RNA molecules. Specific viral mRNA of the '+' type is formed, and other strands of '+' RNA are used as the core of new virus particles.

With a *double-stranded RNA virus*, the situation is essentially the same. The necessary RNA synthetase is first made on the host ribosomes, using one strand of the viral RNA as messenger (Fig. 14–5b).

How viral nucleic acids dominate the metabolism of the host cell is still imperfectly understood. Sometimes normal cell synthesis continues during viral reproduction, but in extreme cases all DNA and RNA synthesis on the host chromosome ceases, the cellular RNA templates are degraded, and all protein synthesis is coded for by the viral nucleic acids. Some viruses, in which the capsid is composed of only a few types of protein, contain relatively small amounts of nucleic acid. With increasing coat complexity, more genetic material is required to code for the increased number of proteins. Sometimes, too, there are complex interactions between the virus and the host cell, and the synthesis of various enzymes is required to produce viral molecules; this too involves a greater amount of nucleic acid in the virus. Viruses that have an envelope of protein and lipid are thought to appropriate membrane material from the host cell.

Reproduction of the virus usually results in the rupture of the host cell, and the release of the many new virus particles. Many bacterial viruses possess a region of DNA which codes for the enzyme lysozyme (present in the phage tail), the production of which helps to rupture the rigid bacterial wall. We do not yet know how viruses break out of animal cells but the release of lysosomal enzymes may play a part.

14–3
Shape of viruses

Viruses, with their simple composition, yet precise and characteristic shapes (Fig. 14–2), are suitable objects for studying how biological materials are put together.

The tobacco mosaic virus has a simple structure and was the first virus to be taken apart and reassembled in the test tube. It is a rod-shaped particle about 3,000 Å in length and 180 Å in diameter. It consists of a long molecule of RNA surrounded by a cylindrical capsid (Fig. 14–6a) composed of helically arranged capsomeres. The virus can readily be dissociated into its nucleic acid and protein components by a number of reagents. Fraenkel-Conrat (1957) used rather gentle treatment to separate the nucleic acid and the coat protein and obtained from the coat some 2,000 sub-units, or capsomeres, per virion, which were all formed from one type of protein. Under suitable conditions of temperature and ionic strength these sub-units reassemble to form rods indistinguishable from the capsid of the intact virus. They are not infectious, however, since they contain no RNA. If recombination of the protein sub-units occurs in the presence of viral nucleic acid, rods are obtained which not only appear to be identical to the intact virus but also produce mosaic disease in plants.

The fact that the protein sub-units alone can spon-

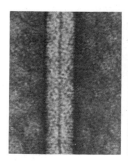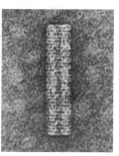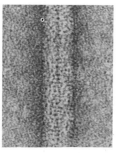

Fig. 14–6 Electron micrographs of tobacco mosaic virus. **(a)** Intact TMV. **(b)** Repolymerized TMV protein —stacked disk rod. **(c)** Repolymerized TMV protein— helical rod Magnification × 400,000. (With permission from J. T. Finch and R. Leberman.)

taneously reassemble into rods means that all of the information required for this process is in the sub-unit itself. It has been found by electron microscopy (using the negative contrast phosphotungstate technique, in which particles are examined against an electron dense matrix) that the asymmetric protein molecules, in the form of elongated sub-units, join side by side at sites of affinity and form disks with a hole in the centre. Under suitable conditions the capsomeres form larger aggregates, one of which is a rod-like structure composed of a stack of two-turn disks (Fig. 14–6b). The other, a helical rod in which the sub-units are packed in the same way as in the virus (Fig. 14–6c and Fig. 14–7), is only obtained under carefully controlled conditions. The bonds between the sub-units in these two structures are not identical, but are probably very similar, especially in the case of the lateral bonds.

How protein molecules assemble to form aggregates must depend not only upon the shape of the molecules, but also upon the position and specificity of the various chemical bonds. In the sort of situation described above, where all of the shape-specifying information is carried in one type of protein sub-unit (and therefore coded for by one gene), the process may be called morphogenesis of the first order.

Most animal viruses, however, are not rod-shaped but nearly spherical. Crick and Watson (1956) suggested that small spherical viruses probably have capsids composed of identical sub-units, linked at specific sites, as this would be the most efficient way of using the small amount of genetic information contained in their nucleic acid. If identical sub-units are bound at specific sites, geometrical considerations indicate that they must be arranged in a regular pattern, called a surface crystal. X-ray diffraction analysis reveals that

Fig. 14–7 Manner of reassembly of the protein sub-units in tobacco mosaic virus. **(a)** Flat disk of protein sub-units linked sideways. **(b)** Increased spiralling growth. **(c)** Cylinder of tobacco mosaic virus protein. The helical coil in the centre indicates the position of RNA in the intact virus

(a) flat disk of protein sub-units
linked sideways

(b) increased spiralling growth

(c) cylinder of tobacco mosaic
virus protein
helical coil in centre indicates
position of RNA in intact virus.

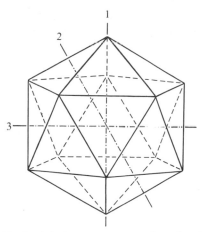

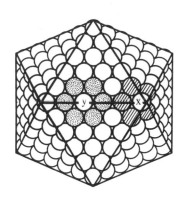

Fig. 14–8 (a) An icosa-hedron showing axes of sym-metry: (1) fivefold, (2) three-fold, (3) twofold. (b) Draw-ing of adenovirus showing how the capsomeres are arranged with icosahedral symmetry. Pentagonal pack-ing round x at vertex; hexa-gonal packing round the remaining capsomeres on edges (y) or faces. (After R. W. Horne.)

nearly spherical viruses have symmetry along three mutually perpendicular axes (cubic symmetry), and that the capsid usually is an icosahedron. (This, as Caspar and Klug concluded in 1962, would also be the most probable shape from energy considerations.) An icosahedron (Fig. 14–8a) is a polyhedron with 20 faces (each an equilateral triangle), 30 edges, and 12 vertices. It has three kinds of axis of symmetry—one fivefold, one threefold, and one twofold.

The regular icosahedral structure of tipula iridescent virus has been demonstrated by the technique of metal 'shadowing'. A stream of heavy-metal atoms is evaporated in a vacuum chamber and allowed to fall on the virus particles at an angle. Electrons are blocked by the metal atoms deposited on the virus particles, but can pass freely through the shadows of the particles where no metal atoms were deposited (Fig. 14–9). By this means the overall shape of the virus was determined, and shown to be an icosahedron.

Adenovirus capsid also is a regular icosahedron. From electron micrographs, the arrangement of the capsomeres is as shown in Fig. 14–8b. There are 252 spherical capsomeres, each of which must be situated on the vertex, side, or face of an icosahedron. Certain of the capsomeres are surrounded by five neighbours (pentagonal packing), which means that they are placed on a vertex; others are surrounded by six neighbours (hexagonal packing) and must lie on either a face or an edge (Fig. 14–8b). Many different solids, some regular and some irregular, can be built up from a combination of pentagonal and hexagonal packing of identical units. This means that symmetrical virus capsids of various shapes can theoretically be constructed from morphogenesis of the first order with each capsid being formed from a single type of protein sub-unit. Very large capsids may not be completely symmetrical; the regular shape may be somewhat distorted and bond angles between the protein building units may vary. Some form of scaffolding may be required during the assembly of these large capsids, so that only the complete shells are stable.

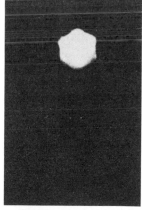

Fig. 14–9 Electron micro-graph of tipula iridescent virus. Heavy-metal shadow-ing shows its icosahedral structure. Magnification ×75,000. (With permission from Robley C. Williams.)

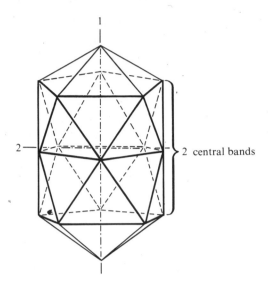

2 central bands

Fig. 14–10   Diagram of the head of T4 bacteriophage with (1) an axis of fivefold symmetry, (2) an axis of twofold symmetry. (After E. Kellenberger.)

The capsids of some viruses are composed of more than one protein building unit, as is shown by the sub-units reacting differently to an antiserum, or moving at different speeds in an electrical field. The more elaborate the shape and structure of the virus, the more genetic information it must carry. One of the most complex viruses is the T4 bacteriophage (one of the T-even bacteriophages Fig. 14–2), which consists of several different components. The genes which control the assembly and formation of the virus have been extensively studied by Edgar and his colleagues (1965 onwards), and others. Kellenberger and Moody (1966) showed that the head capsid is close to icosahedral, but lacks an axis of threefold symmetry. In fact it consists of the two pyramids of an icosahedron separated by two central bands (Fig. 14–10). There must therefore be an 'elongation factor' which imposes the assembly of two bands instead of the usual one, and determines that one vertex shall be the top of the pyramid. This factor may be a component of the viral capsid, or a morphogenetic core, consisting of sub-units arranged in a specific shape, may be built first and then act as a scaffolding by imposing the correct shape on the capsid as it is assembled. In favour of the second possibility is the finding that proteins exist inside the head of the T4 virus; these may be the core proteins.

All the information necessary for producing the various T4 phage components must be in the viral nucleic acid. The products of seven genes are required for the assembly of a complete, stable, virus head. The functions of most of these genes have been precisely determined, such as the production of the elongation factor mentioned above. Specific information seems also to be needed to close off the icosahedral shape and, in the absence of the gene which produces this 'rounding-up factor', long tubular heads called polyheads are formed, much as if the elongation factor were continuing to operate in an uncontrolled form. A gene is required to specify the protein which forms the major sub-unit of the capsid and another gene produces the 'solubility factor' without which this sub-unit protein forms insoluble aggregates and is prevented from participating in capsid assembly. The processes involved in determining the assembly of viral components may be common to all protein structures. Elucidation of the mechanisms involved should help in understanding, for example, the assembly of cell membranes from lipid and protein components.

Another level of arrangement in viruses is the aggregation of intact virus particles in crystals. Plant viruses tend to form true three-dimensional crystals, or sometimes two-dimensional arrays of what are known as paracrystals. The latter can also be obtained from poliomyelitis virus. The types of aggregate formed by different virus particles are indicative of their individual shapes.

14–4 Viruses and cancer

Rous observed in 1910 that a particular type of virus causes tumours to develop in chickens. The virus is an RNA virus now known as the Rous sarcoma virus. More recently, viruses have been associated with several kinds of tumours in mammals, especially in mice. During the last twenty years there has been intensive study of tumour forming or carcinogenic viruses, in particular of polyoma virus (Fig. 14–2) which causes tumours to develop in rodents. The action of polyoma virus on cells *in vitro* has been studied by Stoker (1964) and others. The virus transforms the cells from their normal state to a cancerous state; the transformation is accompanied by changes in the DNA of the infected cell and profound functional disturbances, including changes in cell surface properties and behaviour. *In vivo*, the transformed cells are no longer completely bound by the normal regulatory processes and begin to proliferate.

Although viruses have been isolated from human tumour material, in no case has any human cancer yet been conclusively shown to have been caused by a virus. It is impossible obviously to test viruses isolated from human tumours for their carcinogenicity in man, although some cause tumours in animals.

14–5 Viruses and cell hybridization

Viruses are used to elucidate various cellular functions, particularly the transfer of genetic information from the nucleus to the cytoplasm and the control of RNA synthesis. It was known in the 19th century that many diseases are associated with lesions in which the multinucleate cells occur. More recently, such cells have been found in lesions produced by certain pathogenic viruses, and in the late 1950s it became clear that some of these multinucleate cells were produced by the fusion of two single cells through viral action.

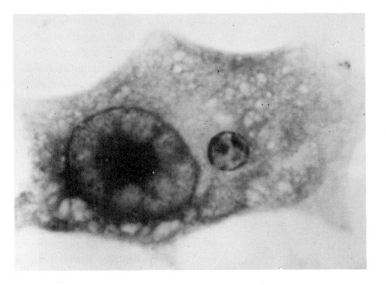

Fig. 14–11 A dikaryon produced by fusion of a hen erythrocyte with a HeLa cell due to the action of inactivated Sendai virus. The hen erythrocyte nucleus has begun to enlarge and the nuclear bodies are less deeply stained and more diffuse than normally. Magnification ca ×2,500. (With permission from H. Harris, *Nucleus and Cytoplasm.* Clarendon Press, Oxford, 1968.)

Harris and his collaborators (1965 onwards) have used an animal virus, inactivated by ultraviolet irradiation (to avoid the complications of infection by living virus), to fuse together cells from different species and so produce artificial hybrids. They used Sendai virus, a member of the para-influenza group of myxoviruses (Fig. 14–3). Other members of this group have also been used, but Sendai virus was found to bring about rapid fusion of animal cells. The first step is the formation of cytoplasmic bridges between two cells at their points of contact. These bridges increase in number and extent until finally the cytoplasms of the two cells coalesce. Harris has produced such hybrids or *heterokaryons* (cells containing nuclei of different kinds) from human tumour and hen erythrocyte cells (Fig. 14–11), among others. The properties and behaviour of such hybrids have been mentioned in Chapter 13.

Viruses are the simplest form of life, and studies of their comparatively simple composition and structure provide an important step towards understanding the arrangement, functioning, and interactions of the much more complex structures of plant and animal cells.

**14–6
Bacteria**

Bacteria, like fungi, are ubiquitous; they occur in both fresh and salt water, in soil, and in plants and animals. The biochemistry of bacteria is equally various; certain soil bacteria fix nitrogen, other bacteria turn milk into cheese by converting lactose into lactic acid. Many micro-organisms present in animals are harmless or even add to the well-being of the host; others are pathogenic.

Van Leeuwenhoek at the beginning of the 18th century observed the bacteria in pond water and in scrapings of tartar from teeth. Little progress in the field was made until the late

19th century, when the outstanding work of Pasteur in particular resulted in the emergence of microbiology as a branch of scientific study.

Pasteur showed in his studies of fermentation that some micro-organisms (*anaerobes*) can exist only in the absence of air (hitherto air had been considered essential to all life). He developed techniques for growing bacteria and devised methods for removing unwanted micro-organisms by heating or filtration. His observations had a profound effect on surgery and were applied by Lister to surgical practice where aseptic methods were previously unknown.

Koch applied staining techniques to various micro-organisms and showed in 1876 that anthrax, an infectious disease of domestic animals which is transmissible to man, is caused by a bacterium. Pasteur, working independently on anthrax, reached the same conclusion but made a significant advance by successfully preparing vaccines from dead or attenuated (weakened in virulence by various treatments) micro-organisms and using them to confer immunity to diseases such as anthrax and rabies.

During the next few decades, work centred on the search for useful vaccines, and it was not until 1909 that Ehrlich, who had observed the selective staining of bacteria in microscopic preparations of host tissues, conceived the idea of looking for compounds which might be selectively toxic to bacterial parasites. After more than 600 attempts, he succeeded in synthesizing an organic compound of arsenic which killed the infective agent of syphilis (a spirochaete) without harming the host.

Further advances in antibacterial therapy came with the discovery of the sulphonamides in the mid-1930s, and of penicillin in 1939. The bactericidal action of penicillin was first observed by Fleming in 1929, but it was more than ten years later that Florey and Chain isolated, purified, and clinically tested the antibiotic. The widespread use of penicillin and other antibiotics has revolutionized medical practice. (Antibiotics are defined as compounds produced by micro-organisms which are capable of inhibiting the growth of other micro-organisms). Antibiotics may now be synthesized in the laboratory, but this is more laborious and expensive than extraction from large-scale cultures of micro-organisms.

Since the 1930s much has been learnt about the chemical composition, structure, and function of bacteria. In 1937 it was observed by light microscopy that some form of nuclear organization is present in bacteria, but little or no structure could be observed in the cytoplasm. The development of high-resolution electron microscopy, combined with sophisticated biochemical techniques for the isolation and analysis of cellular fractions, has given us a much deeper understanding of bacterial structure and function.

14–7
General structure
of bacteria

As shown in Fig. 1–2 most bacteria are in the size range of 0·5–10 μ. Unlike viruses, bacteria contain the necessary materials and structures to support an independent existence. Viruses possess DNA or RNA (never both), whereas bacteria possess and can synthesize both types of nucleic acid, as well as the proteins, lipids, polysaccharides and other materials necessary for independent reproduction, provided that adequate supplies of nutrients are available. Bacteria differ from typical plant and animal cells (Fig. 14–12, cf Fig. 1–3) in that their nuclear material is not separated from the cytoplasmic region by a membrane, nor does their cytoplasm appear to contain an endoplasmic reticulum, Golgi apparatus, or mitochondria. The respiratory activity of bacteria is apparently located in membranous structures in the cytoplasm, called *mesosomes* (Fitz-James, 1960), which are thus the bacterial equivalent of mitochondria. Ribosomes have been identified under the electron microscope, and sometimes seem to be interconnected by fine fibrillar structures. In photosynthetic bacteria the site of the photochemical reactions is a vesicle known as the *chromatophore*, which is simpler both in structure and function than the chloroplast of plant cells; it is covered by a single membrane.

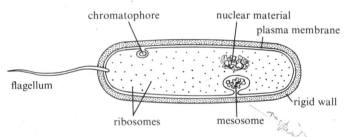

chromatophore   nuclear material
                     plasma membrane

flagellum

Fig. 14–12 Generalized diagram of a eubacterial cell

ribosomes   mesosome

rigid wall

In *eubacteria*, one of the three main groups of bacteria, and the group we chiefly consider in this section, the plasma membrane is surrounded by a thick rigid outer *wall* which, as in plants, confers fixity of shape on the cell. The other two groups of bacteria, myxobacteria and spirochaetes, have a thin flexible wall and change shape considerably, particularly as they are motile. The chemical composition of the rigid wall of eubacteria differs greatly, however, from that of the plant cell wall. Eubacteria are either permanently immotile, or move by using flagella.

Eubacteria may be classified into three categories on the basis of their shape. First are the *cocci*, spherical or ovoid cells, about 1 μ in diameter—for example, the staphylococci. Second are the *bacilli*, cylindrical cells commonly referred to as rods, which are about 1 μ in diameter and 5–8 μ in length —for example, the coliform bacilli. Third, the helically coiled organisms known as *spirilla* may be several times longer than bacilli—for example, *Spirillum volutans*, which lives in fresh water.

Another classification is based on a staining test, the Gram method, named after Christian Gram who developed it in Denmark in 1884. A heat-fixed bacterial preparation is stained with a basic dye, such as crystal violet, and with iodine, and then treated with an organic solvent. The cells of what are known as *Gram-negative* bacteria are rapidly decolorized by the organic solvent, while *Gram-positive* cells remain stained. It was realized that this must indicate some basic chemical difference between the two cell types, but not until the late 1950s, when the cell walls of bacteria were carefully studied, was the difference found to lie in the high lipid content of the walls of Gram-negative cells. The lipid material is readily dissolved by organic solvents, which can then enter the cell and leach out the dye, whereas the walls of Gram-positive bacteria form a barrier that prevents penetration of the solvent.

The most common method of reproduction in eubacteria is by simple cell division (binary fission), but some forms reproduce by budding. The transfer of genetic information by a process of conjugation in *E. coli* has been described in Chapter 10.

14–8
Bacterial nucleus

It is not entirely correct to speak of the bacterial nucleus because there is no limiting membrane to separate the region containing the genetic material from the cytoplasm. The structures formed by the bacterial genome, sometimes known as *nucleoids*, are undoubtedly nuclear in function. They are, however, much simpler in structure and morphology than the nuclei of higher organisms, so it is incorrect to speak of the bacterial chromosome. For convenience, however, we shall use the terms nuclear material and nuclear region, in addition to nucleoid.

Until the 1930s it was hard to demonstrate the presence of DNA in bacteria with the available staining techniques because of the large amounts of RNA in the bacterial cytoplasm. RNA combines strongly with basic dyes, and obscures the staining of the nuclear material. The Feulgen reaction, which is specific to DNA, and other techniques, such as the degradation of RNA by ribonuclease treatment or acid hydrolysis, made clear the existence of a nuclear region.

Since the late 1950s, techniques such as the use of electron microscopy and ultrathin sectioning have been used to study the nucleoids of micro-organisms such as *Bacillus subtilis*, which is a Gram-positive bacterium, and *Escherichia coli*, which is Gram-negative. It is now clear that nucleoids are regions consisting of a single coherent structure of delicate filaments (Fig. 14–13), the actual shape and size of these regions varying from one bacterium to another. The regions contain DNA and apparently no other macromolecules, except in rapidly growing cells, when RNA appears in the nuclear

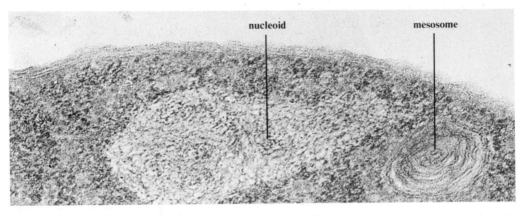

nucleoid   mesosome

Fig. 14–13 Electron micrograph of a section of a mutant form of *E. coli* bacterium. The nuclear region seems to be attached to the base of a mesosome. Magnification ca ×112,000. (With permission from A. Ryter.)

region. Evidence so far points to the absence of any protein comparable to the histones or acidic proteins of plant and animal chromosomes. Studies on *E. coli* indicate that in a slow-growing population the cells contain one or two nucleoids, but at high growth rates they contain between two and four (the latter presumably in cells which have completed nuclear, but not cellular, division).

Genetic studies indicate that in bacteria the DNA has the characteristics of a linear structure closed on itself, like a simple ring. Electron microscopy carried out by Kleinschmidt (1961) on isolated bacterial DNA, which indicated that there are no free ends or sudden irregularities in shape, supports the idea that there must be only one or very few molecular strands. The autoradiographic studies of Cairns (1963) provided further evidence that certainly in *E. coli* the genetic material consists of a single giant DNA molecule, about 1,000 μ in length. Clearly, this long filament of DNA must be closely folded or coiled in some way inside the bacterial cell, since the volume occupied by the nuclear region is of the order of 1 μ³. Electron microscopic studies of serial sections of the nucleoids of *B. subtilis* and *E. coli* (Kellenberger, 1960; Finks, 1965; and others) have revealed an ordered structure, notably a parallel arrangement of DNA fibres folded back and forth to form bundles, some of which contain up to 500 DNA helices. These bundles are usually orientated along the long axis of the bacterium (Fig. 14–14). Sometimes two bundles of DNA fibres are wrapped around each other. Since the patterns differ from cell to cell, analysis is difficult though it is clear that the fibres are arranged in an ordered manner. The degree of regularity in the structural arrangement is probably lower than in the chromosomal structures of higher organisms.

Nucleoid division and separation of the two daughter nucleoids is preceded by replication of the DNA molecule (Section 3–1 and Fig. 3–8). Replication begins at one point on the DNA and continues in the same direction round the circular molecule until the starting point is reached, at which a new cycle begins. But how does the genetic material manage

Fig. 14–14 Electron micrographs of *Bacillus subtilis*; a short series of thick sections showing nucleoids. The nucleoid on the left consists of a bundle originating near the centre of the cell; it proceeds to the left pole of the cell, then turns backwards and seems to end again near the centre. The nucleoid on the right shows similar arrangement. P, possible points of attachment to plasma membrane. Magnification ×61,000. (Reprinted with permission from G. W. Fuhs in *Genetics Today*. Copyright © 1965, Pergamon Press Ltd.)

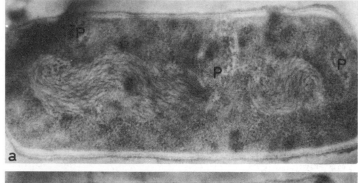

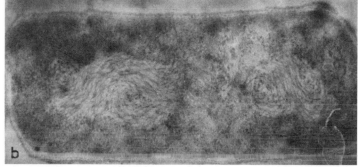

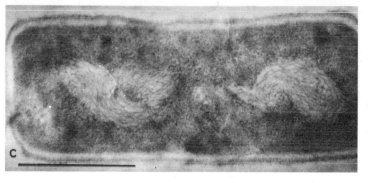

to unwind along its entire length, since it is folded and coiled? As with the chromosomes of plant and animal cells, we must assume a point (or points) of breakage and reunion, allowing sections of the DNA molecule to rotate freely so that the helix may uncoil. The early stages of nucleoid division, as observed by phase-contrast microscopy, are accompanied by rapid changes in nucleoid shape. Occasionally, systems of three circles touching each other in two points and looking very like the picture of bacterial replication obtained from the autoradiographic studies of Cairns (Fig. 3–8) have been observed, but detailed analysis is extremely difficult because of the overlapping of bundles of fibres.

Another aspect of division which is not fully understood is the separation and equal distribution of the replicated genetic material in the daughter cells. There is no sign of any structure corresponding to the mitotic apparatus of higher cells. In 1964, Jacob, Brenner, and Cuzin suggested that the bacterial nucleoid is attached to the plasma membrane and that during cell division synthesis of fresh membrane material at the point of attachment causes separation of the two daughter nucleoids. There is an appreciable amount of morphological evidence in support of a relationship between the nucleoid and the plasma membrane, and studies on isolated membrane fractions have also indicated an association between the two structures. The connection may not be a direct one, but may exist between the nucleoid and a mesosome which is continuous with the plasma membrane. Mesosomes are clearly observable in the cytoplasmic region of Gram-positive bacteria, and in a less well-developed form in Gram-negative cells. Cytochemical studies reveal them to be centres of the electron transport system in bacteria. Frequently the nuclear region appears to be in contact with a mesosome (Fig. 14–13). Probably, therefore, there is a connection either between the plasma membrane or the mesosomal membrane and the DNA, which ensures separation of the two daughter nucleoids after DNA replication.

This process has been studied by electron microscopy of serial sections, and the following picture emerges. When DNA synthesis begins, the nucleoid is connected to a single mesosome, but during synthesis the mesosome splits in two and when replication is complete the two mesosomes move apart, each carrying a daughter nucleoid with it. New mesosomal membrane material is therefore being formed while DNA is being synthesized. It is not easy to distinguish new membrane material from old, but the indications are that new membrane is generated near the point of attachment of the nucleoid to the membrane. There is also evidence that mesosomes are attached to, and responsible for the orderly formation of the cross wall or division septum formed during cell division (see Fig. 14–17).

14–9
Cytoplasm and
plasma membrane

Rapidly dividing bacterial cells would be expected to contain many ribosomes in their cytoplasmic region. The ribonucleoprotein in *E. coli* cells which are dividing every 30 minutes accounts for up to 30 per cent of the cell's dry weight, and the RNA for up to 25 per cent. Electron micrographs of rapidly dividing cells show that the cytoplasm has a fine granular structure (Fig. 14–15) due to many ribosomes. These ribosomes are not independent rounded particles, but part of a complicated network of interconnected linear arrays. The fine, electron-dense fibrillar structure which connects the individual units varies in width, and may be only just visible, being as narrow as 10 Å. The fibrils can sometimes be traced from the plasma membrane to the nuclear region and in some regions are clearly helical. So bacterial ribosomes may be polyribosomes, with the ribosomal particles interconnected by fine fibres of RNA.

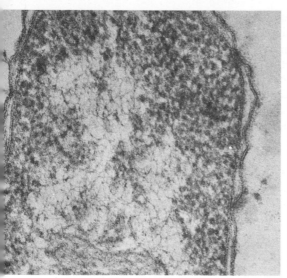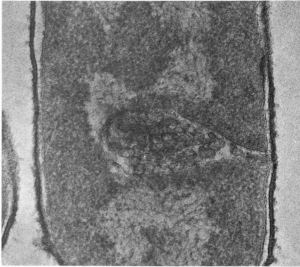

Left: **Fig. 14–15**   Electron micrograph of an *E. coli* bacterium in exponential growth phase. There are no independent rounded ribosomes, but denser areas, which are interconnected. The cytoplasma appears essentially to have a polyribosomal structure. Magnification ca ×130,00. (With permission from W. van Iterson.)

Right: **Fig. 14–16**   Electron micrograph of *Bacillus subtilis* treated with tellurite (highly electron-opaque in its reduced form); the mesosome (chondrioid) is enveloped by a continuation of the plasma membrane. The space inside the mesosome is thus connected with that between the plasma membrane and the cell wall. Magnification ×97,000. (With permission from W. van Iterson.)

The chief membrane-bound structure in the bacterial cytoplasm is the mesosome (Fig. 14–16). Mesosomes form clear well-developed structures in Gram-positive bacteria, but are much more indefinite in Gram-negative forms. They have a typical triple-layered membrane structure, continuous with the plasma membrane, and appear to be formed from a pocket of invaginated plasma membrane. Their internal structure varies from one type of Gram-positive bacterium to another, sometimes appearing as tubules and sometimes resembling a string of beads.

The mesosome, rather than the plasma membrane, seems to be the centre of the electron transport system, and to contain the cytochromes and succinate dehydrogenase. Mesosomes are probably the bacterial equivalent of mitochondria and are sometimes known as chrondrioids. Unlike mitochondria, however, they are not closed systems but extensions of the plasma membrane. Because they are also in close contact with the nuclear material they may play a role in intracellular transport, as well as being a source of energy close to the nucleoid. Because of their close proximity to the genetic material, they may not need to carry DNA, as do the mitochondria of plant and animal cells.

In photosynthetic bacteria, such as *Rhodospirillum*

rubrum, there is another type of membranous structure, the *chromatophore*, which is the site of photosynthesis. It is surrounded by a single membrane continuous with the plasma membrane. Much of the cytoplasm is filled with chromatophores in photosynthetic bacteria, which obscures the presence of ribosomes, since chromatophores may be more than 1,000 Å in diameter. The chief constituents of chromatophores are proteins and lipids. They contain the chlorophyll and carotenoid pigments and the system of enzymes and electron carriers required in photosynthetic phosphorylation.

Reserve materials are stored in the cytoplasm of bacterial cells either as finely dispersed or distinct granules. There are three main types of reserve material. First, there are organic polymers which either serve as reserves of carbon, as does poly-β-hydroxybutyric acid, or as stores of energy, as does a polymer of glucose sometimes called granulose. Second, many bacteria contain large reserves of inorganic phosphate as granules of metaphosphate polymers known as volutin. The granules are highly refractile and can be seen in the light microscope. The third type of material is elemental sulphur, formed by oxidation from hydrogen sulphide. It occurs as an energy reserve in the form of spherical droplets in certain sulphur bacteria.

The bacterial plasma membrane appears to have the usual triple-layered lipoprotein structure possessed by plant and animal cells. How substances are transported across the membrane is important in determining the nutritional specificity of many bacteria. Specific nutritional needs of bacteria require not only the correct intracellular enzymes, but also appropriate and highly specific transport mechanisms. The agents responsible for active transport across the membrane against a concentration gradient are known as permeases. The plasma membrane is so efficient at pumping metabolites into the cell that the intracellular osmotic pressure can reach 20 atmospheres.

14–10
Bacterial cell wall

The plasma membrane alone could not withstand the high internal osmotic pressure, and rupture of the cell is prevented by the tough and rigid outer wall. The function of the wall appears to be purely protective and mechanical; it plays no apparent part in permeability or in metabolism. If the wall is removed or damaged, lysis ensues. Bacterial protoplasts from which the wall has been removed, for instance by treatment with lysozyme, may be kept intact if placed in a medium of the same osmotic pressure as bacterial cytoplasm. Protoplasts always assume a spherical shape; it is the wall which is responsible for the characteristic shape of bacteria.

Cell walls may be isolated by rupturing bacteria and washing away the intracellular material. Isolated walls prepared in this way retain their characteristic shape. In

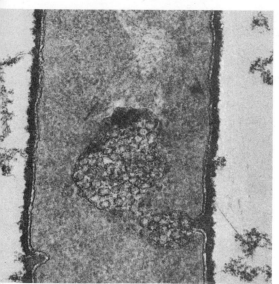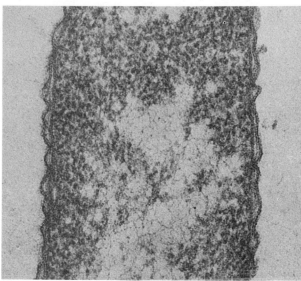

Fig. 14–17 Electron micrograph of *Bacillus subtilis* as an example of the Gram-positive bacterium in which the thick wall and plasma membrane are united along their entire length. The beginnings of cross wall formation can be seen Magnification ×94,000. (With permission from W. van Iterson.)

Fig. 14–18 Electron micrograph of *E. coli B*. as an example of the Gram-negative type of bacterium, with a thin wall which is sinuous and not parallel to the plasma membrane. Magnification ×110,000 (With permission from W. van Iterson.)

Gram-positive bacteria the cell wall is 150–200 Å thick and appears to be attached to the plasma membrane along its entire length (Fig. 14–17). In Gram-negative bacteria, however, the wall is thinner, 75–120 A, and only makes intermittent contact with the plasma membrane (Fig 14–18). Along most of its length the Gram-negative wall forms a sinuous covering clearly separated from the plasma membrane.

The first chemical analyses of bacterial walls were carried out by Salton in the 1950s. Much of the detailed structure was determined by Sharon, Perkins, and others in the 1960s. Bacterial walls are much more complex structures than the cellulose walls of plants. There are consistent major differences between the walls of Gram-positive and Gram-negative bacteria, and there are variations between different types of bacteria in the two groups.

All cell walls, however, contain a macromolecular compound composed of two simple sugars and three or four amino acids. These molecules form the basal structure or mucocomplex which confers structural rigidity on the wall and is unique to bacterial cells. The two sugars are *N*-acetylglucosamine and *N*-acetylmuramic acid (Figs 7–25 and 7–31). The saccharide part of the wall consists of alternating units of these two amino sugars joined by glycosidic linkages (Fig. 14–19). The amino acids are glutamic acid, alanine, glycine,

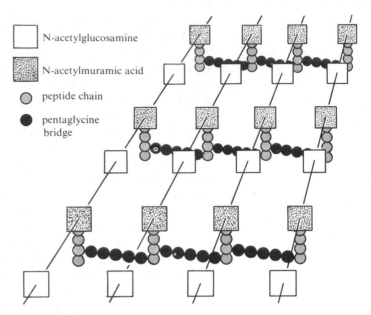

Fig. 14–19 Structure of the bacterial cell wall muco-complex. The polysaccharide chains run from south-west to north-east. Attached to them are peptide chains linked to one another by pentaglycine bridges

and sometimes lysine, though many bacteria have instead of lysine a similar substance known as diaminopimelic acid:

$$HOOC-CH-CH_2-CH_2-CH_2-CH-COOH$$
$$\quad\quad\; | \quad\quad\quad\quad\quad\quad\quad\quad\quad\; |$$
$$\quad\quad NH_2 \quad\quad\quad\quad\quad\quad\quad\; NH_2$$

Much of the glutamic acid and alanine are present as the unusual D-stereoisomers, which never occur in proteins. These few amino acids form short peptide chains which attach to the polysaccharide chains and cross-link to one another (Fig. 14–19). Thus a three-dimensional network of considerable toughness and mechanical strength is formed.

In Gram-positive bacteria there are large amounts of this mucocomplex, which in some cases is the only macromolecular constituent of the wall. In other Gram-positive bacteria there are also simple polysaccharides and teichoic acids, which are polymers of glucose, alanine, and glycerol or ribitol (a five-carbon alcohol).

In Gram-negative bacterial walls much less mucocomplex is present, and, as far as is known, no teichoic acid. There are, however, large amounts of protein, lipid, and polysaccharide. The amount of lipid may range from 20 to 30 per cent. Electron microscopy indicates that there may be several different layers in Gram-negative walls, and *E. coli* has an outer layer of lipoprotein–polysaccharide complex.

Little is known about how bacterial walls increase in length, girth, and mass. Immunofluorescent techniques used by Cole and May in the 1960s to distinguish between old and new wall material have shown that the manner in which new wall is initiated varies from one kind of bacteria to another. For instance, in some bacteria the insertion of new cell wall components into the old wall appears to be evenly distributed, while in others there is a bipolar extension of the wall due to the formation of caps of new material at opposite ends.

Lysozyme disrupts bacterial cells by breaking the linkage between the amino sugar units, as described in Chapter 4. The work of Park and Strominger (1957 onwards) indicates that penicillin kills bacteria by interfering with cell wall synthesis. This explains why the drug is not toxic to animals, because it inhibits the synthesis of material which animals do not possess. It is now known that penicillin inhibits the final stage in the synthesis of the wall mucocomplex, that is the linking of adjacent polysaccharide chains by reaction between their peptide side-chains. Thus penicillin is very toxic to most Gram-positive bacteria, but is toxic to most Gram-negative bacteria only at relatively high concentrations.

Clarke, O'Grady, and coworkers (1969) have studied with the scanning electron microscope the surfaces of bacteria which have been exposed to the antibiotic ampicillin. They find that staphylococci appear to develop multiple points of weakness all over their surface, with no obvious pattern, but in streptococci the main effect seems to be in the division region where the cell wall often collapses completely to give an 'apple core' appearance. Differences of this type would be expected from the observations on variations in sites of wall synthesis mentioned above. Other antibiotics act upon bacteria in different ways.

Cline and Lehrer (1969) have discovered how phagocytic white blood cells kill ingested micro-organisms. The cells contain the enzyme D-amino acid oxidase, which converts the D-amino acids of the ingested bacterial cell wall into the corresponding keto acids. The process is accompanied by the release of hydrogen peroxide, which is believed to be responsible for the anti-bacterial action of white cells.

Even closely related bacteria may be distinguished from one another by their antigenic specificity. An important group of antigens are located in the cell wall and both qualitative and quantitative differences in wall composition play a major part in determining specificity.

14–11
Flagella of
bacteria

Most eubacteria are immotile. In the types which are motile, movement is always due to flagella. In peritrichous types (Fig. 14–20a), the flagella are attached at many points along the sides of the cell, while in polar forms attachment of the flagella (or sometimes a single flagellum) is exclusively to one or both ends of the cell (Fig. 14–20b). Bacteria may be deflagellated by mechanical means so that the flagella break off at or near the surface of the wall. Studies of flagella show that they possess antigenic specificity; each bacterial species, and sometimes different strains of the same species, has a characteristic type of flagellar protein. Flagella originate in the cytoplasm; this is demonstrated by removing the cell wall

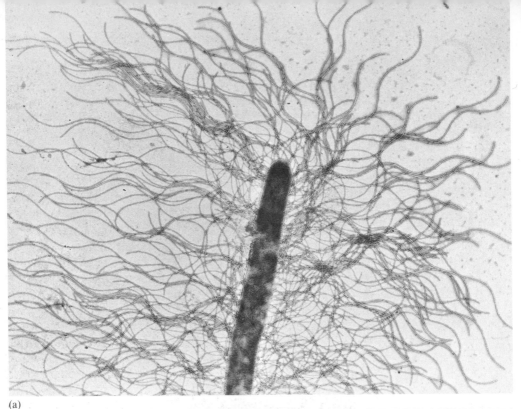

(a)

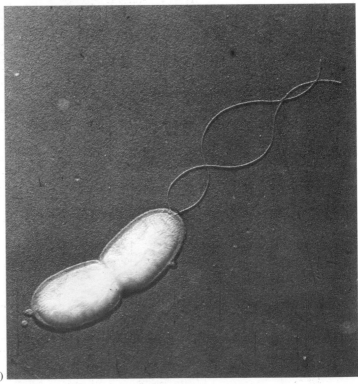

(b)

Fig. 14–20 **(a)** Electron micrograph showing peritrichous flagellation in *Proteus mirabilis.* Magnification ×9,700. (With permission from J. F. M. Hoeniger.) **(b)** Electron micrograph showing polar flagellation in *Pseudomonas fluorescens* after metal shadowing. Magnification ×22,000. (With permission from A. L. Houwink.)

with lysozyme, after which the flagella can be seen to be attached to the protoplast. The flagellated protoplast is not motile, however, because the rigid wall is needed to provide a solid fulcrum.

14–12
Bacterial colonies

When a bacterium reproduces in or on the surface of a solid medium, the progeny form a localized colony. Inside a solid medium such as agar, the shape of the colony largely depends upon how the medium can accommodate the enlarging mass. On the surface of agar, however, the appearance of the colonies is often characteristic for a given species, so that their gross morphology provides a means of identification. The chief criteria are size, shape, and colour. Colonies may be smooth, granular, or filamentous; they may have a smooth or jagged edge; they may be transparent or opaque, and dull or glistening.

The structure of the individual cells and their behaviour during growth determine the appearance of the colonies. For instance, rod-shaped cells which tend to remain attached in chains after cell division usually give a filamentous structure, with curled hair like extensions at the edges. The curl arises because the fission process is slightly asymmetric with respect to the long axis of the cells and this results in a regularly twisted chain of cells. Bacteria with capsules (see below) produce mucoid colonies. Flagellated forms are sometimes so orientated and their flagellar movements so coordinated that the whole colony rotates and moves over the agar surface!

14–13
Mechanisms of survival

How do bacterial cells survive? Obviously the cell wall, which protects the cell against mechanical injury, enables it to survive in an environment which is frequently hypotonic compared to its own cytoplasm. In fact, bacteria possessing a defective cell wall, due for instance to penicillin treatment, can survive if the surrounding medium is balanced osmotically by the addition of salts or serum.

In 1935, a bacterium lacking a cell wall was isolated from a rat. It had apparently arisen spontaneously from a strain of bacillus. The bacterium was called an L-form, after the Lister Institute in London. It was later discovered that most bacteria can give rise to L-forms when subjected to environmental stress, and that they usually revert to the parental form when the stress is removed. L-forms are remarkably similar to mycoplasmas in their lack of a cell wall, the appearance of their colonies, and certain aspects of their metabolism. It has been suggested that mycoplasmas are L-forms of bacteria which have lost their ability to revert to the parent. On the other hand, L-forms are produced in the laboratory as a response to adverse conditions and have never been obtained from animal diseases in the absence of the parent bacterium, so that the evidence is inconclusive.

Many bacteria have the capacity to secrete around themselves, outside the wall, gelatinous gums or slimes, known as capsules or slime layers. The composition of the capsular substances varies with different bacterial types; usually they are polysaccharides, but some are polypeptides. Capsules apparently play no direct role in cellular function, since their removal by specific enzymes is not harmful, nor is the loss of ability to produce capsules as the result of a mutation. However, capsules confer considerable survival value because they protect the bacterium from being engulfed by the phagocytic cells of the host. The slippery surface of the capsule apparently prevents the phagocytic cell from adhering to and ingesting the bacterium.

Some bacteria secrete substances which enhance their invasiveness in an infected host. These substances, known as aggressins, are frequently enzymes. For example, hyaluronidase and collagenase are aggressins which can break down two important constituents of connective tissue, and presumably facilitate the spread of the bacteria. Streptokinase is a bacterial enzyme, produced by streptococci, which dissolves blood clots. Since the latter may play a part in localizing infection, the enzyme is again thought to promote bacterial spread. Some bacteria not only resist phagocytosis but produce substances, known as leucocidins, which kill phagocytes, and others produce toxins which damage or kill host cells. Certain types make an enzyme, penicillinase, which destroys penicillin.

Another survival mechanism possessed by a few bacteria, which is very efficient, is the ability to become a dormant form called a spore. A striking property of spores is their resistance to heat; some can survive for 2 hours at a temperature of 100°C. They are also more resistant than the normal vegetative bacterial cell to dehydration, irradiation, mechanical damage, and toxic chemicals. Spores are metabolically inert, and can remain viable for many years at ordinary temperatures in the dry state.

Spores are formed in response to a nutrient deficiency in the bacterial environment. Spore formation occurs inside the parental vegetative cell and is a remarkable process in that the spore differs from the parent in chemical structure and physiological function. Yet it can revert to being a vegetative cell by germinating under suitable conditions. The formation of a mature spore from a vegetative cell, known as sporulation, appears to involve an ordered sequence of events and serves as a simple model system for studying differentiation.

First, part of the genetic material separates, later to form the spore nucleus. Second, a thin transverse wall develops, separating the part of the cell containing the nuclear material from the rest of the cell. Third, the transverse wall grows around the cytoplasm and nucleus of the future spore, forming the spore protoplast or core. Finally, the spore matures and

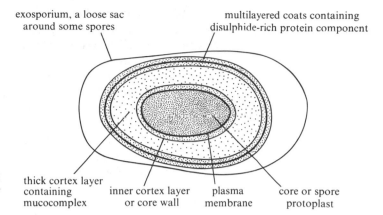

exosporium, a loose sac
around some spores

multilayered coats containing
disulphide-rich protein component

thick cortex layer
containing
mucocomplex

inner cortex layer
or core wall

plasma
membrane

core or spore
protoplast

Fig. 14–21 Representation of the structure of a bacterial spore; specialized layers of material surround and protect the spore. (After G. Gould.)

the mother cell dissolves away, leaving the free spore.

In maturation various layers are formed around the spore wall (Fig. 14–21). There is a thick cortex around the core wall, a major component of which is the polysaccharide and peptide mucocomplex present in vegetative cell walls. Then there are multi-layered protein coats which contain a disulphide rich protein with properties similar to those of keratin. Some spores also have a further loose, thin layer known as the exosporium. During maturation, the spore steadily loses water. The mechanisms involved are not clearly understood, but dehydration is thought to be connected with the synthesis of dipicolinic acid:

$$HO-\underset{O}{\underset{\|}{C}}\quad\overset{}{\underset{N}{\bigcirc}}\quad\underset{O}{\underset{\|}{C}}-OH$$

which is a characteristic component of spores, probably in the form of its calcium salt.

The reverse process, germination, can be initiated by a simple nutrient such as L-alanine. A dramatic and rapid depolymerization of the cortex takes place, and dipicolinic acid, calcium, and mucocomplex leak from the spore, which may lose 30 per cent of its weight in 30 seconds. Outgrowth of the core from within the spore coats and cortex then occurs, followed by cell division. Here also we have a well-ordered sequence of events, and some of the changes resemble those occurring during the initiation of new growth in higher organisms—for instance, in the growth of embryonic or regenerating tissues.

Bibliography

Viruses

Crawford, L. V. and M. G. P. Stoker (eds), *The Molecular Biology of Viruses*, Symposia of the Society for General Microbiology, vol 18. Cambridge University Press, London, 1968.

Fraenkel-Conrat, H., Rebuilding a virus. *Scientific American,* June 1956.

Horne, R. W., The structure of viruses. *Scientific American*, January 1963.

Kellenberger, E., The genetic control of the shape of a virus. *Scientific American*, December 1966.

Lwoff, A. and P. Tournier, The classification of viruses. *Ann. Rev. Microbiol.*, **20**, 45, 1966.

Bacteria

Cole, R. M., Bacterial cell wall replication followed by immunofluorescence. *Bacteriological Revs*, **29**, 3, 326, 1965.

Fuhs, G. W., Fine structure and replication of bacterial nucleoids. *Bacteriological Revs*, **29**, 3, 277, 1965.

Gould, G. and S. Warren, The bacterium's survival pack. *New Scientist*, 10 July 1969.

O'Grady, Francis and D. Greenwood, Antibiotic-induced surface changes in micro-organisms demonstrated by scanning-electron microscope. *Science*, **163**, 1076, 1969.

Ryter, A., Association of the nucleus and the membrane of bacteria: a morphological study. *Bacteriological Revs*, **32**, 1, 39, 1968.

Sharon, N., The bacterial cell wall. *Scientific American*, May, 1969.

Stanier, R. Y., M. Doudoroff, and L. A. Adelberg, *General Microbiology*, 2nd edn. Macmillan, London, 1964.

Van Iterson, W., Bacterial cytoplasm. *Bacterialogical Revs*, **29**, 3, 299, 1965.

15 Origin of life

The problem of the origin of life has interested mankind since earliest times, but the suggestion that all living things come from existing forms of life, as expressed by Virchow in 'Omnis cellula e cellula', was certainly not accepted until the mid-19th century. Aristotle maintained that any dry body becoming moist, or on the other hand any wet body becoming dry, would give rise to animals. Van Leeuwenhoek's discovery of minute organisms, which he called 'very little animals', in decaying substances led many to take the view that life arose spontaneously in decaying matter which had been kept warm. Pasteur showed that adequate sterilization removed all microbes and that sterilized organic matter would remain free from living organisms indefinitely. With the acceptance of the view that all forms of life come from previous life, the problem then became to determine when the first life arose on the earth.

Astronomical evidence indicates that the earth was probably formed about $5,000 \times 10^6$ years ago as a large cloud of extremely hot incandescent gas. Some scientists take the view that the earth was derived from the sun, and others that it was captured later by the sun after formation from a gaseous cloud or from some other star. Recent analysis by O'Hare and Biggar of the rocks brought back from the moon by the *Apollo 11* astronauts indicates that they are of different origin from the earth, which suggests that other bodies in the solar system (and perhaps the earth itself) may not be of the same origin as the sun. The initial temperature of the earth was so high that even heavy metals such as iron were present in the form of vapour, and it would have been impossible for any life to exist. Richter proposed the theory of panspermia, which suggested that everywhere in space are found small particles of solid matter (cosmozoa) which have been cast off by celestial bodies; along with these are carried the germs of micro-organisms. This would mean that life arose on other planets and was carried to the earth by these particles, possibly by becoming trapped at the centre of meteorites (the outer surface of meteorites reaches an enormous temperature on entering the earth's atmosphere, so life could not be carried on their surface, and the cosmozoa themselves would certainly become incandescent). Helmholtz said that he was forced to take up the theory of panspermia because of the complete impossibility of explaining the origin of life scientifically in any other way. However, the view that life did in fact originate on earth itself after it had cooled over a period of many thousands of years is almost universally accepted today.

It must follow, if life originated on the earth, that the geological conditions at the surface and in the atmosphere were, at some time in the earth's history, suitable to provide all of the components needed by the simplest living organisms. An important factor to be taken into account is that almost all life on the earth at the present time is interdependent in a biochemical sense. Green plants generate molecular oxygen from carbon dioxide, and produce glucose, an energy source which is utilized by both plants and animals. Certain bacteria, which are dependent on green plants for energy, are able to fix atmospheric nitrogen in the form of ammonia and proteins. Both green plants and animals are dependent in their turn on these bacteria to maintain the nitrogen cycle. In making hypotheses about the conditions under which the first life could have originated, all these factors must be considered. The authors feel that they cannot do better than refer the reader to the masterly and farsighted essays of Oparin (1925) and Haldane (1929), and to the book of Bernal (1967). Here we give only a brief summary of the main conclusions and fields of speculation, including new information which has become available since these works were written.

Three approaches to a study of the problem can be made:

To examine the earliest records left by living organisms in geological strata. This is the direct approach and is being continuously carried out by palaeontologists.

To carry out experiments in the laboratory in an attempt to simulate the conditions on the earth at the time when the first life probably arose, and to examine the type of organic synthesis that can be initiated. Such investigations can never, however, prove how or when life arose —they can only lead to plausible hypotheses.

To examine in general molecular terms the nature of living organisms today and to speculate as to how these might have arisen on the basis of known physico-chemical phenomena. This again can only lead to plausible hypotheses.

15–1
Geological
findings

We will first describe two of the most recent geological findings, which are illustrated in Figs. 15–1 and 15–2. These records, studied by Cloud and Barghoorn, are more than $1,000 \times 10^6$ years older than the earliest records of other types of organisms (which date from about 800×10^6 to 900×10^6 years ago).

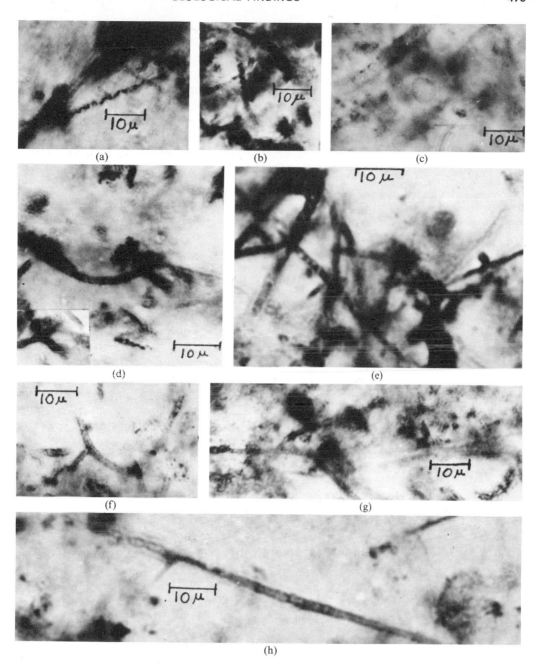

Fig. 15–1 Fossils of photosynthetic algae, ca 2000×10^6 years old, showing septate threads of blue – green algae, resembling *Oscillatoria*, and bacteria resembling the iron-precipitating bacterium *Sphaerotilus*. From the Gunflint chert (Pre-Cambrain). **(a)** and **(b)** Spiral threads resembling degraded filaments of *Oscillatoria* and *Sphaerotilus*. **(c)** Small sphere in centre; *Oscillatoria*-like structure at upper left. **(d)** Looped structure occurring in threads; inset shows septate nature of threads. **(e)** Thread at right has attached lump, possibly a reproductive structure. **(f)** Possible branching in threads? **(g)** Possible reproductive structure attached to threads which are part of meshwork or overlap. **(h)** Thin section of thread. (With permission from P. E. Cloud, Jnr.)

They have been preserved under rather extraordinary conditions in rocks of the Canadian pre-Cambrian shield. At the time that these organisms were alive, there was much silicate in solution and the gelatinous coats of the cell walls were replaced by silica. All pre-Cambrian formations have been subject to numerous subsequent earth movements under intense pressure and heat. Being already impregnated by silica, somewhat in the way that specimens for electron microscopy are impregnated with plastic, these cellular structures have been preserved for this vast period of time. The date of the rocks has been determined radioactively by the potassium–argon technique. The two main types of organism in Fig. 15–1 are:

> Filamentous blue–green algae, seen as 6μ diameter septate threads. These cells closely resemble in form the modern *Oscillatoria*, a freshwater blue–green alga.
>
> Bacteria seen as thin septate threads, $1\cdot5$–2μ in diameter and more than 100μ long. These consist of separate cylindrical units 1–2μ in length. They resemble filamentous iron-precipitating bacteria such as *Sphaerotilus*. Large deposits of oxidized iron have in fact been found in some pre-Cambrian strata; these could have been produced by bacteria of this type. These structures are also rather similar in appearance to nostocalean blue-green algae.

More recent discoveries have been made in the black cherts of the Fig Tree series in Swaziland, Eastern Transvaal. These have been dated at $3,100 \times 10^6$ years old! They have been studied by making polished specimens and etching the surface, followed by platinum–carbon replica microscopy, according to the technique illustrated in Fig. 8–14. These appear to be isolated cells similar to modern bacilli (Fig. 15–2); a cell wall $0\cdot015\mu$ thick, with a two-layered structure, can be seen in some specimens. In these formations are also found filamentous structures resembling plant debris.

Another method of approach is to study hydrocarbon deposits. Most oils are of biological origin, and these have been studied by Abelson and Calvin in particular. They often contain inclusions of complex hydrocarbon molecules, particularly phytane (C_{20}) and pristane (C_{19}). These are both isoprene derivatives, and are components of the chlorophyll molecule. Complex molecules of this type are found in a number of strata which are $3,000 \times 10^6$ years old, including the Fig Tree strata mentioned above.

It is interesting to consider the nature of the earth's crust and atmosphere before the date of these records of living organisms. To a cell biologist it is particularly fascinating to speculate as to what processes occurred over many millions of years in the chemical substances which were present at the time

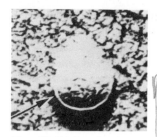

Fig. 15—2 Fossils of bacteria ca 3000×10^6 years old. The circular structure is interpreted as a transverse section showing a cell wall with a two-layered structure 0.015μ thick, similar in size and thickness to modern bacteria. Platinum—carbon replicas and negative prints were used. From the Fig Tree series in South Africa. Magnification $\times 50,000$. (With permission from E. S. Barghoorn.)

of the earth's formation, and how these resulted finally in the formation of the first living cell. As the earth cooled over a period of many thousands of years a certain order probably began to be established. The work of physicists on matter in the vapour state and on the spectral analysis of the sun and other stars indicates that, as cooling occurred, the heavier atoms would begin to sink towards the centre of the cloud of gas due to the action of gravity, and the lighter ones would remain near the surface. As cooling continued, the density of the material must have increased until at a certain stage it changed in the centre of the cloud from gas to liquid. It was still red-hot, however, and was surrounded by a huge envelope of incandescent gas. Possibly the first compounds to exist were the carbides of heavy metals, which are stable in even extreme heat. After a further period of cooling a solid crust formed; around this was an atmosphere consisting of various gases.

Certain conclusions, based partly on studies of other planets in the solar system, can be drawn about the possible composition of the atmosphere at that time. Much steam must have been present, as all of the water now found in the seas, rivers, and lakes of the earth must have existed in the form of gas. Some of this steam may also have been trapped later as water inside the earth's crust, because volcanic eruptions still cause the release of vast amounts of water vapour (up to 70% of the total mass erupted). It is unlikely that any free oxygen was present; Haldane has noted that the amount of available oxygen now is only about enough to combine with all of the coal and other organic remains found below and on the earth's surface. Probably all of this organic carbon (and much of the carbon present in chalk, limestone, etc.) was present as carbon dioxide, some of which was later converted into oxygen by plant photosynthesis.

Thus three of the four chief elements concerned in biosynthesis—carbon, hydrogen, and oxygen—were present. The fourth common element, nitrogen, was probably present in the earth's crust in the form of metallic nitrides from which ammonia was formed by the action of water. Sensitive spectral analysis of interstellar dust and gas clouds from 1951 onwards has shown that hydrogen molecules and hydroxyl radicals are present in these regions, as well as water and ammonia. In 1969, formaldehyde, $H \cdot CHO$, was found to be present; under reducing conditions formaldehyde is readily converted into methane, CH_4.

In addition to carbon, hydrogen, oxygen, and nitrogen, other elements are essential for all existing forms of life: sodium, potassium, chlorine, calcium, magnesium, phosphorus, and sulphur, together with traces of about 20 other elements. These elements are all found today in sea water as inorganic ions, phosphates, sulphates, etc. The hard crust of the earth contained these elements in the form of simple silicates (acidic), basalts (basic), and other igneous rocks. When the temperature of the earth's surface had dropped below 100°C, steam would gradually become liquid, and a hot rain would fall through the atmosphere and collect on the earth's surface. The process of weathering of the igneous rocks on the mountain ranges would then begin and erosion would gradually lead to the solution of sodium, potassium, and all the other elements now found in salt water; the first water would have been free of inorganic salts.

This process of erosion must have taken place for many thousands (possibly many millions) of years before appreciable concentrations of the inorganic salts necessary for life could have been removed from the igneous rocks. Evidence for the date at which liquid water condensed on the earth has been estimated by geologists, because sedimentary rocks formed from sands and clays can only have been formed by the flowing of water into lakes and seas. This is estimated to have taken place about $3,000 \times 10^6$ years ago. By this time, the stage had been set and the necessary elements—carbon, hydrogen, oxygen, nitrogen, phosphorus, etc.—were present either in the atmosphere or in solution in water.

**15–2
Laboratory
investigations**

Modern research has demonstrated the high degree of molecular complexity and integrated function at the biochemical and molecular biological level even with the simplest forms of life. No protozoologist today would name a new species of amoeba as *Chaos chaos*. In studying the origin of life in the laboratory, the problem is to demonstrate certain limited activities as observed in living cells, while utilizing molecular organization of a simpler type.

It is useful at this stage to consider laboratory investigations attempting to simulate conditions which may have existed on the earth at the time that life began. As the temperature of the earth dropped, the various compounds in the earth's atmosphere would interact with each other. Ultraviolet light from the sun is now prevented from penetrating the earth's atmosphere mainly by ozone (in the upper atmosphere) and by oxygen, but probably there was then no oxygen present. (Not all geologists would accept, however, that the atmosphere was a reducing one at that time.) So ultraviolet light from the sun would have been able to act on the mixture in the earth's atmosphere of water, carbon dioxide, ammonia, and methane.

Under these conditions it has been demonstrated that a variety of simple organic molecules are made, including sugars and amino acids. Urey and Miller, in 1953, obtained such molecules by exposing mixtures of water, methane, and ammonia to various types of electrical excitation. Various later workers have carried out similar studies in the dry state. These indicate that certain molecules are formed with greater frequency than others; these include amino acids, lipids, the base adenine, and the pentose sugar ribose and its phosphates (phosphorus and sulphur are two of the most abundantly occurring elements on the earth). This suggests that the formation of these compounds from their primary constituents may depend upon the fact that they possess particular stability relative to various other compounds which could have been formed.

The next stage, the formation of the large polymeric molecules of living cells, such as the nucleic acids and proteins, must have occurred by the coming together of monomeric molecules.

One of the puzzling problems is to explain how the molecules formed in the early stages of biochemical synthesis could have become sufficiently concentrated to interact effectively with each other. Oparin has suggested that this took place by coacervate formation. When two hydrophilic sols carrying opposite charges are mixed, viscous drops known as coacervates often form instead of a continuous liquid phase (Bungenburg de Jong, 1929). It is believed that these drops contain a number of particles of opposite charge, held together by electrostatic attraction, but prevented from coalescing by their shells of tightly bound water molecules. Coacervates, therefore, are regions of high concentrations of the two oppositely charged particles. Oparin, in 1968, obtained evidence that the photosynthetic activity of porphyrin is promoted when it is incorporated in coacervates, so if this situation had occurred in the early stages of biosynthesis it would have resulted in a more efficient utilization of energy from the sun.

Bernal, however, has made an interesting suggestion that interfaces may have been the regions where the necessary concentration of monomeric molecules could have been achieved. Surfaces are well known to act as catalysts; residual forces at interfaces reduce the energy between interacting molecules. Montmorillonite and kaolinite clays contain a layer structure on which organic molecules can be adsorbed. Such clays lying at the bottom of a shallow pond could have provided the conditions suitable for further stages of biochemical synthesis. Alternatively, Bernal has suggested that small molecules could have been adsorbed at the surface of the sea itself, and then become concentrated by being blown on to the beaches in the the form of foam. Foam has a very large surface area per unit volume, and so a very large capacity for concentrating molecules capable of being adsorbed. None of these concentration hypotheses has yet been proved or disproved.

Possible mechanisms of formation of biological polymers have been studied in the laboratory. Fox (1960 onwards) has demonstrated that protein-like polymers, which he has called proteinoids, can be prepared by the condensation of dry amino acids at temperatures above 100°C. The most important property of proteinoids is that their amino acid sequence is not

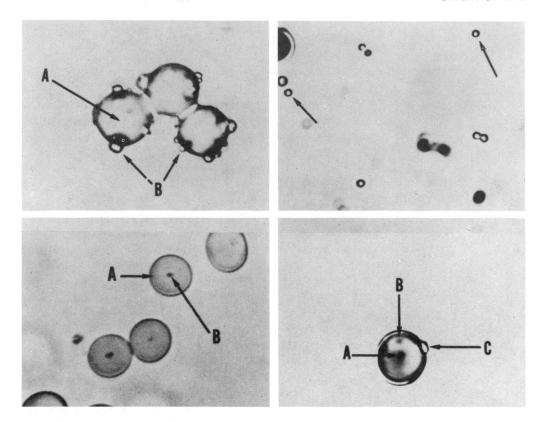

Fig. 15–3 The 'reproduction' of proteinoid microspheres. *Top left:* Microspheres, A, with buds, B. *Top right:* Buds which have been released (arrowed). *Bottom left:* Microspheres grown from stained buds B by accretion. *Bottom right:* Second generation bud, C, on such a microsphere. (With permission from S. Fox.)

random, but shows some form of ordering, only a few relatively uniform and similar fractions being formed. In water or saline solution at high concentrations proteinoids form microspherical units of about 0.5μ to 3μ in diameter. These have a relatively stable structure, and show osmotic and catalytic properties. As Fox states, the process of proteinoid formation has been accompanied by the appearance of order of a new kind. Where there were only molecules in solution before, there is now a system with a unique set of properties. The constituent amino acids must have possessed in their shape and charged groups sufficient potential information to yield proteinoids, which show morphogenic and catalytic properties. Proteinoids are able to "reproduce" in that they can form buds which are released and then grow to form microspheres by a process of accretion (Fig. 15–3). Fox looks upon cells as being composed internally of highly organized interacting micro-systems, ordered structures being obtained from a matrix of ordered molecules.

15–3
Nucleic acids and
proteins

One big problem concerns the interdependence of the two important classes of biological polymers—the nucleic acids and the proteins. Nucleic acids code for the synthesis of protein molecules, yet enzymes are required for the formation of nucleic acids, the exact replication of which is essential for the continuity of any form of life. The two types of polymer are interdependent, so the problem is which came first, and how could the one exist without the other?

Both Crick and Orgel in 1969 suggested that it would be easier for nucleic acids to act as crude catalysts than for proteins to produce identical copies of themselves. It is known that short lengths of nucleic acid may be formed in the laboratory from nucleotides in conditions simulating those thought to have existed in the early stages of biochemical synthesis. This happens, however, less readily than the formation of crude proteins from amino acids. Orgel points out that short lengths of nucleic acid of this kind are able to fold up to form specific structures, and that certain of these structures may have shown catalytic properties.

Crick also postulates that the whole machinery of protein synthesis may have been carried out originally by nucleic acids. Initially, he suggests, only a few amino acids were coded for in a rather imprecise way. Gradually, he supposes, the whole mechanism became more sophisticated; new amino acids were introduced and the coding became more exact. Thus the proteins formed must have at first been crude, with a poorly defined structure, but they gradually became extremely complex. Finally, a stage must have been reached at which the introduction of further amino acids would have disrupted too many proteins, and so the code was 'frozen' as further changes would have been too damaging. Each new amino acid as it was introduced would require a new transfer RNA, which was probably derived from an existing transfer RNA. So the least amount of disruption would be caused if the new amino acid were similar to the old amino acid coded for by the original transfer RNA. In fact, transfer RNA molecules with similar codons do tend to carry similar amino acids.

Crick also suggests that the code must always have been a triplet, since if it had initially been something else, then the change to a triplet would have meant that all previous messages would have become nonsense, and this would probably have been lethal.

Lacy and Pruitt (1969) have obtained evidence supporting the theory that the genetic code originated as a result of specific interactions between helical polypeptides and mononucleotide units, which would mean that proteins were formed before nucleic acids. They have shown from molecular models that stereospecific interactions take place in the ratio of 3 nucleotides to 1 amino acid residue. The attached mononucleotides presumably then polymerized to form a polynucleotide chain, which dissociated from the polypeptide and served in turn as a template for synthesis of the complementary polynucleotide chain.

Another problem is to explain why all structural biochemical molecules are asymmetrical, as shown by the fact that they have the property of optical rotation. For instance, the amino acids used in the formation of proteins are all L forms, which gives a 'left-handedness' to protein structure. Once life had started and was functioning, a molecule of opposite handedness to the existing one could not fit into the appropriate place in the metabolic processes. In this way, asymmetry would become continuous. Whether the first determining step was due to chance, however, or whether there exists some minute free-energy difference between the right-handed and left-handed forms has not yet been resolved.

15–4 Cell formation and energy utilization

A very important step in the formation of a cell must have been the development of lipid membranes. In order that biological systems can function efficiently it is essential that the enzymes connected with successive stages of synthesis of a biochemical pathway should be in close proximity to one another. The necessary conditions for this are obtained in cells by means of lipid membranes, which can maintain local high concentrations of reactants. The presence of hydrocarbons early in the earth's history has already been mentioned. Some cell biologists believe that the polar lipids may have been the first requirement to bring small molecules together within a lipid boundary before any synthesis of nucleic acids or proteins could have commenced. It may have taken millions of years, however, before the assembly of an elementary cellular unit took place. The first successful cell would have plenty of food available, and a great advantage over its competitors.

Life requires for its maintenance a continuous supply of energy. This could have been provided by ultraviolet or visible light from the sun, or possibly partly from the breakdown of unstable free radicals produced in the earth's atmosphere by ultraviolet light. Before the formation of the important energy-transferring compound, ATP, the phosphate part of the molecule was probably present in the form of inorganic phosphates. Laboratory experiments indicate that these may have shown electron transfer properties which were later modified by the addition of the base adenine and the ribose sugar molecule. Photosynthesis probably began very early in the history of life, since geological evidence indicates that molecules similar to the chlorophyll molecule were present in some of the earliest organisms. The production of molecular oxygen from CO_2 probably occurred much later, however. For instance, iron-oxidizing bacteria are far more readily able to oxidize Fe^{2+} to Fe^{3+} than to produce molecular oxygen from water.

15–5 Evolution and protein structure

Modern biochemical techniques have proved of value in tracing evolutionary relationships. Methods of determining protein structure mean that it is possible to compare the composition of protein molecules which carry out similar functions in various widely different organisms. It was mentioned in Chapter 3 that histones show a marked similarity in structure

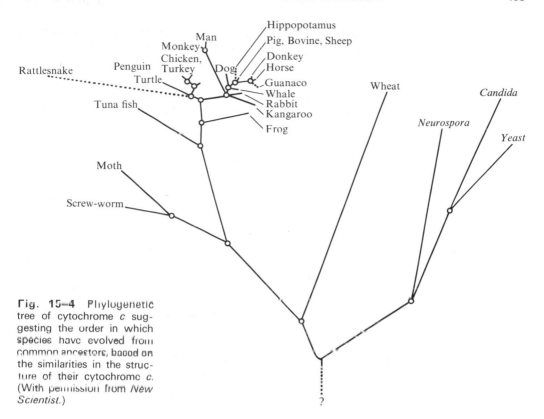

Fig. 15-4 Phylogenetic tree of cytochrome *c* suggesting the order in which species have evolved from common ancestors, based on the similarities in the structure of their cytochrome *c*. (With permission from *New Scientist.*)

in the cells of all species so far studied. This indicates that histone composition must have very great evolutionary stability. Another protein which has been widely studied is the respiratory enzyme, cytochrome *c* This protein molecule is widely distributed in nature, is easily extracted from cells, and contains only about 110 amino acid units. The amino acid sequences of more than thirty cytochrome *c* molecules from a wide range of organisms have now been determined. From these, a phylogenetic tree has been constructed showing the evolutionary relationships between most of these species (Fig. 15-4). The number of amino acid changes between one species and another permits calculation of the time lapse since divergence, by assuming a steady mutation rate, and so the location of the junction points of the branches. Studies were made by Dus and coworkers in 1968 on the amino acid sequence of a cytochrome *c* from a primitive organism, the photosynthetic bacterium *Rhodospirillum rubrum*. They found clearly discernible homologies between the sequence of this bacterial molecule and that of cytochrome *c* in man and mammals, although these organisms are so widely separated on the evolutionary scale. It is anticipated that further studies will enable the structure of the lower part of the tree to be determined in more detail.

15–6
Life on other planets

In conclusion, it is fascinating to speculate on the possibility of life existing on other planets in the solar system. The two most important aspects are those of temperature and availability of water. These together mean that a planet must be at a distance of between 100 and 200 million miles from the sun for life to be possible. If it were nearer, water would evaporate, and if it were farther away, water would freeze. Size is also important—a large planet would tend to retain water, while on a smaller one water would evaporate and escape.

Using these criteria, it can be stated that the outer planets must be far too cold. Mercury is too hot and dry and Venus is probably too hot. Mars is too cold and dry, and the moon is too small and dry. Certainly the moon samples brought back by the *Apollo 11* astronauts in 1969 show that the so-called lunar 'seas' could never have been covered with water. It appears to be much more likely that these large flat areas were the remnants of huge outpourings of lava. The samples contain traces of the carbon compounds CH_4, CO, and CO_2, the presence of which is difficult to explain by terrestrial contamination or by transport by solar winds; a few higher molecular weight compounds, thought perhaps to be amino acids, have also been found (1970). The moon samples indicate that, although the moon originated at approximately the same time as earth, some major catastrophic event may have occurred on the moon between 2,000 and 3,000 million years ago. This could have been damage by a giant meteorite, or by a solar flare. All we can conclude is that the moon, lacking water and having no atmosphere, does not have suitable conditions at the present time for the existence of life.

We have reviewed briefly some of the more recent discoveries relating to the origin of life on earth, and mentioned various recent hypotheses. To quote from Bernal—"while removing most of the mysteries of life . . . this has not reduced in the minds of scientific biologists of today any of the appreciation of its complexity and its beauty". A study of cell biology emphasizes, perhaps as no other study, the essential unity of life. The function of natural science is to hold up a mirror to nature; it is not its function to ask the question why life arose on earth, whether by accident or design. This comes within the province of the individual's philosophy and credo.

For the future, it must be clear that this is a field in which far-reaching discoveries will be made for good or for ill. These can lead almost certainly in the next century to the removal of the major scourges of cancer and certain genetic diseases which at present afflict mankind. Such possibilities promise well for man's future happiness; whether the discoveries will be so used will depend largely on the foresight and dedication of those in whom the responsibility lies for their control.

Bibliography   Barghoorn, E. S. and J. W. Schopf, Micro-organisms three billion years old from the pre-Cambrian strata of South Africa. *Science,* **152,** 758–763, 1966.

Bernal, J. D., *The Origin of Life.* Weidenfeld and Nicholson, London, 1967.

Fox, S. W. (ed.), *The Origins of Prebiological Systems and of their Molecular Structure.* Academic Press, New York, 1965.

Haldane, J. B. S., The origin of life. *The Rationalist Annual,* 1929. (In Bernal, J. D., above, p. 242.)

Oparin, A. I., *The Origin of Life,* translated by Ann Synge from Proiskhozhdenie zhizny, Izd. Moskovski Rabochii, 1924. (In Bernal, J. D., above, p. 199.)

Index